Pauli and the
Spin-Statistics Theorem

Ian Duck
Rice University, Houston, Texas, USA

E C G Sudarshan
University of Texas at Austin, USA

Pauli and the
Spin-Statistics Theorem

World Scientific
Singapore • New Jersey • London • Hong Kong

Published by

World Scientific Publishing Co. Pte. Ltd.

P O Box 128, Farrer Road, Singapore 912805

USA office: Suite 1B, 1060 Main Street, River Edge, NJ 07661

UK office: 57 Shelton Street, Covent Garden, London WC2H 9HE

Library of Congress Cataloging-in-Publication Data
Pauli and the spin-statistics theorem / editors, Ian Duck, E.C.G. Sudarshan.
     p. cm.
    Includes bibliographical references and index.
    ISBN 9810231148
    1. Nuclear spin -- Mathematical models.   2. Nuclear spin -- Statistical methods.
    3. Pauli exclusion principle.   I. Duck, Jan. 1933–
    II. Sudarshan, E. C. G.
    QC793.3.S6P38   1997
    530.13'3--dc21   97-36353
                                                      CIP

British Library Cataloguing-in-Publication Data
A catalogue record for this book is available from the British Library.

First published 1997
First reprint 1998

Printed in Singapore.

Table of Contents

Part II: The Pauli Era

Part III: The Wightman-Schwinger Era

Part IV: The Contemporary Era

Authors

Ian Duck, b. 1933 Kamloops, Canada; BSc 55 Queen's; res asst, 55-56 UBC; PhD 61 Caltech; Research Assoc. 61-63 USC; Research Assoc. to Prof of Physics 63- Rice University; theoretical research in nuclear and particle physics from ev to Gev; radiative capture of light nuclei, muon capture, bremstrahlung in p-p scattering, Faddeev equations, N-Δ process, quark model of nucleon structure, models of color confinement, quark-gluon plasma excitation in 10 Gev $\bar{p}$-nucleus annihilation, fine structure of the baryon octet-decuplet spectrum.

Ennackel Chandy George Sudarshan, b. 1931 Kottayam, India; BSc 51, MA 52 Madras; res asst 52-55 Tata Inst; PhD 58 Rochester; Res Fellow 57-59 Harvard; to Assoc Prof Rochester 59-64; Prof Syracuse 64-69; Prof of Physics, Center for Theoretical Physics, Texas 69- ; concurrent positions at Bern 63-64, Brandeis 64, Madras 70-71, Bangalore 72-. Honors: DSc from Wisconsin, Delhi, Chalmers, Madras, Burdwan, Cochin; Medals: Order of the Lotus, Bose Medal, Sarvadhikari Medal and many others. Research accomplishments: coinventer with R.E. Marshak of the universal V-A weak interaction (1957); published six books and over 350 journal articles in the areas of elementary particle theory, quantum theory, group theory, quantum optics, and the foundations, philosophy and history of contemporary physics.

Acknowledgements

This book would never have been written without the impetus of a question raised by Professor D.E. Neuenschwander in the *American Journal of Physics*. His question expressed the widespread dissatisfaction, shared even by Feynman, with the proof of the Spin-Statistics Theorem (that integral spin particles satisfy Bose-Einstein statistics, half-integral spin particles satisfy Fermi-Dirac statistics and the Pauli Exclusion Principle). The complaint was that the widely accepted Pauli proof depends on complicated techniques in relativistic quantum field theory which mystify most people and offer no physical insight. Neuenschwander asked the physics community for a simple, direct, intuitive, physical proof of the Spin-Statistics Theorem. This book is our response.

The short answer to Neuenschwander's question is our Section 21.2 which contains an "Elementary Proof" completely free of the complications of *relativistic* quantum field theory. An explanation of the spin-statistics connection intended to make the Spin-Statistics Theorem *understandable* is contained in the remainder of our closing Chapter 21.

Our book follows the Spin-Statistics Theorem from its genesis in the Pauli Exclusion Principle ; through the original developments in statistical mechanics, quantum mechanics, and quantum field theory; to a first generation of proofs culminating in Pauli's 1940 proof which has been the standard for almost sixty years; to a second generation of proofs by Lüders and Zumino, and Burgoyne based on work by Wightman; and to the start of a modern era by Schwinger. Next we analyze, criticize and dispose of various intuitive explanations of the spin-statistics connection. In our final chapter the long mystery is resolved in a simple elementary proof, and the fundamental basis for understanding it in an intuitive and physical way.

We wish to dedicate this book to our parents who, simultaneously but a world apart, inspired us with their devotion to education and set us on the paths which led to this collaboration.

Ian Duck and George Sudarshan, March 1997.

Foreword

§1. Introduction.

The principal motivation for writing this book came from a question raised by Neuenschwander in the "Questions and Answers" section of The American Journal of Physics [0.1], whom we quote:

"In the *Feynman Lectures on Physics*, Richard Feynman said [0.2]

'Why is it that particles with half-integral spin are Fermi particles whose amplitudes add with the minus sign, whereas particles with integral spin are Bose particles whose amplitudes add with the positive sign? We apologize for the fact that we cannot give you an elementary explanation. An explanation has been worked out by Pauli from complicated arguments of quantum field theory and relativity. He has shown that the two must necessarily go together, but we have not been able to find a way of reproducing his arguments on an elementary level ⋯. This probably means that we do not have a complete understanding of the fundamental principle involved ⋯.'

Has anyone made any progress towards an "elementary" argument for the Spin-Statistics Theorem?"

Within months a few responses appeared which we will explore in due time.

What Neuenschwander's question did for us was to bring home the blunt realization that we did not "really understand" the original Pauli proof [0.3], and did not understand the too terse comments on the subject that one finds in all textbooks on quantum mechanics and field theory, and especially did not understand the major existing book on the subject by Streater and Wightman [0.4]. In fact Pauli's proof, and Streater and Wightman's explanation and expansion of it, are probably just the complicated arguments which Feynman was apologizing for, and which prompted Neuenschwander's

1

dissatisfaction in the first place.

Another strong influence leading to the format of this book, was a book of the first papers on Quantum Electrodynamics compiled and edited by Julian Schwinger[0.5], which contains a wonderful collection from the genesis of the subject in 1927 to that date. We have long believed that one learns best from the original founders of a subject, in spite of the fact that first papers are frequently difficult and obscure. Of course, just the opposite also occurs. There is also the problem that many of these source papers are inaccessible to those who are not fluent in German. We set out to collect key papers leading to the Spin-Statistics Theorem, in one place, in English, in chronological order, in a unified modern notation, with what clarification and explanation we could offer, and edited with as close a focus on the chosen subject as understanding would allow.

This book is the result. It contains classic achievements of the great contributors to the foundations of quantum mechanics, quantum statistics, and quantum field theory all leading to the classic proofs of the Spin-Statistics Theorem. The introductory explanations make the readings selected for each chapter more accessible than they would be alone. The frequent omissions from the originals are necessitated by the fact that the original authors were usually addressing a broader subject than is done here, and would have led us too far afield. Nonetheless, in certain cases the originals are too beautiful to set aside. For our taste, this happens most frequently in the writings of Dirac, Fermi, and Feynman. Other authors seemed so difficult that it was necessary to cut to the subject at hand. This was frequently true of Pauli himself, of Schwinger who encased his remarks on the Spin-Statistics Theorem within a massive tome on field theory, and of Hall and Wightman, whose mathematics is too demanding for most tastes.

There are no ideas original in this book, so there are no advances in the status of the Spin-Statistics Theorem that did not exist before. But Neuenschwander's question does ultimately find an answer in this book.

The answer is as simple and as direct as we can make it. We are sure that further improvements will be made.

The title includes Pauli's name for the obvious reasons that:

1) he invented the Pauli Exclusion Principle.

2) he dictated the program for proving the Spin-Statistics Theorem and authored the widely recognized 1940 proof.

but most of all,

3) by sheer force of will, by persistent and dogged criticism, and by an arrogant assertion of some proprietary right, he did in effect possess the field, it was his theorem, and he defended it against all incursions as long as he lived, which coincidentally was as long as it took for Lüders, Zumino, Burgoyne, and Hall and Wightman to resolve the flaws of logic and rigor which infected the earlier proofs, Pauli's included.

This book is accessible to all physicists and will be of interest to the many who have an historical bent and subscribe, as we do, to a personality cult which reveres these giants of the recent past. We hope also, that this book will be useful as a text in modern physics classes for students who have had a junior level introduction to quantum mechanics and are concurrently taking a senior level statistical mechanics course.

§2. Preliminary Remarks.

Everyone *knows* the Spin-Statistics Theorem but no one *understands* it.

The key word of course is "understand". The Spin-Statistics Theorem - which states that *identical half-integral spin particles satisfy the Pauli Exclusion Principle and Fermi-Dirac statistics which permit no more than one particle per quantum state; identical integral spin particles do not, but satisfy Bose-Einstein statistics which permits any number of particles in each*

3

quantum state - stands as a fact of nature. The question is whether physics contains this fact, and if so how this comes about; or whether physics is merely consistent with the Spin-Statistics Theorem and that some deeper explanation exists. The Spin-Statistics Theorem could conceivably be an essential ingredient of a more fundamental view of the world, of which the physics of the last seventy years gives the currently observable manifestation. This could be the case in fundamental string theories or their successors, formulated in some extension of four-dimensional Minkowski space which is the customary but suspect venue of physics.

With such a point of view forced upon us, we should modify the meaning of "understand", and at the same time reduce our expectations of any proof of the Spin-Statistics Theorem. What is proved - whether truly or not, whether optimally or not, in an acceptable logical sequence or not - is that the existing theory is consistent with the spin-statistics relation. What is not demonstrated is a reason for the spin-statistics relation.

To belabor the point, it is difficult to imagine a fundamental mechanism for the Pauli Exclusion Principle - upon which *all* depends - which predicates it (looking ahead to the work of Hall and Wightman [0.6]) upon the analyticity properties of vacuum expectation values of products of quantized field operators. Did God - for lack of a better word - build a series of failed worlds which sputtered and died, or exploded and disintegrated, before discovering the stabilizing effect of anticommutation relations for half-integral spin fields? Was this before or after imposing the requirements of Lorentz invariance? Are we the lucky winners of a Monte Carlo simulation in which every choice was tried and one survived?

Must we reduce our demands on physics to require only consistency. Does an understanding of the "Why?" of the spin-statistics relation have no direct answer in physics? Or must physics be formulated to include it. The Pauli result does not explain the spin-statistics relation and cannot. The Neuenschwanders and the Feynmans of the world must remain unsatisfied (Ch.20) because the consistency of relativistic quantum mechanics and quantum field theory with the Pauli Exclusion Principle has every reason to be as complicated as these subjects are, not as simple and direct as the Pauli Exclusion Principle itself.

It must be admitted that the simple, pristine Dirac equation has many marvelous unintended consequences: spin-$\frac{1}{2}$, the anomalous magnetic moment of the electron, the prediction of the positron, the process of pair production and annihilation, and finally the consistency with the Pauli Exclusion Principle. All of these, along with more detailed successes of quantum electrodynamics, combine to give an illusion of absolute truth, the feeling that we are dealing with an ultimate description of nature. This illusion survives through the successes of the Standard Model where the electron-photon system of quantum electrodynamics is extended to include weak, electromagnetic, and strong interactions. However, the illusion begins to fade when we escalate successively through Grand Unified Theories to Supersymmetry to String Theory and beyond.

In the following chapters we examine in detail the genesis of the ideas essential to the discussion of the Spin-Statistics Theorem. These include the Pauli Exclusion Principle, and the discoveries of electron spin, quantum statistics, wave function symmetrization, field theory, anticommutation relations, and the quantization of fields. All these subjects are presented and discussed on the basis of their first published reports. But first, for the sake of orientation, we present very brief, heuristic arguments for the standard choices of

(1) anticommutation relations leading to antisymmetrized many particle wave functions, Fermi-Dirac statistics and the Pauli Exclusion Principle for identical particles of spin-$\frac{1}{2}$ which satisfy the first order relativistic free particle Dirac equation; and

(2) commutation relations which lead to symmetrized many particle wave functions and Bose-Einstein statistics for identical particles of spin-0 which

satisfy the second order relativistic free particle Klein-Gordon equation.

§3. Anticommutation Relations for Dirac Spin-$\frac{1}{2}$ Fields.

The Dirac equation, linear in the space and time derivatives, is

$$i\frac{\partial}{\partial t}\psi = H_D\psi = \left(\vec{\alpha}\cdot\frac{\vec{\nabla}}{i} + \beta m\right)\psi. \tag{1}$$

Dirac found the four 4×4 matrices $\vec{\alpha}$ and β by requiring ψ also to satisfy the relativistic energy-momentum relation $E^2 = p^2 + m^2$. Consider a state of zero momentum with the Dirac wave function

$$\psi = aue^{-imt} + b^*ve^{+imt}, \tag{2}$$

which requires interpretation. The short but confused (and confusing) history of the "negative energy" piece $\sim \exp(imt)$, for which

$$i\frac{\partial}{\partial t}e^{imt} = -me^{imt} \equiv Ee^{imt},$$

is presented in Dirac's original remarks quoted in our Chapter 8.

The four component Dirac spinors u and v are trivial for zero momentum. u has the top two components equal to a Pauli spinor, the bottom two equal to zero, and the opposite for v. They are normalized to

$$u^\dagger u = v^\dagger v = 1; \qquad u^\dagger v = v^\dagger u = 0. \tag{3}$$

where $u^\dagger = (u^*)^\sim$ is the row spinor Hermitian conjugate (complex conjugate transposed) to u.

We begin by interpreting a (and similarly b) as the amplitude to be in the electron (positron) state with energy $+m$, and charge e ($-e$) (more confusion - the charge on the electron is taken to be $e = -|e|$).

Speaking loosely, ψ has a piece aue^{-imt} which is the wave function of an electron of energy m, and another piece b^*ve^{imt} which is the complex conjugate of the wave function of a positron of energy $+m$. The probability for an electron is a^*a and for a positron is bb^*. We are reminded of the raising and lowering operators in the matrix solution of the quantum theory of the harmonic oscillator, where the *operators* a and $a^\dagger$ (b, $b^\dagger$) replace the *complex numbers* a and a^* (b, b^*) and the number operators $N_e = a^\dagger a$ and $N_{\bar{e}} = b^\dagger b$ replace the corresponding probabilities a^*a and bb^*.

Reinterpreting (a^*, a) as raising and lowering (better, creation and annihilation) operators $(a^\dagger, a)$, we have that:

1) $a^\dagger$ creates an electron of energy m, charge e; a annihilates an electron of energy m, charge e. (There is also an unspecified spin component, and usually a momentum which is here set to zero.)

2) similarly, $b^\dagger$ creates a positron of energy m, charge $-e$, and so on.

In the usual harmonic oscillator solution of non-relativistic quantum mechanics, only the operator a and its Hermitian conjugate $a^\dagger$ occur, but not b and $b^\dagger$. There is only one kind of quantum, and no charge, and no antiparticle. For the harmonic oscillator case, the operators a and $a^\dagger$ are linear combinations of the coordinate q and momentum p which satisfy canonical commutation relations, and the a and $a^\dagger$ can also be shown to satisfy the commutation relations

$$[a, a^\dagger]_- = 1, \qquad [a, a]_- = [a^\dagger, a^\dagger]_- = 0. \tag{4}$$

In a standard elementary but fundamental exercise, the matrix representation of the quantum harmonic oscillator follows from this algebra plus the requirement that the positive definite Hamiltonian operator

$$H = a^\dagger a + \frac{1}{2} \tag{5}$$

must have a lowest energy eigenstate $|0\rangle$. We are led to the occupation number operator $N = a^\dagger a$ whose eigenstates $|0\rangle, |1\rangle, \cdots |n\rangle, \cdots$ are characterized by the integer occupation number eigenvalues $n = 0, 1, 2, \cdots$ which are the number of quanta in each state. To complete the solution we have the orthonormal occupation number eigenstates $|n\rangle$, and the matrix elements

$$\langle n-1|a|n\rangle = \sqrt{n}, \quad \langle n|a^\dagger|n-1\rangle = \sqrt{n}, \quad \langle n|a^\dagger a|n\rangle = n, \qquad (6)$$

all others zero, and the energy eigenvalues

$$E_n - \frac{1}{2} = \langle n|H - \frac{1}{2}|n\rangle = \langle n|N|n\rangle = n. \qquad (7)$$

Returning now to the discussion of the zero momentum solutions of the Dirac equation and the interpretation of the $(a, a^\dagger)$, $(b, b^\dagger)$ as annihilation operators and creation operators, we can write

$$
\begin{aligned}
E &= \psi^\dagger \left(i\frac{\partial}{\partial t} \right) \psi \\
&= m(a^\dagger a - bb^\dagger).
\end{aligned}
\qquad (8)
$$

The cross terms disappear because of the orthogonality of u and v. We are now in a quandary. If we attempt to pursue the analogy with the harmonic oscillator solution and assign commutation relations to $(a, a^\dagger)$ and $(b, b^\dagger)$ (which are no longer simply related to coordinate and momentum), then the eigenvalues of $a^\dagger a$ and $b^\dagger b$ (or $bb^\dagger$) would be all possible positive integers, and the energy eigenvalues would range from $+\infty$ to $-\infty$. There would be no lowest energy state but separate a and b oscillators with positive and negative energies.

The way out of this quandary, which is to replace commutation relations for the $(a, a^\dagger)$ and $(b, b^\dagger)$ by *anticommutation* relations, was invented even before the Dirac equation by Jordan, and clearly explained by Jordan and Wigner (our Ch.7) in the context of non-relativistic quantum mechanics. In spite of that fact, it took some five years to understand the problem of "negative energy states" in the Dirac equation and to realize that the "filled Dirac negative energy sea" with "holes", and all the associated circumlocution and confusion could be replaced neatly by the concept of antiparticles. The part of the field operator ψ which $\sim e^{-i\omega t}$ (where we always mean

$\omega \equiv +\sqrt{p^2 + m^2}$) is to be interpreted as a particle annihilation piece, the part $\sim e^{+i\omega t}$ as an antiparticle creation piece. No more talk of negative energies!

With Jordan's anticommutation relations

$$[a, a^\dagger]_+ = [b, b^\dagger]_+ = 1, \tag{9}$$

the situation clarifies remarkably. It will be necessary to extend these anticommutation relations by requiring all others to be zero,

$$[a, b]_+ = [a, a]_+ = \text{ etc. } = 0. \tag{10}$$

The question will arise what happens, for example, between the a for an electron, and the a or $a^\dagger$ for, say, a muon. We will not pursue the various possibilities here, but suffice it to say that they can be chosen to anticommute. A similar question arises between the a for an electron and the a or $a^\dagger$ for, say, a photon, which satisfy commutation relations themselves. Here, it is sufficient to choose the spin-$\frac{1}{2}$ field operators to commute with the spin-1 operators.

Now back to interesting business. With the anticommutation relation for b and $b^\dagger$, the above expression for the energy

$$E = m(a^\dagger a - bb^\dagger) \Rightarrow m(a^\dagger a + b^\dagger b - 1),$$

which can provisionally but correctly be identified in terms of occupation number operators

$$N_e = a^\dagger a, \qquad N_{\bar{e}} = b^\dagger b \tag{11}$$

as

$$E = m(N_e + N_{\bar{e}} - 1) \tag{12}$$

and interpreted as the total energy consisting of the energy (m) of each (at rest) electron multiplied by the positive (or zero) number of electrons N_e, plus the energy (m) of each (at rest) positron multiplied by the positive (or zero) number of positrons $N_{\bar{e}}$.

9

The negative constant $-m$ has an interesting significance. The zero occupation number, zero-point energy for the spin-$\frac{1}{2}$ particle is *minus* $\frac{1}{2}$ multiplied by two for particle and antiparticle as above, and multiplied by another factor of two for the two spin possibilities. This is opposite in sign to the *plus* $\frac{1}{2}$ which occurs in the zero-point energy for each independent mode of an integer spin field.

For our purposes, the energy difference

$$E - E_0 = m\left(N_e + N_{\bar{e}}\right)$$

is positive as required. It is reassuring to calculate the the electric charge

$$Q = e\psi^\dagger\psi = e\left(a^\dagger a + bb^\dagger\right),$$

which becomes

$$Q = e\left(N_e - N_{\bar{e}}\right), \tag{13}$$

in accord with our interpretation of the occupation numbers N_e and $N_{\bar{e}}$ in the expression for the energy (although the zero point charge does need some special attention).

Finally, it is instructive and reassuring to construct a matrix realization of the operator algebra. From the anticommutator $[a, a]_+ = 0$, we must have $aa|\psi\rangle = -aa|\psi\rangle$ for any state $|\psi\rangle$. So either $a|0\rangle = 0$, which defines the state $|0\rangle$; or $a|1\rangle = |0\rangle$, which defines the state $|1\rangle$. Similarly we find $a^\dagger|1\rangle = 0$ and $a^\dagger|0\rangle = |1\rangle$. We find that there are only two states, $|0\rangle$ and $|1\rangle$, for which

$$a|0\rangle = a^\dagger|1\rangle = 0; \quad a^\dagger|0\rangle = |1\rangle \text{ and } a|1\rangle = |0\rangle. \tag{14}$$

Each state is labelled by its eigenvalue of the occupation number operator $N = a^\dagger a$ so

$$N|1\rangle = 1|1\rangle; \qquad N|0\rangle = 0|0\rangle = 0. \tag{15}$$

The 2×2-matrix representation of these operators is simply

$$\langle 0|a|1\rangle = \langle 1|a^\dagger|0\rangle = 1, \quad \langle 1|N|1\rangle = 1, \tag{16}$$

and all other matrix elements are zero. In matrix form

$$|0\rangle = \begin{pmatrix} 1 \\ 0 \end{pmatrix}, \quad |1\rangle = \begin{pmatrix} 0 \\ 1 \end{pmatrix}, \quad a = \begin{pmatrix} 0 & 1 \\ 0 & 0 \end{pmatrix}, \quad a^\dagger = \begin{pmatrix} 0 & 0 \\ 1 & 0 \end{pmatrix}. \tag{17}$$

The number operator

$$N = a^\dagger a = \begin{pmatrix} 0 & 0 \\ 1 & 0 \end{pmatrix} \begin{pmatrix} 0 & 1 \\ 0 & 0 \end{pmatrix} = \begin{pmatrix} 0 & 0 \\ 0 & 1 \end{pmatrix} \tag{18}$$

has eigenvalues 0 and 1, which are the only occupation numbers allowed by the anticommuting field operators, in accord with the Pauli Exclusion Principle.

One more point on which we can instructively reassure ourselves, is that the many-particle (in our example many = 2) non-relativistic wave function is antisymmetric in the exchange of the two identical particles. For this, we need to review the concept of a field operator $\Psi^\dagger(x)$ which creates a particle at position x and its Hermitian conjugate which annihilates it. (We are considering non-relativistic quantum mechanics for the moment, and need only the particle but not the antiparticle degrees of freedom.) We have

$$\Psi(x) = \sum_j a_j \psi_j(x), \qquad \Psi^\dagger(x) = \sum_j a_j^\dagger \psi_j^*(x) \tag{19}$$

where the sum is over a convenient complete orthonormal set of single particle wavefunctions $\psi_j(x)$, and the corresponding (annihilation) operators a_j.

The state vector for a particle localized at x is

$$|x\rangle = \Psi^\dagger(x)|0\rangle \tag{20}$$

in a standard notation. The amplitude for a particle in state s to be at position x is

$$\begin{aligned} \langle x|s \rangle &= \langle 0|\Psi(x)a_s^\dagger|0\rangle \\ &= \sum_j \psi_j(x)\langle 0|a_j a_s^\dagger|0\rangle \\ &= \sum_j \psi_j(x)\delta_{js} = \psi_s(x). \end{aligned} \tag{21}$$

11

Now we can apply the formalism to a two particle state

$$
\begin{aligned}
\langle x, y | s, t \rangle &= \langle 0 | \Psi(x) \Psi(y) a_s^\dagger a_t^\dagger | 0 \rangle \\
&= \sum_{jk} \psi_j(x) \psi_k(y) \langle 0 | a_j a_k a_s^\dagger a_t^\dagger | 0 \rangle \\
&= \sum_{jk} \psi_j(x) \psi_k(y) \left\{ \delta_{ks} \delta_{jt} - \delta_{kt} \delta_{js} \right\} \\
&= \psi_t(x) \psi_s(y) - \psi_s(x) \psi_t(y).
\end{aligned}
\tag{22}
$$

The result is antisymmetric as desired, as a result of the anticommutations required to move the annihilation operators a_j, a_k through the creation operators $a_s^\dagger$, $a_t^\dagger$ where they can then be eliminated against the vacuum state $|0\rangle$. The overall phase and the normalization are a matter of choice.

In summary, this brief and elementary exercise has demonstrated that quantization of the Dirac equation using Jordan-Wigner anticommutation relations (instead of canonical commutation relations) for the field operators gives a theory with a positive energy spectrum, which is naturally interpreted in terms of electrons and positrons of spin-$\frac{1}{2}$ satisfying the Pauli Exclusion Principle (Ch.1) and Fermi-Dirac statistics (Ch.6), and possessing antisymmetric many-particle wave functions in accord with Heisenberg and Dirac (Ch.5).

This situation, which was in place by 1932, could be characterized as a scenario. Probably correct, but not proven. It would occupy some of the best minds in physics over the next generation to satisfy themselves, perhaps, but certainly not everyone, that the matter was closed.

Next we look at the comparable situation for integral spin fields, specifically at the spin-0 relativistic scalar field satisfying the free particle Klein-Gordon wave equation.

§4. Commutation Relations for Klein-Gordon Spin-0 Fields.

Here again, there was a long delay and much confusion before the ostensibly simple spin-0 relativistic scalar wave equation was treated in a systematic and rigorous way by Pauli and Weisskopf (Ch.9). Here we will present an abbreviated version, as we did above for the Dirac field. We find, in contrast to what occurred for the Dirac case, that everything goes smoothly if we choose commutation relations for the field operators, and the resulting Bose-Einstein statistics (Ch.3) for the Klein-Gordon spin-0 particles. On the other hand, if we were to choose anticommutation relations, we would still find a positive energy spectrum. Different arguments must be constructed to rule out this possibility (Ch.10).

The essential reason that no negative energy problems are generated by the solutions $\sim e^{+imt}$, which still occur just as in the Dirac case, is that the energy operator, or more properly the Hamiltonian, or still more correctly the volume integral of the time-time component of the energy-momentum tensor, is bilinear in the time derivatives and produces a factor $+m^2$ regardless of the sign in the exponential. The Dirac Lagrangian, which is linear in the time and space derivatives, leads to the opposite result.

Next, the question arises how we get a quantity with the dimension of an energy from a quantity that goes like m^2, the square of an energy. The answer lies in the normalization of the relativistic wave function, which is dictated by the form of the relativistic (charge-)density, which contains a single time derivative. For the piece of the (zero momentum mode) solution $\sim e^{-imt}$ this derivative produces a factor $+m$, for the piece $\sim e^{+imt}$ a factor $-m$. This sign difference caused the original confusion that the Klein-Gordon equation failed because it did not produce a positive-definite conserved density. Finally, after some eight years and a few precursor comments to the same effect which seem to have been ignored, Pauli and Weisskopf established that these were particle and antiparticle contributions to a charge density and sensibly had opposite signs. Then it becomes a matter of normalizing each term to the number of particles or antiparticles, which required cancelling the factor m in the density of each by dividing the field amplitude by a factor $\sqrt{2m}$. Now the charge densities have factors

$(\pm m/m) = \pm$ times what turns out to be a number operator. In the same way, the energy has a factor $(m^2/m) = m$ times a number operator, which has the appropriate dimension. For modes with momentum $\vec{k}$, we must judiciously replace m by $\omega_k = +\sqrt{k^2 + m^2}$, but everything goes through easily. The factor $1/\sqrt{\omega_k}$ plays an essential role in the manifest Lorentz covariance of all densities, combining with a 3-volume to give a Lorentz invariant measure d^3k/ω_k. A similar factor is extracted in the Dirac case, but is compensated in the normalization of the Dirac spinor $u^\dagger u \Rightarrow \omega/m$ for non-zero momentum, which transforms like the time-component of a four-vector.

Now, we briefly sketch the Pauli-Weisskopf Lagrangian-based canonical quantization of the relativistic scalar Klein-Gordon spin-0 field.

The Lagrangian density which produces the Klein-Gordon equation as the field equation for a complex one-component field $\phi(\vec{x}, t)$ is

$$\mathcal{L} = \frac{\partial \phi^\dagger}{\partial t} \frac{\partial \phi}{\partial t} - \sum_{j=x,y,z} \nabla_j \phi^\dagger \nabla_j \phi - m^2 \phi^\dagger \phi. \tag{23}$$

$\mathcal{L}$ will be Lorentz invariant (as required for a Lorentz invariant action $\mathcal{S} = \int \mathcal{L} d^4 x$) if ϕ is an invariant scalar under Lorentz transformations. We have gone directly to the Hermitian conjugate of the field in $\phi^\dagger$ rather than the complex conjugate ϕ^* anticipating the quantization of the fields which will elevate them from number valued quantities to operators in the Hilbert space of states.

The generalized momentum canonically conjugate to the fields $\phi, \phi^\dagger$ are (abbreviating $\partial \phi / \partial t$ as $\phi_{;t}$ and so on)

$$\Pi_\phi = \frac{\partial \mathcal{L}}{\partial \phi_{;t}} = \frac{\partial \phi^\dagger}{\partial t} \quad \text{and} \quad \Pi_{\phi^\dagger} = \frac{\partial \phi}{\partial t}. \tag{24}$$

The Euler-Lagrange equation

$$\frac{\partial}{\partial t} \frac{\partial \mathcal{L}}{\partial \phi_{;t}^\dagger} + \nabla_j \frac{\partial \mathcal{L}}{\partial \phi_{;j}^\dagger} = \frac{\partial \mathcal{L}}{\partial \phi^\dagger} \tag{25}$$

14

is

$$\frac{\partial^2 \phi}{\partial t^2} - \nabla^2 \phi = -m^2 \phi, \tag{26}$$

just the Klein-Gordon relativistic wave equation $(\Box + m^2)\phi = 0$ as required.

The Hamiltonian density

$$\begin{aligned} \mathcal{H} &= \Pi_\phi \phi_{;t} + \phi_{;t}^\dagger \Pi_{\phi^\dagger} - \mathcal{L} \\ &= \frac{\partial \phi^\dagger}{\partial t} \frac{\partial \phi}{\partial t} + \nabla_j \phi^\dagger \nabla_j \phi + m^2 \phi^\dagger \phi \end{aligned} \tag{27}$$

is positive which guarantees a positive energy spectrum, in contrast to the situation which arises in the Dirac case.

One also finds a conserved four-vector charge-current density

$$\rho = e\phi^\dagger i \frac{\overleftrightarrow{\partial}}{\partial t} \phi, \quad \vec{j} = e\phi^\dagger \frac{\overleftrightarrow{\nabla}}{i} \phi. \tag{28}$$

Expressed in terms of momentum eigenstates

$$\phi = \sum_k \frac{1}{\sqrt{2\omega_k}} \left\{ a_k e^{i(k \cdot r - \omega_k t)} + b_k^\dagger e^{-i(k \cdot r - \omega_k t)} \right\}, \tag{29}$$

where we are still anticipating the interpretation of the expansion coefficients as annihilation operators a_k for a particle of momentum $\vec{k}$ and creation operators $b_k^\dagger$ for an antiparticle also of momentum $\vec{k}$ and the appropriate energy $\omega_k = \sqrt{k^2 + m^2}$. Substituting these expansions of ϕ and $\phi^\dagger$ into $E = \int \mathcal{H} d^3 x$ and $Q = \int \rho d^3 x$ gives the total energy and total charge

$$E = \sum_k \omega_k \left(a_k^\dagger a_k + b_k b_k^\dagger \right) \quad \text{and} \quad Q = \sum_k e(a_k^\dagger a_k - b_k b_k^\dagger). \tag{30}$$

All the cross terms cancel, leaving expressions that are readily interpreted in terms of particle and antiparticle occupation numbers

$$N_e = a^\dagger a, \text{ and } N_{\bar{e}} = b^\dagger b, \tag{31}$$

if the operators $a, a^\dagger$ and $b, b^\dagger$ are assigned the commutation relations

$$[a_k, a_{k'}^\dagger]_- = [b_k, b_{k'}^\dagger]_- = \delta_{k,k'} \tag{32}$$

15

and all others zero.

This is just the desired result and leads to Bose-Einstein statistics with occupation numbers for each mode allowed to take any positive integer value, and to symmetric many particle wave functions.

Without delving into all the pathologies that might develop if we try to quantize Klein-Gordon spin-0 fields with anticommutation relations (which are discussed by Pauli in the first paper exploring the Spin-Statistics Theorem, published in 1936, and presented in our Ch.10), we present an argument due to deWet. deWet's proof (Ch.13) was the subject of his 1939 Princeton PhD thesis. Although incomplete, it is remarkably close to the proof finally constructed by Lüders and Zumino and by Burgoyne (Ch.17) for both integral and half-integral spin fields.

For deWet's discussion we need the (anti-)commutation relations for the full field operators ϕ and $\phi^\dagger$ at different points of space-time, at equal times, and at coincident points of space-time. For example, for the free fields above

$$[\phi(x,t), \phi(x',t)]_- \sim [a + b^\dagger, a + b^\dagger]_- \equiv 0.$$

This is clear from the interpretation of the fields as generalized coordinates q_x and $q_{x'}$, whose Poisson brackets vanish, and whose commutator brackets, after canonical quantization, do also. It follows here because the commutators involved $[a, a]_-, [a, b^\dagger]_-$, and so on, are in the category "all others are zero".

Also, the commutator

$$[\phi(x,t), \phi^\dagger(x',t)]_- = 0.$$

This follows, not quite so trivially as above, from direct calculation. It follows from the classical analog because ϕ and $\phi^\dagger$ are both generalized coordinates q_1, q_2, say, whose Poisson brackets vanish, and whose commutator brackets also vanish. It can also be argued to vanish because these equal time operators are separated by a space-like interval and the two operations cannot be causally connected, therefore the order of the operation cannot matter. The difficulty with this argument is taking the limit as $x \to x'$ in a convincing way.

One more exercise is worth doing to support our choice of a and b. The equal time commutation relation

$$[\phi(x,t), \Pi_{\phi(x',t)}]_- = [\phi(x,t), \phi^\dagger_{;t}(x',t)]_- \tag{33}$$

can easily be shown from the above expansions in terms of the $a, a^\dagger$ and $b, b^\dagger$ to be $i\delta^3(\vec{x} - \vec{x}')$, just the intuitive result for the canonical commutation relation

$$[q_k, \frac{1}{i}\frac{\partial}{\partial q_j}]_- F(q's) = i\delta_{kj} F(q's). \tag{34}$$

Now we sketch deWet's proof that anticommutation relations are untenable for spin-0 fields.

Suppose we try to replace the canonical commutation by an anticommutation, so we investigate the possibility

$$[\phi(x,t), \phi^\dagger(x',t)]_+ \overset{?}{=} 0,$$

which we assume to hold even at $x = x'$. The diagonal matrix element for an arbitrary state $|\mu\rangle$ is

$$\sum_\chi \langle\mu|\phi(x,t)|\chi\rangle\langle\chi|\phi^\dagger|\mu\rangle + \text{hermitian conjugate} = 0$$

where we have inserted a complete orthonormal set of states

$$1 = \sum_\chi |\chi\rangle\langle\chi|.$$

The result is that

$$\sum_\chi |\langle\chi|\phi|\mu\rangle|^2 = 0,$$

17

leading to the conclusion that a scalar field operator satisfying the above anticommutation relation has no matrix elements in the Hilbert space and would have to be the null operator.

More simply, deWet concluded that the anticommutation relation required $|\phi(x,t)|^2 = 0$, which could only be satisfied if $\phi(x,t) = 0$. This non-operator statement is the correct conclusion, but does not make explicit the underlying assumptions on the requirements of the Hilbert space.

The Dirac field escapes this fate. The anticommutator $[\psi, \psi^\dagger]_+$ is not in the category "all the rest are zero", because $i\psi^\dagger$ is the momentum Π_ψ conjugate to ψ, and this anticommutator, by analogy to the canonical commutators, is not zero, but $\delta^3(x - x')$. The above matrix element is not zero but infinite, $\sim \delta^3(0)$, and the Dirac field survives. deWet went on to show that no such exceptions could exist for tensor fields and therefore that it is impossible to quantize integral spin fields with anticommutation relations.

§5. Concluding Remarks.

Our introductory remarks are basically the proofs of Iwanenko and Socolow of 1937 for the spin-$\frac{1}{2}$ Dirac equation (Ch.10), and deWet of 1939 for the spin-0 Klein-Gordon equation. Where did they fall short? Were all further contributions legitimate quarrels or nit-picking quibbles? There is a long list of criticisms, not specifically aimed at these authors because their contributions, which were cited (Iwanenko and Socolow by deWet and deWet by Pauli, at least) so people were aware of them, were almost universally ignored.

Let's list the shortcomings of these simple proofs and others to be presented over the years as they were addressed:

1) Pauli (1940-Ch.14) criticized deWet's proof for being limited to spin-0 and spin-$\frac{1}{2}$, and also for being limited to the canonical formalism which is difficult to carry beyond these low spins because of the need for a prolifer-

ating array of subsidiary conditions.

2) None of the proofs until those finally put forward by Lüders and Zumino and Burgoyne (1958- Ch.17) included the effect of interactions, but dealt with free fields only, or in the case of Feynman (Ch.15) and Schwinger (Ch.16) with free fields interacting perturbatively.

3) The formal requirements underlying the analytic continuations and the manipulations of the usually singular products of operators and their matrix elements were eventually addressed. The formalism reached daunting dimensions in the work of Hall and Wightman (Ch.18) but they substantiate the intuitive conclusions which are attributable most especially to deWet.

Schwinger was the only one to take a less formal view about higher spin particles and interactions. He dismissed the fundamental significance of higher spin particles, and assumed that a model using spin-$\frac{1}{2}$ and spin-0 constituents would give the required generalization with sufficient rigor. Belinfante also constructed product wave functions of free undors (his name for relativistic spinors, see Ch.12) which are very suggestive of todays constituent models. Schwinger seemed also to have an informal view based on perturbation theory and the interaction representation to deal with the (non-)effect of interactions on statistics. His view resembled the original work of Heisenberg and Dirac on the non-relativistic many-body problem, that a (anti-)symmetrized wave function propagated forward in time by a symmetric Hamiltonian necessarily retains its (anti-)symmetry.

Schwinger's opinions on higher spin particles probably correspond to the present view, that spin-0 and spin-$\frac{1}{2}$ solve the basic problem, and that other spins can be understood in principle by imagining that such particles are made up of these either virtually or as constituents, and their statistics determined accordingly. This leaves the truly arduous early work of Fierz (Ch.11), Belinfante, and Pauli himself, based on the spinor representations of the Lorentz group, as an isolated bastion of arcane, and even obsolete and unnecessary, formalism. It is not possible to understand the evolution of the

19

subject without these proofs, and they do provide a consistency backup, but they play no essential role in the much more economical, general, and elegant proofs especially of Burgoyne, which of course has a very deep foundation in Hall and Wightman, as well as deep roots, whether or not they were acknowledged or even known, especially in the work of deWet.

Ultimately we present a much simpler proof, based on work by Schwinger (Ch.19), which frees the Spin-Statistics Theorem from the complicating shackles imposed on it by Pauli, and returns the proof to the much simpler realm of non-relativistic quantum mechanics (Ch.21), and which we hope satisfies Neuenschwander first of all.

We now embark on a tour of what veritably might be called archeological treasures of Physics' own Valley of the Kings.

Bibliography and References.

0.1) D.E. Neuenschwander, Am. J. Phys. **62**(11), 972 (1994).

0.2) R.P. Feynman, R.B. Leighton. and M. Sands, *The Feynman Lectures on Physics, Vol.3* (Addison-Wesley, Reading, MA 1963), Ch.4, Sec.1.

0.3) W. Pauli, Phys. Rev. **56**, 716 (1940); see also our Ch.14.

0.4) R.F. Streater and A.S. Wightman, *PCT, Spin and Statistics, and All That* (Benjamin, NY, 1964), pp.146-161.

0.5) J. Schwinger, *Quantum Electrodynamics* (Dover, New York, 1958).

0.6) A.S. Wightman, Phys. Rev. **101**, 860 (1956); see also our Ch.18.

Chapter 1

Discovery of the Exclusion Principle

Summary: We describe Pauli's deduction of the Exclusion Principle, based on Stoner's explanation based on four quantum numbers of the Periodic Table of the Elements and the magic numbers 2,8,18,32,···.

§1. Introduction.

There is no one fact in the physical world which has a greater impact on the way things *are*, than the Pauli Exclusion Principle. To this great Principle we credit the very existence of the hierarchy of matter, both nuclear and atomic, as ordered in Mendelejev's Periodic Table of the chemical elements, which makes possible all of nuclear and atomic physics, chemistry, biology, and the macroscopic world that we see.

The critical fact originated by Bohr that atomic electrons must occupy only a discrete set of quantum states is necessary but not sufficient to understand the Periodic Table. Bohr himself struggled with the problem of constructing an atomic model which would have the required detailed properties, especially the observed family structure of the elements, but he was forced to invoke *ad hoc* excitations of the inner atomic cores which were not only unappealing but which still failed to produce key features of the Periodic Table [1.1]. Furthermore, they were soon shown by Pauli to be inconsistent with the properties of inert gases. As we discuss at length in this chapter, E.C. Stoner, then a research student at Cambridge, had the remarkable insight to use a new quantum number, introduced by Landé in a different context, with which he was able to modify Bohr's atomic model and to reproduce exactly the observed $2, 8, 18, \cdots$ family structure of the Periodic Table of the Elements [1.2].

Stoner was on the verge of even more remarkable insights which he in fact stated and even emphasized, but never quite explicitly recognized. He was on the verge of discovering the electron spin-$\frac{1}{2}$. He twice stated the Exclusion Principle, in effect if not in generality. He wrote the statistical weight of the individual quantum state (h^3) which is the essence of Bose's contribution (Ch.3).

It was left however to Pauli [1.3] to realize the full implications of Stoner's analysis.

§2. Pauli's Deduction of the Exclusion Principle.

Pauli seized upon the inspired analysis by Stoner to abstract from the closed shell structure of the inert gases the profound generalization:

"There can never be two or more equivalent electrons in an atom. These are defined to be electrons for which - in strong magnetic fields - the value of all quantum numbers n, k_1, k_2, m_1, (or equivalently n, k_1, m_1, m_2), is the same. If, in the atom, one electron occurs which has quantum numbers (in an external field) with these specific values, then this state is occupied."

In contemporary notation these quantum numbers are replaced by n (unchanged, the principal quantum number); l (replacing k_1 or k in Bohr's original model) - the orbital angular momentum quantum number with

$$l = k_1 - 1 < n, \quad = 0, 1, 2 \cdots = s, p, d \cdots; \tag{1}$$

j (replacing k_2 in Pauli which, in turn, was identical to what Stoner called j) - the total angular momentum quantum number

$$j = k_2 - \frac{1}{2} = l \pm \frac{1}{2} \tag{2}$$

- which serves to distinguish the alkali doublets; $m_j \equiv j_z$ (replacing m_1 or m in Bohr's model) - the component of the total angular momentum in the field direction; and m_2, giving the component of the electron's magnetic moment in the field direction, equivalent to the component of the electron's spin,

$$m_2 = l_z + 2s_z = j_z + s_z = m_j \pm \frac{1}{2}. \tag{3}$$

In all that follows, we have taken the liberty of rephrasing the arguments of Stoner and Pauli in terms of (n, l, j, m_j) or (n, l, m_j, m_2) instead of the original (n, k_1, k_2, m_1) or (n, k_1, m_1, m_2). We also denote the total angular momentum of the whole atom by J, to distinguish it from the total angular momentum of the state of an individual electron j.

Pauli eliminated an earlier hypothesis in which the complex optical spectra and the anomalous Zeeman effect were explained by *ad hoc* active properties of the atomic core. He had already pointed out the difficulties of this model, especially regarding the closed shells of inert gases which would be expected also to have such active properties. Both these properties were now explained by Pauli as due to a doubling of the quantum states of the valence electrons which manifests itself in the optical doublets of the alkalis as well as in the relativistic doublets of the X-ray spectrum.

In terms of Stoner's original quantum numbers, the number of states for given (k_1, k_2) is $2k_2 = 2j + 1$ and the number of states for given k_1 is

$$2 \cdot (k_2 = k_1 - 1) + 2 \cdot (k_2 = k_1) = 2 \cdot (2k_1 - 1) = 2 \cdot (2l + 1). \tag{4}$$

Pauli purposely disavows any theoretical model building. "We will seek no further theoretical analysis $\cdots\cdots$ the following deductions $\cdots\cdots$ on the foundation of experimental fact." In doing this Pauli firmly closed his eyes to the two great clues for the electron spin and anomalous moment. "Often the appearance of half-integral (effective) quantum numbers, $\cdots\cdots$ the hypothetical value $g = 2 \cdots\cdots$ are closely connected with the doubling $\cdots\cdots$."

Certainly the triumph he did achieve was great, but he must have been galled by what he missed.

In his analysis of the alkali earths Pauli adds an (n, l) valence electron to an alkali atom with an $(l = 0)$ valence electron to get a total of $2 \cdot (2[l = 0] + 1)$ states from the alkali atom combined with $2 \cdot (2l + 1)$ states of the added valence electron for a total of

$$4 \cdot (2l + 1) = 1 \cdot (2l + 1) + 3 \cdot (2l + 1) \tag{5}$$

states in the singlet and triplet systems. The number of states is clearly understood without the need for any induced states in the core atom as Bohr had suggested. "$\cdots$ the Bohr 'induction' is understood $\cdots$ as the inherent doubling of the quantum-theoretic properties of the individual electrons in the stationary states of the atoms."

Then with uncharacteristic humility, perhaps anticipating Bohr's criticism, he says of his own purely combinatoric explanations "$\cdots$ the explanation put forth here $\cdots\cdots$ cannot be accepted as a sufficiently physical basis $\cdots$ especially because $\cdots$ better explanations have already been given."

In his first step to the Exclusion Principle, Pauli observes that in the alkali earths with two equivalent valence electrons there is a Singlet-S term but no Triplet-S except with larger principal quantum numbers. After some discussion, he raises the crucial question: "$\cdots\cdots$ for those values of $\cdots n$ and l, for which electrons are already present $\cdots$, certain multiplet terms of the spectrum will be absent. Then the question arises, by what quantum-theoretic Rule is this term suppression dictated. $\cdots\cdots$ this question is deeply involved with the $\cdots$ closing of the electron shells $\cdots$ expresses the fact that a shell $\cdots$ can take up no more than $2n^2$ electrons $\cdots$ Stoner $\cdots$ noticed $\cdots$ the number of electrons agreed with the number of $\cdots$ states." Then follows the statement of the Exclusion Principle (italicized above), and the recognition of its fundamental nature: "$\cdots$ we cannot give a deeper explanation $\cdots$."

The moment of Genesis was complete.

Pauli does recognize the power of the Exclusion Principle, not only to

24

explain the period lengths $2, 8, 18, 32 \cdots$ of the Periodic Table, but also to explain the missing spectral terms and to predict the possible multiplets of complex atoms. The remainder of the paper makes application to known and conjectured spectra: "$\cdots$ it is clear that our Rule gives unambiguous answers $\cdots$ the results are in agreement with experiments $\cdots$ in the simplest cases."

Pauli's closing statement was to be a dominant theme of his research for the next thirty years: "The problem of a deeper foundation for the fundamentally general Rule $\cdots$ about the existence of equivalent electrons in atoms, probably $\cdots$ needs first a more profound understanding of the basic principles of the Quantum Theory in order to be be effectively attacked."

In the following chapters, we collect some of the fundamental advances which contributed to this more profound understanding of quantum theory. But first, we return to the earlier work of Stoner which is so important as *the* source of inspiration for Pauli's creation of the Exclusion Principle.

§3. Stoner's Explanation of the Periodic Table.

Only a sampling of Stoner's paper is reproduced here (App.1B), just enough to give on the one hand, the flavor of the confused and incomplete evidence he had to work with; and on the other, the almost outright statements of the Exclusion Principle which he did write down, but left for Pauli to completely explicate, just as Pauli seemingly willfully refused to interpret the quantum doubling and left the electron spin to be discovered by others.

In fact, Stoner quotes Landé as the originator of the inner quantum number k_2 (Stoner's j; Pauli's $k_2 = k_1, k_1 - 1$; now replaced by $k_2 = j + \frac{1}{2} = l + 1, l$ in modern notation). The difficulty solved by Landé was to correct the earlier Bohr model, and replace it by a unified interpretation of the alkali optical doublets and the X-ray doublets as described by a division of the k_1 level into *two* sub-levels labeled by a *two*-valued quantum number k_2.

25

Stoner's great success was to choose for k_2 a number linear in k_1 and directly related to the number of electrons in the sub-level. The choice $k_2 = k_1, k_1 - 1$ and $N(k_1, k_2) = 2k_2$ reproduces the inert gas series $2, 8, 18, 32 \cdots$. Stoner actually contemplated using the half-integral j values now assigned to these levels, but opted instead for the integral valued $k_2 = j + \frac{1}{2}$.

Stoner says in a sadly timorous and self-effacing if not self-defeating way "Without laying too much stress on any definite physical interpretation $\cdots$ it may be suggested that for an inner sub-level $\cdots$ the number of possible orbits is equal to twice the inner quantum number $\cdots$." Then Stoner almost definitively states the Exclusion Principle: "Electrons can enter a group until all the possible orbits are occupied $\cdots$".

Stoner's next remark, about the statistical weight of equivalent electrons in a completed group being unity, is particularly inspired, as well as a precursor although unacknowledged source of inspiration for quantum statistics.

The most direct evidence for the inner electron distribution predicted by Stoner comes from the intensity of X-ray lines, and supports the choice $2k_2 = 2j + 1 = 2, 4$ for the L_{II} and L_{III} sub-levels from Fe(26) to W(74).

Also, Stoner anticipated Pauli's Reciprocity Rule for the equivalent magnetic behavior of levels with x electrons and the closed level missing x electrons.

Pauli was almost effusive in his generous references to Stoner. But there remained serious doubt [1.4] that Stoner was properly credited, even though Pauli made the essential abstraction from Stoner's analysis to the statement of the Exclusion Principle in its full generality.

Twenty one years later Pauli was honored for his achievement with the award of the Nobel Prize. On this occasion the credit he gave to Stoner was ungenerous [1.5]: "$\cdots$ Stoner $\cdots$ improvements in the classification of electrons in subgroups, but also the essential remark that the number of energy levels $\cdots$ is the same as the number of electrons in the closed shells $\cdots$".

§4. Biographical Note on E.C. Stoner.(†)

Edmund Clifton Stoner (1899-1968) was a student at Emmanuel College, Cambridge (1918); a research student at Cavendish, where he found experimental research frustrating and Rutherford's bullying manner disconcerting. After intense reading and discussions in the Kapitza Club (a theoretical seminar), he was prepared for his successful first research (Phil. Mag. **48**, 719 (1924), reproduced in part in our App.1B); this led to a lectureship at Leeds University (1924) where he spent his entire career; with FRS (1937); professor of theoretical physics (1939). He had major research achievements in the theory of magnetism, with pioneering contributions successively in dia-, para-, and ferromagnetism; leading to the books *Magnetism and Atomic Structure* (London, 1926), *Magnetism and Matter* (London, 1934) and the text *Magnetism* (London, 1930; 1948); and in the theory of specific heats; and in the theory of dense stars where his use of Fermi-Dirac statistics to predict the limiting mass of white dwarfs anticipated Chandrasekhar's.

The originality and impact of Stoner's many contributions, especially in the field of magnetism, can not be overstated.

(† - G. Cantor in *Dictionary of Scientific Biography* (Charles Scribner's Sons, New York, 1994), edited by Frederic L. Holmes, pp.876-877; E.P. Wohlfarth in *The University of Leeds Review* **12**, 98-101 (1969); L.F. Bates in *Biographical Memoirs of Fellows of the Royal Society* **15**, 201-237 (1969); private communication from Professor Phillip Rhodes, Leeds.)

Bibliography and References.

1.1) N. Bohr, Zeits. f. Phys. **9**, 1 (1922).

1.2) E.C. Stoner, Phil. Mag. **38**,749 (1924); see our App.1B.

1.3) W.Pauli, Zeits. f. Phys. **31**, 765 (1925); see our App.1A.

1.4) J.L. Heilborn, *The Origins of the Exclusion Principle* in Historical Studies in the Physical Sciences **13**, 261-310 (1983).

1.5) W. Pauli, *Remarks on the History of the Exclusion Principle* in Science **103**,213-215 (1946).

APPENDIX 1.A:

Excerpt from: Zeitschrift für Physik, **31**, 765 (1925)

About the Relationship of the Closing of Electron Shells in Atoms with the Complex Structure of the Spectrum

by W. Pauli jr. in Hamburg

(Received on 16 January 1925.)

Abstract: In the description of the alkali doublets and their anomalous Zeeman effect by the relativistic Millikan-Landé formula, and also in results obtained in other work, there occurs a doubling of the quantum-theoretic properties of the valence electrons which is not describable classically. Without this doubling, the closing of the inert gas configurations of the atomic core in the form of inner electron shells, not as the site of a magnetic anomaly of the atom, is a concern. When this doubling is used as a working hypothesis for the assignment of quantum numbers, it is possible, in spite of its difficulties, to include other atoms than the alkalis in its consequences. In contrast to the usual understanding, the result, at least in the case of a strong external magnetic field where the coupling between the valence electron and the atomic core can be ignored, is that these two particle systems can be described very simply. The number of stationary states, the value of their quantum numbers, and their magnetic energy require no other properties in their description than those of the free atomic core and the valence electron of the alkalis. On the basis of these considerations we arrive at a general classification of each electron in the atom according to the principal quantum number n and two auxiliary quantum numbers l and j, and, in the presence of an external field, a further quantum number m_j. Following new work of E.C. Stoner, this classification leads to a general quantum-theoretic explanation of the closing of electron shells in atoms.

§1. The Permanence of the Quantum Numbers (Constituent Principle) of Complex Structures and the Zeeman Effect.

In an earlier work [1], it was emphasized that the usual idea, that the inner closed electron shells of the atom form a core and are the inherent site of the magnetic anomaly in the complex structure of the optical spectrum and its anomalous Zeeman effect, gives rise to various serious difficulties. Here it will be made clear that this idea is in contrast to another, *that the doublet structure of the alkali spectrum as well as its anomalous Zeeman effect result from a doubling of the quantum-theoretic properties of the valence electrons which cannot be described classically.*[Note: All such italics added.] This observation is supported by the result of Millikan and Landé, that the optical doublets of the alkalis are analogous to the relativistic doublets in the X-ray spectrum and their magnitude is determined by a relativistic formula.

We now discuss this viewpoint further. Following Bohr and Coster for the X-ray spectrum, we label the stationary states of the valence electron corresponding to the emission of the alkali spectrum by the principal quantum number n and two auxiliary quantum numbers l and j. The first quantum number $l \cdots$ has the values $0, 1, 2 \cdots$ for the $s, p, d \cdots$ terms and changes by one unit during transitions; it determines the magnitude of the central force interaction of the valence electron with the atomic core. The second quantum number j is equal to $l - \frac{1}{2}$ and $l + \frac{1}{2}$ for the two terms of a doublet (for example $p^{\frac{1}{2}}$ and $p^{\frac{3}{2}}$); in the transition processes it changes by ± 1 or 0 and determines the magnitude of the relativistic correction (according to Landé, because of the different penetration of the valence electron into the atomic core). If we define - with Sommerfeld - for a particular stationary state of the atom, the total angular momentum quantum number j as the maximum value of the quantum number m_j (usually written m), which corresponds to the angular momentum component parallel to an external field, then for the alkali atoms $j = l \pm \frac{1}{2}$. The number of stationary states in the magnetic field for given l and j is $2j + 1$. The number of these states for both doublet terms of given l is $2(2l + 1)$.

Next consider the strong field case (the Paschen-Back effect). In addition to l and m_j, we introduce instead of j another magnetic quantum number m_2, which gives directly the energy of the atom in the magnetic field. It is the component of the magnetic moment of the valence electron parallel to the field. The two terms corresponding to the doublet have the values $m_2 = m_j + \frac{1}{2}$ and $m_j - \frac{1}{2}$. Just as in the doublet structure of the alkali spectra, the "anomaly of the relativistic correction" to the expression occurs (for whose magnitude a different quantum number is required than for the magnitude of the central force interaction energy of the valence electron and the atomic core). This gives rise to an anomaly of the Zeeman type from the normal Lorentzian triplet which, for the known anomaly, is the analog in appearance to the "magnetic anomaly" (for the magnitude of the magnetic moment of the valence electron still another quantum number is necessary, different from the angular momentum). Often the appearance of half-integral (effective) quantum numbers and, in the same formalism, the need for the hypothetical value $g = 2$ for the splitting factor for the s-term of the alkalis, are closely connected with the doubling of the term levels. *We will seek no further theoretical analysis of this state of affairs.* Instead, the following deductions from the Zeeman effect of the alkalis will be laid down

29

on the foundation of experimental fact.

Without concerning ourselves with the status of our understanding or with the possible difficulties, we attempt the formal classification of the valence electrons using the four quantum numbers n, l, j, m_j, just by transcribing for complex atoms the results for the alkali atoms. Then it turns out, on the basis of this classification and in contrast to the usual assumption about the permanence of the quantum numbers, that we can understand completely both the complex spectra and the anomalous Zeeman effect. This principle of Bohr states that during the addition of further electrons to an atom (loading it in all possible ways), the quantum numbers of the already present electrons keep the same values which they had in the previous stationary states of the free atom.

First we consider the alkali earths. The spectrum here consists of a singlet- and a triplet-system. For a definite value of the quantum number l for the valence electron, there are $1 \cdot (2l + 1)$ singlet states and $3 \cdot (2l + 1)$ triplet states in an external field. Previously one has assumed, that in strong fields in each case there are $2l + 1$ configurations for the valence electron, while the core atom in the first case has one, and in the second case three possible configurations. The number of these configurations is obviously different from the number 2 of configurations of the free core atom (alkali-like s-term) in a field. Bohr called these states "induced", but the analogy cannot be with the effect of an external field [2]. We can now however simply conclude that the total of $4 \cdot (2l + 1)$ states of the atom result from the fact that the core atom has two states in an external field, and the valence electron - as for the alkalis - has $2 \cdot (2l + 1)$ states.

Generally, according to a branching rule given by Heisenberg and Landé [3], a stationary state of a core atom with N states in an external field, by addition of a further electron gives rise to two term systems, which for a definite value of the quantum number l of the added electron correspond to $(N + 1)(2l + 1)$ or $(N - 1)(2l + 1)$ states. According to our interpretation these $2N(2l + 1)$ states of the whole atom arise from N states of the core atom and $2(2l + 1)$ states of the valence electron. From the assumed quantum-theoretic classification of the electrons it is clear that the term multiplicity of the branching rule follows simply as a consequence of the constituent principle. According to the idea proposed here, the Bohr "induction" is understood not as a violation of the Permanence of the Quantum Numbers by the coupling of the valence electrons to the core atom, but *as the inherent doubling of the quantum-theoretic properties of the individual electrons in the stationary states of the atoms.*

Moreover, in agreement with the constituent principle, we can calculate both the number of stationary states and also their energies in the case of strong external fields (the part proportional to the field strength), by adding those of the free core atom and of the valence electron, which are known from the alkali spectrum. In this case, it depends both on the total component $\bar{m}_j$ of the angular momentum of the atom parallel to the field (in units of $\hbar$), and also on the component $\bar{m}_2$ of the magnetic moments of the atom in the same direction (in units of Bohr magnetons). These are the sums of the quantum numbers m_j or m_2 for the individual electrons

$$\bar{m}_j = \sum m_j, \qquad \bar{m}_2 = \sum m_2. \tag{1}$$

The last sum runs over values, independent of one another, which correspond to the values of the angular momentum quantum numbers l and j of the electrons in the particular stationary states of the atom. (Here $\bar{m}_2 h \nu$, with $\nu =$ the Larmor frequency, is the part of the atom's energy proportional to the field strength.)

We consider as an example the two s-terms (singlet and triplet) of the alkali earths. We need to include only two valence electrons, since the contribution of the rest of the electrons to the sums in Eqn.1 vanishes. Each of these two valence electrons can assume the values $m_j = -\frac{1}{2}$, $m_2 = -1$ and $m_j = \frac{1}{2}$, $m_2 = 1$ (independent of the other electrons) of the s-terms of the alkalis. We get from (1) the following values of the quantum numbers $\bar{m}_j$ and $\bar{m}_2$ for the whole atom

$$\bar{m}_j = -\frac{1}{2} - \frac{1}{2}, \quad -\frac{1}{2} + \frac{1}{2}, \quad +\frac{1}{2} - \frac{1}{2}, \quad +\frac{1}{2} + \frac{1}{2}$$
$$\bar{m}_2 = -1 - 1, \quad -1 + 1, \quad +1 - 1, \quad +1 + 1$$

or

$$\bar{m}_j = -1, 0, 0, +1$$
$$\bar{m}_2 = -2, 0, 0, +2$$

(corresponding to a term with total angular momentum J of the whole atom $J = 0$ and a term with $J = 1$ in weak fields) [4]. In order to get the p-,d-$\cdots$ terms of the alkali earths, for a fixed contribution of the first valence electron, one must insert for the second electron the m_j- and m_2-values of the p-,d-,$\cdots$ terms of the alkalis in (1).

In general, the prescription (1) leads exactly to a technique recently given by Landé [5] for the calculation of strong field energy values. As I have pointed out, it also gives the correct result in complicated cases. For example, according to Landé, this prescription gives the Zeeman terms of the neon spectrum (at least for strong fields), when one assumes in the core atom an active electron in a p-term instead of in an s-term [6] and further allows the valence electron to run through the s-,p-,d-,f-,$\cdots$ terms.

This result also requires that each electron in the atom must be characterized by two auxiliary quantum numbers l and j in addition to the principal quantum number n, even in the presence of many equivalent electrons or in closed electron groups. Further we will

31

imagine a magnetic field sufficiently strong that we can label each electron, independent of the others, by two quantum numbers m_j and m_2 in addition to n and l (where m_2 determines the electron's share of the magnetic energy). The relation between l and m_j and between j and m_2 is to be determined as for the alkali spectrum.

Before we apply this quantum-theoretic classification of the electrons to the problem of the closing of the electron groups, we must still address the difficulties which the complex structure and the anomalous Zeeman effect present to the ideas introduced here, and to the limitation of their exact significance.

First, this idea is not dependent on the detailed behavior of the different term systems (for example, of the alkali earth singlet and triplet systems), or on the exact validity of the Landé interval rule. Of course, one cannot assume for the energy splitting of the alkali triplet levels two different causes, for example both the anomaly of the relativistic correction of the valence electrons and also the dependence of the interaction energy with the atom core upon the relative orientation of the two systems.

A still more important difficulty appears in the compatibility of the proposed idea with the correspondence principle, which has been indispensable for the explanation of the selection rules of the quantum numbers l, j, m, as well as for the polarization of the Zeeman components. According to this principle, it is not necessary in a definite stationary state for each electron to be assigned a unique orbit in the sense of ordinary kinematics; however the state of the atom as a whole should correspond to a class of orbits with a definite type of periodicity properties. This leads to the selection- and polarization-rules of the correspondence principle for the types of motion of a central orbit with superposed precession giving an ill-defined orbit around a definite axis of the atom for weak external fields or even a precession in the field direction along an axis through the nucleus.

Such a dynamical explanation of these types of motion of the valence electrons, which depends upon the assumption of a weakening of the central symmetry of the force of the core atom on the electron, seems to be incompatible with the representability of the alkali doublets (and also the magnitude of the corresponding precession frequency) by the relativistic formula. Similar conclusions hold for the motion in strong fields.

There arises here the difficult problem of how the types of motion of the valence electrons, required by the correspondence principle, can be interpreted physically, independent of their previously assumed but no longer supportable special dynamical representation. Also involved with this problem is the question of the magnitude of the term values of the Zeeman effect, in particular the alkali spectrum.

As long as this problem remains unsolved, the explanation put forth here of the

32

complex structure and of the anomalous Zeeman effect certainly cannot be accepted as a sufficiently physical basis for an explanation of these phenomena, especially because in many respects better explanations have already been given. Perhaps in the future a merging of these two viewpoints will be given. For now the question of most interest to us, is to follow the consequences of this new explanation as far as possible. In this sense it will be possible to understand, quite unexpectedly, from the starting point of difficulties opposed to it, the problem of the closing of the electron groups in the atom. To that end, we will discuss only the number of possible stationary states of the atom for the existence of more equivalent electrons, but not the positions and ordering of the term values.

§2. About a General Quantum-Theoretic Rule for the Possibility of the Occurrence of Equivalent Electrons in Atoms.

It is well known that the existence of many equivalent electrons in an atom is possible only under particular conditions which depend on the details of the complex structure of the spectrum. For example, for the alkali earths in the normal state, in which the two valence electrons are equivalent, there is a Singlet-S term; whereas in the usual stationary states of atoms, those valence electrons which are almost equally bound belong to the Triplet system and the large Triplet-S term corresponds to a larger principal quantum number than the normal state. We observe a second example in the neon spectrum. There, two groups of terms occur with different series limits, corresponding to different states of the rest of the atom. The first group, which belongs to electrons with the quantum numbers $l = 1$ and $j = \frac{1}{2}$ separated from the rest of the atom, can be characterized as a Singlet- and a Triplet-system, whereas the second group, electrons with $l = 1$ and $j = \frac{3}{2}$ separated from the rest of the atom, can be characterized as a Triplet- plus a Quintet-system. The ultraviolet resonance lines of neon are still not observed, but there is little doubt that the normal state of the neon atom which combines with the known excited states of the atom must be a p-term. In fact, only one such term gives unequivocally certain agreement with the diamagnetic behavior of the inert gas, and it has the value $J = 0$ [7]. Since the p-term with $J = 0$ is the lowest Triplet term p_0 of the two groups, we can conclude that for Ne the principal quantum number of these two Triplet terms must be $n = 2$ and both term-series must be identical.

Generally, we can expect that for those values of the quantum numbers n and l, for which electrons are already present in the atom, certain multiplet terms of the spectrum will be absent. *Then the question arises, by what quantum-theoretical Rule is this term suppression dictated.*

As already mentioned in the example of the neon spectrum, this question is deeply involved with the problem of the closing of the electron shells in atoms, which are restricted to the numbers $2, 8, 18, 32, \cdots$ the lengths of the periods of the Periodic Table of the Elements. This closure expresses the fact that a shell with principal quantum number n, whether through emission or absorption of radiation or through other external interactions, can take up no more than $2n^2$ electrons.

Bohr, in his theory of the Periodic Table, has given a unified explanation of the spectroscopy and chemical properties, and a quantum-theoretical basis for the behavior of the chemically analogous elements, such as the iron- and platinum-group metals, and the rare-earth salts in the later periods of the system. He does all this by a partitioning of the electron shells into subshells. In his theory, he characterizes each electron in the stationary states of the atom as determined by the stationary states of a central motion designated by a symbol n_k with $k \leq n$. Then he adopts in general, for an electron shell with the principal quantum number n, n subshells. In this way, Bohr was able to construct the scheme of atomic structure for inert gases shown in Table 1. He himself concluded, however, that the assumed equality of the number of electrons in the different subshells of a principal shell was extremely tentative, and that a complete and satisfactory theoretical representation of the closure of the shells in atoms, in particular a representation of the period lengths 2,8,18,32,... in the Periodic Table, had not yet been given [8].

Recently, essential progress has been made on the problem of the Closure of the Electron Shells in Atoms in the work of E.C. Stoner [9]. Stoner has constructed a scheme for the atomic structure of the ideal gases for which, in contrast to Bohr, no opening of a closed subshell by the addition of further electrons of the same principal shell is allowed. Also, the number of electrons in a closed subshell depends only on the value of $l \equiv k - 1$, but not on the value of n, which requires the existence of further subshells of the same principal shell. Stoner's scheme represents things with a great simplicity, and moreover it can fit many different experimental facts. For this, one must assume for $l = 0$ two, for $l = 1$ six, for $l = 2$ ten, and generally for a particular l value $2(2l + 1)$ electrons in closing the states of a particular subshell, in order to obtain agreement with the empirically known electron numbers of the inert gas atoms.

Stoner also noticed that these numbers of electrons agreed with the number of stationary states of the alkali atoms in an external field for a given value of l. He arrived at the analogy to the the stationary states in the alkali spectrum, by assuming that the complex structure of these spectra (and the X-ray spectra) correspond to further division of the subshells into two partial-subshells distinguished by two numbers l, j, where $l + 1$ agrees with Bohr's number k and $j = l - \frac{1}{2}$ or $j = l + \frac{1}{2}$ (except for $l = 0$, where, to agree with the uniqueness of the s-terms, $j = \frac{1}{2}$ only). Corresponding to the number $2j + 1$ of stationary states, into which a stationary state of the alkali atom with given quantum numbers l and j divides in an external field, Stoner assumes $2j + 1$ electrons required to close the corresponding partial-subshells of quantum numbers n, l, j. The scheme for the atomic structure of the inert gases arrived at in this way by Stoner, is shown in Table 2.

We can now summarize and generalize Stoner's idea. We apply the explanation of the complex structure of the spectra and the anomalous Zeeman effect described in the above paragraph to the problem of the existence of equivalent electrons in atoms. We are led, from the possibility of having particular overall quantum numbers, to characterizing each electron in the atom by a principal quantum number n as well as by the two supplementary quantum numbers l and j. In strong magnetic fields an angular momentum quantum number m_j occurs for each electron. And we also introduce, in addition to l and m_j, and instead of j, a magnetic quantum number m_2. Next we see that the use of both quantum numbers l and j for each electron is in best accord with Stoner's partitioning of Bohr's subshell [10]. Now, from the consideration in Stoner's work of the case of strong magnetic fields, where the number of electrons closing a subshell was determined to correspond with the number of terms in the Zeeman effect in the alkali spectrum, we can state the following general Rule about the occurrence of equivalent electrons in atoms:

There can never be two or more equivalent electrons in an atom. These are defined to be electrons for which - in strong magnetic fields - the value of all quantum numbers n, l, j, m_j, (or equivalently n, l, m_j, m_2) is the same. If, in the atom, one electron occurs which has quantum numbers (in an external field) with these specific values, then this state is occupied.

It is essential to keep in mind, that the principal quantum number n is included in this rule; obviously very many more (non-equivalent) electrons can occur in the atom, whose values of the quantum numbers l, j, m_j coincide, but they differ from each other through their principal quantum number n.

We can not give a deeper explanation of this rule, but it does seem to be very natural. It applies, as mentioned, particularly for the case of a strong field. From thermodynamic considerations [11] the number of stationary states of the atom must be the same in strong and weak fields and be given by the values of the numbers l and j for the individual electrons and the value of $\bar{m}_j = \sum m_j$ for the whole atom. Therefore, even in the latter case, we can make a definite statement about the number of stationary states and the value of the total J belonging to them (which gives the number of equivalent electrons for different values of l and j). Then the number of possibilities for the different

35

closed shells can be deduced. Also the question asked at the beginning of this section about the exclusion or suppression of a given multiplet term in the spectrum for values of the principal quantum number for which many equivalent electrons are present, can be answered unequivocally for each individual case. Still we can only say something about the number of terms and the value of their quantum numbers, but nothing about their magnitudes and their interval ratios [12].

Next we have to show that the consequences of our Rule agree with the facts in the simplest cases. Whether this will remain true in complicated cases, or whether changes will be required, remains to be seen and might be determined by further disentangling complicated spectra.

First we see that Stoner's result and with it the period lengths $2, 8, 18, 32, \cdots$ in the Periodic Table are in immediate agreement with our Rule. Obviously, for fixed l and j and no more equivalent electrons than there are values of m_j (namely $2j + 1$), it leads directly to the closed shells with one electron for each of these values of m_j.

Secondly, the absence in alkali-earths of Triplet-S-terms with the same principal quantum number as the normal state follows directly from our Rule. In particular, for the binding of two equivalent electrons in s-terms (here $l = 0$ and j can only have the value $\frac{1}{2}$), then according to our Rule, in strong fields the case with $m_j = \frac{1}{2}$ or $-\frac{1}{2}$ for both electrons is excluded; we can only have $m_j = \frac{1}{2}$ for the first electron and $m_j = -\frac{1}{2}$ for the second or vice versa [13]. As a result, the quantum number $\bar{m}_j = \sum m_j$ of the whole atom must be zero. Also in the weak field (or no field) case, only the value $J = 0$ (Singlet-S-term) is possible.

Next, consider the case where one electron is removed from a closed shell, as happens in the X-ray spectrum. It is possible to have the case where one electron is missing from one of Stoner's partial-subshells, so that no electron has one of the values of m_j; we call this the "unoccupied" value of m_j. The remaining electrons divide themselves among the other values of m_j, so that one electron occupies each value of m_j. The total sum of these remaining values of m_j and therefore the quantum number $\bar{m}_j$ of the whole atom is just the negative of the m_j which is unoccupied. Let us run through all these possible values, and keep in mind that for each partial-subshell, one electron can be missing. Then we see that in strong fields the multiplicity of the unoccupied value of m_j and therefore also of $\bar{m}_j$, is the same as that of the m_j-value of a single electron. Because of the invariance of the statistical weight, it follows also for weak fields that the equality of the number of stationary states and the j-value of the singly ionized closed electron shell (X-ray spectrum) corresponds with the alkali spectrum, as is the case.

This is a special case of a general Reciprocity Principle: To each configuration of the electrons, there is a conjugate configuration, for which the unoccupied m_j and the occupied m_j are interchangeable. This interchange can refer to a single partial-subshell without changing configurations of other partial-subshells, or to one Bohr subshell, or to a total principal shell, since the individual partial-subshells are completely independent of one another. The electron numbers of the two conjugate configurations add up to the number of electrons in the closed state of the particular shell (or subshell). The j-values of the two configurations are equal to one another. It also follows that the sum of the unoccupied m_j is equal but opposite to the sum of the occupied m_j within a configuration. That is, the quantum numbers $\bar{m}_j$ of the whole atom are equal but opposite for conjugate configurations. Since the J-values are defined as the limits of the $\bar{m}_j$-values, they must coincide in the conjugate configurations (see also the examples described below). From this Reciprocity Principle, the behavior at the end of a period of the Periodic Table can be inferred from that at the beginning of the period. It must be noted however, that the Reciprocity Principle can say something only about the number of stationary states of the particular shell and the value of its quantum numbers, but nothing about the magnitude of its energy values and its interval structure [14].

We now discuss the application of our Rule to the special case of the successive construction of the octet-shell (where no electrons with $l > 1$ are allowed in the normal state of this particular principal quantum number), which provides another example of the Reciprocity Principle. The binding of the first two electrons in this shell was already described and in the following we will make the simplification that neither of the two electrons is missing from the $l = 0$ subshell, which is therefore closed (see the above description of Stoner's scheme). Then - according to Stoner - for the elements which close the octet-shell (that is from B to Ne), we must add a p-term to the normal state in order to agree with the known facts. In particular, we get the alkali-like doublet spectrum by adding the third electron to the octet-shell, which exhibits the familiar suppression of the s-terms with the same principal quantum number as the normal state.

We can continue in this way to bind the fourth electron in the octet-shell. In this way we make some progress in understanding the still unanalysed arc-spectrum of carbon and the partially understood arc-spectrum of lead. According to the Landé-Heisenberg Branching Rule (compare the previous section) the resulting spectrum should be of the same general structure as the neon-spectrum, namely consisting of a Singlet-Triplet group and a Triplet-Quintet group with different series limits, which correspond to the $2_p^{\frac{1}{2}}$ and the $2_p^{\frac{3}{2}}$ Doublets with $n = 2, l = 1, j = \frac{1}{2}, \frac{3}{2}$ of the corresponding ions [15]. We next show that according to our Rule, these spectra - in contrast to the analogous structure of their excited states expected from the number and j-value of the p-terms with larger principal quantum number (for C $n = 2$, for Pb $n = 6$) - must differ essentially from the Ne-spectrum (where in addition to the normal state with $J = 0$, no further p-terms with

37

$n = 2$ exist, as stated at the beginning of this paragraph).

Here we distinguish three cases, according to the electron numbers in the two partial-subshells with $l = 1, j = \frac{1}{2}$ and with $l = 1, j = \frac{3}{2}$. We have to assign two electrons (the first two are assumed to be bound in s-terms with $l = 0, j = \frac{1}{2}$):

a) Two equivalent $n_p^{\frac{1}{2}}$-electrons. For this partial-subshell (corresponding to the $p_{\frac{1}{2}}$-term of the alkalis), m_j can assume only two values $m_j = \pm\frac{1}{2}$. These are therefore in the closed state with $\bar{m}_j = 0$ and $J = 0$.

b) One $n_p^{\frac{1}{2}}$-electron and one $n_p^{\frac{3}{2}}$-electron. For the latter partial-subshell corresponding to the p_2-term of the alkalis, m_j can take four values $\pm\frac{1}{2}, \pm\frac{3}{2}$. These can combine directly with the already given values $m_j = \pm\frac{1}{2}$ of the first electron, since the two electrons are in different partial-subshells and are therefore not equivalent [16]. We get

$$
\begin{aligned}
\bar{m}_j &= (-\frac{3}{2}, -\frac{1}{2}, \frac{1}{2}, \frac{3}{2}) + (-\frac{1}{2}, \frac{1}{2}) \\
&= \pm(\frac{3}{2} + \frac{1}{2}), \quad \pm(\frac{3}{2} - \frac{1}{2}), \quad \pm(\frac{1}{2} + \frac{1}{2}), \quad \pm(\frac{1}{2} - \frac{1}{2}) \\
&= \pm2, \quad \pm1, \quad \pm1, \quad 0, \quad 0.
\end{aligned}
$$

One sees that the terms split into two series, one with $|\bar{m}_j| \leq 2$ and one with $|\bar{m}_j| \leq 1$. These correspond in the field free case to two terms, one with $J = 2$ and one with $J = 1$.

c) Two equivalent $n_p^{\frac{3}{2}}$-electrons. Here the m_j values of the two electrons must be different according to our Rule. For the possible values of $\bar{m}_j$ we get:

$$
\begin{aligned}
\bar{m}_j &= \pm(\frac{3}{2} + \frac{1}{2}), \quad \pm(\frac{3}{2} - \frac{1}{2}), \quad (\frac{3}{2} - \frac{3}{2}), \quad (\frac{1}{2} - \frac{1}{2}) \\
&= \pm2, \quad \pm1, \quad 0, \quad 0.
\end{aligned}
$$

In the absence of a magnetic field these give one term with $J = 2$ and one term with $J = 0$.

We get - in total - for the 4-shell, five different p-terms with the same principal quantum number: two with $J = 2$, one with $J = 1$ and two with $J = 0$.

About the grouping of these terms or their magnitudes and their interval behavior, we can say nothing. On the other hand, we can make definite statements about the expected Zeeman splittings of these terms.

By fixing the m_2-value of the individual electrons with the indicated m_j-values (as inferred from the strong-field Zeeman terms of the alkalis), we get the prescription from which the following table of strong-field Zeeman splitting of the five p-terms is constructed:

$$
\bar{m}_j = -2; -1; 0; 1; 2;
$$

38

$$\bar{m}_2 \;=\; -3,-2; \quad -2,-1,-1; \quad 0,0,0,0,0; \quad 1,1,2; \quad 2,3.$$

With the help of the sum rule given by Landé for multiplets of higher rank [17], for the weak-field value of the sum of the g-values for the two $J = 2$ terms ($\sum g_2$) and of the g-value for the $J = 1$ term (g_1), one gets the equations

$$2 \sum g_2 = 2 + 3 = 5, \qquad \sum g_2 + g_1 = 1 + 1 + 2 = 4.$$

The solution is

$$\sum g_2 = \frac{5}{2}, \quad g_1 = \frac{3}{2}.$$

A test of this theoretical result is almost possible for lead. In this case, four p-terms have been reliably determined from observations, whereas the existence of a fifth p-term is still doubtful [18]. Although as yet unpublished, measurements made by E. Back on the Zeeman effect of several lead lines make it very plausible that the just mentioned p-terms have the J-values (2,2,1,0) and that the g-values of these terms are in agreement with the theory [19].

We now discuss further the successive construction of the octet-shell. With the help of the Reciprocity Principle, and using the whole Bohr subshell with $l = 1$, which contains six electrons in the closed state, then we can take over the results from the 4-shell directly to the number of possible configurations of the 6-shell (four electrons with $l = 1$), as for example in oxygen. The cases a), b), c) of the 4-shell are obviously conjugate to the following cases of the 6-shell:

a) Four equivalent $n_p^{\frac{3}{2}}$-electrons (two empty places in the $n_p^{\frac{1}{2}}$-shell). These partial-subshells are closed; therefore we get under a) a term with $J = 0$.

b) One $n_p^{\frac{1}{2}}$-, three equivalent $n_p^{\frac{3}{2}}$- electrons (one empty place in the $n_p^{\frac{1}{2}}$-, one empty place in the $n_p^{\frac{3}{2}}$-shell). We get one term with $J = 2$ and one term with $J = 1$.

c) Two equivalent $n_p^{\frac{1}{2}}$-, two equivalent $n_p^{\frac{3}{2}}$-electrons (two empty places in the $n_p^{\frac{3}{2}}$-shell). The first partial-subshell is closed. We get one term with $J = 2$, one term with $J = 0$.

We have here also, for example for oxygen, five p-terms with the lowest principal quantum number. So far, for oxygen and sulfur only three such terms have been observed, and these have $J = 2, 1, 0$ [20]. It remains to be seen, whether there are two further p-terms of the particular principal quantum number still to be observed, or whether our Rule must be modified for this case.

There are no observations for the 5-shell, so we will only give the results predicted by our Rule. For this shell, there are five p-terms: one with $J = \frac{5}{2}$, three with $J = \frac{3}{2}$, and one with $J = \frac{1}{2}$. For the 7-shell appearing in the X-ray spectrum, there is, as already stated, the alkali-like term.

We will not discuss any other special cases which depend on existing observational results. From the given examples it is clear that our Rule gives unambiguous answers to the question of the possible configurations of each shell, together with the number of equivalent electrons in each case. That the results are in agreement with experiment, can be claimed so far, it is true, only in the simplest cases.

In general it should be remarked, that these considerations, which hold for the transition from strong to weak and even to vanishing fields, rely in principle upon the invariance of the statistical weight of the quantum states. *For the existence of a solution to the problem of the closing of the electron shells in atoms based on the Correspondence Principle, as was conjectured by Bohr, there seems to be no available starting point as a basis for the desired result. The problem of a deeper foundation for the fundamentally important general Rule about the existence of equivalent electrons in atoms, probably first needs a more profound understanding of the basic principles of the Quantum Theory in order to be effectively attacked.*

Footnotes and References.

1) W. Pauli jr., Zeits. f. Phys. **31**, 373 (1925).

2) N. Bohr, Ann. d. Phys. **71**, 228 (1923), especially p.276.

3) W. Heisenberg and A. Landé, Zeits. f. Phys. **25**, 279 (1924). On the question of the limit of validity of this rule and especially on the theoretical explanation of the so-called excluded term $\cdots\cdots$ we will not go into detail here. Suffice it to say, there is an essential disagreement with the usual combination rule.

4) One sees that the two cases, $m_j = -\frac{1}{2}$ for the first electron and $m_j = +\frac{1}{2}$ for the second, and vice versa, must be considered as two different terms (in regard to the field independent part of the energy). This is possibly an incompleteness in the classification carried out here. It will be shown later, that for the equivalence of the inner and outer valence electrons these two terms must in fact be identical.

5) A. Landé, Ann. d. Phys. **76**, 273 (1925); see especially §2.

6) The equivalence of the seven-electron shell (core atom of neon) with an electron will be justified theoretically in the next paragraph.

7) This is defined, as usual, as the value of j, and in the following is fixed by the maximum

value of the quantum number m_j.

8) N. Bohr, Three Essays about Spectra and Atomic Structure, 2. Aufl. Braunschweig 1924, Anhang.

9) E.C. Stoner, Phil. Mag. **48**, 719 (1924). This important article was referred to in the forward to the new edition of Sommerfeld's book "Atombau und Spektrallinien".

10) That this partitioning and the question of the number of electrons in the partial-subshell required for closing the electron shells has merit, arises directly from the Millikan-Landé result about the relativistic doublets in the X-ray spectra. These same numbers occur explicitly in the expression for the energy of the whole shell as a function of the order number, as factors with known values in the screening number (determined by l) and as relativistic corrections (determined by j) in the Mosely-Sommerfeld expression.

11) This invariance is independent of the validity of classical mechanics during the transformation.

12) Comment added in proof: In work by A. Sommerfeld (Zeits. f. Phys. **26**, 70 (1925)) the question of the normal state of the atom was discussed in connection with Stoner's proposal.

13) The reverse case corresponds to an exchange of two electrons and does not give a new quantum state. But in this double realization of a quantum state, one must abstain from multiplying by two for a statistical weight with regard to exchange of the two electrons (compare the observations about statistical weights in the cited work of Stoner).

14) It follows from the equality of the multiplicity of the m_2-values under conjugation, that even in strong fields the g-sums of the corresponding terms (extending over terms with equal j) are equal to one another.

15) This theoretical expectation can not as yet be conclusively verified empirically because of the incomplete disentangling of the spectra involved. A. Fowler (Proc. Roy. Soc. **107**, 31 (1925)) recently discovered in the N^+-spectrum, in addition to forbidden terms (about whose appearance, as yet, nothing theoretical can be said), Singlet and Triplet terms. Furthermore, Kiess (Science **60**, 249 (1924)) also discovered Quintet terms in this spectrum. This result, which arises from excited states of N^+, at least does not contradict the theoretical expectation.

16) On this basis, the case with $m_j = +\frac{1}{2}$ for the first, $m_j = -\frac{1}{2}$ for the second electron, and $m_j = -\frac{1}{2}$ for the first, $m_j = +\frac{1}{2}$ for the second must be counted here as different. See Footnote 13.

17) A. Landé, Ann. d. Phys. **1**, 6 (1925).

18) V. Thorsen, Naturwissenschaften **11**, 78 (1923); W. Grotrian, Zeits. f. Phys. **18**, 169 (1923).

41

19) This information was made available to me through a most friendly communication from Herr Back, who provided me with the results even before their publication. To him, I wish to express my warmest thanks. Herr Back's measurements were concerned with the general question of term analysis in the Lead spectrum, amongst other things.

Comment added in Proof: the term analysis of the spectra of Lead and Tin was meanwhile carried further in work by Frl. Sponer, whose important results are contained in the Sitzung des Gauvereins Niedersachsen in Göttingen on February 9. For Tin, five p-terms were obtained with the largest principal quantum number, whose J-values correspond exactly to those given theoretically. For Lead, the J-values obtained for the known large p-terms were in agreement with those in the Text. The existence of a fifth p-term with $J = 0$, as for Tin, could also be verified for Lead, as an essential criterion.

20) J.J. Hopfield, Astrophys. Journ. **59**, 114 (1924); O. Laporte, Naturw. **12**, 598 (1924). See also A. Sommerfeld, Atombau und Spektrallinien, 4 Aufl. 1924, Ch.8, P.598 and 599.

APPENDIX 1.B:

Excerpt from: Philosophical Magazine 48, 719 (1924)

The Distribution of Electrons Among Atomic Levels

By EDMUND C. STONER, B.A.

Emmanuel College, Cambridge†

§1. Introduction.

The scheme for the distribution of electrons among the completed sub-levels in atoms proposed by Bohr [1] is based on somewhat arbitrary arguments as to symmetry requirements; it is also incomplete in that all the sub-levels known to exist are not separately considered. It is here suggested that the number of electrons associated with a sub-level is connected with the inner quantum number characterizing it, such a connection being strongly indicated by the term multiplicity observed for optical spectra. The distribution arrived at in this way necessitates no essential change in the process of atom-building pictured by Bohr; but the final result is somewhat different, in that a greater concentration of electrons in outer sub-groups is indicated, and the inner sub-groups are complete at an earlier stage. The available evidence as to the final distribution is discussed, and is not unfavorable to the scheme proposed.

§2. Classification and Number of X-ray Levels.

The X-ray atomic levels may be conveniently classified by means of three quantum numbers - n (total), k_1 (azimuthal (Note added: $k_1 = l + 1$ in modern notation, which will be used here for the rest of Stoner's paper)), and k_2 (inner (Note added: $k_2 = j + \frac{1}{2}$)), as shown in Table I. This classification has been put forward by Landé [2]. In contradistinction to the older schemes, such as that of Sommerfeld, it gives a satisfactory selection principle (l changes by 1, j by 1 or 0), and at the same time brings out clearly the analogy between X-ray and optical spectra. The sub-levels may, in fact, be regarded as corresponding to σ, π, δ $\cdots$ doublet-series terms, as for alkali metal arc spectra, in the way indicated in the last row of the table.

The main criticism to be advanced against the classification is that it invalidates the interpretation of terms such as L_{II}-L_{III} as simple relativistic doublets, although their

43

separation is given accurately by Sommerfeld's formula. Bohr [3], however, has shown that although L_I-L_{III} was to be regarded as a relativistic + screening doublet, the further subdivision into separate relativistic (L_{II}-L_{III}) and screening doublets (L_I-L_{II}) was not justifiable. · · · · · · Landé [4] · · · · · · shows · · · beyond doubt that the two types of doublet are essentially similar in origin.

Now, observations on the anomalous Zeeman effect show that the optical doublets must have a magnetic origin; · · · · · · it is justifiable to apply those ideas which have coordinated the optical data to the case of X-rays - in particular in the assignment of inner quantum numbers, as in the above scheme.

· · · · · · experiments thus support the view that for X-ray spectra, as for optical doublet series, the number of j sub-levels into which a given l level subdivides is restricted to two.

§3. Suggested Distribution of Electrons.

In the classification adopted, the remarkable feature emerges that the number of electrons in each completed level is equal to double the sum of the inner quantum numbers assigned, there being in the K, L, M, N levels when complete,

$$2, 8(2 + 2 + 4), 18(2 + 2 + 4 + 4 + 6), 32 \cdots$$

electrons. It is suggested that the number of electrons associated with each sub-level separately is also equal to double the inner quantum number $2k_2 = 2j+1$. The justification for this is discussed below. A summarized periodic table (Table II.) is given, which shows the nature of the distribution suggested. · · ·

§4. Comparison with Bohr's Distribution.

· · · · · ·(Note added: See the tables in Pauli's article for this comparison.)

The present scheme, then, accounts well for the chemical properties; it differs from Bohr's in the final distribution suggested, and in the fact that inner sub-groups are completed at an earlier stage, subsequent changes being made by simple addition of electrons to outer sub-levels without reorganization of the group as a whole.

§5. Significance of Inner Quantum Numbers.

From a physical point of view, the real significance of inner quantum numbers, especially when applied to inner X-ray atomic levels, is very problematical. Evidence based on the analogous optical spectra, however, provides strong justification for the idea of the number of electrons in a sub-group being related to the inner quantum number in the way assumed.

The case of the doublet series of the alkali metals only need be considered. In the atoms there is one electron external to a core of electronic groups. Observations on the Zeeman effect can be correlated by assigning inner quantum numbers corresponding to $j = l - \frac{1}{2}$ and $l + \frac{1}{2}$ to the atom with the electron in an l level. For π terms ($l = 1$), $j = \frac{1}{2}$ or $\frac{3}{2}$, giving π_2 and π_1; for δ terms ($l = 2$), $j = \frac{3}{2}$ or $\frac{5}{2}$. There is actually a certain degree of arbitrariness as to the absolute value given to j; for instance, values 1 and 2 for π_2 and π_1, 2 and 3 for δ_2 and δ_1 can be made to fit the facts equally well [5]. This difficulty, however, is irrelevant, for the reasons given below. The inner quantum number is usually interpreted as giving the magnetic moment of the atom as a whole, and the number of possible energy states of the atom in a weak external magnetic field, in which core and light electron are not separately affected, is attributed to the number of possible orientations of the of the atom in virtue of space-quantization.

The actual number of possibilities is given by the multiplicity of terms in the anomalous Zeeman effect, and can be deduced very straight-forwardly (in simple cases at least) from observations on the behavior of the lines. The point which it is desired to emphasize here is that, however the inner quantum numbers are interpreted, if they are given the values above ($j = l + \frac{1}{2}$ and $l - \frac{1}{2}$), twice the inner quantum number $k_2 = j + \frac{1}{2}$ does give the observed term multiplicity as revealed by the spectra in a weak magnetic field. (Thus in a weak magnetic field there are 2σ, $2\pi_2$, and $4\pi_1$ terms.) In other words, the number of possible states of the (core + electron) system is equal to twice the inner quantum number, these $2j + 1$ states being always possible and equally probable, but only manifesting themselves in the presence of the external field.

At present it is not clear whether the number of equally probable states indicates that the atom as a whole is always the same as concerns relative orientation of core and outer electron orbit, and can take up $2j + 1$ different orientations relative to the (weak) field; or that the core takes up a definite orientation relative to the field, and the outer electron orbit can take up $2j + 1$ different orientations relative to the core [6]. (The mutual influence becomes of less relative importance as the strength of the field increases, so that ultimately the field affects the electron and core separately and the Paschen-Back effect is obtained.) There may be some quite different interpretation.

The spectral term-values themselves, in so far as they are altered by external fields, would seem to depend primarily on the outer electron orbit itself (and not so much on electron + core, as do the magnetic properties of the atom); and remembering this, it seems reasonable to take $2j + 1$ as the number of possible equally probable orbits.

Without laying too much stress on any definite physical interpretation, or pressing the analogy too far, it may be suggested that for an inner sub-level, in a similar way,

45

the number of possible orbits is equal to twice the inner quantum number, these orbits differing in their orientation relative to the atom as a whole. *Electrons can enter a group until all the possible orbits are occupied* [Note: All such italics added.], when the atom will possess a symmetrical structure.

That the inner quantum numbers are analogous to those for the alkali metals, is presumably connected with the fact that the building-up can always be regarded as occurring on a sub-structure of the inert gas atom type with completed group systems. The complicated optical multiplet series occur when there are light-electrons moving externally to incompleted groups.

In brief, then, it is suggested that, corresponding roughly to the definite indication in the optical case, the number of possible states is equal to $2j + 1$; so, for the X-ray sub-levels, $2j + 1$ gives the number of possible orbits differing in orientation relative to the atom as a whole; *and that electrons can enter a sub-level until all the orbits are occupied.*

§6. Statistical Weight of Electrons Bound in Atoms.

If electrons in the atom are distributed according to the present scheme, the interesting point is suggested that all electrons bound in the atom forming constituents of completed groups are to be regarded as having the same statistical weight, namely unity (or h^3); *for there is then one electron in each possible equally probable state.*

§7. Evidence as to Electron Distribution.

$\cdots$ in the present state of the theory, a definite test $\cdots$ is difficult. $\cdots$

(a)Intensities of X-ray Lines.

$\cdots$ In one case very accurate measurements have been made, namely for

$$K\alpha_1(L_{III} \rightarrow K) \quad \text{and} \quad K\alpha_2(L_{II} \rightarrow K).$$

The results are as follows:-

	Fe	Cu	Za	Mo	W
$\alpha_2/\alpha_1 \cdots$	.50	.51	.50	.52	.50

The ratio α_1/α_2 is thus practically constant and equal to 2/1 from Fe(26) to W(74).

This result can at once be explained on the assumption that there are twice as many electrons in the L_{III} as in the L_{II} sub-level. Now in Bohr's scheme four electrons are

46

assigned to the (2,2) level, so that 4 electrons have to be divided between L_{II} and L_{III} - a ratio 3/1 or 1/1 for the α lines would be expected, certainly not 2/1.

$\cdots\cdots$ The results $\cdots$ do give definite support to the allocation of 2 electrons to the II sub-levels and 4 to the III sub-levels, in the L, M, and probably N groups.

(b) Absorption of X-rays.

$\cdots\cdots$ Owing to experimental difficulties, data are meagre, Dauvillier's for gold being the only direct ones available [7]. From his curves $\cdots$, applying de Broglie's correction, he obtains

$$N(L_{II})/N(L_{III}) = .495, \qquad N(L_I)/N(L_{II}) = .78,$$

a conclusion that may be taken to support the $2, 2, 4$ distribution of electrons.

$\cdots\cdots$

(c) Magnetic Properties.

$\cdots\cdots$ in agreement with the M group by the simple addition of 10 similar (n, l) orbits; the presence of x electrons in the $(M_{IV}$ and $M_V)$ levels, superposed on completed groups, may be expected (by a sort of Babinet principle!) to produce the same paramagnetic properties as (10-x) electrons, for with the latter number the group diverges from non-paramagnetic completeness in the absence of x electrons. $\cdots\cdots$

(d) Chemical Properties.

$\cdots\cdots$

(e) Optical Spectra.

$\cdots$ the doublet spectrum of carbon (CII) [8] can at once be explained if C_+ has 2(2,1) electrons and 1(2,2); whereas it cannot be fitted easily into a 4(2,1) scheme for neutral carbon. The spectra of silica $\cdots$ fit $\cdots$ as also the recently analyzed oxygen spectrum [9]. $\cdots$ many unsolved problems for the future.

This section may be briefly summarized. The X-ray emission-line intensities seem to provide conclusive evidence for the presence of 2 and 4 electrons in L_{II} and L_{III} as inner atomic sub-levels over a wide range of atomic numbers, a subdivision also indicated for M_{II} and M_{III}. The absorption measurements confirm this, and suggest a distribution of the $2, 2, 4, 4, 6$ type for the $5M$ and the first $5N$ sub-levels. The chemical evidence is strongly in support of the up-building of the $I, II, III L, M, \cdots P$ sub-levels as suggested

47

by the final 2, 2, 4 distribution, and this is confirmed in an important case by the optical spectra. The chemical properties also indicate strongly the number 6 as characterizing the $M, N,$ and $O\ V$ sub-groups, and magnetic considerations suggest 10 as the number of electrons in the completed M_{IV} and M_V sub-levels.

While evidence based on experiment is inadequate to provide quantitative proof of the correctness of the whole system of electronic distribution proposed, it seems conclusive as to the simpler sub-groupings, and collectively does lend strong support to a scheme in itself simple and consistent.

Summary.

A distribution of electrons in the atom is proposed, according to which the number in a sub-group is simply related to the inner quantum number characterizing it. A formal justification of the connection is given. The suggested number of electrons in the completed $K, L, M \cdots$ groups are (2), (2,2,4), (2,2,4,4,6) $\cdots$ respectively. The scheme is compared to that of Bohr. It enables all the essential features involved in Bohr's picture of atom-building to be retained, and so is equally in accord with general chemical and spectroscopic evidence; but it differs in the distribution in the completed groups, and in indicating a somewhat simpler mode of development. Evidence based on considerations of intensities of X-rays, chemical and magnetic properties, and optical spectra is discussed and shown to give considerable support to a distribution of the kind put forward.

I would like to thank Mr. R.H. Fowler for helpful criticism and discussion.

Footnotes and References.

†) Communicated by R.H. Fowler, M.A.

1) N. Bohr, Zeits. f. Phys. **9**, 1 (1922); or 'The Theory of Spectra and Atomic Constitution', Essay III (Cambridge, 1922).

2) A. Landé, Zeits. f. Phys. **16**, 391 (1923).

3) N. Bohr, Zeits. f. Phys. **12**, 342 (1922).

4) A. Landé, Zeits. f. Phys. **24**, 88 (1924); **25**, 96 (1924).

5) A. Sommerfeld, Ann. der Phys. **70**, 32 (1923); **73**, 39 (1924). See also A. Landé, Zeits. f. Phys. **19**, 112 (1924).

6) N. Bohr, Ann. der Phys. **71**, 228 (1923).

7) A. Dauvillier, Comptes Rendus **178**, 476 (1924).

8) R.H. Fowler, Proc. Roy. Soc. **105**, 228 (1924).

9) J.J. Hopfield, Phys. Rev. **21**, 710 (1923).

Cavendish Laboratory, July 1924.

Chapter 2

The Discovery of the Electron Spin-$\frac{1}{2}$

Summary: The discovery of the spin of the electron as the physical interpretation of the fourth quantum number introduced by Stoner is followed through the credited discovery work of Goudsmit and Uhlenbeck, back to a precursor in the suggestion of Compton, and finally to Kronig, whose primary work never reached publication.

§1. Introduction.

The story of the discovery of the electron spin-$\frac{1}{2}$ is a fascinating tale of the various ways that science attains its intended goal. The credited discovery by Goudsmit and Uhlenbeck [2.1] is well known, and will be discussed here in close reference to their original papers. Less well known, and briefly included here, is a precursor to that discovery in a qualitative proposal by A.H. Compton [2.2] that the electron "spinning like a tiny gyroscope" could be the source of magnetization needed to explain many magnetic properties of matter. Apparently Compton's proposal had no direct impact on the ultimate discovery of electron spin. Nonetheless, he does deserve credit for his early proposal deduced from macroscopic properties of magnetic materials, long before the classification of atomic spectra made the quantitative discovery of a spinning electron almost mandatory.

There is also a dark side to the history of the discovery of the electron [2.3]. Kronig [2.4], as a brilliant neophyte student, had the idea at least a full six months before Goudsmit and Uhlenbeck but was discouraged from taking it seriously by Pauli and others including Bohr and Heisenberg. In closing, we include a brief comment about Kronig and his work on spin.

§2. Goudsmit and Uhlenbeck's Discovery of Spin.

Goudsmit and Uhlenbeck were led to the necessity of a fourth, internal, essentially quantum degree of freedom for the electron to correspond to the fourth of the four quantum numbers (n, l, j, m) or (n, l, m, m_2) used by Pauli and Stoner to classify the atomic electrons. In Goudsmit and Uhlenbeck's first paper, the fourth quantum number is ascribed "to an eigen-rotation of the electron" instead of just ascribing it, as Pauli did, to a "classically undescribable double-valuedness". They also pointed out that Abraham's rotating spherical electron (with only a surface charge) has the desired gyromagnetic ratio for spin which is twice that for orbital motion. They offered no excuses for the fact that the classical Abraham electron also requires a surface velocity greater than the speed of light.

Two months later they supported their initial report with a paper in Nature, which expanded on their suggestion and developed the spectroscopic consequences which, in every case, amounted to a physical basis for the results of Pauli, Stoner, and Landé [2.5].

Their first application is to the fine structure of hydrogen-like spectra in Fig 1. The quantum number then in use was k (Pauli and Stoner's k_1) with $l = k_1 - 1 = 0, 1, 2 \cdots$ (or $(s, p, d \cdots)$) in modern notation) which was replaced in their view by

$$K = k_1 - \frac{1}{2} = l + \frac{1}{2} = \frac{1}{2}, \frac{3}{2}, \frac{5}{2}, \cdots \tag{1}$$

which split to states specified by J (Pauli and Stoner's $k_2 = k_1, k_1 - 1$) with

$$\begin{aligned} J &= K \pm \frac{1}{2} = (1,1), (2,2) \cdots \quad \text{(doublets)} \\ &= k_1, k_1 - 1 = l + 1, l \quad \text{but not 0 when} \quad l = 0 \\ &= j + \frac{1}{2} \Rightarrow j = l \pm \frac{1}{2} \end{aligned} \tag{2}$$

in modern notation. For reasons left unexplained, the doublets

$$(S_{\frac{1}{2}}, P_{\frac{1}{2}}), \quad (P_{\frac{3}{2}}, D_{\frac{3}{2}})$$

and so on, are nearly degenerate. This modification resolved paradoxes connected with $(\Delta L = 1, \Delta J = 0)$ transitions which had previously been thought to be $\Delta L = 0$, for example

$$4D_{\frac{3}{2}} \rightarrow 3P_{\frac{3}{2}}.$$

In addition, the two phenomena of screening doublets and relativistic doublets are seen clearly to have the same origin, as different manifestations of spin doublets. Screening doublets are nearly degenerate electron states with the same J but different L which have different penetration of the inner core removing the degeneracy. For example

$$nS_{\frac{1}{2}}, nP_{\frac{1}{2}}.$$

Relativistic doublets are states with the same L and the same screening, but different J, which are approximately degenerate. An example is

$$nP_{\frac{1}{2}}, nP_{\frac{3}{2}}.$$

Goudsmit and Uhlenbeck interpret the alkali-spectra and the complex spectra in terms not different from Pauli. And finally the anomalous Zeeman effect and the high field Paschen-Back effect are given a fully unified dynamical explanation in terms of the anomalous gyromagnetic ratio for the electron spin.

Initially, Goudsmit and Uhlenbeck met with much negative criticism of their idea from Lorentz and, as recounted by Uhlenbeck in his 1955 inaugural lecture as Lorentz professor at Leiden, they were so discouraged that they asked their mentor Ehrenfest to withdraw their paper. Ehrenfest's great kindness - accompanied by the remark "I have already sent off your paper; besides, you are both young enough that you are permitted to make dumbheads of yourselves." - assured its publication.

§3. Kronig's Frustrated Discovery of the Electron Spin.

The opposite fate befell the twenty year old Ralph de Laer Kronig who actually anticipated the discovery of the electron spin (in January 1925, only a little more than a week after arriving to begin his graduate education in Germany, and some eight months before the work of Goudsmit and Uhlenbeck in September 1925). The tragic obstruction of the young Kronig by the criticism of Pauli, Heisenberg (themselves only 25) and Bohr is documented in meticulous detail by B.L. van der Waerden in his chapter "Exclusion Principle and Spin" included in the Pauli Memorial Volume published in 1960. The same volume includes Kronig's own recollection of events, including a 1955 statement of Uhlenbeck that he and Goudsmit had heard a hint from Heisenberg who could not remember where he had heard the idea, and also a very contrite letter from Bohr to Kronig in 1926 to the same effect.

The detailed history of these events is definitively documented with first person quotations and letters, in van der Waerden's eulogy to Pauli. There is no better source for an intimate understanding of the origin of the ideas or of the interplay of the participants in this great drama, than van der Waerden's authoritative article from which we quote his last words on the subject absolving Pauli - "We have seen that the doubts of Pauli, Heisenberg and Kronig were justified in many respects. Lorentz, too, had serious doubts, which were well founded at that time. In my opinion, Pauli and Heisenberg cannot be blamed for not having encouraged Kronig to publish his hypothesis."

Kronig, who served after the resolution of the electron spin problem as Pauli's assistant, closes his paper (written before Pauli's sudden death) with the generous remark - "··· After our first meeting, Pauli's orbit and my own have crossed repeatedly and in various ways, leaving me with the feeling of having received more than I ever could give in the vital atmosphere which surrounds him."

Kronig's belated article on spin is included to get a flavor of his thinking also. As we now know, all the difficulties have been resolved: the anomalous gyromagnetic ratio in the classical theory by Thomas precession; the concern with an extended electron by the Dirac point electron with the correct magnetic property; the absence of electron magnetic moments in nuclei by the discovery of the neutron and the invention of field theory. Finally, Kronig likens the physics of Goudsmit and Uhlenbeck to hiding ghosts rather than truly exorcising them.

§4. Biographical Notes.

a) Profile of A. H. Compton.(†)

Arthur H. Compton (1892-1962) was educated at the College of Wooster, with a PhD from Princeton (1916); subsequent appointments were at Minnesota, Cambridge, Oxford, Washington University, and the University of Chicago. He was Wayman Crowe Professor of Physics and Chairman of the Physics Department at Washington University (1921-23); Professor of Physics (1923-), Chairman of the Physics Department and Dean of Natural Sciences (1940-), and Director of the Mannhattan District's Plutonium Research Project (1942-) at the the University of Chicago (1923-45); Chancellor (1945-53), Distinguished Professor of Natural Philosophy (1954-61), and Professor-at-large (1961-62) at Washington University. His wartime research experiences are described in his book *Atomic Quest, a Personal Narrative (1956)*.

Compton won the 1927 Nobel Prize in Physics (shared with C.T.R. Wilson) for his experiments on the elastic scattering of photons by electrons. In earlier work, he was first to succeed in finding diffraction patterns with X-rays at grazing incidence on ruled gratings. This permitted successively: 1) standardization of X-ray wavelengths; 2) determination of lattice constants using Bragg diffraction; and 3) a determination of Avogadro's number N^* and the electron charge e.

He was President of the American Physical Society, was awarded some twenty medals and prizes, and twenty-five honorary degrees. Compton's older brother Karl T. Compton, who is referred to in his paper suggesting a spinning electron, was later President of the A.P.S. and President of MIT.

(† - from PHYSICS TODAY, May 62, p.88; also, D.L. Livesey, *Atomic and Nuclear Physics* (Blaisdell, Waltham, 1966), pp.34,95,107.)

b) Profile of S.A. Goudsmit.(††)

Samuel A. Goudsmit (1902-1978), born in The Hague, educated at the University of Leiden, received his PhD in 1927. Already having published his first paper, at the age of 19 he had established himself as "house theoretician" with a special talent for solving puzzles of all kinds. His and Uhlenbeck's personal reminiscences appear in PHYSICS TODAY, June 76, p.40. He claimed that his determination with Ernst Back of the first nuclear spin by the analysis of the hyperfine structure of Bi^{209} was more thrilling than the discovery of the spin of the electron. With R.F. Bacher, he introduced fractional parentage coefficients. With Bacher, he coauthored the book *Atomic Energy States* (McGraw-Hill, New York 1932); with Linus Pauling, *The Structure of Line Spectra* (McGraw-Hill, New York, 1930).

Goudsmit was a professor at the University of Michigan (1927-41); worked on radar research at the MIT radiation Laboratory (1941-43); was a professor at Northwestern University (1946-48); and a member of Brookhaven National Laboratory (1948-70) where he was Chairman of the Brookhaven Physics Department (1952-60) during a vital period in its development; visiting professor at Rockefeller University (1960-70); professor at the University of Nevada (1971-1977).

Goudsmit had a remarkable episode in his career when he was detailed to the Army on a scientific intelligence mission to determine the status of the Nazi atomic bomb effort. His experiences are reported in his book *Alsos* (H. Shuman, New York, 1947, out of print), with the code name of his mission. He had the task of de-briefing Heisenberg as a post-war detainee, which must have been made difficult by deep respect for Heisenberg's profound place in physics; and even moreso because Goudsmit's parents had been killed in the Holocaust.

Goudsmit was Editor-in-Chief of the American Physical Society (1951-74) and founding editor of *Physical Review Letters* (1958). His editorial broadsides directed sometimes at the writing abilities of physicists, sometimes at the hyperactivity of theorists, were filled with exasperation tempered with good humor. One from Phys. Rev. Lett. **15**, 543 (1965): "··· a few theorists ··· want a central register of preprints and other unpublished reports ··· The next step might be to equip theorists with portable recorders so that all their statements about physics, including those uttered in their sleep, ··· would be transmitted to interested colleagues ··· computers coded with key words could scan the tapes ··· result in such chaos as to make priority assignments impossible ··· theoretical physics would become anonymous, just like ··· art of ancient Egypt."

Goudsmit's many honors include; the Max Planck Medal of the German Physical Society (1965); the Karl T. Compton Award AIP (1974); Commander of the Order of Orange-Nassau (1977); and the National Medal of Science (1977).

(†† - from Maurice Goldhaber, PHYSICS TODAY, April 79, p.71.)

c) Profile of G. E. Uhlenbeck.(†∗, † † ∗)

George Eugene Uhlenbeck (1900-1988) was one of the last survivors of the *Wunderkind* who established quantum theory in the years 1924-26. After early studies in engineering and then mathematics and physics, Uhlenbeck entered graduate school at the University of Leiden in 1920. After a period of indecision from 1922-25 - during which he travelled, taught in Italy, learned Italian, met and studied with Fermi, passed his doctoral exams, and published three papers - he settled down at Leiden where in a period of weeks he had shared in the discovery of the electron spin.

His PhD thesis under Paul Ehrenfest was on statistical mechanics, and his life is characterized as the conscience of that subject.

Uhlenbeck was a professor at the University of Michigan (1927-35; 39-60); head of the theory division at the MIT Radiation Laboratory (1942-45); Henry Carhart Professor of Physics at Michigan (1954); and was founding professor of theoretical physics at Rockefeller University (1960-74).

Uhlenbeck sought to systematize and organize Statistical Mechanics, and one of his proudest accomplishments, with Soon Tok Choh, was to derive the virial expansion of the transport coefficients for dense gases using Bogoliubov's method.

Uhlenbeck won many awards, including the Research Corporation Award (1953), the Oersted Medal of the AAPT (1955), the Max Planck Medal (1964), the Lorentz Medal of the Royal Netherland Academy of Science (1970), and the National Medal of Science (1977). He insisted on sharing the 1979 Wolf Prize, awarded for the discovery of the electron spin, with Irene Goudsmit, the widow of his lifelong friend Sam Goudsmit.

One of us (ID) once asked Uhlenbeck what year he had won the Nobel Prize for spin. Maybe it was some consolation to both Goudsmit and Uhlenbeck that so many people thought that they deserved it, and a few just naturally assumed that they had won it.

(†* - from Max Dresden, PHYSICS TODAY, Dec 89, p.93; ††* - D.E. Newton in *Notable Twentieth-Century Scientists* (Thomson Publishing Co., 1995), edited by E.L.McMurray pp.2061-2.)

d) Profile of R. de L. Kronig.

The drama and disappointment of his wonderful debut at age twenty on January 7, 1925 into the midst of the intense European theoretical physics scene did not seem to discourage Ralph de Laer Kronig (1904-1995). Thirty years later, he tells († * *) of his education at Columbia (BA21, MA24, PhD25) where he got a thorough classical education but was largely self-taught in atomic physics from Sommerfeld's *Atombau und*

Spektrallinien. He had the good fortune to meet Ehrenfest who invited him to visit Leiden. There, in a few weeks, he had written a paper with Goudsmit (*Naturwissenschaften* **13**, 90 (1925)) where they analyzed the relative intensities of Zeeman components in terms of polynomials in J and M. This propelled him to visit Landé in Tübingen, where Pauli had just written Landé that he would arrive the next day. The great drama was set in motion. Landé handed Pauli's letter to Kronig.

Immediately upon reading Pauli's letter, Kronig realized that the $s = \frac{1}{2}$ "might be considered as an intrinsic angular momentum of the electron. $\cdots\cdots$ the same afternoon $\cdots$ I succeeded in deriving the so-called relativistic doublet formula."

The next day Pauli's reaction was "Das ist ja ein ganz witziger einfall." That is really a very amusing idea. But Pauli did not believe the suggestion had any connection with reality. They proceeded with the scheduled discussion of the spectrum of lead.

Kronig went on to a prolific and varied career with fundamental contributions in dispersion theory (*J. Opt. Soc. Amer.* **12**, 547 (1926); band theory of solids (the Kronig-Penney one dimensional periodic square-well lattice, R.deL. Kronig and W.G. Penney, *Proc. Roy. Soc.* A**130**, 409 (1931); also, R.deL. Kronig, *Band Theory and Molecular Structure* (Cambridge, New York, 1930); valence theory, (*The Optical Basis of the Theory of Valency* (Macmillan, New York, 1935); causality in the dispersion relation for the S-matrix (1946); and many other subjects; editor of *Textbook of Physics* (Pergamon Press, New York, 1959) with contributions from many prominent Dutch physicists. He was Assistant Professor at Columbia (1925); Pauli's assistant at ETH Zurich (27); Lecturer at Groningen (30-39); Professor to Rector to Emeritus at Delft (39-69).

It is impossible to do justice to Kronig's achievement without reading his own account in Pauli's memorial volume, and also the meticulously documented account contained in van der Waerden's *Sources of Quantum Mechanics.*

(† * * - from R. Kronig, "The Turning Point" in *Theoretical Physics in the Twentieth Century: A Memorial Volume to Wolfgang Pauli* (Interscience, New York, 1960), edited by M. Fierz and V.F. Weisskopf, pp.5-39. Also, B.L. van der Waerden, *Sources of Quantum Mechanics* (North Holland, Amsterdam, 1967), pp.23-25. Also Max Dresden, PHYSICS TODAY Mar 97, pp.97-98.)

Bibliography and References.

2.1) G.E. Uhlenbeck and S. Goudsmit, Die Naturwissenschaften **47**, 953 (1925); Nature **117**, 264 (1926); see our App.2A, 2B.

2.2) A.H. Compton, Journal of the Franklin Institute **192**-2, 145 (1921); see our App.2C.

2.3) B.L. van der Waerden, in *Theoretical Physics in the Twentieth Century: A Memorial Volume to Wolfgang Pauli* (Interscience, New York, 1960), edited by M. Fierz and V.F. Weisskopf, pp.199-244.

2.4) R. Kronig, in *Theoretical Physics in the Twentieth Century: A Memorial Volume to Wolfgang Pauli* (Interscience, New York, 1960), edited by M. Fierz and V.F. Weisskopf, pp.5-39.

2.5) A. Landé, Zeits. f. Phys. **16**, 342 (1923).

APPENDIX 2.A:

Excerpt from: Die Naturwissenschaften 47, 953 (1925)

Replacement of the Hypothesis of a Non-Mechanical Constraint by the Requirement of an Internal Property of Each Individual Electron

§1. As is well known, one can describe in great detail the structure and the magnetic behavior of the spectra with the help of Landé's vector model R, K, J and m [1]. Here R is the angular momentum of the core atom (that is the atom without the light emitting electron), K the angular momentum of this electron (Note added: actually $K = l + \frac{1}{2}$), J their resultant (again, $J = j + \frac{1}{2}$), and m the projection of J on the direction of the external magnetic field, all in the quantum unit $\hbar$. One must further assume in this model:

a) that for the core atom, the ratio of the magnetic moment to the angular momentum is twice as great as expected classically.

b) wherever R^2, K^2, J^2 appear, one must replace them by $R^2 - \frac{1}{4}, K^2 - \frac{1}{4}, J^2 - \frac{1}{4}$ (Heisenberg's Rule [2]).

This model has proved useful and has led to the interpretation of even very complicated spectra.

§2. However, one runs into difficulties when one tries to interpret Landé's vector model in terms of the electronic structure of atoms. For example:

a) Pauli [3] has just shown that for alkali-atoms the core must be magnetically inactive, since otherwise the effect of the relativistic corrections would produce a dependence of the Zeeman effect on the nuclear charge which is not observed in these spectra.

b) In the Landé model, one should not identify the angular momentum of the core atom with that of the residual positive ion, but one must allow it to depend on the definition of the core atom. (the Branching Rule of Landé-Heisenberg [4]- a non-mechanical constraint).

c) And recently, spectra (e.g. Vanadium, Titanium) analyzed with the help of Landé's scheme, show that the K of the groundstates do not have values expected from the Bohr-

59

Stoner model of the Periodic Table.

§3. All the above difficulties point in the same direction, that the understanding of the Landé vectors is probably not correct. Pauli [5] has recently adopted a new way, which addresses the difficulty (a). Here one must uniquely assign all quantum numbers of the emitting-electron from the alkali-spectrum. According to Pauli, in a magnetic field, each electron gets a unique set of four independent quantum numbers. With the help of Bohr's Structure Rules and other Rules, we can derive Landé's results in a simple way. The construction of the Bohr-Stoner Periodic Table follows, and it opens up a new starting point [7].

§4. In seeking the explanation of the behavior of the relativistic doublets in the X-ray and the alkali-spectra, there remains a puzzle. In order to explain the facts, one has had to assume a classically undescribable double-valuedness in the quantum theoretic properties of the electrons [8].

§5. It occurred to us that there is another way. Pauli himself did not suggest a model interpretation. The four quantum numbers describing each electron have lost their original Landé significance. It is apparent that each electron with its four quantum numbers, should also have four degrees of freedom. One can then give the following explanation of the quantum numbers: "n" and "k" remain, as before, the principal- and azimuthal-quantum numbers of the electron in its orbit.

R however is ascribed to an eigen-rotation of the electron [9]. (Italics added.)

The other quantum numbers retain their old significance. Throughout our discussion the formal explanations of Landé and Pauli combine with one another and retain all their former successes [10]. The electron must now be given the previously unrecognized property of an angular momentum (mentioned in §1a) , which Landé ascribed to the core atom. The quantitative investigation of this idea will depend strongly upon the details of the electron model. In order to agree with the facts, one must adjust the model for the following results:

a) The ratio of the magnetic moment of the electron to its mechanical angular momentum must be twice as great for the eigen-rotation as for the orbital-motion [11].

b) The different orientations of R with respect to the orbital-plane (or K) of the electron must be able to explain the relativistic doublets, in order to agree with the Heisenberg-Wentzel prescription [12].

60

G.E. Uhlenbeck and S. Goudsmit.

Leiden, 17 October 1925.

Instituut voor Theoretische Natuurkunde.

......

Footnotes and References.

1) See E. Back and A. Landé, Zeeman Effekt und Multiplettstruktur der Spektrallinien.

2) W. Heisenberg, Zeits. f. Phys. **31**, 291 (1925).

3) W. Pauli Jr., Zeits. f. Phys. **31**, 373 (1925).

4) E. Back and A. Landé, loc. cit. P.55 ff.

5) W. Pauli Jr., Zeits. f. Phys. **31**, 765 (1925).

6) Compare: S. Goudsmit, Zeits. f. Phys. **32**, 794 (1925); W. Heisenberg, Zeits. f. Phys. **32**, 841 (1925); F. Hund, Zeits. f. Phys. **33**, 345 (1925).

7) See the work cited in [5].

8) W. Heisenberg, Zeits. f. Phys. **32**, 841 (1925).

9) Note that one must infer the quantum numbers of the electrons in the alkalispectrum. Thus $R = 1$ for each electron (in Landé's standard notation).

10) For example, now the significance of Heisenberg's Model III is understandable, where one must combine both the R of the whole atom and the K of the electrons.

11) For example, for a rotating spherical electron with surface charge one can read off from Abraham's formulas (Ann. d. Phys. **10**, 105 (1903)):

Rotational energy: $\frac{1}{9}\frac{e^2 a}{c^2}\dot{\phi}^2$ (where a is the electron radius),

Also: $P_\phi = \frac{2}{9}\frac{e^2 a}{c^2}\dot{\phi}$,

Magnetic moment: $\mu = \frac{1}{3}\frac{e^2 a}{c^2}\dot{\phi}$,

Mass: $m = \frac{2}{3}\frac{e^2}{c^2 a}$.

Therefore $\frac{\mu}{P_\phi} = \frac{3}{2}\frac{ac}{e} = 2 \times \frac{e}{2mc}$,

which is in fact twice as great as for the orbital-motion. Note however, when one quantizes this rotational motion the peripheral surface velocity of the electron would be greater than the speed of light.

12) W. Heisenberg, l.c.; G. Wentzel, Ann. d. Phys. **76**, 803 (1925).

APPENDIX 2.B:

Excerpt from: Nature 117, 264 (1926)

Spinning Electrons and the Structure of Spectra

So far as we know, the idea of a quantized spinning electron was put forward for the first time by A.H. Compton (*Journ. Frankl. Inst.*, Aug. 1921, p.145), who pointed out the possible bearing of this idea on the origin of the natural unit of magnetism. Without being aware of Compton's suggestion, we have directed attention in a recent note (*Naturwissenschaften*, Nov. 20, 1925) to the possibility of applying the spinning electron to interpret a number of features of the quantum theory of the Zeeman effect, which were brought to light by the work especially of van Lohuizen, Sommerfeld, Landé and Pauli, and also of the analysis of complex spectra in general. In this letter we shall try to show how our hypothesis enables us to overcome certain fundamental difficulties which have hindered the interpretation of the results arrived at by those authors.

To start with, we shall consider the effect of the spin on the stationary states which correspond to motion of an electron around the nucleus. On account of its magnetic moment, the electron will be acted on by a couple just as if it were placed at rest in a magnetic field of magnitude equal to the vector product of the nuclear electric field and the velocity of the electron relative to the nucleus, divided by the speed of light. This couple will cause a slow precession of the spin axis, the conservation of the angular momentum of the atom being ensured by a compensating precession of the orbital plane of the electron. This complexity of the motion requires that, corresponding to each stationary state of an imaginary atom in which the electron has no spin, there shall in general exist a set of states which differ in the orientation of the spin axis relative to the orbital plane, the other characteristics of the motion remaining unchanged. If the spin corresponds to a one-quantum rotation there will be in general two such states. Further, the energy difference of these states will, as a simple calculation shows, be proportional to the fourth power of the nuclear charge. It will also depend on the quantum numbers which define the state of motion of the non-spinning electron in a way very similar to the energy differences connected with the rotation of the orbit in its own plane arising from the relativistic variation of the electronic mass. We are indebted to Dr. Heisenberg for a letter containing some calculations on the quantitative side of the problem.

This result suggests an essential modification of the explanation given for the fine structure of the hydrogen-like spectra. As an illustration we may consider the energy

63

levels corresponding to electronic orbits for which the principal quantum number is three. The scheme on the left side of the accompanying figure (Fig. 1) corresponds to the results to be expected from Sommerfeld's theory. The so-called azimuthal quantum number k is defined by the quantity of moment of momentum of the electron around the nucleus, $k\hbar$, where $k = 1, 2, 3$. According to the new theory, depicted in the scheme [1] on the right, this moment of momentum is given by $K\hbar$, where $K = \frac{1}{2}, \frac{3}{2}, \frac{5}{2}$. The total angular momentum of the atom is $J\hbar$, where $J = 1, 2, 3$. The symbols K and J correspond to those used by Landé in his classification of the Zeeman effects of the optical multiplets. The letters S, P, D also relate to the analogy with the structure of optical spectra which we consider below. The dotted lines represent the position of the energy levels to be expected in the absence of the spin of the electron. As the arrows indicate, this spin now splits each level into two, with the exception of the level with $K = \frac{1}{2}$, which is only displaced.

In order to account for the experimental facts, the resulting levels must fall in just the same places as the levels given by the older theory. Nevertheless, the two schemes differ fundamentally. In particular, the new theory explains at once the occurrence of certain components in the fine structure of the hydrogen spectrum and of the helium spark spectrum which according to the old scheme would correspond to transitions where K remains unchanged. Unless these transitions could be ascribed to the action of electric forces in the discharge which would perturb the electronic motion, their occurrence would be in disagreement with the correspondence principle, which only allows transitions in which the azimuthal quantum changes by one unit. In the new scheme we see that, in the transitions in question, K will actually change by one unit and only J will remain unchanged. Their occurrence is, therefore, quite in conformity with the correspondence principle.

The modification proposed is specially important for explaining the structure of X-ray spectra. These spectra differ from the hydrogen-like spectra by the appearance of the so-called "screening" doublets, which are ascribed to the interaction of the electrons within the atom, effective mainly through reducing the effect of the nuclear attraction. In our view, these screening doublets correspond to pairs of levels which have the same angular momentum J but different azimuthal quantum numbers K. Consequently, the orbits will penetrate to different distances from the nucleus, so that the screening of the nuclear charge by the other electrons in the atom will have different effects. This screening effect will, however, be the same for a pair of levels which have the same K but different J's and correspond to the same orbital shape. Such pairs of levels were, on the older theory, labelled with values of k differing by one unit, and it was quite impossible to understand why these so-called "relativistic" doublets should appear separately from the screening doublets. On our view, the doublets in question may more properly be termed "spin" doublets, since the sole reason for their appearance is the difference in orientation of the spin axis relative to the orbital plane. It should be emphasized that our interpretation is in complete accordance with the correspondence principle as regards the rules of combination of X-ray levels.

The assumption of the spinning electron leads to a new insight into the remarkable analogy between the multiplet structure of the optical spectra and the structure of the X-ray spectra, which was emphasized especially by Landé and Millikan. While the attempt to refer this analogy to a relativistic effect common to all structures was most unsatisfactory, it obtains an immediate explanation on the hypothesis of the spin electron. If, for example, we consider the spectra of the alkali-type, we are led to recognize in the well-known doublets regular spin doublets of the character described above. In fact, this enables us to explain the dependence of the doublet width on the effective nuclear charge and the quantum numbers describing the orbit, as well as the rules of combination.

The simplicity of the alkali-spectra is due to the fact that the atom consists of an electron revolving around an atomic-core which contains only completed electronic groups, which are magnetically inert. When we pass to atoms in which several electrons revolve around a core of this kind we meet with new features, since we have to take account of other influences on the spin axis of each electron besides the couple due to its own motion in the electric field. Not only does this enable us to account for the appearance of multiplets of higher complexity, but it also seems to throw light on so-called "branching" of spectra, which usually accompanies the adding of a further electron to the atom, and for which no satisfactory explanation had been given. In fact, it seems that the introduction of the spinning electron makes it possible to maintain the principle of the successive building up of the atoms used by Bohr in his general discussion of the relations between spectra and the Periodic Table of the elements. Above all, it may be possible to account for the important results arrived at by Pauli, without having to assume an unmechanical "duality" in the binding of the electrons.

So far, we have not mentioned the Zeeman effect, although the introduction of the spinning electron was primarily suggested by the analysis of the anomalous Zeeman effect. From the point of view of the correspondence principle, this effect shows that the influence of a magnetic field on the motion of the atom differs from that if the electron had no spin. In fact, from the Larmor theorem, we would expect the effect on any spectral line to be of the simple Lorentz type, quite independently of the character of the multiplet structure. Therefore the anomalous Zeeman effect has presented very grave difficulties. These difficulties disappear at once when the electron has a spin and the ratio between magnetic moment and angular momentum of this spin is different from that for the revolution of the electron in an orbit. On this assumption the spin axis of an electron would precess with a frequency different from the Larmor rotation. It is easily shown that the resultant motion of the atom for magnetic fields of small intensity will be of just the type revealed by Landé's analysis. If the field is so strong that its influence on the precession of the spin axis is comparable with that due to the orbital motion in the atom, this motion will be changed in a way which directly explains the gradual transformation of the multiplet structure for increasing fields known as the Paschen-Back effect.

It seems possible on these lines to develop a quantitative theory of the Zeeman effect, if it is assumed that the ratio between the magnetic moment and the angular momentum due to the spin is twice the ratio corresponding to an orbital revolution. At present, however, it seems difficult to reconcile this assumption with a quantitative analysis of our explanation of the fine structure of levels. In fact it leads, in a preliminary calculation, to widths of the spin doublets just twice as large as those required by observation. It must be remembered, however, that we are here dealing with problems which for their final solution require a closer study of quantum mechanics and perhaps also of questions concerning the structure of the electron.

In conclusion, we wish to acknowledge our indebtedness to Prof. Niels Bohr for an enlightening discussion, and for criticisms which helped us to distinguish between the essential points and the more technical details of the new interpretation.

G.E. UHLENBECK and S. GOUDSMIT.

Instituut voor Theoretische Naturkuunde,

Leyden, December, 1925.

Footnote.

1) Quite independently of the ideas discussed here, a scheme of levels corresponding to this figure has been previously proposed by the writers (*Physica*, **5**, 266 (1925)), on the ground of the formal analogy between spectral structures. From similar formal considerations, this scheme has been recently arrived at by J.C. Slater (*Proc. Washington Acad.*, December 1925).

APPENDIX 2.C:

Excerpt from: Journal of the Franklin Institute, 192-2, 145 (1921)

THE MAGNETIC ELECTRON[†]

BY

ARTHUR H. COMPTON, Ph.D.

Washington University, St. Louis

$\cdots$ many magnetic phenomena (show) that matter contains a large number of minute elementary magnets. The theories of para- and ferro-magnetism $\cdots$, though based upon the hypothesis of such ultimate magnetic particles, makes no assumption concerning their nature. The explanation of diamagnetism $\cdots\cdots$ this effect owes its origin to the circulation of electrons in resistanceless paths. $\cdots$ Let us see if it is possible to identify these elementary magnets with any of the fundamental divisions of matter.

$\cdots$ It is difficult to imagine what mechanism could reasonably give to a group of atoms, such as the chemical molecule, the properties of a single magnetic particle. $\cdots$ if on magnetization such a group of atoms should actually turn around within the crystal $\cdots$ should produce a change in the crystal form $\cdots$. $\cdots$ magnetic fields effect no such change in the form of a magnetic crystal.

$\cdots$ the most generally accepted view of the nature of the elementary magnet, is that the revolution of electrons in orbits within the atom give to the atom as a whole the properties of a tiny permanent magnet. Support of this view is found in the quantitative explanation which it affords of the Zeeman effect. It seems but a step from the explanation of this effect to Langevin's explanation of diamagnetism as another result of the induced electronic currents within the atom. On Langevin's view the electronic orbits act as resistanceless circuits in which an external magnetic field induces changes of current. By Lenz's law these induced currents $\cdots$ opposite to the applied field, thus accounting for the atom's diamagnetic properties. This theory offers a satisfactory qualitative explanation $\cdots$ is independent of temperature. But quantitatively it is inadequate. $\cdots$ one must suppose a number of electrons several times the atomic number, or the distance $\cdots$ must be several times as great $\cdots$. $\cdots$ experiments $\cdots$ show that the ratio of charge to mass $\cdots$ is appreciably greater $\cdots$. But perhaps a more serious difficulty $\cdots$ the induced change in magnetic moment of the electronic orbit involves also a change in its angular momentum.

··· according to classical theory ··· any angular momentum induced ··· rapidly disappears so that diamagnetism should be merely a transient effect. Let us assume with Bohr that if each electron has some definite angular momentum such as $\hbar$, no radiation occurs. ··· the induced change in angular momentum will put the electrons in an unstable condition. ··· soon be dissipated. ··· diamagnetism, accounted for by the induced magnetic moment of electrons revolving in orbits, can be (no) more than a transient phenomenon.

··· many of the magnetic properties of matter receive a satisfactory explanation on Parson's hypothesis [1], that the electron is a continuous ring of negative electricity spinning rapidly about an axis perpendicular to its plane, and therefore possessing a magnetic moment as well as an electric charge. Thus, for example, the fact that such a ring can rotate without radiating enables this hypothesis to account for diamagnetism as a permanent instead of a transient effect. While retaining Parson's view of a magnetic electron of comparatively large size, we may suppose with Nicholson that instead of being a ring of electricity, the electron has a more nearly isotropic form with a strong concentration of electric charge near the center and a diminution of electric density as the radius increases. It is natural to suppose that the mass of such an electron is concentrated principally near its center and that the ratio of charge to mass of its external portions will be greater than that for the electron as a whole. While the explanation of the inertia of such a charge of electricity is perhaps not obvious, it is at least consistent with our usual conceptions and it has the advantage of offering an explanation for the large value of e/m observed in Barnett and Stewart's experiment [2]. It also makes possible an explanation of the relatively large induced currents required to account for diamagnetism without introducing the assumption of a prohibitively large radius for the electric charge.

A series of experiments has recently been performed ······ [3]. It is difficult to avoid the conclusion that the elementary magnet is not the atom as a whole.

Since neither the molecule nor the atom gives a satisfactory explanation of these experiments ······ Let us see then if we can find any positive evidence for the existence of an electron with a magnetic moment.

··· we should expect the electron ··· to possess thermal energy of rotation motion···. On Planck's more recent quantum hypothesis, however, at absolute zero temperature each particle of matter - including the electron - should retain an average amount of energy $\frac{1}{2}h\nu$ for each degree of freedom. For a rotating system this corresponds to an angular momentum of $\hbar$. Thus ··· the thermal motions of the electron will give it an appreciable magnetic moment. ··· nearly the same at different temperatures - a property characteristic of the elementary magnets. It is interesting to notice, also, that the magnitude of the magnetic moment of an electron spinning with an angular momentum $\hbar$ is of the proper order to account for the ferro-magnetic properties, being about one-third the magnetic moment of the iron atom.

⋯⋯ On the present view we may well suppose that the electron is spinning like a gyroscope and on traversing matter is set into nutational oscillations, resulting in the observed radiation.

Let us then review the different lines of evidence that have given us information concerning the nature of the elementary magnet. In the first place, the Richardson-Barnett effect shows that magnetism is due chiefly to the circulation of negative electricity whose charge to mass ratio is not greatly different from that of the electron. In the second place, experiments on the diffraction of X-rays by magnetic crystals indicate that the elementary magnet is not any group of atoms, such as the chemical molecule, nor even the atom itself; but lead rather to the view that it is the electron rotating about its own axis which is responsible for the ferro-magnetism. ⋯⋯ *May I then conclude that the electron itself, spinning like a tiny gyroscope, is probably the ultimate magnetic particle.* (Italics added.)

Footnotes and References.

†) Based on a paper read before Section B of the American Association for the Advancement of Science, December 27, 1920.

1) A.L. Parson, Smithsonian Misc. Collections, 1915.

2) S.J. Barnett, Phys. Rev. **6**, 240 (1915); J.Q. Stewart, Phys. Rev. **11**, 100 (1918).

3) K.T. Compton and E.A. Trousdale, Phys. Rev. **5**, 315 (1915).

4) A.H. Compton and O. Rognley, Phys. Rev. **16**, 464 (1920).

APPENDIX 2.D:

Excerpt from: Nature 117, 550 (1926)

Spinning Electrons and the Structure of Spectra

Recently Uhlenbeck and Goudsmit (Nature, February 20, p.264; see also *Naturw.*, November 20 1925) have directed attention to the fact that a number of features of multiplet structure and the anomalous Zeeman effect can be described by assuming the electrons in the atom to possess an inherent magnetic moment, pictured as being due to a spinning motion of the electron about an axis of symmetry. This moment must be considered $\cdots$ 1 or 2 Bohr magnetons $\cdots$. The above-named authors discuss the advantages which such a view brings with it, but fail to point out some serious difficulties.

If it is permissible at all to use pictorial concepts such as the word 'spinning' evidently implies, it must be permissible to speak of the 'dimensions' of an electron. These dimensions would then have to be taken as of order 10^{-13} cm. To give a magnetic moment around 1 Bohr magneton, the internal velocities would have to be exceedingly close to that of light. Now the elementary unit of magnetic moment, the Bohr magneton, is derived from considerations of the *orbital* motions of electrons with velocities *much smaller* than that of light (W. Pauli, jun., *Zeit. f. Phys.* **31**, 373 (1925)), in which the internal structure of these electrons does not enter at all, so that they can be regarded as point charges. It is hard to see, then, why this elementary unit should also be characteristic of the internal motion of the electrons in spite of the high velocities involved, and that with a precision which would have to be considerable, if the measurements on the anomalous Zeeman effect were to be explained in this way.

Quite aside from this objection, the validity of which could perhaps be questioned, since we may not be justified in applying the classical concepts of kinematics to the case of the structure of the electron, even if we only wish to obtain rough estimates, the following difficulty arises. In order to account for the observed Zeeman effect by the hypothesis of Uhlenbeck and Goudsmit, it is necessary to assume that an orbital electron always has the same magnetic moment, of the order of a Bohr magneton, no matter in what orbit or in what atom. One is thus led to expect that this also remains true when an electron forms part of the nuclear structure. But then the nucleus, too, will have a magnetic moment of the order of a Bohr magneton, unless the magnetic moments of all the nuclear electrons just happened to cancel. For such an additional moment of the nucleus there is no place in the theory of the Zeeman effect, and the probability that in all atomic nuclei the magnetic moments of the electrons neutralize seems *a priori* to be very small.

The new hypothesis, therefore, appears rather to effect the removal of the family ghost from the basement to the sub-basement, instead of expelling it definitely from the house.

R. DE L. KRONIG.

Columbia University, New York.

(Submission undated; publication April 17, 1926)

Chapter 3

Bose-Einstein Statistics

Summary: The fundamental derivation of the Planck blackbody distribution in the work of Bose, and the immediate recognition by Einstein of the possibility of an ideal gas satisfying the same statistics, with the resulting consequence of Bose-Einstein condensation are presented with the original literature.

§1. Bose's Quantum Derivation of Planck's Distribution.

In yet another classic drama, the unknown outsider Satyendra Nath Bose of Dacca, India (now Bangladesh), sent to Einstein his manuscript with the key normalization of the phase space for a gas of light quanta. By dividing the phase space into cells of volume h^3 (h = Planck's constant = $2\pi\hbar$), Bose, for the first time, was able to derive *ab initio* the probability distribution for a photon gas. It was his declared aim to make the derivation without using Correspondence Principle connections to the classical theory which had been an essential feature in Einstein's derivation, but which Bose considered a logical defect in the theory. Writing in early 1924, Bose - although he does not say so explicitly, and in fact gives no formal references - uses the Bohr-Sommerfeld quantization condition

$$\int dpdq = Nh$$

to give the volume in phase space of allowed particle orbits in three dimensions as integer multiples of h^3.

After this crucial step, Bose makes contact with the Einstein derivation of the Planck distribution, where the number of possible partitions does not include permutations of identical quanta in the individual cells in phase space. Bose's derivation of Planck's formula using the Bohr-Sommerfeld quantization condition and Einstein's analysis of the statistical weights of identical quanta was recognized immediately as an important advance by Einstein, who translated the manuscript from English into German and forwarded it to Zeitschrift für Physik for publication [3.1]. Here it is translated back to English.

§2. Einstein's Applications of Bose's New Insight.

There followed three papers by Einstein which make essential use of Bose's idea, in which Einstein generalized the statistical analysis from the photon gas to the ideal gas of identical molecules [3.2] and then discovered the theoretical phenomenon of Bose-Einstein condensation [3.3].

Einstein's three papers are classic and fabulously interesting in their own right, but a diversion from our primary concern with the Spin-Statistics Theorem. Nonetheless, we include them in their entirety because of their interest and also because of their relative inaccessibility.

In what he characterizes as a "striking impact" of Bose's idea, Einstein calculates the number of cells Δs in an element of phase space volume corresponding to a gas of molecules of mass m and energy in the interval E to $E + \Delta E$, as (equation numbers from Einstein's paper 1)

$$\Delta s = 2\pi \frac{V}{h^3} (2m)^{3/2} E^{1/2} \Delta E. \tag{2}$$

Then - following the same development as for the photon gas - he derives the Bose-Einstein probability distribution

$$n_s = \frac{1}{e^{\alpha_s} - 1} \tag{11}$$

with $\alpha_s = \alpha + \beta E^s$.

The total number of particles

$$n = \sum_s n_s \tag{6}$$

and the total energy

$$\bar{E} = \sum_s n_s E^s \tag{7}$$

determine the parameters α and β. From the entropy and the entropy increase in an infinitesimal heating at constant volume $d\bar{E} = TdS$, one gets the Lagrange multiplier in terms of the temperature, $\beta = 1/kT$ (13). The equation of state $\bar{E} = 3/2pV$ and particle number in Eqns.18,19 can be expanded in powers of the dimensionless parameter $\lambda \equiv e^{-\alpha}$, which, in turn, can be calculated from the inversion of Eqn.6 to be a function of the dimensionless parameter

$$\lambda_0 = \frac{h^3 n}{V}(2\pi mkT)^{-3/2} \tag{16}$$

giving

$$\frac{\bar{E}}{n} = \frac{3}{2}kT \left(1 - \lambda/2^{5/2} + \cdots\right) \tag{22}$$

and

$$n_s = \text{const} \cdot e^{-E^s/kT} \left(1 + \lambda e^{-E^s/kT} + \cdots\right), \tag{23}$$

which display the leading quantum corrections to the classical Maxwell gas, for λ set to λ_0.

In the second of the papers inspired by Bose's idea, Einstein resolves a paradoxical situation that can arise if one increases the density of the gas keeping the temperature fixed. The analytic solution for the distribution function (Eqn.18 of the first paper)

$$n = \sum_s \frac{1}{e^{\alpha_s} - 1}$$

with $\alpha_s = \alpha + \beta E^s$ requires $\alpha > 0$ in order to include the $E^s = 0$ cell in the sum. In this case the parameter $\lambda = e^{-\alpha}$ must be < 1 and the series expansion (Eqns.18b,c in paper 1).

$$\frac{n}{V} = \frac{(2\pi mkT)^{3/2}}{h^3} \sum_{\tau=1}^{\infty} \lambda^\tau \tau^{-3/2}$$

predicts a maximum density that cannot exceed

$$\frac{n}{V} \leq \frac{(2\pi mkT)^{3/2}}{h^3} \sum_{\tau=1}^{\infty} \tau^{-3/2}.$$

Einstein avoids this paradox by treating the $E^s = 0$ cell separately. He demonstrates that at the critical density (for fixed temperature) the excess particles condense into the $E^s = 0$ cell and the rest are distributed over the remaining cells with $\lambda = 1$ giving them their maximum (sub-)density. He proves that there is thermodynamic equilibrium between the two components of the ideal gas: the condensate with zero energy and density variable as required, and the "saturated" ideal gas possessing the density for a given cell which is the maximum possible, determined by $\lambda = 1$.

In the remainder of the paper, Einstein demonstrates that the quantum gas obeys Nernst's theorem simply by reason of the hypothesized counting of the groundstate of many identical quanta as unique; that the fluctuation phenomena in the quantum gas (Eqn.34) are just what one would expect from the analogy of de Broglie wave interference compared to that of radiation; and, finally, that the equation of state of the saturated ideal gas, as applied to electrons in metals, leads to unacceptable requirements for the mean free path.

In the third and last paper of the series [3.4], Einstein uses general arguments based on dimensional analysis to determine the most general form for the equation of state of an ideal gas. The techniques were familiar from the thermodynamic derivation of Wien's law in radiation theory.

He summarizes the arguments - based on Nernst's theorem (entropy $S = 0$ at $T = 0$), Boltzmann's rule for the entropy ($S = k \log W$ with W the probability of the macrostate under consideration) and Planck's prescription ($W = N$, where N is the number of microstates corresponding to a particular macrostate) - that the classical equations of state must fail and that a quantum theory of the ideal gas is required. The classical equations of state inevitably lead to negative entropy states realizable at least in principle. In contrast, the quantum theory, in particular Bose's prescription for counting the groundstate of a many-quantum system as unique, satisfies Nernst's theorem in a simple but profound way.

In the distribution function

$$dn = \frac{V}{h^3}\rho d\Phi$$

the density function ρ must be a dimensionless function of the dimensionless parameters

$$v = \frac{L}{kT} \quad \text{and} \quad u = \frac{h^3}{VN}(2\pi mkT)^{-3/2},$$

with L the kinetic energy per molecule. Einstein shows that the density ρ is an adiabatic invariant, and that the entropy density can only be a function of ρ, $s(\rho)$. Minimizing the entropy while holding fixed the number of quanta and the total energy leads to (14)

$$\frac{\partial s}{\partial \rho} = \alpha + \beta L$$

which itself is a function only of ρ, so ρ can be expressed as a function

$$\rho = \Psi(\alpha + \beta L).$$

β is determined to be $1/kT$ by an infinitesimal heating at constant volume giving the change in energy $DE = TDS$. So far, Einstein has expressed the density as

$$\rho = \psi(v, u) = \psi(v + \phi(u))$$

a universal function ψ of v plus another universal function ϕ of u. The same result is deduced by an independent argument using a potential Π to confine the gas.

Finally, Einstein specializes to the classical Maxwell result, and to the Bose-Einstein distribution by requiring the dependence on Planck's constant h to cancel out of the density ρ.

§3. Biographical Note on S.N. Bose.(†)

Satyendra Nath Bose (1894-1974), was a brilliant student at Calcutta, became a lecturer in Applied Mathematics at Calcutta (1916), then reader in Physics at Dacca (1921). After his bombshell paper to Einstein, he visited Europe but his visit was not a success and he seems to have been sidetracked in Paris with the status of a graduate student. Returning to Dacca, his career was blocked because he had no PhD. A postcard from Einstein to the vice-chancellor of Dacca cleared the way for his appointment as Professor and Chairman (1927-45).

Bose was Guprasad Singh and Khaira Professor of Physics at Calcutta (1945-58); FRS (1958); and Vice-Chancellor of Vishwabharati University (1958-). His research was in unified field theory and number theory. Bose was revered as kindly, generous and inspirational by his colleagues and students.

He never met Einstein.

(† - from Jagadish Sharma, PHYSICS TODAY, April (1974), p.129-130; London Times, 5Feb74, p.14g; P.T. Landsburg, London Times, 8Feb74, p.18g.)

Bibliography and References.

3.1) Bose, Zeits. f. Phys. **26**, 178 (1924); see our App.3A.)

3.2) A. Einstein, S.B.d. Preuss. Akad. Wiss. Ber. **22**,261 (1924); see our App.3B.

3.3) A. Einstein, S.B.d. Preuss. Akad. Wiss. Ber. **1**, 3 (1925); see our App.3C.

3.4) A. Einstein, S.B.d. Preuss. Akad. Wiss. Ber. **3**, 18 (1925); see our App.3D.

APPENDIX 3.A:

Excerpt from: Zeitschrift für Physik 26, 178 (1924)

Planck's Law and the Light Quantum Hypothesis

by Bose (Dacca University, India)

(Received on 2 July 1924)

The phase space of a light quantum in a given volume is expressed in cells of size $(h)^3$. The number of possible configurations of light quanta in a macroscopic field among these cells determines the entropy and thereby all thermodynamic properties of the radiation.

Planck's formula for the distribution of the energy in black body radiation is the starting point for the Quantum Theory which has been developed in the last twenty years and which has made rich contributions in all areas of physics. Since its publication in 1901 many derivations of this law have been put forth. It is recognized that the fundamental postulates of the Quantum Theory are incompatible with the laws of classical electrodynamics. Previously, all derivations made use of the relation

$$\rho_\nu d\nu = \frac{8\pi\nu^2 d\nu}{c^3} E,$$

between the radiation density ρ_ν and the average energy E of an oscillator of frequency ν. They make assumptions about the number of degrees of freedom of the ether, which enter the above equation (the first factor on the right side). This factor, however, can only be inferred from the classical theory. This is the unsatisfactory starting point in all derivations, and it cannot be assumed that further developments, starting from this assumption, are free of this logical defect.

A remarkably elegant derivation has been given by Einstein. This has the logical defect of all previously known derivations and tries to deduce the formula from the classical theory. From very simple assumptions about the energy exchange between the molecules and the radiation field, he finds the relation

$$\rho_\nu = \frac{\alpha_{mn}}{e^{\frac{\varepsilon_m - \varepsilon_n}{kT}} - 1}.$$

However, in order to bring this formula into agreement with Planck's Law, he has to make use of Wien's Displacement Law and Bohr's Correspondence Principle. Wien's Law is based on the classical theory, and the Correspondence Principle assumes that the Quantum Theory coincides with the classical theory in certain limiting cases.

In all cases it seems to me the derivations are not logically sufficient. On the other hand it seems to me that the light quantum hypothesis in combination with statistical mechanics (as it has been adapted from Planck to the needs of the Quantum Theory) is sufficient for the derivation of this law independent of the classical theory. In the following I will briefly sketch the method.

Let the radiation be enclosed in the volume V and let its total energy be E. It consists of different sorts of quanta characterized by their number N_s and energy $h\nu_s$ ($s = 0$ to ∞). The total energy is then

$$E = \sum_s N_s h\nu_s = V \int \rho_\nu d\nu. \tag{1}$$

The solution of the problem requires the determination of the N_s, which define the ρ_ν. When we can specify the probability for each distribution characterized by the particular N_s, then the solution will be determined by the requirement that this probability should be a maximum subject to the restriction that the energy of Eqn.1 is constant. We will now derive this probability.

The quantum has a momentum of magnitude $\frac{h\nu_s}{c}$ in the direction of its motion. The momentum state of a quantum is characterized by its coordinates x, y, z and its corresponding momenta p_x, p_y, p_z; these six quantities can be interpreted as coordinates in a six-dimensional space, where we have the relation

$$p_x^2 + p_y^2 + p_z^2 = \frac{h^2 \nu^2}{c^2},$$

according to which the given point is required to move on spherical surface determined by the frequency of the quantum. The frequency interval $d\nu_s$ corresponds in this sense to the phase space

$$\int dx\,dy\,dz\,dp_x\,dp_y\,dp_z = V 4\pi \left(\frac{h\nu}{c}\right)^2 \frac{h\,d\nu}{c} = 4\pi \frac{h^3 \nu^2}{c^3} \nu\,d\nu.$$

If we divide the phase volume into cells of size h^3, then

$$4\pi V \frac{\nu^2}{c^3} d\nu$$

cells correspond to the frequency interval $d\nu$. In regard to the nature of these distributions, nothing more definite can be said. However the total number of cells in the given volume must be recognized as the number of possible configurations of a quantum in the given volume. In order to include the states of polarization, it is necessary to multiply this number by two, so that we obtain the number of cells corresponding to $d\nu$ as

$$8\pi V \frac{\nu^2 d\nu}{c^3}.$$

Now it is simple to calculate the thermodynamic probability of a (macroscopically defined) state. Let N^s be the number of quanta corresponding to the frequency interval $d\nu_s$. In how many ways can these be distributed over the cells corresponding to $d\nu_s$? Let p_0^s be the number of vacant cells, p_1^s the number of those which contain one quantum, p_2^s the number of those with two, and so on. The number of possible partitions is then

$$\frac{A^s!}{p_0^s! p_1^s! p_2^s! \ldots},$$

where $A^s = 8\pi\nu_s^2 d\nu_s / c^3$ and where $N^s = 0 \cdot p_0^s + 1 \cdot p_1^s + 2 \cdot p_2^s + \ldots$ is the number of quanta in the interval $d\nu_s$.

The probability of the states defined by all p_r^s is obviously

$$W = \prod_s \frac{A^s!}{p_0^s! p_1^s! \ldots}.$$

With the understanding that we can treat the p_r^s as large numbers, we have

$$\log[W] = \sum_s A^s \log[A^s] - \sum_{rs} p_r^s \log[p_r^s]$$

where

$$A^s = \sum_r p_r^s.$$

This expression should be a maximum under the restriction

$$E = \sum_s N^s h\nu_s; \qquad N^s = \sum_r r p_r^s.$$

Carrying out the variation yields the relations

$$\sum_{rs} \delta p_r^s \left(1 + \log[p_r^s]\right) = 0, \qquad \sum_s \delta N^s \cdot h\nu_s = 0,$$

$$\sum_r \delta p_r^s = 0, \qquad \delta N^s = \sum_r r \cdot \delta p_r^s.$$

It follows that

$$\sum_{rs} \delta p_r^s \left(1 + \log[p_r^s] + \lambda_s + \frac{r \cdot h\nu_s}{\beta}\right) = 0.$$

80

If we define

$$p_r^s = B^s e^{-\frac{rh\nu_s}{\beta}},$$

then

$$A^s = \sum_r B^s e^{-\frac{rh\nu_s}{\beta}} = B^s \left(1 - e^{-\frac{h\nu_s}{\beta}}\right)^{-1},$$

so

$$B^s = A^s \left(1 - e^{-\frac{h\nu_s}{\beta}}\right).$$

Further one has

$$N^s = \sum_r r p_r^s = \sum_r r A^s \left(1 - e^{-\frac{h\nu_s}{\beta}}\right) e^{-\frac{rh\nu_s}{\beta}}$$

$$= \frac{A^s e^{-\frac{h\nu_s}{\beta}}}{1 - e^{-\frac{h\nu_s}{\beta}}}.$$

Inserting the value of A^s

$$E = \sum_s \frac{8\pi h\nu_s^3 d\nu_s}{c^3} V \frac{e^{-\frac{h\nu_s}{\beta}}}{1 - e^{-\frac{h\nu_s}{\beta}}}.$$

Using these results one finds the entropy

$$S = k \left[\frac{E}{\beta} - \sum_s A^s \log\left(1 - e^{-\frac{h\nu_s}{\beta}}\right)\right],$$

and using $\frac{\partial S}{\partial T} = \frac{1}{T}$, it follows that $\beta = kT$. If one inserts this in the above expression for E, one gets

$$E = \sum_s \frac{8\pi h\nu_s^3}{c^3} V \frac{1}{e^{\frac{h\nu_s}{kT}} - 1} d\nu_s$$

which is equivalent to Planck's formula.

(Translated by A. Einstein)

Comment of the Translator: In my opinion, Bose's derivation of Planck's formula constitutes an important advance. The method used here also yields the quantum theory of the ideal gas, as I will show in another place.

APPENDIX 3.B:

Excerpt from: S. B. d. Preuss. Akad. Wiss. Ber. 22, 261 (1924)

Quantum Theory of the Monatomic Ideal Gas

By A. EINSTEIN

A Quantum Theory of the monatomic ideal gas free of arbitrary prescriptions did not exist before now. This defect will be filled here on the basis of a new analysis developed by Bose, which this author has described as a most remarkable derivation of the Planck radiation formula [1].

What follows can be characterized as a striking impact of Bose's method. The phase space of an elementary particle (here a monatomic molecule) in a given (three-dimensional) volume will be divided into "cells" of size h^3. If there are many elementary particles present, then, for thermodynamic purposes, they are characterized in their microscopic distribution by the number and way with which the elementary particles are distributed over these cells. The "Probability" of a macroscopically defined state (in Planck's sense) is equal to the number of different microscopic states by which the macroscopic state can be realized. The entropy of the macroscopic state and therefore the statistical and thermodynamic behavior of the system are then determined by Boltzmann's rule.

§1. The Cells.

The phase space volume belonging to a given region of the coordinates (x, y, z) and the corresponding momenta (p_x, p_y, p_z) of a monatomic molecule, is given by the integral

$$\Phi = \int dx\,dy\,dz\,dp_x\,dp_y\,dp_z. \tag{1}$$

If V is the volume in which the molecule is confined, then the phase volume of all states whose energy $E = (p_x^2 + p_y^2 + p_z^2)/2m$ is less than some definite value E, is given by

$$\Phi = V \cdot \frac{4}{3}\pi(2mE)^{\frac{3}{2}}. \tag{1}$$

The number of cells Δs which correspond to an energy interval ΔE is

$$\Delta s = 2\pi \frac{V}{h^3}(2m)^{\frac{3}{2}} E^{\frac{1}{2}} \Delta E. \tag{2}$$

82

For arbitrarily small $(\Delta E)/E$ one can still choose V so large that Δs is a very large number.

§2. State-Probability and Entropy.

We now define the macroscopic states of the gas.

Suppose the volume V to contain Vn molecules of mass m. Δn of these have energy between E and $E + \Delta E$. These divide themselves among the Δs cells. Among the Δs cells, suppose $p_0 \Delta s$ have no molecules, $p_1 \Delta s$ have one molecule, $p_2 \Delta s$ have two molecules, and so on. The probabilities p_r for these Δs cells are obviously functions of the cell type s and the index r, and will be designated p_r^s. It is obvious that for all s

$$\sum_r p_r^s = 1. \tag{3}$$

For given p_r^s and given Δn, the number of possible arrangements of the Δn molecules among the Δs cells of a given energy equals

$$\frac{\Delta s!}{\prod_{r=0}^{r=\infty} (p_r^s \Delta s)!},$$

which, using Stirling's approximation and Eqn.3, can be replaced by

$$\frac{1}{\prod_r (p_r^s)^{\Delta s p_r^s}},$$

which one can also replace by the product over all r and permitted s

$$\frac{1}{\prod_{rs} (p_r^s)^{p_r^s}}. \tag{4}$$

If one extends the product to all values of s from 1 to ∞, then (4) obviously represents the total number of configurations - that is, the probability in Planck's sense - of a macroscopic state of the gas defined by the p_r^s. For the entropy S of this state, Boltzmann's rule gives

$$S = -k \log \sum_{rs} (p_r^s \log p_r^s). \tag{5}$$

§3. Thermodynamic Equilibrium.

For thermodynamic equilibrium S is a maximum, but besides (3) the supplementary conditions must be satisfied that the total number of atoms n as well as the total energy $\bar{E}$ must be fixed at their given values. These conditions are obviously expressed in the two equations [2]

$$n = \sum_{sr} r p_r^s \tag{6}$$

$$\bar{E} = \sum_{sr} E^s r p_r^s, \tag{7}$$

where E^s is the energy of a molecule in a phase cell labelled s. From (1) it follows that

$$E^s = cs^{\frac{2}{3}}$$
$$c = (2m)^{-1}h^2 \left(\frac{4}{3}\pi V\right)^{-\frac{2}{3}}. \tag{8}$$

By carrying out the variation with respect to the p_r^s, one finds for an acceptable solution that the constants β_s, A and B must satisfy

$$p_r^s = \beta_s e^{-\alpha_s r}$$
$$\alpha_s = \alpha + \beta c s^{2/3}. \tag{9}$$

Using (3),

$$\beta_s = 1 - e^{-\alpha_s}. \tag{10}$$

From these we get the average number of molecules per cell

$$
\begin{aligned}
n^s &= \sum_r r p_r^s = \beta_s \sum_r r e^{-\alpha_s r} \\
&= -\beta_s \frac{d}{d\alpha_s}\left(\sum_r e^{-\alpha_s r}\right) = -\beta_s \frac{d}{d\alpha_s}\left(\frac{1}{1-e^{-\alpha_s}}\right) \\
&= \frac{1}{e^{\alpha_s} - 1}.
\end{aligned}
\tag{11}
$$

Eqns.6,7 become

$$n = \sum_s \frac{1}{e^{\alpha_s} - 1} \tag{6}$$

$$\bar{E} = c \sum_s \frac{s^{2/3}}{e^{\alpha_s} - 1}, \tag{7}$$

which together with α_s determine the constants α and β. In this way, the law for the macroscopic distribution of states in thermodynamic equilibrium is completely determined.

Substituting the results of these paragraphs into (5) gives the equilibrium entropy

$$S = -k\left\{\sum_s \left[\log\left(1 - e^{-\alpha_s}\right)\right] - \alpha n - \beta \bar{E}\right\}. \tag{12}$$

We can now calculate the temperature of the system. To do so, use the defining equation of the entropy in an infinitesimal heating at constant volume to get

$$d\bar{E} = TdS = -kT\left\{\sum_s \frac{d\alpha_s}{1 - e^{\alpha_s}} - nd\alpha - \bar{E}d\beta - \beta d\bar{E}\right\},$$

which with substitutions from (9), (6) and (7) gives

$$d\bar{E} = kT\beta d\bar{E}$$

84

or

$$\frac{1}{kT} = \beta. \tag{13}$$

In this way the temperature is expressed indirectly in terms of the energy and the other parameters. From (12) and (13) it follows that the free energy F is given by

$$F = \bar{E} - TS = kT \left\{ \sum_s \left[\log \left(1 - e^{-\alpha_s} \right) \right] - \alpha n \right\}. \tag{14}$$

The pressure p is [3]

$$p = - \left(\frac{\partial F}{\partial V} \right)_T = -kT \sum_s \beta E^s \left(\frac{1}{E^s} \frac{\partial E^s}{\partial V} \right) = -\bar{E} \frac{\partial \log E^s}{\partial V} = \frac{2}{3} \frac{\bar{E}}{V}. \tag{15}$$

This gives the remarkable result, that the relation between the kinetic energy and the pressure is exactly the same as in the classical theory, where it was derived from the Virial Theorem.

§4. The Classical Theory as Limiting Case.

If we neglect unity compared to the e^{α_s}, then we get the results of the classical theory; in the following, we describe under what conditions this neglect is correct. According to (11), (9), (13) the mean number of molecules per cell is then

$$n_s = e^{-\alpha_s} = e^{-\alpha} e^{-E^s/kT}. \tag{11}$$

The number of molecules whose energy lies in the interval dE^s is given according to (8) by

$$\frac{3}{2} c^{-3/2} e^{-\alpha} e^{-E/kT} E^{1/2} dE,$$

in agreement with the classical theory. Eqn.6 leads in the same approximation to

$$e^\alpha = \frac{V}{nh^3} (2\pi m kT)^{3/2}. \tag{16}$$

For hydrogen at atmospheric pressure this quantity is approximately $6 \cdot 10^4$, and clearly much greater than one. Here the classical theory is certainly correct to a good approximation. The error gets important however with increasing density and decreasing temperature. For helium in the region of the critical point, it is genuinely important; of course, then there can no longer be any talk of an ideal gas.

Now we calculate from (12) the entropy for the limiting case. In (12), if one replaces $\log(1 - e^{-\alpha_s})$ by $-e^{-\alpha_s}$ and this by $-1/(e^{\alpha_s} - 1)$, one gets from (6)

$$S = \nu R \log \left[e^{5/2} \frac{V}{h^3 n} (2\pi m kT)^{3/2} \right], \tag{17}$$

where ν is the number of moles, and R the ideal gas constant. This result for the absolute value of the entropy agrees with the well known result of quantum statistics.

85

Nernst's theorem is satisfied for the ideal gas according to the theory developed here. In fact our formula cannot be applied directly at extremely low temperatures, because we have assumed for its derivation that the p_r^s change relatively only an infinitesimal amount when s changes by one. However one sees immediately that the entropy at absolute zero must vanish. Then all molecules must be in the first cell; for this state there is only a single configuration of molecules in the sense of our enumeration. The correctness of our assertion follows directly.

§5. The Difference of our Gas Equation from the Classical Theory.

Our results for the equation of state are contained in the following equations:

$$n = \sum_s \frac{1}{e^{\alpha_s} - 1} \qquad \text{(see Eqn.6a)} \qquad (18)$$

$$\bar{E} = \frac{3}{2}pV = c\sum_s \frac{s^{2/3}}{e^{\alpha_s} - 1} \quad \text{(see Eqns.7a,15)} \qquad (19)$$

$$\alpha_s = \alpha + \frac{cs^{2/3}}{kT} \qquad \text{(see Eqns.9,13)} \qquad (20)$$

$$c = \frac{E^s}{s^{2/3}} = \frac{h^2}{2m}\left(\frac{4}{3}\pi V\right)^{-2/3}. \qquad \text{(see Eqn.8)} \qquad (21)$$

We will now discuss these results. Sec.(4) states that the quantity $e^{-\alpha}$, abbreviated here as λ, is smaller than 1. It is a measure of the "degeneracy" of the gas. We can now write (18) and (19) in the form of double sums

$$n = \sum_{s\tau} \lambda^\tau e^{-cs^{2/3}\tau/kT} \qquad (18)$$

$$\bar{E} = c\sum_{s\tau} s^{2/3}\lambda^\tau e^{-cs^{2/3}\tau/kT}, \qquad (19)$$

where τ is summed over all values from 1 to ∞ for all s.

We can replace the sum over s by an integration from 0 to ∞. This is permitted because of the slow variation of the exponential with s. We get:

$$n = (2\pi mkT)^{3/2}\frac{V}{h^3}\sum_\tau \tau^{-3/2}\lambda^\tau, \qquad (18)$$

$$\bar{E} = \frac{3}{2}kT(2\pi mkT)^{3/2}\frac{V}{h^3}\sum_\tau \tau^{-5/2}\lambda^\tau. \qquad (19)$$

Eqn.18b determines the degeneracy parameter λ as a function of V, T, n, and from (19b), the energy and through that also the pressure of the gas.

86

The general discussion of these equations can be done by dividing (19b) by (18b) to get

$$\frac{\bar{E}}{n} = \frac{3}{2}kT\frac{\sum_{\tau=1}^{\infty}\tau^{-5/2}\lambda^{\tau}}{\sum_{\tau=1}^{\infty}\tau^{-3/2}\lambda^{\tau}}. \tag{22}$$

The average energy (or the pressure) of the gas molecules at a given temperature is seen to be less than the classical value, and the reduction factor is the smaller, the larger is the degeneracy parameter. This itself, from (18b) and (21), is seen to be a function only of the variable

$$\lambda_0 = \frac{nh^3}{V}(2\pi mkT)^{-3/2}.$$

If λ is small, then λ^2 may be ignored compared to 1, and one gets [4]

$$\frac{\bar{E}}{n} = \frac{3}{2}kT\left[1 - \lambda_0/2^{5/2} + \cdots\right]. \tag{22}$$

We next show how the Maxwell distribution is influenced by the quantum theory. If one expands (11) with (20), in powers of λ, one gets

$$n^s = \text{const} \cdot e^{-E^s/kT}\left(1 + \lambda_0 e^{-E^s/kT} + \cdots\right). \tag{23}$$

The parentheses contain the first order effect of the quantum theory on Maxwell's distribution. One sees that the slow molecules, in comparison to the fast, are more numerous than would be expected from Maxwell's distribution.

To close I might call attention to a paradox for which I have no solution. There is no difficulty to handle, by the method developed here, the case of the mixing of two different gases. In this case, each molecular type has its own "cells". This leads to the additivity of the entropy of the components of the mixture. Each component has the same molecular energy, pressure and statistical distribution, as when it existed alone. A mixture of the molecular numbers n_1, n_2, whose molecules of the first and second kind differ from one another arbitrarily little (in particular, perhaps in the masses m_1, m_2), has therefore at a given temperature a different pressure and a different distribution than a single gas of molecular number $n_1 + n_2$ of practically the same molecular mass and the same volume. But this seems almost impossible.

Footnotes and References:

1) Soon to appear in Zeits. f. Physik.

2) $n^s = \sum_r rp_r^s$ is the average number of molecules in the sth cell.

3) *Note added*: Here we make use of the fact that E^s has dimension $V^{-2/3}$, terms involving $\partial\alpha_s/\partial V$ cancel, and $\beta = kT$ is independent of V.

4) *Note added in translation*: The coefficient "$1/2^{5/2} = .1768$" was ".0318" in the original. The corrected value is taken from Eqn.44 of the second paper, which even there appears with a typographical error in Eqn.22 as ".186".

Session of 10 July 1924.
Published on 20 September 1924.

APPENDIX 3.C:

Excerpt from: S. B. d. Preuss. Akad. Wiss. Ber. 1, 3 (1925)

Quantum Theory of the Monatomic Ideal Gas
Part Two

By A. EINSTEIN

Recently in this Journal [XXII 1924, p.261], a paper was published on an application of a method developed by Bose for the derivation of Planck's radiation formula, giving a theory of the "Degeneracy of an Ideal Gas". The interest in this theory lies in the fact that it is based on the hypothesis of a far-reaching formal relationship between the radiation and the gas. According to this theory, the degeneracy affects the statistical mechanics of the gas in a way analogous to that of radiation according to Planck's Law compared to radiation according to Wien's Law. When the Bose derivation of Planck's radiation formula is taken seriously, then one is not permitted to ignore it as a theory of the ideal gas; when it is correctly applied, the radiation is recognized as a gas of quanta, so the analogy between the gas of quanta and the gas of molecules must be a complete one. In the following the earlier development will be supplemented by something new, which seems to me to increase the interest in the subject. For convenience I write the following formally as a continuation of the cited paper.

§6. The Saturated Ideal Gas.

From the theory of the ideal gas it seems a self-evident requirement that the volume and temperature of an amount of gas can be given arbitrarily. The theory then determines the energy and also the pressure of the gas. The study of the Equation of State contained in Eqns.18,19,20,21 however shows that for a given molecular number n and given temperature T, the volume cannot be made arbitrarily small. Eqn.18 requires that for all s, $\alpha_s \geq 0$, which according to (20) means that $\alpha \geq 0$. This means that in Eqn.18b, which is valid in this case, $\lambda = e^{-\alpha}$ must lie between 0 and 1. From (18b) it follows that the number of molecules in such a gas for given volume V cannot be greater than

$$n = \frac{(2\pi m k T)^{3/2} V}{h^3} \sum_{\tau=1}^{\infty} \tau^{-3/2}. \tag{24}$$

What happens now however, when at fixed temperature I let n/V increase (by isothermal

89

compression) to more than the density of the substance?

I maintain that in this case, with the total density fixed, an increasing number of molecules goes into the quantum state numbered 1 (the state without kinetic energy), leaving the remaining molecules distributed with the parameter value $\lambda = 1$. This assertion also requires that something similar occurs as during the isothermal compression of a vapor beyond the saturation volume. A separation occurs; a part "condenses", the rest remains a "saturated ideal gas" ($\alpha = 0$, $\lambda = 1$).

That the two parts are in thermal equilibrium can be shown, because the condensed substance and the saturated ideal gas have the same Planck function per mole

$$\Phi = S - \frac{\bar{E} + pV}{T}.$$

For the condensed substance Φ vanishes, since S, $\bar{E}$, and V vanish individually [1]. For the saturated gas, from (12) and (13) for $\alpha = 0$, one has

$$S = -k \sum \log(1 - e^{-\alpha_s}) + \bar{E}/T. \tag{25}$$

One can write the sum as an integral and recast it by a partial integration, to obtain

$$\sum_s = -\int_0^\infty s \cdot \frac{1}{e^{cs^{\frac{2}{3}}/kT} - 1} \cdot \frac{2}{3} \frac{cs^{-\frac{1}{3}}}{kT} ds,$$

or from (8), (11) and (15)

$$\sum_s = -\frac{2}{3} \int_0^\infty n_s E^s ds = -\frac{2}{3} \frac{\bar{E}}{kT} = -\frac{pV}{kT}. \tag{26}$$

From (25) and (26) for a saturated ideal gas

$$S = \frac{\bar{E} + pV}{T}$$

or - as is necessary for the coexistence of the saturated ideal gas with the condensate -

$$\Phi = 0. \tag{27}$$

We get the prescription:

According to the derived equation of state the ideal gas at each temperature has a maximum density for finding the molecules in motion. By exceeding this density the excess number of molecules fall out motionless (condense without attractive forces). The remarkable fact is, that the equation of state of the saturated ideal gas also gives the maximum possible density of the state of moving molecules to be that density at which the gas is in thermodynamic equilibrium with the condensate. An analogy with the "supersaturated vapor" therefore does not exist for the ideal gas.

§7. Comparison of the Derived Gas Theory with that which follows from the Opposing Hypothesis of the Statistical Independence of the Gas Molecules.

Ehrenfest and others have reported that in the Bose Theory of Radiation and in my analog of the ideal gas, the quanta or molecules do not act in a manner statistically independent of one another, without which, in our treatment of the situation, things would be singularly different. This is entirely correct. When one treats the quanta as statistically independent of one another in their allocation, one arrives at Wien's displacement law; when one treats the gas molecules analogously, one arrives at the classical equation of state of the ideal gas, just as when one approximates the exact result, as Bose and I have done. I will here compare the two analyses with one another, in order to make the distinction completely clear, and so that we can easily compare our result with that of the theory of independent molecules.

According to both theories, the number of cells z_ν, which correspond to the infinitesimal interval ΔE of the molecular energy (in the following called "elementary interval"), is given by

$$z_\nu = 2\pi \frac{V}{h^3} (2m)^{3/2} E^{1/2} \Delta E. \tag{2}$$

The state of the gas may be defined (macroscopically) when it is given how many molecules n_ν are in each cell. One should calculate the number W of realizable possibilities (Planck's probability) of the thus defined state

a) according to Bose:

A state is microscopically defined by the number of molecules in each cell (a configuration). The number of configurations for the νth elementary interval is

$$\frac{(n_\nu + z_\nu - 1)!}{n_\nu!(z_\nu - 1)!}. \tag{28}$$

By taking the product over all elementary intervals one gets the total number of configurations of a state and from that according to Boltzmann's rule the entropy is

$$S = k \sum_\nu \left\{ (n_\nu + z_\nu) \log(n_\nu + z_\nu) - n_\nu \log(n_\nu) - z_\nu \log(z_\nu) \right\}. \tag{29}$$

That this method of calculating partitions of the molecules among the cells does not treat them as statistically independent, is easy to see. It is connected to the fact that what is here called "configurations", according to the hypothesis of independent partition of individual molecules among the cells would not be considered as cases of equal probability. This counting of these configurations of different probability would then give the incorrect entropy if in fact the molecules were statistically independent. The formula therefore expresses indirectly an implicit hypothesis about the mutual influence of the molecules of a totally new and mysterious kind, which just depends on the cases being defined here as configurations of equal statistical probability.

b) according to the hypothesis of the statistical independence of the molecules:

A state is defined microscopically, when it is specified in which cell each molecule resides (a configuration). How many configurations belong to a macroscopically defined state? I can distribute n_ν distinct molecules in

$$(z_\nu)^{n_\nu}$$

different ways among the z_ν cells of the νth elementary interval. If the arrangement of the molecules is labelled in a definite way, then there is a total of

$$\prod (z_\nu)^{n_\nu}$$

different partitions of the molecules over all cells. In order to obtain the number of configurations in the defined sense, this expression must be multiplied by the number

$$\frac{n!}{\prod n_\nu!}$$

of possible orderings of all molecules in the elementary interval for given n_ν. Boltzmann's rule gives the entropy

$$S = k \left\{ n \log(n) + \sum_\nu \left(n_\nu \log(z_\nu) - n_\nu \log(n_\nu) \right) \right\}. \tag{30}$$

The first term of this expression does not depend on the choice of the macroscopic distribution, but only on the total number of molecules. For the comparison of the entropy of different macroscopic states of the same gas this term plays the role of an unimportant constant which we can ignore. We must omit it, when we - as is customary in thermodynamics - wish to calculate that the entropy for a given internal state of the gas should be proportional to the number of molecules. We have therefore set

$$S = k \sum_\nu n_\nu \left(\log(z_\nu) - \log(n_\nu) \right). \tag{29}$$

One usually neglects this factor $n!$ in W for the gas because one considers such configurations, which differ from one another merely by the exchange of the same kind of molecules, as not different and counts them only once.

Now we have to find, for both cases, the maximum of the entropy under the restrictions

$$
\begin{aligned}
\bar{E} &= \sum E_\nu n_\nu = \text{const} \\
n &= \sum n_\nu = \text{const.}
\end{aligned}
$$

92

In Case a) this gives

$$n_\nu = \frac{z_\nu}{e^{\alpha + \beta E_\nu} - 1},$$ (30)

which agrees with Eqn.13. In Case b) it gives

$$n_\nu = z_\nu e^{-(\alpha + \beta E_\nu)}.$$ (30)

In both cases $\beta k T = 1$.

One further sees, that in Case b) Maxwell's distribution is obtained. The quantum structure has no effect here (at least not for infinitely large total volume of the gas). One easily sees now, that Case b) is incompatible with Nernst's theorem. In particular to calculate the value of the entropy at absolute zero temperature for this case, one has to evaluate (29c) at absolute zero. For this all molecules will be in the first quantum state. We have to set

$$
\begin{aligned}
n_\nu &= 0, \text{ for } \nu \neq 1 \\
n_1 &= 1 \\
z_1 &= 1.
\end{aligned}
$$

Eqn.29c for $T = 0$ gives

$$S = -n \log(n).$$ (31)

Therefore calculation (b) results in a contradiction with Nernst's theorem. On the other hand, calculation (a) agrees with Nernst's theorem, as one sees, when one realizes that at absolute zero, in the sense of (a), there is only a single configuration ($W = 1$). Calculation (b) requires either a rejection of Nernst's theorem or a rejection of the requirement that the entropy of a given inner state must be proportional to the molecular number. On this basis I believe that calculation (a) (that is, the prescription of Bose's statistics) must be preferred. Furthermore, the choice of this method over others can even be made a priori. This result also constitutes essential support for the conception of the deep relationship between radiation and the molecular gas, in that the same statistical method which leads to Planck's formula, in its application to the ideal gas produces agreement of the theory of gases with Nernst's theorem.

§8. The Fluctuation Properties of the Ideal Gas.

Suppose a gas of volume V communicates with one of the same kind of infinitely large volume. Both volumes should be separated by a wall which transmits molecules in an energy interval ΔE, but reflects molecules of other kinetic energies. The fiction of such a wall is the analog of the quasi-monochromatic filter in the subject of radiation theory. The question is, what is the fluctuation Δ_ν in the number of molecules n_ν in the energy interval ΔE? It is assumed that an energy exchange between the molecules in different

93

energy intervals within V cannot take place, so no fluctuations of the molecular numbers belonging to other energy intervals are possible.

Let n_ν be the average number of molecules belonging to ΔE, and $n_\nu + \Delta_\nu$ the instantaneous value. Then (29a) gives the value of the entropy as a function of Δ_ν, if one replaces n_ν in this equation by $n_\nu + \Delta_\nu$. If one expands to quadratic terms, one obtains

$$S = \bar{S} + \overline{\frac{\partial S}{\partial \Delta_\nu}} \Delta_\nu + \frac{1}{2} \overline{\frac{\partial^2 S}{\partial \Delta_\nu{}^2}} {\Delta_\nu}^2.$$

A simple relation holds for the infinitely large external-system, namely

$$S^0 = \overline{S^0} - \overline{\frac{\partial S^x}{\partial \Delta_\nu}} \Delta_\nu.$$

The quadratic term is infinitesimally small for the infinite sized external-system. If one designates the total entropy by $\Sigma(= S + S^0)$, then

$$\overline{\frac{\partial \Sigma}{\partial \Delta_\nu}} = 0,$$

because it exists in mean equilibrium. By addition of these equations, one gets the total entropy

$$\Sigma = \overline{\Sigma} + \frac{1}{2} \overline{\frac{\partial^2 S}{\partial \Delta_\nu{}^2}} \Delta_\nu^2. \tag{32}$$

According to Boltzmann's rule one gets from this the probability of Δ_ν

$$dW = \text{const} \cdot e^{S/k} d\Delta_\nu = \text{const} \cdot e^{\frac{1}{2k} \overline{(\partial^2 S/\partial \Delta_\nu{}^2)} \Delta_\nu^2} d\Delta_\nu.$$

From this follows the mean square fluctuation

$$\overline{\Delta_\nu{}^2} = k / \overline{\left(-\frac{\partial^2 S}{\partial \Delta_\nu{}^2} \right)}. \tag{33}$$

Using (29a) for the entropy S,

$$\overline{\Delta_\nu{}^2} = n_\nu + n_\nu^2/z_\nu. \tag{34}$$

This fluctuation law is the perfect analog of the quasi-monochromatic Planck's radiation result. We write it in the form

$$\overline{\left(\frac{\Delta_\nu}{n_\nu} \right)^2} = \frac{1}{n_\nu} + \frac{1}{z_\nu}. \tag{34}$$

The mean square relative fluctuation of the molecules of this kind is the sum of two terms. The first would occur alone if the molecules were independent of one another. In addition to this, there is a part of the mean square fluctuation which is totally independent of the mean molecular density and is determined solely by the energy interval ΔE and the volume. It corresponds to the interference fluctuations in radiation. One can also understand it for the gas in a corresponding way, because one analyzes the gas in the

94

same way as a radiative process and computes these interference fluctuations. I look closer at this explanation, because I believe that it is more than just an analogy, since a material particle or a system of them can be represented by a (scalar) wave field, as deBroglie stated in his remarkable thesis [2]. A material particle of mass m corresponds first of all to a frequency ν_0 given by

$$mc^2 = h\nu_0. \tag{35}$$

Suppose the particle is at rest in a galilean system K', in which we think of synchronous vibrations of frequency ν_0. Relative to a system K in which K' with the mass m is moving with the velocity v along the (positive) x-axis, there is a wave occurring of the form

$$sin\left(2\pi\nu_0 \frac{t - vx/c^2}{\sqrt{1 - v^2/c^2}}\right).$$

The frequency and phase velocity U of this wave are given by

$$\nu = \frac{\nu_0}{\sqrt{1 - v^2/c^2}}, \tag{36}$$

$$U = \frac{c^2}{v}. \tag{37}$$

v is then - as deBroglie has shown - also the group velocity of this wave. It is of further interest, that the energy of the particle $mc^2/\sqrt{1 - v^2/c^2}$ is, according to (35) and (36), just equal to $h\nu$ in agreement with the basic relation of quantum theory.

One sees now that such a gas can be assigned a scalar wave field, and I have myself shown by direct calculation that $1/z_\nu$ is the mean square fluctuation of this wave field, just as is predicted by our above investigation on the energy interval ΔE.

These considerations throw light on the paradox which is referred to at the end of my first paper. In order that two wave trains can noticeably interfere, both their U and ν must agree very closely. In addition, according to (35), (36), (37) it should be noted, that v as well as m for both gases should be closely the same. The waves corresponding to noticeably different molecular masses, therefore, cannot significantly interfere with one another. From that one concludes that according to the theory developed here the entropy of a gas mixture is exactly additive like that of a mixture of particles according to the classical theory, at least so long as the molecular weight of the components differs from one another in some degree.

§9. Remark about the Viscosity of the Gas at Low Temperatures.

According to the analysis of the preceding paragraphs, it seems that an oscillating field should be associated with each moving process, just as the optical oscillating field is associated with the motion of the light quantum. This oscillating field - whose physical nature is still obscure - must, in principle, permit itself to be demonstrated by the diffraction phenomena corresponding to the motion. So a beam of gas molecules travelling through an opening must experience a bending, which is analogous to that of a beam of light. In order that such a phenomenon should be observable, the wavelength λ must be comparable with the dimension of the opening. From (35), (36), and (37) it follows that for velocity v small compared to c

$$\lambda = \frac{h}{mv}. \tag{38}$$

This λ for gas molecules moving with thermal velocities, is extraordinarily small, most importantly even smaller than the molecular dimension σ. Therefore it follows right away, that for the observation of this diffraction it does not do to think of a manufactured opening or grid.

It appears however, that at low temperatures, for the gases hydrogen and helium λ is of the order of magnitude of σ, and it seems in this case, that the effect depends on a coefficient of viscosity which we must evaluate from the theory.

Suppose in particular that a swarm of molecules moving with the speed v hits another molecule, which for convenience we imagine to be unmovable, so the situation is comparable to a wave-train of wavelength λ hitting a foil of diameter 2σ. It undergoes a (Fraunhofer-like) diffraction, which is the same as that which would occur from an equal sized opening. Large angle diffraction occurs when λ is of the order of or greater than σ. At some critical temperature, this becomes bigger than the mechanical process of impulsive scattering following the (not yet calculable) inelastic scattering of molecules of higher frequency which occurs first, reducing their wavelength. In the neighborhood of that temperature, there sets in a rather sudden acceleration of the decrease of viscosity with decreasing temperature. An estimate of that temperature assuming $\lambda = \sigma$ gives $56°$ for H_2, $40°$ for He. Naturally these are completely crude estimates; similar results can however be obtained by more detailed calculations. These involve a new experimental result on the temperature dependence of the viscosity of hydrogen obtained by P. Gunther, for whose explanation Nernst has already devised a quantum theoretic analysis [3].

§10. Equation of State of the Saturated Ideal Gas.
Comments on the Theory of the Equation of State of the
Gas and on the Electron Theory of Metals.

In §6 it was shown, that for an ideal gas in equilibrium with its condensate the degeneracy parameter $\lambda = 1$. The concentration, energy and pressure of those molecules which are still moving are determined by the temperature T alone using (18b), (22) and

96

(15). The equations are

$$\eta = \frac{n}{NV} = \frac{2.615}{Nh^3}(2\pi mkT)^{3/2} = 1.12 \cdot 10^{-15}(MRT)^{3/2} \tag{39}$$

$$\frac{E}{n} = \frac{1.348}{2.615} \cdot kT \tag{40}$$

$$p = \frac{1.348}{2.615}RT\eta. \tag{41}$$

Here η is the concentration in moles, N the number of molecules in moles, and M the molecular weight.

One finds with the help of (39), that the actual gas never attains the value of the density that would saturate the corresponding ideal gas. Nevertheless the critical density of helium is only some five times smaller than the saturation density η of the ideal gas of the same temperature and the same molecular weight. For hydrogen the corresponding ratio is some 26. Since the real gas exists at densities near the saturation density then according to (41) the degeneracy considerably influences the pressure. Consequently, when the underlying theory is applicable, it predicts a significant quantum influence on the equation of state. In particular, one must investigate whether the deviation from van der Waal's equation of the corresponding state can be explained [4].

In addition one must also expect that the diffractive phenomenon discussed in the foregoing paragraphs, might influence the equation of state at low temperatures.

There is a situation in which, as far as possible, nature has essentially realized the saturated ideal gas, namely for the conduction electrons inside metals. The electron theory of metals has basically given the connection between electric and thermal conductivity with remarkable success (the Drude-Lorentz formula) under the assumption that inside the metal, free electrons exist which conduct both electricity and heat. In spite of this great success however, such a theory at present does not have the capacity to fit, among other things, the fact that the free electrons contribute no significant share of the specific heat of the metal, and therefore the theory cannot be correct. This difficulty disappears when one uses the above theory as the foundation. In particular, it follows from (39) that the saturation concentration of the (moving) electrons at the normal temperature is some $5.5 \cdot 10^{-5}$, so that only a vanishingly small part of the electrons can contribute a share to the thermal energy. The average thermal energy from the electrons participating in the thermal motion is thereby some half as great as according to the classical molecular theory. When only very small forces are involved, which leave the non-moving electrons in their rest positions, then it is also conceivable that these do not participate in the electrical conductivity. It is even possible that suppression of this weak binding force for very low temperatures could cause superconductivity. In addition, on the basis of this theory, the thermoelectric force would not be understandable as long as one treated the electron gas as an ideal gas. Naturally such an electron theory of metals would not lead to Maxwell's velocity distribution, but to that of the saturated ideal gas as in the above theory; from (8), (9), (11) it gives for this special case:

$$dW = \text{const} \cdot \frac{E^{1/2} dE}{e^{E/kT} - 1}. \tag{42}$$

Upon thinking through this theoretical possibility one comes to the difficulty that, to explain the measured conductivity of the metal for heat and electricity compared to the very small density of the electrons which, according to our understanding, contribute to the thermal motion, one must assume very long mean free paths (of the order 10^{-3} cm). Neither does it appear possible on the basis of this theory, to understand the behavior of the metals for reflection and emission of infrared radiation.

§11. Equation of State of the Unsaturated Gas.

We shall now consider more exactly the deviation of the equation of state of the ideal gas in the unsaturated domain, from the classical equation of state. We use here the equations (15), (18b) and (19b).

We define the abbreviations

$$\sum_{1}^{\infty} \tau^{-\frac{3}{2}} \lambda^{\tau} = y(\lambda)$$

$$\sum_{1}^{\infty} \tau^{-\frac{5}{2}} \lambda^{\tau} = z(\lambda)$$

and substituting one in the other, express z as a function of y ($z = \Phi(y)$). The solution of this equation, for which I thank J. Grommer, depends upon the following general rule (Lagrange):

Under the conditions of our case, that y and z vanish for $\lambda = 0$, and that y and z are regular functions of λ in some region of the origin, then for sufficiently small y, the Taylor series holds

$$z = \sum_{\nu=0}^{\infty} \left(\frac{d^{\nu} z}{dy^{\nu}} \right)_{\lambda=0} \frac{y^{\nu}}{\nu!}, \tag{43}$$

where the coefficients can be derived from the functions $y(\lambda)$ and $z(\lambda)$ by the recursion formula

$$\frac{d^{\nu} z}{dy^{\nu}} = \frac{d}{d\lambda} \left(\frac{d^{\nu-1} z}{dy^{\nu-1}} \right) \frac{d\lambda}{dy}. \tag{44}$$

One gets convergence in our case for $\lambda < 1$ and calculation easily gives

$$z = y - 0.1768y^2 - .0034y^3 - .0005y^4.$$

98

We now introduce the notation

$$\frac{z}{y} = F(y).$$

For the unsaturated ideal gas (that is, between $y = 0$ and $y = 2.615$) we get

$$\frac{\overline{E}}{n} = \frac{3}{2}kTF(y) \tag{19}$$

$$p = RT\eta F(y); \tag{22}$$

where

$$y = \frac{h^3}{(2\pi mkT)^{3/2}}\frac{n}{V} = \frac{h^3 N\eta}{(2\pi M RT)^{3/2}}. \tag{18}$$

From (19b) one gets the molar specific heat at constant volume c_v:

$$c_v = \frac{3}{2}R\left(F(y) - \frac{3}{2}yF'(y)\right) = \frac{3}{2}RG(y).$$

$\ldots\ldots$

If one bears in mind the approximate linear behavior of $F(y)$, then to a good approximation

$$F(y) \approx (1 - .177y) \qquad \text{and} \qquad G(y) \approx (1 + .088y),$$

giving

$$p \approx RT\eta\left[1 - 0.177\frac{h^3 N\eta}{(2\pi M RT)^{3/2}}\right]. \tag{22}$$

Footnotes and References:

1) The condensed part of the substance occupies no particular volume, since it does not contribute to the pressure.

2) Louis de Broglie, Thesis, Paris. (Edit. Musson & Co.), 1924. In this dissertation, one also finds a very remarkable geometric interpretation of the Bohr-Sommerfeld quantization rule.

3) W. Nernst, Sitzungsber. 1919, VIII, p.118. P. Günther, Sitzungsber. 1920, XXXVI, p.720.

4) This is not the case as I have subsequently found out by comparison with experiment. The conjectured influence is masked by molecular interactions of another kind.

Dated December 1924.
Session of 8 January 1925.
Published on 9 February 1925.

APPENDIX 3.D:

Excerpt from: S. B. d. Preuss. Akad. Wiss. Ber. 3, 18 (1925)

On the Quantum Theory of the Ideal Gas

By A. Einstein

Prompted by a derivation due to Bose of Planck's radiation formula, which is based on the Light Quantum Hypothesis, I have recently formulated a quantum theory of the ideal gas [1]. This theory assumes that a light quantum (aside from its polarization property) differs essentially from a monatomic molecule only in that the rest-mass of the quantum is vanishingly small. The postulate based on this analogy is by no means approved by all, and furthermore the statistical method used by Bose and by me is not free of all doubt, but it does appear *a posteriori*, from its success in the case of radiation, to be correct. Consequently I, along with others, have sought a quantum theory of the ideal gas which is as free as possible of arbitrary hypotheses. These considerations are set forth in the following report. It gives an efficient structure for future investigations of the theory, when the necessary knowledge of the yet incomplete laws of such a theory are known. It deals with an area in the theory of gases, whose methods and results have far reaching analogs which lead in the subject of radiation theory to Wien's Displacement Law.

§1. Statement of the Problem.

Consider one mole of an ideal gas, whose molecules have mass m, and which is contained in volume V at temperature T. We ask, from the statistical laws of probability distributions, for the analog of the Maxwell distribution formula. We seek an equation of the form

$$dn = \rho(L, kT, V, m)V\frac{dp_1\,dp_2\,dp_3}{h^3}. \tag{1}$$

Here dn is the number of molecules whose momentum components (p_1, p_2, p_3) lie in the interval (dp_1, dp_2, dp_3). L is the kinetic energy of the molecule

$$L = \frac{1}{2m}(p_1^2 + p_2^2 + p_3^2);$$

because of the obvious isotropy, (p_1, p_2, p_3) can appear in ρ only in the combination L. ρ is an as yet unknown function of the four variables indicated. If the density function ρ is known, then naturally the equation of state is also known, because there is no doubt that

100

the pressure follows from a standard calculation of the mechanical impact of the molecules with the wall. On the other hand we need not assume that the collisions of the molecules with one another follow the laws of mechanics; otherwise we would simply arrive at the Maxwell distribution law and the classical gas equations.

§2. Why is the Classical Equation of State Not Appropriate in the Quantum Theory?

Ever since the first of Planck's works on quantum theory, one understood the W in Boltzmann's rule

$$S = k \log W$$

to be a whole number. It gives the number of discrete ways (in the sense of quantum theory) in which the state of entropy S can be realized. In most cases it is not possible to calculate W without theoretical arbitrariness. Thus it is seemingly impossible to count all conceivable ways with enough conviction that S contains no arbitrary additive constant. Nonetheless, in the sense of quantum theory the entropy is completely determined and must always be positive. Planck's conception is almost a necessity to understand Nernst's theorem. At absolute zero all disorder due to thermal agitation stops, and the state can be realized in only one way ($W = 1$), which means that Nernst's theorem ($S = 0$ for $T = 0$) holds.

This simple explanation of Nernst's theorem from Planck's understanding of Boltzmann's rule arises from the general countability, in this conception, of the number of states. It leads us in particular to the result that the entropy can never be negative.

According to the classical equation of state of an ideal gas, the entropy of a mole contains the additive term $R \log V$, which expresses its dependence on the volume at constant temperature. This term can be made arbitrarily negative. Now admittedly this value of V for the interacting gas must be divided by the critical volume of the gas, so reaching a negative entropy value for the interacting gas can be avoided on this basis. However we still need to understand how the fictional ideal gas, which closely approximates gases which actually occur in nature, should not lead to a violation of a general thermodynamic law. According to the classical equation of state, however, it would necessarily lead to a negative entropy for a state realizable in principle. Therefore we must reject on principle the classical equation of state and regard it as other than a fundamental law, like perhaps Wien's displacement law.

§3. Dimensional Considerations Used in the Following Method.

It follows from (1) that ρ is dimensionless. We can therefore reach certain conclusions about the structure of the function ρ, provided that we assume that ρ has no dimensional constants other than Planck's constant h. Then ρ must be of the form

$$\rho = \psi \left(\frac{L}{kT}, \frac{m(V/N)^{2/3} kT}{h^2} \right),\tag{2}$$

where ψ is an unknown universal function of two distinct dimensionless variables. The function ψ is normalized so that

$$\frac{V}{h^3} \int \rho d\Phi = N,\tag{3}$$

where

$$d\Phi = \int_L^{L+dL} dp_1 dp_2 dp_3 = 2\pi (2m)^{\frac{3}{2}} L^{\frac{1}{2}} dL.\tag{4}$$

No more can be said just from dimensional considerations. The two variable function ψ can be further restricted so that only a function of one variable remains undetermined. This can be done in two independent ways, by making one of the following assumptions:

1) The entropy of a gas does not change for infinitely slow adiabatic compression.

2) For an ideal gas in a conservative external field, the stationary states have everywhere the same probability distribution.

These two propositions should hold in the absence of collisions of the molecules with one another. Because of the neglect in principle of the collisions, the only question is of the validity of the two non-demonstrable hypothesis; these are however very natural, and their accuracy is made plausible by the fact that they both lead to the Maxwell distribution in the limiting case of vanishing quantum effects.

§4. Adiabatic Compression.

We suppose the gas to be enclosed in a rectangular box of dimensions l_1, l_2, l_3. The probability distribution should be isotropic, or chosen so. The collisions with the walls are elastic. Then the distribution function does not change with time. Let it be given by

$$dn = \frac{V}{h^3} \rho d\Phi,\tag{5}$$

where ρ is an arbitrary function of L.

When we infinitely slowly and adiabatically move the walls, in such a way that

$$\frac{\Delta l_1}{l_1} = \frac{\Delta l_2}{l_2} = \frac{\Delta l_3}{l_3} = \frac{1}{3} \frac{\Delta V}{V},\tag{6}$$

then the distribution remains isotropic, of the form (5). How does the distribution change?

Let $|p_1|$ be the magnitude of p_1 of a molecule. From the law of elastic collisions, one easily gets

$$\Delta|p_1| = -|p_1|\frac{\Delta l_1}{l_1}, \tag{7}$$

and similarly for $\Delta|p_2|$, $\Delta|p_3|$. From this one gets

$$\Delta L = \frac{1}{m}\left(|p_1|\Delta|p_1| + \cdots\right) = -\frac{2}{3}L\frac{\Delta V}{V}. \tag{8}$$

It follows from (4) that

$$\Delta d\Phi = 2\pi(2m)^{\frac{3}{2}}\left(L^{\frac{1}{2}}\Delta dL + \frac{1}{2}L^{-\frac{1}{2}}\Delta L dL\right),$$

or using (8)

$$\Delta d\Phi = -d\Phi\frac{\Delta V}{V}, \tag{9}$$

that

$$\Delta(V d\Phi) = 0. \tag{10}$$

In all these formulas Δ is the change in the considered quantity due to the adiabatic change in the volume.

Because the number of molecules does not change

$$0 = \Delta dn = \Delta(V\rho d\Phi)$$

or from (10)

$$\Delta\rho = 0. \tag{11}$$

We now consider the entropy of the gas whose distribution function is given by (5). For that we assume that the entropy is the sum of the parts which correspond to the energy increase dL. This hypothesis is the analog of that in the theory of radiation, where the entropy of the radiation is the sum of that for the quasi-monochromatic constituents. It is equivalent to the assumption made for different molecular velocity distribution intervals when one introduces semi-permeable walls [2]. According to this hypothesis if we have a gas whose molecules are isotropically distributed, and whose momentum interval $d\Phi$ changes, the entropy attributed to it is

$$\frac{dS}{k} = \frac{V}{h^3}s(\rho, L)d\Phi, \tag{12}$$

where s is an as yet unknown function of the two variables.

For the adiabatic compression the entropy must remain unchanged; that is

$$\Delta dS = 0$$

103

so from (7) and (10)

$$0 = \Delta s = \frac{\partial s}{\partial \rho} \Delta \rho + \frac{\partial s}{\partial L} \Delta L.$$

From (11) it follows that

$$\frac{\partial s}{\partial L} = 0, \tag{13}$$

so s is a function of ρ alone.

Now we deduce the requirement which the velocity distribution of a gas in thermodynamic equilibrium must satisfy. For that, the entropy

$$\frac{S}{k} = \frac{V}{h^3} \int s \, d\Phi$$

must be a maximum with respect to all variations of ρ which satisfy the two restrictions

$$\delta \left\{ \frac{V}{h^3} \int \rho \, d\Phi \right\} = 0$$

and

$$\delta \left\{ \frac{V}{h^3} \int L\rho \, d\Phi \right\} = 0.$$

Carrying out the variation yields the requirement

$$\frac{\partial s}{\partial \rho} = \alpha + \beta L \tag{14}$$

where α and β are independent of L. Since s, and therefore also $\partial s / \partial \rho$, are functions only of ρ, this equation can be solved for ρ giving

$$\rho = \Psi(\alpha + \beta L) \tag{15}$$

where Ψ is an unknown function. α and β, naturally, can depend on kT, V/N, m and h.

The quantity β can be determined from the entropy law by carrying out an infinitesimal heating of the gas at constant volume. With E the energy of the gas, and D the changes which occur in various quantities, one gets

$$DE = \frac{V}{h^3} \int L D\rho \, d\Phi \quad = TDS = \frac{kTV}{h^3} \int Ds \, d\Phi.$$

Then using (14)

$$Ds = D\rho(\alpha + \beta L)$$

and holding constant the number of molecules

$$\int D\rho \, d\Phi = 0$$

104

one gets

$$\int L D \rho d\Phi (1 - kT\beta) = 0$$

or

$$\beta = \frac{1}{kT}.$$

then (15)gives

$$\rho = \Psi(\frac{L}{kT} + \alpha). \tag{15}$$

§5. Gas in a Conservative Force Field.

Suppose the gas is in dynamic equilibrium under the influence of an external conservative field. The potential energy Π of a molecule is a function of its spatial coordinates only. The density function ρ must be defined in the six-dimensional phase space of the molecule. We will ignore collisions and assume that the motion of the individual molecules under the influence of the external field follows classical mechanics. The requirement that the motion should be stationary yields the equation

$$\sum_i \frac{\partial(\rho x_i)}{\partial x_i} + \frac{\partial(\rho p_i)}{\partial p_i} = 0. \tag{16}$$

From the equations of motion

$$\dot{x}_i = p_i/m, \qquad \dot{p}_i = -\frac{\partial \Pi}{\partial x_i},$$

it follows in the known way that

$$\frac{\partial \rho}{\partial x_i}\dot{x}_i + \frac{\partial \rho}{\partial p_i}\dot{p}_i = 0. \tag{16}$$

Therefore ρ is constant along an orbit. Since from the isotropy of the equilibrium distribution, ρ can depend on the p_i only in the combination L, then ρ must have the form

$$\rho = \Psi^\star(L + \Pi)). \tag{17}$$

Since the equilibrium density distribution holds in the different regions of our gas, the different values of V correspond to the same temperature, so Equation (17) expresses at one and the same time the form of the dependence of the phase density upon V in that Π is a function of V.

§6. Inferences about the Equation of State of the Ideal Gas.

If we summarize the discussions of the last two paragraphs as applied to the problem of the Equation of State, then we must write in place of (15) and (17):

$$\rho = \Psi(h, m, \frac{L}{kT} + \alpha), \tag{15}$$

and

$$\rho = \Psi^{\star}(h, m, kT, L + \Pi). \tag{17}$$

α and Π are as yet unknown universal functions of h, m, kT, V. Ψ and $\Psi^{\star}$ are dimensionless universal functions. Each of these has the property that it must depend in the following special way upon the dimensionless combinations given in Eqn.2:

$$\rho = \psi\left(\frac{L}{kT} + \phi\left(\frac{h^3 N}{V}(2\pi mkT)^{-3/2}\right)\right). \tag{18}$$

Here ψ and ϕ are two universal functions of one dimensionless variable. The two functions ψ and ϕ are connected, so that in fact the result depends only on the unknown function ψ. From (2), (3) and (4) one obtains the specific relation

$$\int_{v=0}^{\infty} \psi(v + \phi)v^{\frac{1}{2}}\, dv = \frac{\sqrt{\pi}}{2}\frac{h^3 N}{V}(2\pi mkT)^{-3/2}. \tag{19}$$

If the function ψ is given, then for each value of ϕ the right side of the equation can be calculated; by inversion one obtains ϕ as a function of the right hand side. The problem is then reduced to the question of the function ψ.

§7. Relation of this Result of the Classical Theory to that of my Quantum Theory of the Ideal Gas.

We investigate the case where the constant h cancels out of the distribution function. For abbreviation, we set

$$u = \frac{h^3 N}{V}(2\pi mkT)^{-3/2}, \qquad v = \frac{L}{kT}.$$

From (1) and (18) one sees that h can only be eliminated from the expression for dn, when ψ/u is independent of u. We will in this case call this function $\overline{\psi}(v)$. By an appropriate choice of the function ϕ, we can require

$$\psi(v + \phi(u)) = u\overline{\psi}(v). \tag{20}$$

If one takes the logarithm of this equation and differentiates it twice (by u and v) then one sees that $\log \psi$ must be a linear function. Also ϕ is then easily found. It turns out that ψ must in fact be an exponential function (Maxwell's velocity distribution).

The classical theory corresponds to the prescription

$$\psi(v) = e^{-v}, \tag{21}$$

and from my derivation of the statistical theory the prescription is

$$\psi(v) = \frac{1}{e^v - 1}. \tag{22}$$

106

The Planck Function [3] replaces the exponential with the negative exponent. The difference of the prescription (22) from (21) is enough to lead to Nernst's theorem, as I have already shown in a recently published work.

Two objectives have been reached in the foregoing investigation. First, a general formulation (Eqn.18) is found, which must be satisfied by any theory of the ideal gas. Second, from the above it has been shown that the equation of state which I derived by adiabatic compression is not affected by a conservative external force field.

Footnotes and References:

1) Diese Ber. XXII P.261, 1924.

2) One can think of realizing such semi-permeable walls by a conservative force field.

3) This follows easily from (18), (20) and (21) of the above cited work.

Session of 29 January 1925.

Chapter 4

Wave Function of State of Many Identical Particles

Summary: The discovery by Heisenberg and by Dirac that the wave functions for identical particles satisfying Bose-Einstein statistics must be symmetric in the interchange of any two particles, and conversely that those satisfying Fermi-Dirac statistics must be antisymmetric, is presented from its origin.

§1. Introduction.

Heisenberg and Dirac independently and almost simultaneously discovered that the wave functions of states of two or more identical particles must be symmetrized or antisymmetrized in the coordinates of any two particles, depending on whether the particles obey Bose-Einstein or Fermi-Dirac statistics. These conclusions seem almost trivial today but in the first instance even these inventors of quantum mechanics seem to have agonized over the result, which immediately defined forever the way we think about the the quantum mechanics of identical particles.

§2. Heisenberg's Deduction of the Symmetrization Requirement.

Heisenberg's first exploration of the problem [4.1], progressing from the simplest model to the most profound conclusion, demonstrated that taken by themselves, the postulates of quantum mechanics alone require one choice or the other - symmetric or antisymmetric - for the wave function of identical particles.

The Hamiltonian for identical coupled oscillators

$$H = \frac{p_1{}^2}{2m} + \frac{m\omega^2}{2}q_1{}^2 + \frac{p_2{}^2}{2m} + \frac{m\omega^2}{2}q_2{}^2 + m\lambda q_1 q_2 \tag{1}$$

is decoupled by the transformation

$$q_1' = (q_1 + q_2)/\sqrt{2}, \quad q_2' = (q_1 - q_2)/\sqrt{2}. \tag{2}$$

With the definitions

$$\omega_1'^2 = \omega^2 + \lambda, \quad \omega_2'^2 = \omega^2 - \lambda, \tag{3}$$

the energy eigenvalues are

$$E(n_1', n_2') = \hbar\omega_1' \left(n_1' + \frac{1}{2} \right) + \hbar\omega_2' \left(n_2' + \frac{1}{2} \right), \tag{4}$$

where n_1' and n_2' are integer excitation numbers for the symmetric (q_1') and the antisymmetric (q_2') oscillators.

Heisenberg gives a detailed analysis - based on an analog to dipole and quadrupole electromagnetic couplings, which are necessarily symmetric in the coordinates q_1 and q_2 - and concludes that there can be no "crossover-transitions" between energy levels with n_2' even and with n_2' odd.

He first considers lowest order perturbation theory. Dipole transitions require a matrix element of $q_1 + q_2 \sim q_1'$, so only transitions with

$$\Delta n_1' = \pm 1, \quad \Delta n_2' = 0 \tag{5}$$

are allowed. Quadrupole transitions depend on a matrix element of $q_1^2 + q_2^2 \sim q_1'^2 + q_2'^2$, with only

$$\Delta n_1' = \pm 2, 0, \quad \Delta n_2' = \pm 2, 0. \tag{6}$$

Even in higher orders of perturbation theory, and including higher multi-poles, still only transitions with

$$\Delta n_2' = \text{even}, \quad \Delta n_1' = 0, \pm 1, \pm 2, \cdots, \tag{7}$$

are allowed.

The energy levels separate into two non-connected spectra, $(+)$ with $n_2' =$ even, and $(-)$ with $n_2' =$ odd. Only by introducing interactions unsymmetrical in the coordinates of the identical oscillators (q_1, q_2) can transitions be induced between the $(+)$ and $(-)$ spectra.

Heisenberg concludes that this "doubling" of the quantum mechanical solution is the essential result of the analysis of this basic example. It remained to be proved, first, that the symmetric states are always separate from the antisymmetric ones in the general case, and, second, that the symmetric (antisymmetric) states should be identified with systems obeying Bose-Einstein (Fermi-Dirac) statistics.

In §2 the first step is done in a rather cumbersome way using matrix mechanics. Heisenberg generalizes the coupled oscillators to the case of two identical sub-systems a and b with interaction-free Hamiltonian H and interaction Hamiltonian H^1. He shows that the energy levels are non-degenerate and separated, as in the earlier example, into two spectra - those even under the exchange of a, b and those odd under the exchange - just like the separate spectra $(+)$ and $(-)$ above. Then follows his decisive result, that a coupling F - symmetric in the systems a and b - cannot induce transitions between these separate spectra, which must therefore be viewed as separate and independent solutions of the quantum mechanical problem. He follows this laborious development with the equivalent and much more transparent statement in the Schrödinger formulation of quantum mechanics. In his Eqn.17, it is immediately obvious that the expectation value of a symmetric operator between a symmetric state and an antisymmetric state necessarily vanishes, so that all couplings also vanish and the two different sets of states never mix.

Next, in §4, Heisenberg generalizes his conclusions from 2 to n identical sub-systems, with $n!$ partitions and $n!$ states with their degeneracies removed by interactions. Of these, there is one state - totally antisymmetric in the n sub-systems, with no two sub-systems in the same configuration - which satisfies the Pauli Exclusion Principle. This state must be unique among all those permitted *a priori* by the quantum mechanics. The statistical weight of the state is reduced from the naive $n!$ to 1.

Heisenberg concludes by constructing the totally antisymmetric state in the matrix mechanics (19), and the determinantal many-body Schrödinger wave function (20).

§3. Dirac's Introduction of the Symmetrization Postulate.

Dirac's discovery of the (anti)symmetrization requirement of the many-body wave function describing identical particles is characteristically direct, elegant and succinct $\cdots\cdots$ "one would expect only symmetrical functions of the coordinates of all the electrons to be capable of being represented by matrices $\cdots$ obtain two solutions $\cdots$ the theory is incapable of deciding which is the correct one $\cdots$ one $\cdots$ leads to Pauli's principle $\cdots$ the other $\cdots$ to the Bose-Einstein statistical mechanics."

These brilliant deductions are all contained in §3 of Dirac's prolific paper [4.2] where it is just one of four great advances in one historic paper.

Dirac formulates the question for an atom with two electrons in states m and n. He concludes that the Heisenberg matrix mechanics which determines observable quantities (and only these) requires that the configurations (m, n) and (n, m) should be counted only once. Then the transitions $mn \rightarrow m'n'$ and $mn \rightarrow n'm'$, which are physically indistinguishable, will appear as a single element in the transition matrix. This conclusion leads to a further paradox if he tries to assign a matrix to any quantity such as x_1 or x_2 which is not symmetric in the variables of the identical particles. His conclusion is that only symmetric functions can be represented as matrices, in accord with Heisenberg's matrix mechanics of observable quantities.

Dirac next constructs the Schrödinger wave functions $\psi_{m,n}(1, 2)$ as either symmetric or antisymmetric functions of the coordinates of the two electrons

(1, 2), in order that a single wave function should describe both (m, n) and (n, m). The symmetric functions alone or the antisymmetric functions alone satisfy the requirements. He points out that a choice must be made which is not dictated *a priori* by quantum mechanics. The extension of the symmetric and antisymmetric wave functions to an arbitrary number of non-interacting identical particles is made directly. The fact that (symmetrical) interactions of the identical particles preserves the symmetry is pointed out, as is the fact that the Pauli Exclusion Principle is expressed in the totally antisymmetric wave function.

§4. Epilog.

More than fifty years later, prompted by a number of proposals to search for small violations of the Pauli Exclusion Principle, Amado and Primakoff [4.3] reiterated, with emphasis, a point made by Heisenberg and Dirac: There can be no *small* violations of particle identity.

A typical experimental test of the Pauli Exclusion Principle might be thought to consist of looking for K-shell X-rays from heavy stable atoms. The X-rays are supposed to be evidence of outer shell electrons making radiative transitions to the filled innermost $1S$ electron shell, in violation of the requirement that the resulting wave function should be antisymmetric and subject to the Exclusion Principle. If no X-rays are seen from an Avogadro number N_A of atoms for a time T_{EXP} of one year, for a transition that normally has a lifetime $\tau \sim 10^{-17}$ seconds, it might naively be said that the Pauli Exclusion Principle has inhibited the transition by a factor

$$N_A \times T \div \tau = 6 \times 10^{23} \times 3 \times 10^7 \div 10^{-17} = \sim 10^{48}!$$

Amado and Primakoff point out, as was clearly stated in the original work of both Heisenberg and Dirac, that this remarkable stability factor has a different origin:

If the electrons are *identical*, then the interaction Hamiltonian *must be*

112

symmetric in the electron coordinates and *cannot* lead from an antisymmetric initial state to anything but an antisymmetric final state.

So it is impossible to simultaneously maintain particle identity and to violate the Pauli Exclusion Principle by changing the wavefunction antisymmetry.

One could postulate a Hamiltonian that somehow differentiates amongst "electrons" but only at the expense of them ceasing to be electrons and ceasing to be subject to the Pauli Exclusion Principle.

Amado and Primakoff point out that such experiments do have a separate motivation, which is to test the stability of the electron. But any conceivable mechanism to violate the Pauli Exclusion Principle by a symmetry violating transition necessarily sacrifices particle identity at the outset and completely voids the requirements of the Pauli Exclusion Principle and its quantum mechanical realization by the Heisenberg-Dirac wave function antisymmetrization. Needless to say, there is not the least shred of evidence for, and mountains of evidence, both direct and circumstantial, against the existence of any faux-electron.

Nor do such arguments discourage further tests of the efficiency of the Pauli Exclusion Principle, both experimental and theoretical [4.4]. But we will leave the subject here.

§5. Biographical Note on Heisenberg. (†, ††, †∗)

Werner Karl Heisenberg (1901-1976), a student of Sommerfeld at Munich, PhD(1923) on hydrodynamics; collaborator (1922-) and assistant to Born at Göttingen (1923-24); visitor at Copenhagen (1925-27); introduced the principle of observables as dynamical variables, formulated the matrix theory as the first breakthrough to modern quantum mechanics (1925); with Born and Jordan, developed the Matrix Theory of Quantum Mechanics (1925); treated effects of particle identity (1926); invented the Uncertainty Principle as the epitome and essence of Quantum Mechanics (1927); Heisenberg ferromagnetism (1928); hole theory in solid state physics (1931); Nobel Prize (1932); invented isotopic spin (1932); S-matrix theory (1943); non-linear unified field theory of elementary particles

(1950-). He was Professor at Liepzig (1927-41); director of the Kaiser Wilhelm Institute for Physics in Berlin (1941-45), and of the Max Planck Institute in Göttingen, and then in Munich; President of the Alexander von Humboldt Foundation (1953-).

Heisenberg has been much analyzed and criticized for his role in the Nazi atom bomb project. The tape recordings of his reaction while in confinement at the time of the Hiroshima blast are interpreted as evidence of confusion and incompetence. Without pretending any expertise on the particular events, it seems to us that Heisenberg's capabilities at age 40-44 must have been what they had always been - monumental; and he accomplished just what he set out to accomplish, which is to say - nothing. If he deserves any criticism, it is from whatever Nazis are left.

The rest of us should remember him at his best, as Born first saw him (††) : 'He looked like a simple peasant boy, with short, fair hair, clear bright eyes and a charming expression. ⋯ His incredible quickness and acuteness of apprehension has always enabled him to do a colossal amount of work without much effort ⋯ .' or as Kramers described both him and Bohr (†*): 'tough, hard nosed, uncompromising, and indefatigable.'

(† - E.P. Wigner, PHYSICS TODAY, April 1976, p.86; The London Times, Feb2, 1976, p.26; †† - B.L. van der Waerden, in *Sources of Quantum Mechanics* (North Holland, Amsterdam, 1967), ed. B.L. van der Waerden, p.19; †* - A. Pais, in *Niels Bohr's Times, In Physics, Philosophy, and Polity* (Clarendon Press, Oxford, 1991), pp.21, 275-279.)

Bibliography and References.

4.1) W. Heisenberg, Zeits. f. Phys. **38**, 411 (1926); see also our App.4A.

4.2) P.A.M. Dirac, Proc. Roy. Soc. (London) **A112**, 661 (1926); see also our App.4B.

4.3) R.D. Amado and H. Primakoff, Phys. Rev. **C22**, 1338 (1980).

4.4) O.W. Greenberg and R.N. Mohapatra, Phys. Rev. **D39**, 2032 (1989); K. Deilamian, J.D. Gillaspy, and D.E. Kelleher, Phys. Rev. Lett. **74**, 4787 (1995).

APPENDIX 4.A:

Excerpt from: Zeitschrift für Physik 38, 411 (1926)

Many-body Problem and Resonance in Quantum Mechanics

by **W. Heisenberg** in Copenhagen

(Received on 11 June 1926)

This paper attempts to give a foundation for the quantum mechanical treatment of the many-body problem. To this end, a characteristic resonance phenomenon in the quantum mechanics of the many-body problem is investigated and an understanding on the basis of this investigation is obtained. The result produces both the Bose-Einstein Statistics and the Pauli Exclusion Principle.

Quantum mechanics has as yet only been applied to the system of a single moving mass point. This limitation was, in the first instance, due to mathematical difficulties, because a calculation of the individual amplitudes still stood in the way. Recently, extraordinary progress has been made on this problem by noteworthy research in which Schrödinger [1], starting from the de Broglie wave theory of matter [2], has given a mathematically new, essentially easy, approach to the field of Quantum Mechanics.

......*(We go directly to the discussion of the Many-body Problem.)*

The aim of our investigation is the Quantum Mechanical treatment of a system which contains several particles. Such a treatment at first appears beset by considerable difficulties: The particles of the de Broglie wave theory, which leads to Bose-Einstein Statistics [3], possess no clear analog in Quantum Mechanics; additional rules, such as the Pauli Exclusion Principle [4], have in this form, no place in the mathematical scheme of Quantum Mechanics. One could therefore imagine a failure of Quantum Mechanics for the problem of equivalent electrons. Finally, we might also remember the well known difficulty of a quantitative representation of the spectra: The splitting between the singlet- and triplet-system in the spectra of the Alkali-earths and in the Helium spectrum is an order of magnitude too wide, so that two rotating electrons could be said simply to differ in their magnetic interaction energy.

The aim of the following investigation is a general analysis of the assertions that one can make about the many-body problem, purely as a consequence of the application of Quantum Mechanics. It is necessary to assume that the above difficulties can be completely solved by this analysis, which leads us to a connection between Bose-Einstein Statistics and Quantum Mechanics.

§1. The simplest imaginable many-body problem is a system of two coupled oscillators. Such a system can be solved provided the interaction energy is a quadratic function of the coordinates of the two uncoupled oscillators. Without doubt, this problem can be handled in Quantum Mechanics without new assumptions; the first result of this example is the characteristic eigenvalues of the quantum-theoretic many-body problem. From the analysis of this simple model, all results are available, and are later used in the explanation of the spectrum. The oscillator example has the further advantage that no differences exist between the treatment of the classical theory, the pre-existing quantum theory, and the present Quantum Mechanics; for each quantum mechanical result, there is a simple classical-mechanical analog. We will later be obliged to give up more or less of these classical-mechanical analogs, depending on the treatment of the more general quantum mechanical system.

It is a characteristic property of atomic systems that the electrons are identical and experience identical forces. In order to express this property in our example, we take the Hamiltonian as

$$H = \frac{p_1^2}{2m} + \frac{m}{2}\omega^2 q_1^2 + \frac{p_2^2}{2m} + \frac{m}{2}\omega^2 q_2^2 + m\lambda q_1 q_2. \tag{1}$$

The frequency ω and mass m of the two coupled oscillators are taken as equal. In Eqn.1, q_1, q_2 represent the coordinates, p_1, p_2 the momenta, and λ the interaction constant. By the transformation:

$$q_1' = (q_1 + q_2)/\sqrt{2}; \quad q_2' = (q_1 - q_2)/\sqrt{2}, \tag{2}$$

Eqn.1 goes into

$$H = \frac{p_1'^2}{2m} + \frac{m}{2}\omega_1'^2 q_1'^2 + \frac{p_2'^2}{2m} + \frac{m}{2}\omega_2'^2 q_2'^2, \tag{3}$$

where

$$\omega_1'^2 = \omega^2 + \lambda; \quad \omega_2'^2 = \omega^2 - \lambda. \tag{4}$$

H is now additive in the two oscillator energies, which correspond to the two "principal vibrations". If only the first q_1' is excited, then the two particles oscillate with equal amplitude and equal phase; if only q_2' is excited, then they oscillate with equal amplitude but opposite phase or, in other words, with a phase difference π.

The energies of the stationary states of the whole system are

$$H = \hbar\omega_1'(n_1' + \frac{1}{2}) + \hbar\omega_2'(n_2' + \frac{1}{2}), \tag{5}$$

where n_1' and n_2' are whole numbers. $\cdots$ one represents an energy eigenvalue by the symbols (n_1', n_2'), $\cdots$ then one obtains the energy-levels $\cdots$

We next investigate the possible transition processes. In order to make the closest possible analogy to an atomic system, we assume that the two oscillators consist of charged particles which can oscillate along the same line, parallel or antiparallel to each other. The electric dipole moment is given essentially by $q_1 + q_2$. It follows $\cdots$ that only the first principal vibration has an electric dipole moment. In this approximation, the electric dipole interaction can change only n_1' and that only by one. Therefore $\cdots$ the system of coupled oscillators can only make such $\Delta n_1' = \pm 1$ transitions. However, the radiation is only given in first approximation by the dipole moment. The quadrupole and higher poles can give lesser contributions to the radiation. These higher terms in the radiation are given by homogeneous, symmetrical terms of second, third, etc. order in q_1 and q_2 and their time derivatives. If one replaces q_1 and q_2 by q_1' and q_2' from Eqn.2, the terms remain homogeneous functions of second, third, etc. order in q_1' and q_2'. Because of the symmetry in q_1 and q_2, the coordinate q_2', which comes from time derivatives, can only occur in even powers. This means that, even in a higher order calculation of the radiation, n_2' can only change by an even number. The energy-levels $\cdots$ can therefore be split into two parts, $(+)$ and $(-)$, such that transitions occur only within the $(+)$-part, or only within the $(-)$-part, but never between the $(+)$-part and the $(-)$-part. Not even by collision processes is a transition from one to the other parts possible, since the probability of a transition through a collision can be expressed as a combination of excitations due to all (dipole, quadrupole, etc.) moments. But the non-existence of such "crossover-transitions" is best demonstrated in the original identity of the coupled oscillators. *As soon as a small difference in the mass or frequency between the oscillators is introduced, crossover-transitions do occur with strength proportional to these differences. But we always assume the perfect identity of the oscillators.* [Note: All such italics have been added in translation.] Now a quandary occurs which is characteristic of the quantum mechanical treatment of such problems. In nature, are both $(+)$- and $(-)$-energy level series realized, or only the $(+)$, or only the $(-)$? For a quantum mechanical solution to this problem, it can only be hoped that appropriate terms be "forbidden"; that is, that all and only those terms - which can combine somehow with each other - do so, and that all transition probabilities to non-occurring terms should vanish. *It is therefore sufficient that the $(-)$-system, as well as the $(+)$-system should each be regarded as a quantum mechanical solution to the problem of Eqn.1, just as well as the combination of the two systems. This doubling of the quantum mechanical solution is the essential result of our investigation.* It gives just the same number of degrees of freedom, those which obey the Bose-Einstein statistics, and those the Pauli Exclusion Principle which can be fit unforced into the Quantum Mechanics of the system. For the full solution of this situation, it must be shown that the coupled system always behaves in Quantum Mechanics as it does in the example above.

§2. In the classical theory, two periodically oscillating systems can only be in a true resonance when the frequency of the individual systems is independent of the energy of the systems and is nearly equal for both systems. One can speak of resonance in this sense only for harmonic oscillators. In Quantum Mechanics - in accord with general experience - two atomic systems come into resonance when the absorption frequency of one system agrees with the emission frequency of the other and vice versa; because quantum mechanical equations are inherently linear, the resonance phenomenon in Quantum Mechanics is as general a phenomenon as in the classical theory.

For our investigation we assume two completely identical systems a and b, each with f degrees of freedom, which are coupled through an interaction energy λH^1 which is symmetric in the two systems. The individual systems a and b should not be degenerate. The energy of their stationary states are given by H_n^a and H_m^b.

If one first considers the two systems together without including the interaction energy, then the total energy of the stationary states "n, m" is

$$H_{nm} = H_n^a + H_m^b. \tag{6}$$

The total system possesses the requirements for resonance: because

$$H_{nm} = H_{mn}, \tag{7}$$

each eigenstate is a doublet with the exception of that with $n = m$. In other words: There is always a resonance when the two systems are not originally in the same state; then there can be the exchange of the same energy between the two systems; only for equivalent states of the two parts of the system does the resonance (or the splitting) fail to occur. $\cdots\cdots$ In every system distorted by the interaction, the degeneracy is broken. It corresponds to secular beats, in which the energy of the two particle system pulses back and forth. Formally, it has the following description: The energy W^1 of the distorted system is given, in first approximation, by the time average of H^1 over the undisturbed motion. In general, this average value will contain other terms which correspond to the transitions in which the systems a and b exchange places. One must therefore make a canonical transformation, so that W^1 is a diagonal matrix. The calculation is given in the paper [5] Quantum Mechanics II, P.589. The canonical transformation is

$$W^1 = S^{-1} \bar{H}^1 S, \tag{8}$$
$$q' = S^{-1} q S. \tag{9}$$

S is a matrix which, like $\bar{H}^1$, contains only diagonal terms and terms which correspond to transitions between states of the same energy. For the non-degenerate states the diagonal

118

terms are 1. There are now two linear equations to solve for the two unknown S_{nm}, S_{mn}:

$$
\begin{aligned}
W^1 S_{nm} - H^1(nm, nm) S_{nm} - H^1(nm, mn) S_{mn} &= 0, \\
-H^1(mn, nm) S_{nm} + W^1 S_{mn} - H^1(nm, mn) S_{mn} &= 0.
\end{aligned} \tag{10}
$$

Here the symmetry of H^1 in the systems a and b requires

$$
H^1(nm, nm) = H^1(mn, mn), \quad H^1(nm, mn) = H^1(mn, nm). \tag{11}
$$

It is useful to enumerate with both mn and nm the solutions for W^1 which follow from the zeros of the determinant of Eqn.10. It must be kept in mind that these numbers correspond to the quantum numbers n' of the "principal vibrations" of §1, and do not characterize the states of the individual systems. Furthermore, they lead in each state to equal amplitudes (in different phase) for each sub-system. The solutions of Eqn.10, using the restriction from Eqn.11, read

$$
\begin{aligned}
W^1_{nm} &= H^1(nm, nm) + H^1(nm, mn), \\
S_{nm,nm} &= 1/\sqrt{2}, \quad S_{mn,nm} = 1/\sqrt{2}. \\
W^1_{mn} &= H^1(nm, nm) - H^1(nm, mn), \\
S_{nm,mn} &= 1/\sqrt{2}, \quad S_{mn,mn} = -1/\sqrt{2}.
\end{aligned} \tag{12}
$$

The effect of the coupling is that all energy eigenvalues are different. $\cdots\cdots$

The level spectrum can further - *and this is the decisive result* - be divided into two spectra which can never combine with one another (the $(+)$ and $(-)$ $\cdots$). In general, the radiation is again represented by a function of the p and q, which does not change under the interchange of the two sub-systems. Let this function be F, then after carrying out the canonical transformation of Eqn.8:

$$
F' = S^{-1} F S, \tag{13}
$$

for which, because of the symmetry of F:

1. When $n_1 \neq m, \quad n_2 \neq m, \quad n \neq m_1, \quad n \neq m_2$,

$$
\begin{aligned}
F'_{n_1 m, n_2 m} &= \frac{1}{2}(F_{n_1 m, n_2 m} + F_{mn_1, mn_2} + F_{n_1 m, mn_2} + F_{mn_1, n_2 m}) \\
&= F_{n_1 m, n_2 m} + F_{n_1 m, mn_2}, \\
F'_{nm_1, nm_2} &= \frac{1}{2}(F_{nm_1, nm_2} + F_{m_1 n, m_2 n} - F_{nm_1, m_2 n} - F_{m_1 n, nm_2}) \\
&= F_{nm_1, nm_2} - F_{nm_1, m_2 n}, \\
F'_{n_1 m, mn_2} &= \frac{1}{2}(F_{n_1 m, n_2 m} - F_{mn_1, mn_2} + F_{mn_1, n_2 m} - F_{n_1 m, mn_2}) \quad = 0, \\
F'_{nm_1, m_2 n} &= \frac{1}{2}(F_{m_1 n, m_2 n} - F_{nm_1, nm_2} + F_{m_1 n, nm_2} - F_{nm_1, m_2 n}) \quad = 0. \tag{14}
\end{aligned}
$$

119

2. For transitions with states in which the sub-systems are in "equivalent states" [6]:

$$F'_{n_1 m, mm} = \frac{1}{\sqrt{2}}(F_{n_1 m, mm} + F_{mn_1, mm}) = \sqrt{2} F_{n_1 m, mm},$$

$$F'_{mm_1, mm} = \frac{1}{\sqrt{2}}(F_{m_1 m, mm} - F_{mm_1, mm}) = 0. \tag{15}$$

The crossover-transitions between the level-spectra $(+)$ and $(-)$ therefore vanish. The line intensities within the two spectra are the same in first order as those between the original levels of the individual sub-systems (a and b); amplitudes of the type $F_{n_1 m, mn_2}$, which represent transitions between the two spectra, vanish in first approximation. According to Eqn.15, for transitions to states which have equivalent states for the sub-systems (a and b), the intensity is twice as great as the original system, since the other spectrum contains no such states. The "total intensity" is, in first approximation, not affected by the resonance.

The division into the two non-combining spectra cannot be altered by external interactions. Obviously there is exactly the same behavior here as in the special example of §1. The non-occurrence of crossover-transitions is based on the symmetry of the two coupled systems; as soon as one system is distinguished from the other, crossover-transitions occur. Furthermore, the quantum mechanical solution has the above noted degree of indeterminacy: Both the $(+)$- and the $(-)$-system, and also any combination of both, can be considered as a complete solution to the problem.

The above resonance problem can also be described by Schrödinger's method. In the original sub-systems the normalized Schrödinger eigenfunctions are ϕ_n^a and ϕ_m^b, which correspond to the states with energies H_n^a and H_m^b. Then, for example, for the $n_1 n_2$ matrix element of the coordinate q_k^a of the system a:

$$q_k^a(n_1, n_2) = \int \cdots \int q_k^a \phi_{n_2}^{a\,\dagger} \phi_{n_1}^a dq_1^a \cdots dq_f^a.$$

(The symbol $\dagger$ on ϕ indicates complex conjugate.) If one considers both systems, the eigenfunction $\phi_n^a \phi_m^b$ belongs to the energy $H_{nm} = E_n^a + E_m^b$. The eigenfunctions belonging to the solutions of Eqn.12 for the coupled system develop from those of the uncoupled system through a linear transformation with the matrix S. Then the eigenfunction belonging to W_{nm}^1 is

$$\frac{1}{\sqrt{2}}(\phi_n^a \phi_m^b + \phi_m^a \phi_n^b),$$

and to W_{mn}^1 is

$$\frac{1}{\sqrt{2}}(\phi_n^a \phi_m^b - \phi_m^a \phi_n^b). \tag{16}$$

The non-occurrence of the crossover-transitions between the two systems now follows simply from the fact that the integral of a function F symmetric in a and b with a symmetric eigenfunction and an antisymmetric eigenfunction, of the type

$$\int F \frac{1}{2} (\phi_n^a \phi_m^b + \phi_m^a \phi_n^b)^\dagger (\phi_n^a \phi_m^b - \phi_m^a \phi_n^b) dq_1^a \cdots dq_f^b, \tag{17}$$

changes sign when a and b are exchanged, but keeps the same value. It must therefore be zero.

§3. In the following, as an example, the application of this general theory to the Helium atom will be sketched briefly; I will return to the quantitative discussion later.

1. We assume that electrons are pointlike without magnetism or spin. $\cdots\cdots$ The two systems into which the spectrum splits are Para- and Ortho-helium. Crossover-transitions between them are, for the moment, not possible. Since the energy difference between two corresponding levels of Para- and Ortho-helium can be reduced to a coulomb repulsion of the electrons in the zeroth-order resonance motion, it is understandable that in general it has the same order of magnitude as the smearing corrections of the corresponding hydrogen spectrum. It also follows $\cdots$ that the $1S$-level is present in only one of the two systems. This system is the one whose energy values in general lie higher, according to the calculation. The transition probabilities within the Ortho- or Para-helium spectra are, in first order, the same as the corresponding ones in hydrogen. The exceptional Para-helium transition to the $1S$-level should be twice as fast, in crudest approximation.

To make a somewhat clearer picture of the quantum mechanical solution for the motion of the electrons in the atom, one must somehow represent the two electrons exchanging their places periodically in a continuous way, in analogy to the energy beats of the previously described oscillator example. The period of these beats is given by the splitting of the Ortho-helium from the corresponding Para-helium levels.

2. In the Compton-Goudsmit-Uhlenbeck Hypothesis [7] of the electrons as small magnetic tops, we have to prescribe a definite axial-direction for the two different electrons. There is no qualitative change in the energy levels, but a weak crossover-transition occurs between Ortho- and Para-helium, induced by the interaction between the spin and the orbital motion. The two electrons are no longer equivalent.

3. We can choose the direction of the electron magnetic moment arbitrarily. Then the calculation shows that each level of the former system is split into four levels, corresponding to the statistical weight of the electrons. Furthermore, the spectrum splits into two completely separate parts, as before; now the electrons are once again identical. However, the division is different from before $\cdots$. One system displays a spectrum where the Ortho-helium is a Triplet- and the Para-helium a Singlet-system, and the other a spectrum for which the opposite occurs. Crossover-transitions between Para- and Ortho-helium do

occur depending on the strength of the spin-orbit interaction. Even so, there are no crossover-transitions between the (+)- and the (−)-spectra. One can briefly characterize the physical situation as follows: There is a strong resonance between the two electrons; these are distorted by the spin-interaction, which induces the transition of Ortho- to Para-helium. But, because of the the spin, there exists a further exact resonance of the electrons, which is the reason for re-division into the two spectra described above. The calculation leading to the above results will be explained in a separate paper to appear soon.

§4. It is an empirical fact of the Helium spectrum, that only the one system occurs, and, as far as we can see, at least qualitatively, that it agrees with the Helium spectrum; The other system is not realized in nature. This fact seems to me - when we assume that our result derived for two systems can be generalized to the case of arbitrarily many systems - to signify the real relationship between the previously mentioned quantum mechanical indeterminacy on the one hand, and the Pauli Exclusion Principle on the other. When only the one (−) of the two systems occurs in nature, then on the one hand there occurs a reduction of the statistical weight even in the sense proposed by Bose; on the other hand, for the right choice of system, the Pauli Exclusion Principle is satisfied. The generalization to systems which consist of n identical particles, can therefore be conjectured in the following way: The $n!$ partitions of n systems without interactions correspond in general to $n!$ degenerate systems with equal eigenvalues. Through the interaction, the degeneracy is broken, and the degenerate system splits into $n!$ sub-systems. Among these is one system which contains no equivalent orbits and which cannot make transitions to the other systems. This system occurs alone in nature and represents the actual solution. At the same time, it corresponds to the reduction of the statistical weight of the Bose-Einstein enumeration from $n!$ to 1. This formalism goes beyond the Bose-Einstein enumeration, and prescribes the choice of a definite system from the $n!$ solutions, namely that one system which contains no equivalent sub-systems, and which satisfies the Pauli Exclusion Principle. A principle which directly selects only this one system from all possible quantum mechanical solutions, cannot be given from the simple quantum mechanical calculation alone. *It seems to me an important result of this investigation, that the Pauli Exclusion Principle and the Bose-Einstein Statistics have the same origin, and that Quantum Mechanics cannot say more about it.*

Also one of Einstein's many concrete paradoxes has an analog in our considerations: *When the sub-systems coupled to one another are different, then classical statistics must hold; in principle this is true for infinitely small differences. Nevertheless the enumeration of identical systems is entirely different.* For different systems according to our calculational techniques, there must still be the classical number, since transitions between the $n!$ systems do occur; therefore no sub-system can be excluded. The transitions for vanishing differences of the particles will always be rare. The amplitudes of the transitions will be smaller, after all, than the quantities defined by the sharpness of the involved states, so there exists the possibility that the transition is completely excluded and the enumeration changed. It should be noted however, that for the proposed transitions to alter the enumeration, a finite interaction between the systems is a necessary assumption. When the period of the energy pulses corresponding to the resonance effect is longer than the lifetime, then the sense of the above described calculation is lost.

We have above stated that it is possible to specify the exclusion of all systems except one without infringing on the laws of Quantum Mechanics. This exclusion involves some characteristic restrictions. It means, for example, that it makes no physical sense to speak about the motion of a single electron or about the matrix representing this motion, or about the matrix corresponding to a non-symmetric function of the electrons in an atomic system. Such a matrix contains terms corresponding to transitions from one system to another, and therefore also to the non-occurring systems. Also, as another example, exchange operations in their original form can be given no physical meaning. They occur here only as a formal contribution, when any observable quantity is calculated from symmetrical functions of the electrons.

The consistency portrayed here must be retained as an important requirement when the mathematical treatment of the system of n identical particles is carried through. I might give a brief supplement here, to show how that one system, which has no equivalent sub-systems and which should survive as a unique solution, can in general be unambiguously constructed.

Choose the n completely identical sub-systems to be in the stationary state identified by the quantum numbers $m_1 m_2 \cdots m_n$. Then all the states of the total system which are obtained by permutations among the $m_1 \cdots m_n$, have the same energy. The interaction energy H^1 will, in first approximation, contain only terms which correspond to transitions of at most two sub-systems provided that H^1 itself is the sum of interactions between just two sub-systems. In order to calculate the time-average value of the perturbation energy W^1 as a diagonal matrix, a canonical transformation must be performed as in §2. In order to find it, we must - in the known way - solve a system of $n!$ equations with $n!$ unknowns S_k. We designate that term in H^1 which corresponds to no transitions among the sub-system as H_α^1; those terms, which involve interchanging two of the numbers $m_1 \cdots m_n$ (there are $n(n-1)/2$ such terms), with $H_\beta^1, H_\gamma^1 \cdots H_\nu^1$; then we order the $n!$ states of equal energy so that in the first place the state $m_1 \cdots m_n$ occurs; then all those connected to the first by one transposition; then all those connected to the first by two transpositions, and so on. Then the system of $n!$ linear equations has a characteristic determinant of the form

$$
\begin{bmatrix}
W - H_\alpha^1 & -H_\beta^1 & -H_\gamma^1 & \cdots & -H_\nu^1 & \cdots \\
-H_\delta^1 & W - H_\alpha^1 & 0 & \cdots & 0 & -H_\epsilon^1 & \cdots \\
-H_\eta^1 & 0 & W - H_\alpha^1 & 0 & \cdots & 0 & -H_\xi^1 & \cdots \\
\cdots & & & & & & \\
\cdots & & & & & &
\end{bmatrix}.
\tag{18}
$$

A solution of these equations is:

$$
W^1 = H_\alpha^1 - H_\beta^1 - \cdots H_\nu^1; \quad S^k = (-1)^{\delta_k}/\sqrt{n!},
\tag{19}
$$

where δ_k will be defined in a moment. The Schrödinger eigenfunction for the solution k is

$$
\phi = \frac{1}{\sqrt{n!}} \sum_k (-1)^{\delta_k} \phi_1(m_\alpha^k) \cdots \phi_n(m_\nu^k),
\tag{20}
$$

where now δ_k is the number of transpositions required to bring the series $m_1 m_2 \cdots m_n$ into agreement with $m_\alpha^k m_\beta^k \cdots m_\nu^k$. This function ϕ has the property that it is antisymmetric under the interchange of any two sub-systems, that is, the sign changes. It follows that states characterized by Eqn.20 cannot combine with states for which two or more sub-systems are in equivalent states. The eigenfunction ψ of such a state must be invariant under exchange of the equivalent sub-systems. If F represents the radiation and is thus a symmetric function (or operator) of the coordinates of the sub-systems, then

$$
\int F\psi\phi d\Omega \equiv 0,
\tag{21}
$$

since it must change sign under the interchange of two equivalent sub-systems, without changing its value.

One can see by induction (replacing n by $n+1$) that the term Eqn.20 actually forms a closed system, that is it combines with no other terms, and that the state given by Eqn.20 is unique, in that it contains no equivalent states of the sub-systems. To be sure, it seems to me a not yet rigorously proven result of this analysis. From these considerations, we conclude that Eqn.20 is the definitive construction of the solution representing the state required by the Pauli Exclusion Principle.

After this mathematical excursion, I come to the physical interpretation of this investigation. The relationship between Bose-Einstein Statistics of identical particles and Quantum Mechanics, is in the choice of a definite quantum mechanical solution from among many possible solutions. Such a choice represents - for vanishingly small interactions - essentially a phase-separation between the two types of sub-systems or particles. Possibly one could, through an approximate investigation of this phase-separation, come to a result which can be interpreted directly, analogous to the interference of the de Broglie waves. It might lead, even in the face of the immense physical difficulty which occurs for the coupled problem, to an approximate basis for choosing this solution. However, to me it seems to follow from the present investigation, that we probably do not have to solve this difficulty in order to calculate the spectra of atoms with many electrons. These spectra might be determined much more by the Quantum Mechanics alone.

Footnotes and References.

1) E. Schrödinger, Ann. d. Phys. **79**, 361, 489, 734 (1926).

2) L. de Broglie, Ann. de phys. **3**, 22 (1925).

3) N.S. Bose, Zeits. f. Phys. **26**, 178 (1924). The application of this statistics to the subject at hand is given by A. Einstein, Sitzungsber. der preuss. Akad. d. Wiss. 1924, P.261; 1925, P.3,8.

4) W. Pauli, Zeits. f. Phys. **31**, 765 (1925).

5) M. Born, W. Heisenberg and P. Jordan, Zeits. f. Phys. **35**, 557 (1926).

6) Here and in the rest of the paper, the expression "in equivalent states" refers to the non-interacting system. In the interacting system, of course, the sub-systems still carry out the same motions but with different phase.

7) A.H. Compton, Journ. Frankl. Inst. **192**, 145 (1924); E. Uhlenbeck and S. Goudsmit, Naturwiss. **13**, 953 (1925).

125

APPENDIX 4.B:

Excerpt from: Proc. Roy. Soc. (London) A112, 661 (1926)

On the Theory of Quantum Mechanics

By P.A.M. DIRAC, St. John's College, Cambridge

(Communicated by R.H. Fowler, F.R.S.- Received August 26, 1926)

§1. Introduction and Summary.

The new mechanics of the atom introduced by Heisenberg [1] may be based on the assumption that the variables that describe a dynamical system do not obey the commutative law of multiplication, but satisfy instead certain quantum conditions. One can build up a theory without knowing anything about the dynamical variables except the algebraic laws that they are subject to, and can show that they may be represented by matrices whenever a set of uniformizing variables for the dynamical system exists [2]. It may be shown, however (see §3), that there is no set of uniformizing variables for a system containing more than one electron, so that the theory cannot progress very far on these lines.

A new development of the theory has recently been given by Schrödinger [3]. Starting from the idea that an atomic system cannot be represented by a trajectory, *i.e.*, by a point moving through the co-ordinate space, but must be represented by a wave in this space, Schrödinger obtains from a variation principle a differential equation which the wave function ψ must satisfy. This differential equation turns out to be very closely connected with the Hamiltonian equation which specifies the system, namely, if

$$H(q_r, p_r) - W = 0$$

is the Hamiltonian equation of the system, where the q_r, p_r are canonical variables, then the wave equation for ψ is

$$\left\{ H\left(q_r, -i\hbar \frac{\partial}{\partial q_r}\right) - W \right\} \psi = 0, \tag{1}$$

$\cdots\cdots$. Each momentum p_r in H is replaced by the operator $-i\hbar\partial/\partial q_r$, and is supposed to operate on all that exists on its right-hand side in the term in which it occurs. Schrödinger takes the values of the parameter W for which there exists a ψ satisfying (1) that is

126

continuous, single-valued and bounded throughout the whole of q-space to be the energy levels of the system, and shows that when the general solution of (1) is known, matrices to represent the p_r and q_r may be easily obtained, satisfying all the conditions that they have to satisfy according to Heisenberg's matrix mechanics, and consistent with the energy levels previously found. The mathematical equivalence of the theories is thus established.

In the present paper, Schrödinger's theory is considered in §2 from a slightly more general point of view, in which the time t and its conjugate momentum W are treated from the beginning on the same footing as the other variables. A more general method, requiring only elementary symbolic algebra, of obtaining matrix representations of the dynamical variables is given.

In §3 the problem is considered of a system containing several similar particles, such as an atom with several electrons. If the positions of two of the electrons are interchanged, the new state of the atom is indistinguishable from the original one. In such a case one would expect only symmetrical functions of the co-ordinates of all the electrons to be capable of being represented by matrices. It is found that this allows one to obtain two solutions of the problem satisfying all the necessary conditions, and the theory is incapable of deciding which is the correct one. One of the solutions leads to Pauli's principle that not more than one electron can be in any given orbit, and the other, when applied to the analogous problem of the ideal gas, leads to the Bose-Einstein statistical mechanics.

......*(We go directly to the discussion of identical particles in §3.)*

§3. Systems containing Several Similar Particles.

In Heisenberg's matrix mechanics it is assumed that the elements of the matrices that represent the dynamical variables determine the frequencies and intensities of the components of radiation emitted. The theory thus enables one to calculate just those quantities that are of physical importance, and gives no information about quantities such as orbital frequencies that one can never hope to measure experimentally. We should expect this very satisfactory characteristic to persist in all future developments of the theory.

Consider now a system that contains two or more similar particles, say, for definiteness, an atom with two electrons. Denote by (mn) that state of the electron in which one electron is in an orbit labelled m, and the other in the orbit labelled n. The question arises whether the two states (mn) and (nm), which are physically indistinguishable as they differ only by the interchange of the two electrons, are to be counted as two different states or as only one state, *i.e.*, do they give rise to two rows and columns in the matrices or to only one? If the first alternative is right, then the theory would enable one to calculate the intensities due to the two transitions $(mn) \rightarrow (m'n')$ and $(mn) \rightarrow (n'm')$ separately, as the amplitude corresponding to either would be given by a definite element in the matrix representing the total polarization. The two transitions are, however, physically indistinguishable, and only the sum of the intensities for the two together could be determined experimentally. Hence, in order to keep the essential characteristic of the theory that it shall enable one to calculate only the observable quantities, one must adopt the second alternative that (mn) and (nm) count as only one state.

This alternative, though, also leads to difficulties. The symmetry between the two electrons requires that the amplitude associated with the transition $(mn) \rightarrow (m'n')$ of x_1, a coordinate of one of the electrons, shall equal the amplitude associated with the transition $(nm) \rightarrow (n'm')$ of x_2, the corresponding co-ordinate of the other electron, i.e.,

$$x_1(mn; m'n') = x_2(nm; n'm').\tag{2}$$

If we now count (mn) and (nm) as both defining the same row and column of the matrices, and similarly for $(m'n')$ and $(n'm')$, Eqn.2 shows that each element of the matrix x_1 equals the corresponding element of the matrix x_2, so that we should have the matrix equation

$$x_1 = x_2.$$

This relation is obviously impossible, as, amongst other things, it is inconsistent with the quantum conditions. We must infer that unsymmetrical functions of the co-ordinates (and momenta) of the two electrons cannot be represented by matrices. Symmetrical functions, such as the total polarization of the atom, can be considered to be represented by matrices without inconsistency, and these matrices are by themselves sufficient to determine all the physical properties of the system.

One consequence of these considerations is that the theory of uniformizing variables introduced by the author can no longer apply.······

If we neglect the interaction between the two electrons, then we can obtain the eigenfunctions for the whole atom simply by multiplying the eigenfunctions for one electron when it exists alone in the atom by the eigenfunctions for the other electron alone, and taking the same time variable for each [4]. Thus if $\psi_n(x, y, z, t)$ is the eigenfunction for a single electron in the orbit n, then the eigenfunction for the whole atom in the state (mn) is

$$\psi_m(x_1, y_1, z_1, t)\psi_n(x_2, y_2, z_2, t) = \psi_m(1)\psi_n(2),$$

······. The eigenfunction $\psi_m(2)\psi_n(1)$, however, also corresponds to the same state of the atom if we count the (mn) and (nm) states as identical. But two independent eigenfunctions must give rise to two rows and columns in the matrices. If we are to have only one row and column in the matrices corresponding to both (mn) and (nm), we must find a set of eigenfunctions ψ_{mn} of the form

$$\psi_{mn} = a_{mn}\psi_m(1)\psi_n(2) + b_{mn}\psi_m(2)\psi_n(1),$$

where the a_{mn}'s and b_{mn}'s are constants, which set must contain only one ψ_{mn} corresponding to both (mn) and (nm), and must be sufficient to obtain the matrix representing any symmetrical function A of the two electrons. This means the ψ_{mn}'s must be chosen such that A times any chosen ψ_{mn} can be expanded in terms of the chosen ψ_{mn}'s in the form

$$A\psi_{mn} = \sum_{m'n'} \psi_{m'n'} A_{m'n',mn}, \tag{3}$$

where the $A_{m'n',mn}$'s are constants or functions of the time only.

There are two ways of choosing the set of ψ_{mn}'s to satisfy the conditions. We may either take $a_{mn} = b_{mn}$, which makes ψ_{mn} a symmetrical function of the two electrons so that the left-hand side of (3) is symmetrical and only symmetrical eigenfunctions will be required for its expansion, or we may take $a_{mn} = -b_{mn}$, which makes ψ_{mn} antisymmetrical, so that the left-hand side of (3) is antisymmetrical and only antisymmetrical eigenfunctions will be required for its expansion. Thus the symmetrical eigenfunctions alone or the antisymmetrical eigenfunctions alone give a complete solution of the problem. The theory at present is incapable of deciding which solution is the correct one. We are able to get complete solutions of the problem which make use of less than the total number of possible eigenfunctions at the expense of being able to represent only symmetrical functions of the two electrons by matrices.

These results may evidently be extended to any number of electrons. For r non-interacting electrons $\cdots\cdots$, the symmetrical eigenfunctions are

$$\sum_{\alpha_1\alpha_2\cdots\alpha_r} \psi_{n_1}(\alpha_1)\psi_{n_2}(\alpha_2)\cdots\psi_{n_r}(\alpha_r), \tag{4}$$

where $\alpha_1, \alpha_2, \cdots \alpha_r$ are any permutation of the integers $1, 2, \cdots r$, while the antisymmetrical ones may be written in the determinantal form

$$\begin{vmatrix} \psi_{n_1}(1) & \psi_{n_1}(2) & \cdots & \psi_{n_1}(r) \\ \psi_{n_2}(1) & \psi_{n_2}(2) & \cdots & \psi_{n_2}(r) \\ \cdots & \cdots & \cdots & \\ \psi_{n_r}(1) & \psi_{n_r}(2) & \cdots & \psi_{n_r}(r) \end{vmatrix}. \tag{5}$$

If there is interaction between the electrons, there will still be symmetrical and antisymmetrical eigenfunctions, although they can no longer be put in these simple forms. In any case the symmetrical ones alone or the antisymmetrical ones alone give a complete solution of the problem.

An antisymmetrical eigenfunction vanishes identically when two of the electrons are in the same orbit. This means that in the solution of the problem with antisymmetrical eigenfunctions there can be no stationary states with two or more electrons in the same orbit, which is just Pauli's principle [5]. The solution with symmetrical eigenfunctions, on the other hand, allows any number of electrons in the same orbit, so that this solution cannot be the correct one for the problem of electrons in an atom [6].

· · · · · ·(§4 *of Dirac's paper, with the title "Theory of the Ideal Gas", follows Fermi's paper in Chapter 4. The last section will not be excerpted here.*)

Footnotes and References.

1) See various papers by Born, Heisenberg and Jordan, Zeits. f. Phys., vol.**33** onwards.

2) Proc. Roy. Soc. Proc. **A110**, 561 (1926).

3) See various papers in Ann. d. Phys., beginning with vol.**79**, 361 (1926).

4) The same time variable t must be taken in each owing to the fact that we write the Hamiltonian equation for the whole system: $H(1) + H(2) - W = 0$, where $H(1)$ and $H(2)$ are the Hamiltonians for the two electrons separately, so that there is a common time t conjugate to minus the total energy W.

5) Pauli, Zeits. f. Phys. **31**, 765 (1925).

6) Prof. Born has informed me that Heisenberg has independently obtained results equivalent to these. (Added in proof) - see Heisenberg, Zeits. f. Phys. **38**, 411 (1926).

Chapter 5

Fermi-Dirac Statistics

Summary: The derivation by Fermi and by Dirac of the Statistical Mechanics for an ideal gas of identical particles satisfying the Pauli Exclusion Principle is developed.

§1. Introduction to Fermi's Derivation

The invention of the statistical mechanics for an ideal gas of particles satisfying the Pauli Exclusion Principle was made first by Fermi [5.1], then independently by Dirac [5.2]. Fermi's derivation is notable for its novelty, elegance, rigor, and completeness. He considers a gas confined in a harmonic oscillator potential, rather than the now usual technique of confining the gas in a box with periodic boundary conditions. Each molecule is uniquely labelled by integer quantum numbers s_1, s_2, s_3 and has energy

$$E = h\nu(s_1 + s_2 + s_3) = h\nu s. \tag{1}$$

The number of molecular states Q_s for a given energy $E = sh\nu$ is the number of partitions of the integer s into integers s_1, s_2, s_3

$$Q_s = (s+1)(s+2)/2, \tag{2}$$

giving one molecule possible with energy (in units $h\nu$) $s = 0$, three with $s = 1$, six with $s = 2$, and so on. The groundstate energy at absolute zero temperature Fermi gives as $1(0) + 3(1h\nu) + 6(2h\nu)+$ etc.

The distribution of N particles with

$$N = \sum_{s=0}^{\infty} N_s \tag{3}$$

and energy

$$E = \sum_{s=0}^{\infty} s N_s \tag{4}$$

is determined from the number of ways

$$P = \Pi_{s=0}^{\infty} \frac{Q_s!}{N_s!(Q_s - N_s)!} \tag{5}$$

of arranging N_s identical molecules in Q_s states, leaving $Q_s - N_s$ empty, for each value of s. Maximizing P by varying N_s subject to the constraints of constant N and E gives

$$\frac{N_s}{Q_s} = \frac{\alpha e^{-\beta s}}{1 + \alpha e^{-\beta s}} \tag{6}$$

as the most probable distribution. It would be easiest to identify the constants α and β by requiring the distribution to be Maxwellian for large s and find $\beta = h\nu/kT$, but Fermi followed a more rigorous derivation made possible by his use of the harmonic oscillator potential for confinement. At large values of $r \to \infty$, the density of the gas is small, the effects of degeneracy are ignorable and the gas becomes Maxwellian. Fermi characterizes it as using a gas thermometer with an infinitely dilute ideal gas to measure the temperature of the degenerate gas.

Fermi further analyzes the properties of the gas in terms of a pair of functions $F(A)$ in Eqn.22 and $G(A)$ in Eqn.24. $F(A)$ is invertible (numerically or, in Fermi's treatment, asymptotically) to express A in terms of the temperature T and density n of the gas. The energy (23) and the pressure (24) are expressible in terms of $F(A)$ and $G(A)$. In a marvelous tour de force, Fermi obtains asymptotic expressions, valid in the limit of weak (high T, low n) or strong degeneracy, for the equation of state, average energy and specific heat, including $T = 0$ limits of the pressure and specific heat. He concludes with a calculation of the absolute entropy.

§2. Dirac's Derivation of Fermi-Dirac Statistics

Five months later, as part of a longer paper in which he

132

(1)refined quantum mechanics,

(2)introduced symmetrization/antisymmetrization of the wave function of identical particles,

(3)introduced time-dependent perturbation theory, and

(4)the interaction representation as the foundation for quantum mechanical and even for the quantum field theoretic calculations which he was soon to invent. Finally, Dirac also

(5)deduced the statistical mechanics for identical particles satisfying the Pauli Exclusion Principle. His derivation is extremely brief and elegant. If he was aware of Fermi's previous work, he did not acknowledge it. He was almost dismissive about the physical consequences of his results. He mentioned that the specific heat went to zero at zero temperature, and that there was no condensation phenomenon like that found by Einstein for the Bose-Einstein gas, but he quoted no analysis.

§3. Biographical Note on Enrico Fermi (†).

Enrico Fermi (1901-54) was largely self taught as a pre-adolescent, using Sommerfeld's *Atombau und Spektrallinien* for modern physics; then student at Rome and Scuola Normale, Pisa; PhD (1922); Fermi-Dirac statistics (1926); Professor at Rome (1927); Thomas-Fermi statistical model of the atom (1928); quantum electrodynamics (1930); theory of beta-decay (1933); experiments on neutrons (1934-); Nobel Prize for neutron studies (1938); at Columbia University (1939-); Charles H. Swift Distinguished Service Professor of Physics at the Institute for Nuclear Studies of the University of Chicago (1942-); led the development of the first nuclear reactor (1942); at Los Alamos, a leader in the development of the atom bomb (1945); led the development of meson research at the Chicago synchrocyclotron (1946-).

Fermi's brilliance was similar to that of Dirac or Feynman; his way of doing physics was comparably pure, simple, elegant, powerful and inspired. But he was much more intellectually gregarious and he had superb leadership ability which enabled him to inspire the coherent effort of brilliant people in large projects; he was able to lead their scientific effort by the example of his own unmatched energy, creativity and intellect.

133

Fermi was revered by his peers and his many protegés in what comes down to us as a completely positive and loving way. So much has been written (†∗, † ∗ ∗ that we can best defer to some of Fermi's own words (††) where his friendly, humorous, energetic and optimistic spirit shines through.

(† - S.K. Allison, E. Segré, and H.L. Anderson, PHYSICS TODAY Jan 1955, pp.9-13); †∗ - L. Fermi, *Atoms in the Family* (University of Chicago Press, Chicago, 1954) ; † ∗ ∗ - E. Segré, *Enrico Fermi - Physicist* (University of Chicago Press, Chicago, 1970); †† - N. Metropolis, PHYSICS TODAY Nov 1955, pp.10-16.)

Bibliography and References.

5.1) E. Fermi, Zeits. f. Phys. **36**, 902 (1926); see our App.5A.

5.2) P.A.M. Dirac, Proc. Roy. Soc. (London) **A112**, 661 (1926); see our App.5B.

APPENDIX 5.A:

Excerpt from: Zeitschrift für Physik **36**, 902 (1926)

On the Quantization of the Ideal Monatomic Gases

by E. Fermi in Florence

(Received on 24 March 1926)

Because the Nernst Heat Law should hold even for the ideal gas without forces, one must assume that the Ideal Gas Law deviates at low temperatures from the classical. The cause of this difference is to be sought in the quantization of the molecular motions. For all theories, the difference is always more or less determined by assumptions made about the statistical behavior of the molecules, or about their quantization. In the following work the only assumption used was that first deduced by Pauli based on numerous spectroscopic facts: in a system no two identical particles can coexist, whose quantum numbers are completely equal. With this hypothesis, the equation of state and the internal energy of the ideal gas are derived. The entropy for high temperatures agrees with the Stern-Tetrode value.

In classical thermodynamics the molar heat capacity at constant volume is

$$C_v = \frac{3}{2}k. \qquad (1)$$

However, if one wants to apply the Nernst Heat Law to the ideal gas, then one must take Eqn.1 as merely an approximation for high temperatures, since C_v must vanish in the limit of $T = 0$. It is therefore necessary to assume that the motion of the ideal gas molecules should be quantized. This quantization manifests itself at low temperatures through the degeneracy phenomenon, so that both the specific heat and also the equation of state must be changed from their classical expressions.

The object of the following work [1] is to construct a method for the quantization of the ideal gas, which is as independent as possible of arbitrary assumptions about the

135

statistical behavior of the gas molecules.

Recently, a number of attempts have been made to determine the equation of state of the ideal gas [2]. The equations of state of various authors, and the differences among them and with the classical equation of state $pV = NkT$, arise only from terms which are important at very low temperatures and large pressures; furthermore the deviations of the real gas from the ideal case under these conditions, for the most part, can not be distinguished from the usual not unimportant approximations. It is possible that a deeper understanding of the equation of state of the gas is available, which separates the degeneracy effect from the above deviations from the equation $pV = NkT$, so that an experimental distinction between the different degeneracy theories will be possible.

In order to apply the quantization rules to the motion of the molecules of our ideal gas, one can proceed in different ways; the end result remains the same. For example we can think of enclosing the molecules in a parallelepiped vessel with elastically reflecting walls; then the motion of the molecules flying back and forth between the walls will be periodic and can be quantized accordingly; more generally one can imagine enclosing the molecules in a kind of external force-field, so that their motion will be periodic; the assumption that the gas is ideal allows us to ignore the mechanism of interaction between the molecules, so that their motion is carried out only under the influence of the external force. It is also evident, that the quantization of the molecular motion, carried out under the assumption of the complete independence of the molecules upon one another, is not sufficient to give us an account of the expected degeneracy. One sees this best from the example of a molecule enclosed in a container: the linear dimensions of the quantized states of each individual molecule will always be so small, that for a container of macroscopic dimensions the influence of the discreteness of the energy values practically vanishes. The influence of the volume of the container depends also on whether the number of molecules enclosed in the container is changed, so that their density remains constant.

From a quantitative calculation of these effects [3], one gets a degeneracy of the desired magnitude only when the container is so small that it contains just one molecule.

We conjecture then, that for the quantization of the ideal gas, a principle supplementary to Sommerfeld's Quantization Condition must be important.

Recently, in response to a paper by Stoner [4], Pauli [5] has stated the Rule that, when an electron in an atom has quantum numbers (including the magnetic quantum number) of a definite value, no other electron can exist in the atom with the same quantum numbers. In other words, each quantum state (in an external magnetic field) is already completely occupied by one single electron.

136

Since this Pauli Principle has also been fruitful in the explanation of other spectroscopic facts [6], we wish to investigate whether or not it might also be useful for the problem of the Quantization of the Ideal Gas.

We will show that this is in fact the case, and that the application of the Pauli Principle allows us to derive a completely consistent theory of the degeneracy of the ideal gas.

We will therefore assume that at most one molecule with a particular set of quantum numbers can be contained in our gas; included in the set of quantum numbers are not only the quantum numbers which determine the inner motions of the molecule, but also those which determine its translational motion.

First we must confine our molecule in a suitable external force-field, so that its hypothetical motion will be periodic. This can be done in an infinite number of ways; however since the result is not dependent on the choice of force-field, we will subject the molecule to an elastic force centered at the origin of coordinates "O", so that each molecule will constitute an harmonic oscillator. This central force confines our gas in the neighbourhood of "O"; the gas density will vary with the distance from "O" and will vanish for infinite distance. Let ν be the eigenfrequency of the oscillators. Then the force acting on the molecule is

$$4\pi\nu^2 mr,$$

where m is the mass of the molecule and r its distance from "O". The potential energy of the attractive force is

$$U = 2\pi^2\nu^2 mr^2.$$

The quantum numbers of a molecule forming the oscillator are s_1, s_2, s_3. We ignore the inner motions and just consider the molecules as mass points. The Pauli Principle reads for our case: There can only be, in the whole gas, at most one molecule with the given quantum numbers s_1, s_2, s_3.

The total energy of each molecule is given by

$$w = h\nu(s_1 + s_2 + s_3) = h\nu s. \tag{2}$$

Its total energy can therefore be expressed as integer multiples of $h\nu$. Each realizable possibility corresponds to a solution of the equation

$$s = s_1 + s_2 + s_3 \tag{3}$$

where s_1, s_2, s_3 can take the values $0, 1, 2, 3 \cdots$. It is known that Eqn.3 has

$$Q_s = \frac{(s+1)(s+2)}{2} \tag{4}$$

137

solutions. The zero energy case in particular, can be realized in only one way, the energy $h\nu$ in three, the energy $2h\nu$ in six and so on. A molecule with the energy $sh\nu$ will be called simply an "s-molecule".

According to our assumptions, in the whole gas at most Q_s "s-molecules" can occur; therefore at most one molecule with energy zero, three with energy $h\nu$, six with energy $2h\nu$ and so on. In order to see clearly the consequences of these facts, we will calculate the extreme case, that the absolute temperature of our gas is zero. Let N be the number of molecules. At the absolute zero point the gas must be in the state of lowest energy. If there were no restriction for the number of molecules of a given energy, then each molecule would be found in the state of zero energy ($s_1 = s_2 = s_3 = 0$). According to the above considerations however, at most one molecule can occur with the energy zero; if therefore $N = 1$, there would be a single molecule at the absolute zero point in the state of energy zero; if N were 4, then one molecule would be in the zero energy state, the other three take the places with the energy $h\nu$; if N were 10, then one molecule would be found in the zero energy state, three in the three places with energy $h\nu$, and six others in the six places with the energy $2h\nu$, and so on.

At the absolute zero point the molecules of our gas therefore assume a kind of shell-like structure, in perfect analogy to the shell-like ordering of the electrons in an atom with many electrons.

We will now investigate how a fixed energy

$$W = Eh\nu \tag{5}$$

(E = a whole number) divides among our N molecules.

Let N_s be the number of "s-molecules", with

$$N_s \leq Q_s. \tag{6}$$

One has

$$\sum_s N_s = N, \tag{7}$$

and

$$\sum_s sN_s = E, \tag{8}$$

which state that the total number of molecules is N and their total energy is $Eh\nu$.

Now we want to calculate the number P of such orderings of our N molecules, such that the number N_0 occupy places of zero energy, N_1 places of energy $h\nu$, N_2 places of energy $2h\nu$, and so on. Two such orderings should be considered the same, if the places occupied by the molecules are the same; two orderings which differ only through a permutation of the molecules in their places, are therefore regarded as a single ordering. If one were to regard two such orderings as different, then one should multiply P by the constant $N!$; one can easily see that this would have no influence on the following. In this sense, the number of orderings of N_s molecules in the Q_s places of energy $s h \nu$ is

$$\binom{Q_s}{N_s} = \frac{Q_s!}{N_s!(Q_s - N_s)!}.$$

We therefore get for P

$$P = \binom{Q_0}{N_0}\binom{Q_1}{N_1} \cdots = \prod_s \binom{Q_s}{N_s}. \tag{9}$$

One obtains the most probable value of the N_s by calculating the maximum of P with the restrictions (7) and (8). Using Stirling's formula with the usual approximation, one can write

$$\log[P] = \sum \log \binom{Q_s}{N_s} = -\sum \left(N_s \log \left[\frac{N_s}{Q_s - N_s} \right] - Q_s \log \left[\frac{Q_s}{Q_s - N_s} \right] \right). \tag{10}$$

We seek the value of N_s which satisfies (7) and (8) and for which P is a maximum. One finds

$$\alpha e^{-\beta s} = \frac{N_s}{Q_s - N_s}$$

where α and β are constants. The above equation gives

$$N_s = Q_s \frac{\alpha e^{-\beta s}}{1 + \alpha e^{-\beta s}}. \tag{11}$$

The value of α and β can be determined using Eqns.7 and 8, or alternatively one can assume α and β as given; then Eqns.7 and 8 determine the total number and the total energy of our molecules. Explicitly we find

$$N = \sum_0^\infty Q_s \frac{\alpha e^{-\beta s}}{1 + \alpha e^{-\beta s}},$$

$$\frac{W}{h\nu} = E = \sum_0^\infty s Q_s \frac{\alpha e^{-\beta s}}{1 + \alpha e^{-\beta s}}. \tag{12}$$

The absolute temperature T of the gas is a function of N and E or of α and β. This function can be determined in two ways. One knows for example from Boltzmann's Principle that the entropy satisfies

$$S = k \log[P]$$

and then the temperature is computed from

$$T = \frac{dW}{dS}.$$

139

This method has the same disadvantage as do all methods based on Boltzmann's Principle, that one has to take a more or less arbitrary prescription for the probability of the state. We prefer to proceed in the following way: we observe that the density of our gas is a function of the radius, which vanishes for infinite separation. For infinite values of r therefore, the degeneracy effect vanishes and the statistics of our gas goes over into the classical. In particular, for $r = \infty$, the mean kinetic energy of the molecule must be $\frac{3}{2}kT$, and its velocity distribution must become Maxwellian. We can therefore determine the temperature from the velocity distribution in the limit of infinitely small density; and since the whole gas is at constant temperature, we will at the same time know the temperature in the region of higher density. We are using, so to say, a gas thermometer with an infinitely dilute ideal gas.

First we must calculate the density of the molecules with a kinetic energy between L and $L + dL$ at the separation r. The total energy of these molecules according to Eqn.1 lies between

$$L + 2\pi^2 \nu^2 m r^2 \quad \text{and} \quad L + 2\pi^2 \nu^2 m r^2 + dL.$$

Now the total energy of a molecule equals $sh\nu$. For our molecule s must lie between s and $s + ds$ so

$$s = \frac{L}{h\nu} + \frac{2\pi^2 \nu m}{h} r^2, \quad ds = \frac{dL}{h\nu}. \tag{13}$$

We consider now a molecule, whose motion is characterized by the quantum numbers s_1, s_2, s_3. Its coordinates are given as functions of time by

$$
\begin{aligned}
x &= \sqrt{H s_1} \cos(2\pi\nu t - \alpha_1), \\
y &= \sqrt{H s_2} \cos(2\pi\nu t - \alpha_2), \\
z &= \sqrt{H s_3} \cos(2\pi\nu t - \alpha_3).
\end{aligned}
\tag{14}
$$

Here we have set

$$H = \frac{h}{2\pi^2 \nu m}; \tag{15}$$

$\alpha_1, \alpha_2, \alpha_3$ are phases which take every possible value with equal probability. From Eqn.14 it follows that $|x| \le \sqrt{H s_1}$, $|y| \le \sqrt{H s_2}$, and $|z| \le \sqrt{H s_3}$, and that the probability that x, y, z lie between x and $x + dx$, y and $y + dy$, z and $z + dz$, is

$$\frac{dx\,dy\,dz}{\pi^3 \sqrt{(H s_1 - x^2)(H s_2 - y^2)(H s_3 - z^2)}}.$$

When we know not the individual values of s_1, s_2, s_3, but only their sum, then the probability is

$$\frac{1}{Q_s} \frac{dx\,dy\,dz}{\pi^3} \sum \frac{1}{\sqrt{(H s_1 - x^2)(H s_2 - y^2)(H s_3 - z^2)}}; \tag{16}$$

the sum is over all integer solutions of Eqn.3, which satisfy the inequalities

$$H s_1 \ge x^2, \quad H s_2 \ge y^2, \quad H s_3 \ge z^2.$$

140

When we multiply the probability Eqn.16 by the number N_s of "s-molecules", it becomes the number of "s-molecules" in the volume element $dx\,dy\,dz$. Using Eqn.11, we find the density of "s-molecules" at the point x, y, z as

$$n_s = \frac{\alpha e^{-\beta s}}{1 + \alpha e^{-\beta s}} \frac{1}{\pi^3} \sum \frac{1}{\sqrt{(Hs_1 - x^2)(Hs_2 - y^2)(Hs_3 - z^2)}}.$$

For sufficiently large s one can replace the sum by a double integral; carrying out the integrations we find

$$n_s = \frac{2}{\pi^2 H^2} \frac{\alpha e^{-\beta s}}{1 + \alpha e^{-\beta s}} \sqrt{Hs - r^2}.$$

Using Eqns.13 and 15 we find the density of molecules with a kinetic energy between L and $L + dL$ at the point x, y, z

$$n(L)dL = n_s ds = \frac{2\pi(2m)^{3/2}}{h^3} \sqrt{L}dL \frac{\alpha e^{-2\pi^2 \nu m \beta r^2 / h} e^{-\beta L / h\nu}}{1 + \alpha e^{-2\pi^2 \nu m \beta r^2 / h} e^{-\beta L / h\nu}}. \tag{17}$$

This formula must reduce to the classical Maxwellian distribution

$$n^*(L)dL = K\sqrt{L}dL e^{-L/kT},$$

in the limit $r \to \infty$. Eqn.17 coincides with this provided we set

$$\beta = \frac{h\nu}{kT}. \tag{18}$$

Then Eqn.17 can be written

$$n(L)dL = \frac{(2\pi)(2m)^{3/2}}{h^3} \sqrt{L}dL \frac{Ae^{-L/kT}}{1 + Ae^{-L/kT}}, \tag{19}$$

where

$$A = \alpha e^{-2\pi^2 \nu^2 m r^2 / kT}. \tag{20}$$

The total density of the molecules at r is now

$$n = \int_0^\infty n(L)dL = \frac{(2\pi mkT)^{3/2}}{h^3} F(A), \tag{21}$$

where

$$F(A) = \frac{2}{\sqrt{\pi}} \int_0^\infty \frac{Ax^{1/2}e^{-x}}{1 + Ae^{-x}}dx. \tag{22}$$

The mean kinetic energy of the molecules at r is

$$\bar{L} = \frac{1}{n} \int_0^\infty Ln(L)dL = \frac{3}{2}kT \frac{G(A)}{F(A)} \tag{23}$$

where

$$G(A) = \frac{4}{3\sqrt{\pi}} \int_0^\infty \frac{Ax^{3/2}e^{-x}}{1 + Ae^{-x}}dx. \tag{24}$$

141

From Eqn.21 one can determine A as a function of temperature and density; when one inserts the value found into Eqns.19 and 23, one gets the velocity distribution function and the mean kinetic energy of the molecules as functions of the density and temperature.

To derive the equation of state we use the Virial Theorem, which gives

$$p = \frac{2}{3}n\bar{L} = nkT\frac{G(A)}{F(A)};$$ (25)

the value of A is taken from Eqn.21 as a function of temperature and density.

Before going further, we demonstrate some mathematical properties of the functions $F(A)$ and $G(A)$.

For $A \leq 1$ one can expand the functions in convergent series

$$
\begin{aligned}
F(A) &= A - \frac{A^2}{2^{3/2}} + \frac{A^3}{3^{3/2}} - \cdots, \\
G(A) &= A - \frac{A^2}{2^{5/2}} + \frac{A^3}{3^{5/2}} - \cdots.
\end{aligned}
$$ (26)

For large A one has the asymptotic expression

$$
\begin{aligned}
F(A) &= \frac{4}{3\sqrt{\pi}}(\log A)^{3/2}\left[1 + \frac{\pi^2}{8(\log A)^2} + \cdots\right], \\
G(A) &= \frac{8}{15\sqrt{\pi}}(\log A)^{5/2}\left[1 + \frac{5\pi^2}{8(\log A)^2} + \cdots\right].
\end{aligned}
$$ (27)

It yields the relation

$$\frac{dG(A)}{F(A)} = d(\log A).$$ (28)

We must introduce another function $\mathcal{P}(\Theta)$ as

$$\mathcal{P}(\Theta) = \Theta\frac{G(A)}{F(A)}, \quad F(A) = \frac{1}{\Theta^{3/2}}.$$ (29)

For very large or very small Θ, $\mathcal{P}(\Theta)$ can be calculated from the approximate formulas

$$
\begin{aligned}
\mathcal{P}(\Theta) &= \Theta\left[1 + \frac{1}{2^{5/2}\Theta^{3/2}} + \cdots\right], \quad \Theta \to \infty, \\
&= \frac{3^{2/3}\pi^{1/3}}{5 \cdot 2^{1/3}}\left[1 + \frac{5 \cdot 2^{2/3}\pi^{4/3}}{3^{7/3}}\Theta^2 + \cdots\right], \quad \Theta \to 0.
\end{aligned}
$$ (30)

Using Eqns.27,28,29 one gets

$$\int_0^\infty \frac{d\mathcal{P}(\Theta)}{\Theta} = \frac{5}{3}\frac{G(A)}{F(A)} - \frac{2}{3}\log A.$$ (31)

We are now in a position to eliminate the parameter A from Eqns.23 and 25, and find the pressure and mean kinetic energy of the molecules as explicit functions of density and temperature:

$$p = \frac{h^2 n^{5/3}}{2\pi m} \mathcal{P}\left(\frac{2\pi mkT}{h^2 n^{2/3}}\right), \tag{32}$$

and

$$\bar{L} = \frac{3}{2} \frac{h^2 n^{2/3}}{2\pi m} \mathcal{P}\left(\frac{2\pi mkT}{h^2 n^{2/3}}\right). \tag{33}$$

In the limiting case of weak degeneracy (T large and n small) the equation of state becomes

$$p = nkT\left(1 + \frac{1}{16}\frac{h^3 n}{(\pi mkT)^{3/2}} + \cdots\right). \tag{34}$$

The pressure is therefore greater than the classical equation of state ($p = nkT$). For an ideal gas with the atomic weight of helium, at $T = 5°K$ and a pressure of $10 Atm.$ the difference is some 15%.

In the case of large degeneracy Eqns.32 and 33 assume the form

$$p = \frac{1}{20}\left(\frac{6}{\pi}\right)^{2/3}\frac{h^2 n^{5/3}}{m} + \frac{2^{4/3}\pi^{8/3}}{3^{5/3}}\frac{mn^{1/3}k^2T^2}{h^2} + \cdots, \tag{35}$$

$$\bar{L} = \frac{3}{40}\left(\frac{6}{\pi}\right)^{2/3}\frac{h^2 n^{2/3}}{m} + \frac{2^{1/3}\pi^{8/3}}{3^{2/3}}\frac{mk^2T^2}{h^2 n^{2/3}} + \cdots. \tag{36}$$

One sees that the degeneracy gives rise to a zero point pressure and a zero point energy.

From Eqn.36 one calculates the specific heat at low temperature as

$$C_v = \frac{d\bar{L}}{dT} = \frac{2^{4/3}\pi^{8/3}}{3^{2/3}}\frac{mk^2T}{h^2 n^{2/3}} + \cdots. \tag{37}$$

One sees that the specific heat at absolute zero vanishes, and that at low temperatures it is proportional to the absolute temperature.

Finally we will show that our theory leads to the Stern-Tetrode value for the absolute entropy of the gas. From Eqn.33 one finds in fact

$$S = n\int_0^T \frac{d\bar{L}}{T} = \frac{3}{2}nk\int_0^\Theta \frac{\mathcal{P}'(\Theta)d\Theta}{\Theta}.$$

Eqn.31 gives

$$S = nk\left(\frac{5}{2}\frac{G(A)}{F(A)} - \log A\right), \tag{38}$$

where the value of A is again taken from Eqn.21. For high temperatures, using Eqn.26, we find

$$A = \frac{nh^3}{(2\pi mkT)^{3/2}}, \qquad \frac{G(A)}{F(A)} = 1.$$

143

.

Eqn.38 then gives

$$
\begin{aligned}
S &= nk\left(\log\frac{(2\pi mkT)^{3/2}}{nh^3} + \frac{5}{2}\right) \\
&= nk\left(\frac{3}{2}\log T - \log n + \log\frac{(2\pi mk)^{3/2}e^{5/2}}{h^3}\right),
\end{aligned}
$$

which agrees with the entropy value of Stern and Tetrode.

Footnotes and References.

1) See the preliminary version, Lincei Rend.(6) **3**, 145 (1926).

2) See for example A. Einstein, Berl. Ber. 1924, S.261; 1925, S.318; M. Planck, ibid. 1925, S.49. Our method is analogous to Einstein's in that both give up the assumption of statistical independence of the gas molecules. However the nature of the statistical dependence for our case differs completely from Einstein's, and the final result for the deviation from the classical equation of state is even found to be the opposite.

3) E. Fermi, Nuovo Cim. **1**, 145 (1924).

4) E.C. Stoner, Phil. Mag. **48**, 719 (1924).

5) W. Pauli jr., Zeits. f. Phys. **31**, 765 (1925).

6) See for example B.F. Hund, Zeits. f. Phys. **33**, 345 (1925).

APPENDIX 5.B:

Excerpt from: Proc. Roy. Soc. (London) A112, 661 (1926)

On the Theory of Quantum Mechanics

By P.A.M. DIRAC, St. John's College, Cambridge

(Communicated by R.H. Fowler, F.R.S.- Received August 26, 1926)

§1. Introduction and Summary.

······(*Dirac's paper is excerpted at greater length with reference to the symmetrization or antisymmetrization of the wave functions of identical particles. Here we go directly to the discussion of the ideal gas in his §4.*)

§4. Theory of the Ideal Gas.

The results of the preceding section apply to any system containing several similar particles, in particular to an assembly of gas molecules. There will be two solutions of the problem, in one of which the eigenfunctions are symmetrical functions of the co-ordinates of all the molecules, and in the other antisymmetrical.

The wave equation for a single molecule of rest mass m moving in free space is

$$\left\{ p_x^2 + p_y^2 + p_z^2 - W^2 + m^2 \right\} \psi = 0$$

$$\left\{ \frac{\partial^2}{\partial x^2} + \frac{\partial^2}{\partial y^2} + \frac{\partial^2}{\partial z^2} - \frac{\partial^2}{\partial t^2} - m^2 \right\} \psi = 0,$$

and its solution is of the form

$$\psi_{\alpha_1 \alpha_2 \alpha_3} = \exp \cdot i(\alpha_1 x + \alpha_2 y + \alpha_3 z - Et), \tag{16}$$

where $\alpha_1, \alpha_2, \alpha_3$ and E are constants satisfying

$$\alpha_1^2 + \alpha_2^2 + \alpha_3^2 - E^2 + m^2 = 0.$$

The eigenfunction (16) represents an atom having momentum components $\alpha_1, \alpha_2, \alpha_3$ and the energy E.

We must now obtain some restriction on the possible eigenfunctions due to the presence of boundary walls. It is usually assumed that the eigenfunction, or wave function

145

associated with a molecule, vanishes at the boundary, but we should expect to be able to deduce this, if it is true, from the general theory. We assume, as a natural generalization of the methods of the preceding section, that there must be only just sufficient eigenfunctions for one to be able to represent by a matrix any function of the co-ordinates that has a physical meaning. Suppose for definiteness that each molecule is confined between two boundaries at $x = 0$ and $x = 2\pi$. Then only those functions of x that are defined only for $0 < x < 2\pi$ have a physical meaning and must be capable of being represented by matrices. (This will require fewer eigenfunctions than if every function of x had to be capable of being represented by a matrix.) These functions $f(x)$ can always be expanded as Fourier series of the form

$$f(x) = \sum_n a_n e^{inx},\qquad(17)$$

where the a_n's are constants and the n's are integers. If we choose from the eigenfunctions (16) those for which α_1 is an integer, then $f(x)$ times any chosen eigenfunction can be expanded as a series in the chosen eigenfunctions whose coefficients are functions of t only, and hence $f(x)$ can be represented by a matrix. Thus these chosen eigenfunctions are sufficient, and are easily seen to be only just sufficient, for the matrix representation of any function of x of the form (17). Instead of choosing those eigenfunctions with integral values of α_1, we could equally well take those with α_1 equal to half an odd integer, or more generally with $\alpha_1 = n + \epsilon$, where n is an integer and ϵ is any real number. The theory is incapable of deciding which are the correct ones. For statistical problems, though, they all lead to the same results.

When y and z are also bounded by $0 < y < 2\pi$, $0 < z < 2\pi$, we find for the number of waves associated with molecules whose energies lie between E and $E + dE$ the value

$$4\pi(E^2 - m^2)^{\frac{1}{2}} E dE.$$

This value is in agreement with the ordinary assumption that the wave function vanishes at the boundary. It reduces, when one neglects relativity mechanics, to the familiar expression

$$2\pi(2m)^{\frac{3}{2}} E_1^{\frac{1}{2}} dE_1,\qquad(18)$$

where $E_1 = E - m$ is the kinetic energy. For an arbitrary volume of gas V the expression must be multiplied by $V/(2\pi)^3$.

To pass to the eigenfunctions for the assembly of molecules, between which there is assumed to be no interaction, we multiply the eigenfunctions for the separate molecules, and then take either the symmetrical eigenfunctions, of the form (14), or the antisymmetrical ones, of the form (15). We must now make the new assumption that all stationary states of the assembly (each represented by one eigenfunction) have the same *a priori* probability. If now we adopt the solution of the problem that involves symmetrical eigenfunctions, we

146

should find that all values for the number of molecules associated with any wave have the same *a priori* probability, which gives just the Bose-Einstein statistical mechanics [7]. On the other hand, we should find a different statistical mechanics if we adopted the solution with the antisymmetrical eigenfunctions, as we should then have either 0 or 1 molecule associated with each wave. The solution with symmetrical eigenfunctions must be the correct one when applied to light-quanta, since it is known that Bose-Einstein statistical mechanics leads to Planck's law of black-body radiation. The solution with antisymmetrical eigenfunctions, though, is probably the correct one for gas molecules, since it is known to be the correct one for electrons in an atom, and one would expect molecules to resemble electrons more closely than light-quanta.

We shall now work out, according to well-known principles, the equation of state of the gas on the assumption that the solution with antisymmetrical eigenfunctions is the correct one, so that not more than one molecule can be associated with each wave. Divide the waves into a number of sets such that the waves in each set are associated with molecules of about the same energy. Let A_s be the number of waves in the sth set, and let E_s be the kinetic energy of a molecule associated with one of them. Then the probability of a distribution (or the number of antisymmetrical eigenfunctions corresponding to distributions) in which N_s molecules are associated with waves in the sth set is

$$W = \prod_s \frac{A_s!}{N_s!(A_s - N_s)!},$$

giving for the entropy

$$S = k \log W = k \sum_s \left\{ A_s(\log A_s - 1) - N_s(\log N_s - 1) - (A_s - N_s)[\log(A_s - N_s) - 1] \right\}.$$

This is to be a minimum, so that

$$0 = \delta S = k \sum_s \left\{ -\log N_s + \log(A_s - N_s) \right\} \delta N_s = k \sum_s \log(A_s/N_s - 1) \cdot \delta N_s,$$

for all variations δN_s that leave the total number of molecules $N = \sum_s N_s$ and the total energy $E = \sum_s E_s N_s$ unaltered, so that

$$\sum_s \delta N_s = 0, \qquad \sum_s E_s \delta N_s = 0.$$

we thus obtain

$$\log(A_s/N_s - 1) = \alpha + \beta E_s,$$

where α and β are constants, which gives

$$N_s = \frac{A_s}{e^{\alpha + \beta E_s} + 1}. \tag{19}$$

147

By making a variation in the total energy E and putting $\delta E/\delta S = T$, the temperature, we readily find that $\beta = 1/kT$, so that (19) becomes

$$N_s = \frac{A_s}{e^{\alpha + E_s/kT} + 1}.$$

This formula gives the distribution in energy of the molecules. On the Bose-Einstein theory the corresponding formula is

$$N_s = \frac{A_s}{e^{\alpha + E_s/kT} - 1}.$$

If the sth set of waves consists of those associated with molecules whose energies lie between E_s and $E_s + dE_s$, we have from (18) [where E_s now means the E_1 of equation (18)],

$$A_s = 2\pi V (2m)^{\frac{3}{2}} E_s^{\frac{1}{2}} dE_s/(2\pi)^3,$$

where V is the volume of the gas. This gives

$$N = \sum_s N_s = \frac{2\pi V (2m)^{\frac{3}{2}}}{(2\pi)^3} \int_0^\infty \frac{E_s^{\frac{1}{2}} dE_s}{e^{\alpha + E_s/kT} + 1}$$

and

$$E = \sum_s E_s N_s = \frac{2\pi V (2m)^{\frac{3}{2}}}{(2\pi)^3} \int_0^\infty \frac{E_s^{\frac{3}{2}} dE_s}{e^{\alpha + E_s/kT} + 1}.$$

By eliminating α from these two equations and using the formula $PV = \frac{2}{3}E$, where P is the pressure, which holds for any statistical mechanics, the equation of state may be obtained.

The saturation phenomenon of the Bose-Einstein theory does not occur in the present theory. The specific heat can be easily shown to tend steadily to $\cdots\cdots\cdots$

Footnotes and References:

$\cdots\cdots\cdots$

7) Bose, Zeits. f. Phys. **26**, 178 (1924); Einstein, Sitzungsb. d. Preuss. Ac., p.261 (1924) and p.3 (1925).

Chapter 6

Dirac's Invention
of Quantum Field Theory

Summary: Quantum field theory is introduced following the original paper of Dirac, where Dirac invents creation and annihilation operators and a number of other fundamental concepts, and applies them to the problem of emission and absorption of photons.

§1. Introductory Remarks.

Quantum field theory was invented by Dirac although there were precursors by Born, Heisenberg, and Jordan [6.1]. In his aetherial style, Dirac created the subject whole from the fundamental requirement of canonical commutation relations and Bose-Einstein statistics which he applied first to a quantum gas and then to a photon gas. Dirac finally closed the circle and confirmed his result by quantizing the classical Hamiltonian of an atomic source interacting with a radiation field. In the process, in one grand *tour de force* [6.2], Dirac invented

1) creation and annihilation (raising and lowering) operators;

2) the interaction representation Schrödinger equation;

3) the Fock representation and

4) the occupation number representation of the many particle states, and their connection;

5) perturbation theory in the Hilbert space of these states; and

6) derived the Einstein A and B coefficients for absorption and emission of photons.

149

§2. Dirac's Invention of Second Quantization.

Dirac first considers a quantum gas with a fixed number N of identical Bose-Einstein particles without interaction, each in one of S stationary single particle states r. The stationary states of the gas are described by two equivalent basis states. First he defines a Fock state

$$|r_1 r_2 \cdots r_n \cdots r_N\rangle$$

which corresponds to particle 1 in state r_1 and so on. In order to describe identical Bose-Einstein particles, the state vector must be symmetrized on the N particles and normalized. These states are related to the occupation number state (the N-states)

$$|N_1, N_2 \cdots N_S\rangle$$

corresponding to N_1 particles in state 1, etc. Dirac derives the proportionality factor between these states by an elegant combinatorial argument in Eqn.16. The states satisfy a Schrödinger equation with a Hamiltonian operator which Dirac eventually constructs by imposing canonical commutation relations on suitably chosen variables. His first choice, the wave function expansion coefficient a_r of Eqn.4, obeys a Schrödinger equation with a time dependent interaction Hamiltonian H_1, in what is now called the interaction representation. The expansion coefficients a_r satisfy the equation

$$i\hbar \dot{a}_r = \sum_s V_{rs} a_s \tag{1}$$

and

$$-i\hbar \dot{a}_r^* = \sum_s a_s^* V_{sr}$$

where $V_{rs}^* = V_{sr}$. With the identification $Q_r = a_r$ and $P_r = i\hbar a_r^*$, these have the Hamiltonian form with

$$H_1 = \sum_{rs} -\frac{i}{\hbar} P_r V_{rs} Q_s, \tag{2}$$

150

and

$$\dot{Q}_r = \partial H_1 / \partial P_r, \qquad \dot{P}_r = -\partial H_1 / \partial Q_r. \tag{3}$$

Interpreted as c-number amplitudes, the probability of the system being in state 'r' is $|a_r|^2$ which, properly normalized, is the occupation number N_r. Dirac next introduces new real canonical variables - N_r and a phase $-\phi_r$ - by a classical contact transformation

$$Q_r = N_r^{\frac{1}{2}} e^{-i\phi/\hbar}, \qquad P_r = i\hbar N_r^{\frac{1}{2}} e^{i\phi/\hbar}. \tag{4}$$

The Hamiltonian in terms of these variables is

$$H_1 = \sum_{rs} V_{rs} N_r^{\frac{1}{2}} N_s^{\frac{1}{2}} e^{i(\phi_r - \phi_s)/\hbar}, \tag{5}$$

and the equations of motion for the new variables N_r and $-\phi_r$ are canonical

$$\dot{N}_r = -\frac{\partial H_1}{\partial \phi_r}, \qquad \dot{\phi}_r = \frac{\partial H_1}{\partial N_r}. \tag{6}$$

For convenience, Dirac restores the trivial time dependence by defining the operators

$$b_r = a_r e^{-iW_r t/\hbar} \tag{7}$$

which satisfy a Schrödinger equation (5)

$$i\hbar \dot{b}_r = \sum_s H_{rs} b_s \tag{8}$$

where the H_{rs} are time independent when H is. The $b_r, ib_r{}^*$ are classically canonical variables which Dirac quantizes in the unnumbered equations preceding (10). These are the usual annihilation and creation operators with the commutation relations

$$[b_r, b_s^*]_- = \delta_{rs}, \tag{9}$$

and all others zero.

By a change of variables, Dirac introduces as canonical variables the occupation number N_r and the phase θ_r of the corresponding state. N_r

151

is diagonal in the occupation number representation and $\theta_r = i\frac{\partial}{\partial N_r}$ is (the negative of) its canonically conjugate momentum in this representation. The Hamiltonian (7) in the now familiar form $H = \sum_{rs} b_r{}^* H_{rs} b_s$, is replaced by the Hamiltonian (9,11) in the occupation number representation,

$$H = \sum_r W_r N_r + \sum_{rs} v_{rs} N_r^{\frac{1}{2}} N_s^{\frac{1}{2}} e^{i(\theta_r - \theta_s)}, \qquad (10)$$

The non-diagonal phase factor

$$e^{i(\theta_r - \theta_s)} \equiv e^{-\partial/\partial N_r} e^{+\partial/\partial N_s} \qquad (11)$$

acts to raise N_s by one and lower N_r by one, reading from left to right, as Dirac does in his Eqn.13. The current practice, in accord with the name 'annihilation' or 'lowering' operator for b_s for example, is to read from right to left, in which case N_s is lowered by one. The accompanying amplitude $N_r^{\frac{1}{2}}(N_s + 1)^{\frac{1}{2}}$ is just what is required for the Einstein A and B coefficients. This result is familiar to us from the more usual analysis of the matrix elements of the creation-annihilation combination $b_r{}^* b_s$.

Dirac then goes on to confirm his derivation by comparing to the intuitively obvious case when only one particle is present, and specializes Eqn.13 to Eqn.5; and to deduce the matrix element with the Einstein factors in the occupation number representation from the (again, yet to be invented) Fock representation (15). To do this he examines the number of equivalent Fock states $|r_1 \cdots r_N\rangle$ corresponding to one occupation number state $|N_1 \cdots N_S\rangle$, and from the combinatorial factor in (16) recovers the Einstein coefficient in the occupation number representation of the Hamiltonian in Eqns.17,19,20.

§3. Application to Photon Emission and Absorption.

Still exploring the new formalism, Dirac goes on to apply it to a photon gas. He has to invent an artifice to deal with an interaction Hamiltonian corresponding to the usual electromagnetic interaction which is linear in the vector potential and therefore in the photon creation and annihilation operators, rather than bilinear as was the case in his introductory example of an external potential. He introduces a "zero" state for invisible photons of zero energy and momentum for this purpose and invents the Hamiltonian (21) with creation *or* annihilation of photons (as well as a term with creation *and* annihilation) from which he deduces Einstein's statistical factors.

Finally, Dirac closes the argument in §7 by starting with the classical Hamiltonian for the Fourier components of the unquantized radiation (23) which he then writes as a real q-number in terms of the now quantum operators N_r, θ_r in the Hamiltonian (24) from which he identifies the coefficients introduced in the photon gas derivation (21), and can derive the Einstein A and B coefficients.

Bibliography and References.

6.1) M. Born, W. Heisenberg, and P. Jordan, Zeits. F. Phys. **35**, 606 (1925).

6.2) P.A.M. Dirac, Proc. Roy. Soc. (London) **A114**, 243 (1927); see our App.6A.

APPENDIX 6.A:

Excerpt from: Proc. Roy. Soc. Lond. A, Vol. 114, 243 (1927)

The Quantum Theory of the Emission and Absorption of Radiation

by P.A.M. Dirac
St. John's College, Cambridge, and Institute for
Theoretical Physics, Copenhagen

(Communicated by N. Bohr, For.Mem.R.S. - Received February 2, 1927)

§1. *Introduction and Summary.*

The new quantum theory, based on the assumption that the dynamical variables do not obey the commutative law of multiplication, has by now developed sufficiently to form a fairly complete theory of dynamics. One can treat mathematically the problem of any dynamical system composed of a number of particles with instantaneous forces acting between them, provided it is describable by a Hamiltonian, and one can interpret the mathematics physically by a quite definite general method. On the other hand, hardly anything has been done up to the present on quantum electrodynamics. The questions of the correct treatment of a system in which the forces are propagated with the velocity of light instead of instantaneously, of the production of an electromagnetic field by a moving electron, and of the reaction of this field on the electron have not yet been touched. In addition, there is a serious difficulty in making the theory satisfy all the requirements of the restricted principle of relativity, since a Hamiltonian can no longer be used. This relativity question is, of course, connected with the previous ones, and it will be impossible to answer any one question completely without at the same time answering them all. However, it appears to be possible to build up a fairly satisfactory theory of the emission of radiation and of the reaction of the radiation field on the emitting system on the basis of a kinematics and dynamics which are not strictly relativistic. This is the main object of the present paper. The theory is non-relativistic only on account of the time being counted throughout as a c-number, instead of being treated symmetrically with the space coordinates. The relativistic variation of mass with velocity is taken into account without difficulty.

The underlying ideas are very simple. Consider an atom interacting with a field of radiation, which we may suppose for definiteness to be confined in an enclosure so

154

as to have only a discrete set of degrees of freedom. Resolving the radiation into its Fourier components, we can consider the energy and phase of each of the components to be dynamical variables describing the radiation field. Thus if E_r is the energy of a component labelled r and θ_r is the corresponding phase (defined as the time since the wave was in a standard phase), we can suppose E_r and θ_r to form a pair of canonically conjugate variables. In the absence of any interaction between the field and the atom, the whole system of field plus atom will be describable by the Hamiltonian

$$H = \sum_r E_r + H_0 \qquad (1)$$

equal to the total energy, H_0 being the Hamiltonian for the atom alone. The variables E_r, θ_r obviously satisfy their canonical equations of motion

$$\dot{E}_r = -\frac{\partial H}{\partial \theta_r} = 0, \quad \dot{\theta}_r = \frac{\partial H}{\partial E_r} = 1.$$

When there is interaction between the field and the atom, it could be taken into account on the classical theory by the addition of an interaction term to the Hamiltonian (1), which would be a function of the variables E_r, θ_r that describe the field. The interaction term would give the effect of the radiation on the atom, and also the reaction of the atom on the radiation field.

In order that an analogous method may be used on the quantum theory, it is necessary to assume that the variables E_r, θ_r are q-numbers satisfying the standard quantum conditions

$$\theta_r E_r - E_r \theta_r = i\hbar,$$

etc., like other dynamical variables of the problem. This assumption immediately gives light-quantum properties to the radiation [1]. For if $\omega_r/2\pi$ is the frequency of the component r, $\omega_r \theta_r$ is an angle variable, so that its canonical conjugate E_r/ω_r can only assume a discrete set of values differing by multiples of $\hbar$, which means that E_r can change only by integral multiples of the quantum of energy $\hbar\omega_r$. If we now add an interaction term (taken over from the classical theory) to the Hamiltonian (1), the problem can be solved according to the rules of quantum mechanics, and we would expect to obtain the correct results for the action of the radiation and the atom on one another. It will be shown that we actually get the correct laws for the emission and absorption of radiation, and the correct values for Einstein's A's and B's. In the author's previous theory [2], where the energies and phases of the radiation were c-numbers, only the B's could be obtained, and the reaction of the atom on the radiation could not be taken into account.

It will also be shown that the Hamiltonian which describes the interaction of the atom and the electromagnetic waves can be made identical with that for the interaction of the atom with a gas of particles moving with the speed of light and satisfying Bose-Einstein

155

statistics, by a suitable choice of the interaction energy for the particles. The number of particles having any specified direction of motion and energy, which can be used as a dynamical variable in the Hamiltonian for the particles, is equal to the number of quanta of energy in the corresponding wave in the Hamiltonian for the waves. There is thus a complete harmony between the wave and the light-quantum *(Note added: from now on, Dirac's expression 'light-quantum' will be replaced by the yet to be coined word 'photon'.)* descriptions of the interaction. We shall actually build up the theory from the photon point of view, and show that the Hamiltonian transforms naturally into a form which resembles that for waves.

.... *(We omit some general developments and go to the quantization of the electromagnetic field, but first there is one more classic paragraph from Dirac's introduction.)*

It should be observed that there is a difference between a light-wave and the de Broglie or Schrödinger wave associated with the photon. Firstly, the light wave is always real, while the de Broglie wave associated with a photon moving in a definite direction must be taken to involve an imaginary exponential. A more important difference is that their intensities are to be interpreted in different ways. The number of photons per unit volume associated with a monochromatic light-wave equals the energy per unit volume of the wave divided by the energy $\hbar\omega$ of a single photon. On the other hand a monochromatic de Broglie wave of amplitude a (multiplied into the imaginary exponential factor) must be interpreted as representing a^2 photons per unit volume for all frequencies. ....
The wave whose intensity is to be interpreted in the first of these two ways appears in the theory only when one is dealing with a beam of the associated particles satisfying Bose-Einstein statistics. There is thus no such wave associated with electrons.

§2. *The Perturbation of a Gas of Independent Particles.*

We shall now consider the transitions produced in an atomic system by an arbitrary perturbation. The method we shall adopt ··· leads in a simple way to equations which determine the probability of the system being in any stationary state of the unperturbed system at any time [2]. This, of course, gives immediately the probable number of particles in that state at that time for a gas of the particles that are independent of one another and are all perturbed in the same way. The object of the present section is to show that the equations for the rates of change of these numbers can be put in the Hamiltonian form in a simple manner, which will enable further developments in the theory to be made.

Let H_0 be the Hamiltonian for the unperturbed particle and V the perturbing energy ······, so that the total Hamiltonian for the perturbed particle is $H = H_0 + V$. The

156

eigenfunctions for the perturbed particle must satisfy the wave equation

$$i\hbar \partial \psi / \partial t = (H_0 + V)\psi,$$

where $H_0 + V$ is an operator. If $\psi = \sum_r a_r \psi_r$ is the solution of this equation that satisfies the proper initial conditions, where the ψ_r's are the eigenfunctions for the unperturbed particle, each associated with the stationary state labelled r, and the a_r's are functions of the time only, then $|a_r|^2$ is the probability of the particle being in the state r at any time. The a_r's must be normalized initially, and will then always remain normalized. $\cdots\cdots$

The equation determines that the rate of change of the a_r's is [2]

$$i\hbar \dot{a}_r = \sum_s V_{rs} a_s, \tag{4}$$

where the V_{rs}'s are the matrix elements of V. The Hermitian conjugate equation is

$$- i\hbar \dot{a}_r^* = \sum_s a_s^* V_{sr}, \tag{4}$$

with $V_{rs}^* = V_{sr}$. If we regard a_r and $i\hbar a_r^*$ as canonical conjugates, equations (4) and (4') take the Hamiltonian form with the Hamiltonian $H_1 = \sum_{rs} a_r^* V_{rs} a_s$, namely

$$\frac{da_r}{dt} = \frac{1}{i\hbar}\frac{\partial H_1}{\partial a_r^*}, \quad i\hbar \frac{da_r^*}{dt} = -\frac{\partial H_1}{\partial a_r}.$$

We can transform to the canonical variables N_r, ϕ_r by the contact transformation

$$a_r = N_r^{\frac{1}{2}} e^{-i\phi_r/\hbar}, \quad a_r^* = N_r^{\frac{1}{2}} e^{i\phi_r/\hbar}.$$

This transformation makes the new variables N_r and ϕ_r real, N_r being equal $|a_r|^2$, the probable number of particles in the state r, and $\phi_r/\hbar$ being the phase of the eigenfunction that represents them. The Hamiltonian H_1 now becomes

$$H_1 = \sum_{rs} V_{rs} N_r^{\frac{1}{2}} N_s^{\frac{1}{2}} e^{i(\phi_r - \phi_s)/\hbar},$$

and the equations that determine the rate at which transitions occur have the canonical form

$$\dot{N}_r = -\frac{\partial H_1}{\partial \phi_r}, \quad \dot{\phi}_r = \frac{\partial H_1}{\partial N_r}.$$

A more convenient way of putting the transition equations in the Hamiltonian form may be obtained with the help of the quantities

$$b_r = a_r e^{-iW_r t/\hbar}, \quad b_r^* = a_r^* e^{iW_r t/\hbar},$$

157

W_r being the energy of the state r. We have $|b_r|^2 = |a_r|^2$, the probable number of particles in the state r. For $\dot{b}_r$ we find

$$i\hbar\dot{b}_r = W_r b_r + \sum_s V_{rs} b_s e^{i(W_s - W_r)t/\hbar}$$

using (4). If we put $V_{rs} = v_{rs} e^{i(W_r - W_s)t/\hbar}$, so that v_{rs} is a constant when V does not involve the time explicitly, this reduces to

$$i\hbar\dot{b}_r = \sum_s H_{rs} b_s, \tag{5}$$

where $H_{rs} = W_r \delta_{rs} + v_{rs} \cdots\cdots$ is a constant when H does not involve the time explicitly. Equation (5) is of the same form as equation (4), and may be put in the Hamiltonian form in the same way.

$\cdots\cdots$

We now take b_r and ib_r^* to be canonically conjugate variables instead of a_r and ia_r^* (*Note added: So far, these are c-number variables. In the next section they are elevated to be q-number operators. Also note that from now on, we set $\hbar = 1$.*) Equation (5) and its Hermitian conjugate will now take the Hamiltonian form with the Hamiltonian

$$H = \sum_{rs} b_r^* H_{rs} b_s. \tag{7}$$

Proceeding as before, we make the contact transformation

$$b_r = N_r^{\frac{1}{2}} e^{-i\theta_r}, \quad b_r^* = N_r^{\frac{1}{2}} e^{i\theta_r}, \tag{8}$$

to the new canonical variables N_r, θ_r, where N_r is, as before, the probable number of particles in the state r, and θ_r is a new phase. The Hamiltonian H will now become

$$H = \sum_{rs} H_{rs} N_r^{\frac{1}{2}} N_s^{\frac{1}{2}} e^{i(\theta_r - \theta_s)}.$$

and the equations $\cdots$ take the canonical form

$$\dot{N}_r = -\frac{\partial H}{\partial \theta_r}, \quad \dot{\theta}_r = \frac{\partial H}{\partial N_r}.$$

The Hamiltonian may be written

$$H = \sum_r W_r N_r + \sum_{rs} v_{rs} N_r^{\frac{1}{2}} N_s^{\frac{1}{2}} e^{i(\theta_r - \theta_s)}. \tag{9}$$

The first term is the total proper energy of the gas of particles, and the second may be regarded as the additional energy due to the perturbation. If the perturbation is zero, the phases θ_r would increase linearly with the time, while the previous phases ϕ_r would in this case be constants.

§3. *The Perturbation of a Gas of Particles Satisfying Bose-Einstein Statistics.*

According to the preceding section we can describe the effect of a perturbation on a gas of independent particles by means of canonical variables and Hamiltonian equations of motion. The development of the theory which naturally suggests itself is to make these canonical variables q-numbers satisfying the usual quantum conditions instead of c-numbers, so that their Hamiltonian equations of motion become true quantum equations.

The Hamiltonian will now provide a Schrödinger wave equation, which must be solved and interpreted in the usual manner. The interpretation will give $\cdots$ the probable number of particles in any state $\cdots$ equal to the square of the modulus of the normalized solution of the wave equation that satisfies the appropriate initial conditions. We could, of course, calculate directly $\cdots$ the probability $\cdots$ when the particles are independent $\cdots$. $\cdots$ In the general case it will be shown that the wave equation leads to the correct value for the probability of any given distribution when the particles obey Bose-Einstein statistics instead of being independent.

We assume the variables b_r, ib_r^* of §2 to be canonical q-numbers satisfying the quantum conditions

$$b_r \cdot ib_r^* - ib_r^* \cdot b_r \equiv i[b_r, b_r^*]_- = i$$

or

$$[b_r, b_s^*]_- = \delta_{rs},$$

and

$$[b_r, b_s]_- = [b_r^*, b_s^*]_- = 0.$$

The transformation equations (8) must now be written in the quantum form

$$b_r = (N_r + 1)^{\frac{1}{2}} e^{-i\theta_r} = e^{-i\theta_r} N_r^{\frac{1}{2}}$$
$$b_r^* = N_r^{\frac{1}{2}} e^{i\theta_r} = e^{i\theta_r} (N_r + 1)^{\frac{1}{2}}, \tag{10}$$

in order that N_r, θ_r may also be canonical variables. These equations show that the N_r can have only integral eigenvalues not less than zero [3], which provides us with a justification for the assumption that the variables are q-numbers in the way we have chosen. The numbers of particles in the different states are now ordinary quantum numbers.

The Hamiltonian (7) now becomes

$$H = \sum_{rs} b_r^* H_{rs} b_s = \sum_{rs} N_r^{\frac{1}{2}} e^{i\theta_r} H_{rs} (N_s + 1)^{\frac{1}{2}} e^{-i\theta_s}$$

159

$$= \sum_{rs} H_{rs} N_r^{\frac{1}{2}} (N_s + 1 - \delta_{rs})^{\frac{1}{2}} e^{i(\theta_r - \theta_s)} \tag{11}$$

in which the H_{rs} are still c-numbers. We may write this H in the form corresponding to (9)

$$H = \sum_r W_r N_r + \sum_{rs} v_{rs} N_r^{\frac{1}{2}} (N_s + 1 - \delta_{rs})^{\frac{1}{2}} e^{i(\theta_r - \theta_s)} \tag{11}$$

$\ldots\ldots$.

The wave equation written in terms of the variables N_r is [4]

$$i\frac{\partial}{\partial t}\psi(N_1', N_2' \cdots) = H\psi(N_1', N_2' \cdots), \tag{12}$$

where H is an operator, each θ_r occurring in H being interpreted to mean $i\partial/\partial N_r'$. If we apply the operator $\exp(\pm i\theta_r)$ to any function $f(N_1' \cdots N_r' \cdots)$ the result is

$$e^{\pm i\theta_r} = e^{\mp\partial/\partial N_r'} f(N_1' \cdots N_r') = f(N_1' \cdots N_r' \mp 1, \cdots).$$

If we use this rule in equation (12) and use (11) for H we obtain [5]

$$i\frac{\partial}{\partial t}\psi(\cdots N_r', \cdots N_s', \cdots) =$$
$$\sum_{rs} H_{rs} N_r'^{\frac{1}{2}} (N_s' + 1 - \delta_{rs})^{\frac{1}{2}} \psi(\cdots N_r' - 1, \cdots N_s' + 1, \cdots). \tag{13}$$

We see that $\cdots$ the term in H involving $\exp i(\theta_r - \theta_s)$ will contribute only to $\cdots$ transitions in which N_r decreases by unity and N_s increases by unity $\cdots\cdots$. If we find a solution $\psi(N_1', N_2' \cdots)$ of (13) that is normalized, $i.e.$, one for which

$$\sum_{N_1', N_2' \cdots} |\psi(N_1', N_2' \cdots)|^2 = 1,$$

and which satisfies the proper initial conditions, then $|\psi(N_1', N_2' \cdots)|^2$ will be the probability of that distribution in which N_1' particles are in state 1, N_2' in state 2, $\cdots$ at any time.

Consider first the case when there is only one particle in the gas. The probability of its being in the state q is determined by the eigenfunction $\psi(N_1', N_2', \cdots)$ in which all the N's are put equal to zero except N_q' which is put equal to unity. This eigenfunction we shall denote by $\psi\{q\}$. When it is substituted in the left-hand side of (13), all terms on the right-hand side vanish except those for which $r = q$, and we are left with

$$i\frac{\partial}{\partial t}\psi\{q\} = \sum_s H_{qs}\psi\{s\},$$

which is the same equation as (5) with $\psi\{q\}$ playing the part of b_q. This establishes the fact that the present theory is equivalent to that of the preceding section when there is only one particle in the gas.

160

Now take the general case of an arbitrary number of particles in the gas, and assume that they obey Bose-Einstein statistics. This requires that, in the ordinary treatment of the problem, only those eigenfunctions that are symmetrical between all the particles must be taken into account, these eigenfunctions being by themselves sufficient to give a complete quantum solution of the problem [2]. We shall now obtain the equation for the rate of change of one of these symmetrical eigenfunctions, and show that it is identical with equation (13).

If we label each particle with a number $n(= 1 \cdots N)$ with N the number of particles, then the Hamiltonian for the gas will be $H_A = \sum_n H(n)$, where $H(n)$ is the H of §2 (equal to $H_0 + V$) expressed in terms of the variables of the nth particle. A stationary state of the gas is defined by the numbers $r_1, r_2 \cdots r_n \cdots r_N$ which are the labels of the stationary states in which the separate particles exist. The Schrödinger equation for the gas in a set of variables that specify the stationary states will be of the form $\cdots$ of equation (5) thus:

$$i\dot{b}(r_1 r_2 \cdots r_N) = \sum_{s_1, s_2 \cdots} H_A(r_1 r_2 \cdots r_N; s_1, s_2 \cdots) b(s_1, s_2 \cdots s_N), \tag{14}$$

where $H_A(r_1 r_2 \cdots ; s_1 s_2 \cdots)$ is the general matrix of H_A [with the time factor removed]. This matrix element vanishes when more than one s_n differs from the corresponding r_n; equals $H_{r_m s_m}$ when s_m differs from r_m and every other s_n equals r_n; and equals $\sum_n H_{r_n r_n}$ when every s_n equals r_n. Substituting these values in (14), we obtain

$$i\dot{b}(r_1 r_2 \cdots r_N) = \sum_m \sum_{s_m \neq r_m} H_{r_m s_m} b(r_1 \cdots s_m \cdots r_N) + \sum_n H_{r_n r_n} b(r_1 r_2 \cdots r_N). \tag{15}$$

We must now restrict $b(r_1 r_2 \cdots)$ to be a symmetrical function of the variables $r_1, r_2 \cdots$ in order to obtain Bose-Einstein statistics. This is permissible since if $b(r_1 r_2 \cdots)$ is symmetrical at any time, then equation (15) shows that $\dot{b}(r_1 r_2 \cdots)$ is also symmetrical at that time, so that $b(r_1 r_2 \cdots)$ will remain symmetrical.

Let N_r denote the number of particles in the state r. Then a stationary state of the gas describable by a symmetrical eigenfunction may be specified by the numbers $N_1, N_2 \cdots N_r \cdots N_S$ (with S the number of states) just as well as by the numbers $r_1, r_2 \cdots r_n \cdots r_N$, and we shall be able to transform equation (15) to the variables $N_1, N_2 \cdots$. We cannot actually take the new eigenfunctions $b(N_1, N_2 \cdots N_S)$ equal to the previous one $b(r_1 r_2 \cdots r_N)$, but must take one to be a numerical multiple of the other in order that each may be correctly normalized with respect to its respective variables. We must have, in fact,

$$\sum_{r_1, r_2 \cdots} |b(r_1 r_2 \cdots r_N)|^2 = 1 = \sum_{N_1, N_2 \cdots} |b(N_1, N_2 \cdots N_S)|^2,$$

161

and hence we must take $|b(N_1 N_2 \cdots)|^2$ equal to the sum of $|b(r_1 r_2 \cdots)|^2$ for all values of the numbers $r_1, r_2 \cdots$ such that there are N_1 of them equal to 1, N_2 equal to 2, etc. There are $N!/N_1! N_2! \cdots$ terms in this sum, where $N = \sum_r N_r$ is the total number of particles, and they are all equal, since $b(r_1 r_2 \cdots)$ is a symmetrical function of its variables $r_1, r_2 \cdots$. Hence we must have

$$b(N_1, N_2 \cdots) = [N!/N_1! N_2! \cdots]^{\frac{1}{2}} b(r_1 r_2 \cdots).$$

If we make this substitution in equation (15), the left-hand side will become

$$[N_1! N_2! \cdots /N!]^{\frac{1}{2}} i\dot{b}(N_1, N_2 \cdots).$$

The term $H_{rs} b(r_1 r_2 \cdots s \cdots)$ in the first summation on the right-side will become

$$[N_1! N_2! \cdots (N_r - 1)! \cdots (N_s + 1) \cdots /N!]^{\frac{1}{2}} H_{rs} b(N_1, N_2 \cdots N_r - 1 \cdots N_s + 1 \cdots) \quad (16)$$

where we have written r for r_m and s for s_m. This term must be summed for all values of s except r, and then must be summed for r taking each of the values $r_1, r_2 \cdots$. Thus each term (16) gets repeated by the summation process until it occurs a total of N_r times, so that it contributes

$$N_r [N_1! N_2! \cdots (N_r - 1)! \cdots (N_s + 1)! \cdots /N!]^{\frac{1}{2}} H_{rs} b(\cdots N_r - 1 \cdots N_s + 1 \cdots)$$
$$= N_r^{\frac{1}{2}} (N_s + 1)^{\frac{1}{2}} (N_1! N_2! \cdots /N!)^{\frac{1}{2}} H_{rs} b(\cdots N_r - 1 \cdots N_s + 1 \cdots)$$

to the right-hand side of (15). Finally the term $\sum_n H_{r_n r_n} b(r_1 r_2 \cdots)$ becomes

$$\sum_r N_r H_{rr} \cdot b(r_1 r_2 \cdots) = \sum_r N_r H_{rr} \cdot [N_1! N_2! \cdots /N!]^{\frac{1}{2}} b(N_1, N_2 \cdots).$$

Hence equation (15) becomes, with the removal of the factor $[N_1! N_2! \cdots /N!]^{\frac{1}{2}}$,

$$i\dot{b}(N_1, N_2 \cdots) = \sum_r H_{rr} b(N_1, N_2 \cdots)$$
$$+ \sum_r \sum_{s \neq r} N_r^{\frac{1}{2}} (N_s + 1)^{\frac{1}{2}} H_{rs} b(N_1, N_2 \cdots N_r - 1 \cdots N_s + 1 \cdots), \quad (17)$$

which is identical with (13) [except for the fact that in (17) the primes have been omitted from the N's, which is permissible when we do not need to refer to the N's as q-numbers]. We have thus established that the Hamiltonian (11) describes the effect of a perturbation on a gas of particles satisfying Bose-Einstein statistics.

§4. *The Reaction of the Particles on the Perturbing System.*

(This section involves developments which will be omitted as not essential to the Spin-Statistics Theorem.)

162

...... the Schrödinger equation corresponds to the Hamiltonian

$$H = H_P(J) + \sum_{r,s} H_{rs} N_r^{\frac{1}{2}} (N_s + 1 - \delta_{rs})^{\frac{1}{2}} e^{i(\theta_r - \theta_s)}, \tag{19}$$

in which H_{rs} is now a function of the J's and w's, $\cdots$. (It should be noticed that H_{rs} still commutes with the N's and θ's.)

$\cdots$the matrix element $H(J'r; J''s)$ is the sum of two parts, one from the proper energy H_0 which equals W_r when $J'' = J'$ and $s = r$ and vanishes otherwise, and one from the interaction energy V, which may be denoted by $v(J'r; J''s)$. Thus we shall have

$$H_{rs} = W_r \delta_{rs} + v_{rs},$$

where v_{rs} has the $(J'J'')$ matrix element $v(J'r; J''s)$, so (19) becomes

$$H = H_P(J) + \sum_{r} W_r N_r + \sum_{r,s} v_{rs} N_r^{\frac{1}{2}} (N_s + 1 - \delta_{rs})^{\frac{1}{2}} e^{i(\theta_r - \theta_s)}. \tag{20}$$

The Hamiltonian is thus the sum of the proper energy of the perturbing system $H_P(J)$, the proper energy of the perturbed system $\sum_{r} W_r N_r$ and the perturbation energy

$$\sum_{r,s} v_{rs} N_r^{\frac{1}{2}} (N_s + 1 - \delta_{rs})^{\frac{1}{2}} e^{i(\theta_r - \theta_s)}.$$

§5. *Theory of Transitions* $\cdots$.

(This section involves technical developments which are not essential to our main goal and will be omitted.)

§6. *Application to Photons.*

We shall now apply the theory of §4 to the case when the particles of the gas are photons, the theory being applicable to this case since photons obey Bose-Einstein statistics and have no mutual interaction. A photon is in a stationary state when it is moving with constant momentum in a straight line. Thus a stationary state r is fixed by the three components of momentum of the photon and a variable that specifies its state of polarization. We shall work on the assumption that there are a finite number of these stationary states, lying very close to one another, as it would be inconvenient to use continuous ranges. The interaction of the photons with an atom will be described by a Hamiltonian of the form (20), in which $H_P(J)$ is the Hamiltonian for the atom alone, and the coefficients v_{rs} are for the present unknown. We shall show that this form for the Hamiltonian, with v_{rs} arbitrary, leads to Einstein's laws for the emission and absorption of radiation.

163

The photon has the peculiarity that it apparently ceases to exist when it is in one of its stationary states, namely, the zero state, in which its momentum, and therefore also its energy, are zero. When a photon is absorbed it can be considered to jump into this zero state, and when one is emitted it can be considered to jump from the zero state to one in which it is physically in evidence, so that it appears to have been created. Since there is no limit to the number of photons that may be created in this way, we must suppose that there are an infinite number of photons in the zero state, so that N_0 of the Hamiltonian (20) is infinite. We must now have θ_0, the variable canonically conjugate to N_0, a constant, since

$$\dot{\theta}_0 = \partial H/\partial N_0 = W_0 + \quad \text{terms} \sim N_0^{-\frac{1}{2}} \quad \text{or} \sim (N_0 + 1)^{-\frac{1}{2}}$$

and W_0 is zero. In order that the Hamiltonian (20) may remain finite it is necessary for the coefficients v_{r0}, v_{0r} to be infinitely small. We shall suppose that they are infinitely small in such a way as to make $v_{r0} N_0^{\frac{1}{2}}$ and $v_{0r} N_0^{\frac{1}{2}}$ finite, in order that the transition probability coefficients may be finite. Thus we put

$$v_{r0}(N_0 + 1)^{\frac{1}{2}} e^{-i\theta_0} = v_r, \quad v_{0r} N_0^{\frac{1}{2}} e^{i\theta_0} = v_r{}^*,$$

where v_r and its complex conjugate $v_r{}^*$ are finite. We may consider the v_r and $v_r{}^*$ to be functions of the J's and w's of the atom, since their factors $(N_0 + 1)^{\frac{1}{2}} e^{-i\theta_0}$ and $N_0^{\frac{1}{2}} e^{i\theta_0}$ are practically constants, the rate of change of N_0 being very small compared with N_0. The Hamiltonian (20) now becomes

$$H = H_P(J) + \sum_r W_r N_r + \sum_{r\neq0}[v_r N_r^{\frac{1}{2}} e^{i\theta_r} + v_r{}^*(N_r + 1)^{\frac{1}{2}} e^{-i\theta_r}]$$
$$+ \sum_{r\neq0}\sum_{s\neq0} v_{rs} N_r^{\frac{1}{2}}(N_s + 1 - \delta_{rs})^{\frac{1}{2}} e^{i(\theta_r - \theta_s)}. \tag{21}$$

The probability of a transition in which a photon in the state r is absorbed is proportional to the square of the modulus of that matrix element of the Hamiltonian which refers to this transition. This matrix element must come from the term $v_r N_r^{\frac{1}{2}} e^{i\theta_r}$ in the Hamiltonian, and must therefore be proportional to $N_r'^{\frac{1}{2}}$ where N_r' is the number of photons in state r before the process. In the same way the probability of a photon in state r being emitted is proportional to $(N_r' + 1)$, and the probability of a photon in state r being scattered into state s is proportional to $N_r'(N_s' + 1)$. Radiative processes of the more general type considered by Einstein and Ehrenfest [6], in which more than one photon take part simultaneously, are not allowed in the present theory.

To establish a connection between the number of photons per stationary state and the intensity of radiation, we consider an enclosure of finite volume, V say, containing the radiation. The number of stationary states for photons of a given type of polarization whose energy ($\omega = 2\pi \times$ the frequency ν) lies in the range ω_r to $\omega_r + d\omega_r$ and whose direction of motion lies in the solid angle $d\Omega_r$ about the direction of motion for state r will now be $V\omega_r{}^2 d\omega_r d\Omega_r / (2\pi)^3$. The energy of the photons in these stationary states is thus

$$N_r' \cdot \omega_r \cdot V\omega_r{}^2 d\omega_r d\Omega_r / (2\pi)^3.$$

This must be equal to $V I_r d\nu_r d\Omega_r$, where I_r is the energy per unit volume per unit frequency interval of the radiation about the state r. Hence

$$I_r = N_r' 2\pi\nu_r{}^3, \tag{22}$$

so that N_r' is proportional to I_r and $(N_r' + 1)$ is proportional to $I_r + 2\pi\nu_r{}^3$. We thus obtain that the probability of an absorption process is proportional to I_r, the incident intensity per unit frequency interval, and that of an emission process is proportional to $I_r + 2\pi\nu_r{}^3$, which are just Einstein's laws [7]. In the same way the probability of a process in which a photon is scattered from a state r to a state s is proportional to $I_r[I_s + 2\pi\nu_s{}^3]$, which is Pauli's law for the scattering of radiation by an electron [8].

§7. *The Probability Coefficients for Emission and Absorption.*

We shall now consider the interaction of an atom and radiation from the wave point of view. We resolve the radiation into its Fourier components, and suppose that their number is very large but finite. Let each component be labelled by a suffix $r, \cdots\cdots$. Each component r can be described by a vector potential A_r chosen to make the scalar potential zero. The perturbation term to be added to the Hamiltonian will now be, according to the classical non-relativistic theory, $\sum_r A_r \dot{X}_r$, where X_r is the component of the total polarization of the atom in the direction of A_r, which is the direction of the electric vector of the component r.

We can, as explained in §1, suppose the field to be described by the canonical variables N_r, θ_r of which N_r is the number of photons of the component r, and θ_r is its canonically conjugate phase, equal to ω times the θ of §1. We shall now have $A_r = a_r \cos\theta_r$, where a_r is the amplitude of A_r which must be connected to N_r as follows: The flow of energy per unit area per unit time for the component r is $a_r{}^2 \omega_r{}^2 / 8\pi$. Hence the intensity per unit frequency range of the radiation in the neighbourhood of the component r is $I_r = a_r{}^2 \omega_r{}^2 / 8\pi$. Comparing this with equation (28), we obtain $a_r = (2\omega_r/\pi)^{\frac{1}{2}} N_r{}^{\frac{1}{2}}$, and hence

$$A_r = (2\omega_r/\pi)^{\frac{1}{2}} N_r{}^{\frac{1}{2}} \cos\theta_r.$$

The Hamiltonian for the whole system of atom plus radiation would now be, according to the classical theory,

$$H = H_P(J) + \sum_r N_r \omega_r + \sum_r (2\omega_r/\pi)^{\frac{1}{2}} \dot{X}_r N_r{}^{\frac{1}{2}} \cos\theta_r, \tag{23}$$

where $H_P(J)$ is the Hamiltonian of the atom alone. On the quantum theory we must make the variables N_r and θ_r canonical q-numbers like the variables J_k, w_k that describe the atom. We must now replace the $N_r^{\frac{1}{2}} \cos\theta_r$ in (29) by the real q-number

$$\frac{1}{2}\{N_r^{\frac{1}{2}} e^{i\theta_r} + e^{-i\theta_r} N_r^{\frac{1}{2}}\} = \frac{1}{2}\{N_r^{\frac{1}{2}} e^{i\theta_r} + (N_r + 1)^{\frac{1}{2}} e^{-i\theta_r}\}$$

so that the Hamiltonian (29) becomes

$$H = H_P(J) + \sum_r N_r \omega_r + \sum_r (\omega_r/2\pi)^{\frac{1}{2}} \dot{X}_r \{N_r^{\frac{1}{2}} e^{i\theta_r} + (N_r + 1)^{\frac{1}{2}} e^{-i\theta_r}\}. \quad (24)$$

This is of the form (27), with

$$v_r = v_r^* = (\omega_r/2\pi)^{\frac{1}{2}} \dot{X}_r$$
$$v_{rs} = 0 \qquad (r, s \neq 0). \quad (25)$$

The wave point of view is thus consistent with the photon point of view and gives values for the unknown interaction coefficients v_{rs} in the photon theory. $\cdots\cdots$

(We omit the remainder of this section which reproduces results of §6.)

The present theory, since it gives a proper account of spontaneous emission, must presumably give the effect of radiation reaction on the emitting system, and enable one to calculate the widths of spectral lines, if one can overcome the mathematical difficulties involved in the general solution of the wave problem corresponding to the Hamiltonian (30). Also the theory enables one to understand how it comes about that there is no violation of energy conservation when, say, a photo-electron is emitted from an atom under the action of extremely weak incident radiation. The energy of interaction of the atom and the radiation is a q-number that does not commute with the first integrals of the motion of the atom alone or with the intensity of the radiation. Thus one cannot specify this energy by a c-number at the same time that one specifies the stationary state of the atom and the intensity of the radiation by c-numbers. In particular, one cannot say that the interaction energy tends to zero as the intensity of the incident energy tends to zero. Thus there is always an unspecifiable amount of interaction energy which can supply the energy for the photo-electron.

I would like to thank Prof. Niels Bohr for his interest in this work and for much friendly discussion about it.

Summary.

The problem is treated of a gas of identical particles satisfying Bose-Einstein statistics, which interact with another different system, a Hamiltonian being obtained to describe

166

the motion. The theory is applied to the interaction of a gas of photons with an ordinary atom, and it is shown that it gives Einstein's laws for emission and absorption of radiation.

The interaction of an atom with electromagnetic waves is then considered, and it is shown that if one takes the energies and phases of the waves to be q-numbers satisfying the proper quantum conditions instead of c-numbers, the Hamiltonian takes the same form as in the photon treatment. The theory leads to the correct expressions for Einstein's A's and B's.

Footnotes and References.

1) Similar assumptions have been used by Born and Jordan [Zeits. f. Phys. **34**, 886 (1925).] for the purpose of taking over the classical formula for the emission of radiation by a dipole into the quantum theory, and by Born, Heisenberg and Jordan [Zeits. f. Phys. **35**, 606 (1925)] for calculating the energy fluctuations in a field of black-body radiation.

2) P.A.M. Dirac, Proc. Roy. Soc. A**112**, 661 (1926).

3) See §8 of the author's paper Proc. Roy. Soc. A**111**, 281 (1926). What are there called the c-number values that a q-number can take are here given the more precise name of the eigenvalues of that q-number.

4) We are supposing for definiteness that the label r of the stationary states takes the values $1, 2, 3, \cdots$.

5) When $s = r$, $\psi(N_1', N_2' \cdots N_r' - 1 \cdots N_s' + 1)$ is taken to mean $\psi(N_1', N_2' \cdots N_r' \cdots)$.

6) A. Einstein and P. Ehrenfest, Zeits. f. Phys. **19**, 301 (1923).

7) The ratio of stimulated to spontaneous emission in the present theory is just twice its value in Einstein's. This is because in the present theory either polarized component of the incident radiation can stimulate only radiation polarized in the same way, while in Einstein's the two polarized components are treated together. This remark applies also to the scattering process.

8) W. Pauli, Zeits. f. Phys. **18**, 272 (1923).

Chapter 7

The Jordan-Wigner Invention of
Anticommutation for Fermi-Dirac Fields

Summary: The idea of anticommuting operators for quanta satisfying the Pauli Exclusion Principle is followed from Jordan's original proposal to Jordan and Wigner's final complete exposition.

§1. Introduction.

The germ of the idea of anticommuting operators to represent quanta satisfying the Pauli Exclusion Principle and Fermi-Dirac statistics first appeared in a paper by Jordan [7.1]. Here for the first time he broke through the inhibiting constraint of Heisenberg and Dirac's prescription of canonical commutation relations by the simple but profound idea of replacing the commutator $ab - ba$ by the anticommutator $ab + ba$, although it was not yet explicitly stated in these terms. He immediately made the idea completely explicit in a second paper [7.2], and finally in a marvelously complete discussion with Wigner [7.3].

Jordan labored under an apparent reluctance to simply accept the different algebra, and spent considerable effort to find a representation related in some way to the quantum amplitudes for Bose-Einstein quanta based on fields satisfying commutation relations, and evolved from the Poisson bracket formalism of classical mechanics. By contrast, the anticommutation relations are a completely quantum device which have no precursors in the classical theory.

In the first paper, Jordan claims that a paradox exists in the canonical commutation relations $qp - pq = i\hbar$ as applied to the "spinning magnetic electron". He states that there is no definable quantum analog to a classical angle θ_z conjugate to the spin angular momentum s_z, for the simple reason that the "perfect" spinning electron does not permit the introduction of any variables except the Pauli spins: "··· the magnetic electron is a real example for which the existing canonical commutation relations fail absolutely ···. According to the classical analogy, there is, as canonical conjugate to the z-component of spin, an angle around the z-axis. ··· that is however impossible" (intuitively because there is no marker on the spinning electron to make an angular displacement measurable; mathematically because the finite dimensional spin matrix cannot have a continuous conjugate without contradictions, which Jordan explores in his first paper on this subject.) ···"We wish absolutely to make the starting point for our deliberations the assumption that one can measure no other quantities for the magnetic electron, than the spin component in a given direction $\hat{z}$. ······ As conjugate to the momentum s_z there must be a component $s_{z'}$ ··· $\hat{z}'$ must be perpendicular to $\hat{z}$ ···. We have here a very obvious example of the general quantum mechanical rule, that one cannot observe two conjugate quantities at the same time: one cannot quantize the magnetic electron simultaneously in two different directions."

So not ϕ_z (which cannot be operationally defined) but s_y should be considered, in some sense, the quantity conjugate to s_z!

The above considerations suggested the possibility of abandoning canonical commutation relations as the sole basis for defining dynamical variables, and led Jordan to consider operators with a discrete, finite set of eigenvalues as in Hypothesis B of his second paper. Even here, there is a reluctance to simply write down anticommutation relations for the Fermi-Dirac operators, although they might be suggested as the algebra for his generalized "canonically conjugate" pair

$$[\sigma_\alpha, \sigma_\beta]_+ = 2\delta_{\alpha\beta}.$$

The main inspiration which Jordan derived from this first paper seems to have been the idea to use spinor ($\sim$ quaternion) operators with a 2×2 matrix representation and a discrete, finite spectrum of eigenvalues to replace those introduced in Dirac's quantization of the Bose-Einstein gas.

§2. Jordan's Introduction of Anticommutation Relations.

Jordan follows Dirac by introducing annihilation and creation operators b_r and $b_r^\dagger$ which form an occupation number operator $N_r = b_r^\dagger b_r$ with the eigenvalues $N_r' = 0, 1$. Jordan actually constructs "conjugate" operators N_r, Θ_r as Dirac did, but it would have been more simple to go directly to a 2×2 matrix representation with

$$b_r = \begin{pmatrix} 0 & 1 \\ 0 & 0 \end{pmatrix}, \quad b_r^\dagger = \begin{pmatrix} 0 & 0 \\ 1 & 0 \end{pmatrix}; \quad N_r = b_r^\dagger b_r = \begin{pmatrix} 0 & 0 \\ 0 & 1 \end{pmatrix}. \tag{1}$$

The matrix representation suggests the anticommutation relations

$$[b_r, b_s^\dagger]_+ = \delta_{rs}, \quad [b_r, b_s]_+ = [b_r^\dagger, b_s^\dagger]_+ = 0. \tag{2}$$

From this he could construct an antisymmetric two particle state following Heisenberg and Dirac, which satisfies the Pauli Exclusion Principle:

$$|r, s\rangle = b_r^\dagger b_s^\dagger |0\rangle = -b_s^\dagger b_r^\dagger |0\rangle = -|s, r\rangle. \tag{3}$$

The (non-relativistic) field operators

$$\Psi(x) = \sum_r b_r \psi_r(x), \quad \text{and} \quad \Psi^\dagger(x) = \sum_r \psi_r^\dagger(x) b_r^\dagger \tag{4}$$

give the two particle configuration space state

$$|x_1, x_2\rangle = \Psi^\dagger(x_1)\Psi^\dagger(x_2)|0\rangle \tag{5}$$

and the antisymmetric two particle wave function

$$\psi(x_1, x_2) = \langle x_1, x_2 | r, s \rangle = \psi_r(x_1)\psi_s(x_2) - \psi_s(x_1)\psi_r(x_2), \tag{6}$$

as required.

These are the operations that Jordan eventually makes in his Eqns.18,25; resulting in the N, Θ representation of b and $b^\dagger$ of Eqns.13,15 in Jordan's §2, and in Eqns.5-16 of Jordan and Wigner.

The identities involving N and Θ for Bose-Einstein and for Fermi-Dirac operators follow easily. For example, for Bose-Einstein operators

$$\Theta = i\frac{\partial}{\partial N}. \tag{7}$$

For small ϵ, we get

$$
\begin{aligned}
e^{-i\epsilon\Theta}F(N)G(N) &= e^{\epsilon\frac{\partial}{\partial N}}F(N)G(N) \\
&\approx (1 + \epsilon\frac{\partial}{\partial N})F(N)G(N) \\
&\approx (FG + \epsilon\frac{\partial F}{\partial N}G + \epsilon F\frac{\partial G}{\partial N}) \\
&\approx (F + \epsilon\frac{\partial F}{\partial N})(G + \epsilon\frac{\partial G}{\partial N}) \\
&\approx F(N + \epsilon)e^{\epsilon\frac{\partial}{\partial N}}G(N), \\
\Rightarrow e^{-i\Theta}F(N)G(N) &= F(N+1)e^{-i\Theta}G(N), \tag{8}
\end{aligned}
$$

by repeated application.

For Fermi-Dirac operators,

$$\Theta = \frac{\pi}{2}\sigma_x, \tag{9}$$

so

$$e^{-i\Theta} = \cos\frac{\pi}{2} - i\sigma_x\sin\frac{\pi}{2} = -i\sigma_x, \tag{10}$$

and

$$N = \frac{1 - \sigma_z}{2} = \begin{pmatrix} 0 & 0 \\ 0 & 1 \end{pmatrix}. \tag{11}$$

Finally, for example,

$$
\begin{aligned}
e^{-i\Theta}N &= (-i\sigma_x)\frac{1-\sigma_z}{2} = \frac{1+\sigma_z}{2}(-i\sigma_x) = (1-N)(-i\sigma_x) \\
&= (1-N)e^{-i\Theta}, \tag{12}
\end{aligned}
$$

and so on.

§3. Jordan and Wigner's Algebraic Developments.

In Jordan and Wigner's Eqn.13 the anticommutation relations for b_r and $b_s^\dagger$ are written down for the first time for a single state $r = s$. Finally, after an analysis based on the antisymmetry of the Heisenberg-Dirac determinantal wave-function for many identical Fermi-Dirac particles, Jordan and Wigner write down the full anticommutation relations. Eqn.40 shows that these also anticommute for creation-annihilation operators for identical Fermi-Dirac particles but different states (here labelled q' and q''). The anticommutation for different Fermi-Dirac particles is not established. A final summary, following Eqn.44, makes the comparison of Fermi-Dirac and Bose-Einstein (anti)commutation relations explicit.

Jordan and Wigner demonstrate that anticommuting Fermi-Dirac operators are a necessity if general unitary transformations of the states are to respect *a priori* phase conventions of the many body determinantal wave functions. Recall that the phases depend upon a choice of the ordering of operator eigenvalues. The phases can be explicitly readjusted as in Eqn.32, and, in this sense, commuting operators $b, b^\dagger$ retained but at the expense of some cumbersome algebra. The compelling alternative exists, however, of the near perfect parallel between the Bose-Einstein case and the Fermi-Dirac case following Eqns.43,44 provided that anticommuting operators are chosen as the basic elements of the Fermi-Dirac quantization.

The unitary transformation from one set of $b's$ to another in Eqn.30

$$b_\alpha(\beta') = \sum_{q',p} \Phi_{\alpha p}(\beta'q')b_p(q'), \qquad (13)$$

(written in more familiar notation as

$$\langle \alpha | \beta' \rangle = \langle \alpha | p \rangle \langle p | q' \rangle \langle q' | \beta' \rangle,$$

with

$$\Phi_{\alpha p}(\beta'q') = \langle \alpha | p \rangle \langle q' | \beta' \rangle$$

and $b_p(q') = \langle p | q' \rangle$) holds without complication in the Bose-Einstein case but not in the Fermi-Dirac case.

The reason is that the effect of $b^\dagger(q')$ on an antisymmetrized state depends on the number of occupied states which 'precede' it in the determinantal wave function of the many body state, following the established order of the 'q'. So on the right there must be a conventional phase

$$(-1)^{\sum N_{q''}} = \prod(1 - 2N_{q''}) \tag{14}$$

(where the sum or product extends over all $q'' \leq q'$), which can be absorbed into the operator $a^\dagger(q')$, or kept explicit in an operator $b^\dagger(q')$ if we choose to commute it through others to its appointed position. The multiplication and transformation properties of the anticommuting operators $(a, a^\dagger)$ are seen in Eqns.30-44 to be much more economical, and in fact a near perfect analog of the Bose-Einstein properties. For these reasons the anticommuting $(a, a^\dagger)$ are considered as the fundamental quantities of the theory of fields satisfying the Pauli Exclusion Principle by way of the antisymmetrized many body wave functions of Heisenberg and Dirac.

The identities following Eqn.34 follow easily. For example,

$$
\begin{aligned}
b(q')v(q'') &= b(q') \prod_{q \leq q''} (1 - 2N_q) \\
&= v(q'')b(q'), \quad \text{trivially for } q' > q'' \\
&= v(q'')(1 - 2N_{q'})b(q')(1 - 2N_{q'}) \quad \text{for } q' \leq q'' \\
&= 0 \text{ operating on states for which } N_{q'}|\rangle = 0 \\
&= v(q'')(1 - 0)b(q')(1 - 2) = -v(q'')b(q') \quad \text{for } N_{q'}|\rangle = 1|\rangle \\
&= -v(q'')b(q') \quad \text{generally, for } q' \leq q''.
\end{aligned}
$$

Most of this effort by Jordan and Wigner seems directed towards becoming comfortable with the novelty of anticommuting operators, and abandoning the classical connection provided by canonical commutation relations. They finally state "$\cdots$ the multiplication rules of the $a, a^\dagger$ not only determine the eigenvalues of the $N(\beta)$, but also remain invariant under a canonical transformation of the matrix representation."

§4. Biographical Note on E.P. Jordan.(†)

Ernst Pascual Jordan (1902-80) owes his given name to his father's Spanish ancestry. He read *Introduction to the Mathematical Methods of Natural Science* by Walther Nernst and Arthur Schönflies when he was only sixteen; graduated from the Technische Hochschule in Hanover (1921), where he attended mathematics lectures by Heinrich Müller and by Georg Prange, and electrical engineering lectures by Friedrich Kohlrausch. Dissatisfied by the low level of the physics offerings, he read Sommerfeld's *Atombau und Spektrallinien* and Moritz Schlick's *Space and Time in Modern Physics* by himself. He moved after two semesters to Göttingen where he soon met Bohr, Courant and Born. His exceptional talents were immediately recognized by Born, who introduced him to the current problems of quantum physics. The next five years of Jordan's life were an explosion of creativity that would vault him into the company of Pauli, Heisenberg, Dirac and Born.

As a student, Jordan assisted Born on a paper on lattice dynamics (1924); helped Courant edit *Methods of Mathematical Physics* (1924); co-authored *Impulsive Excitation of Quantum Oscillators* with Franck (eventually published in 1926). Then, strangely, he submitted a PhD thesis - immediately disproved by Einstein himself - which claimed a continuous spectrum of radiation in Compton scattering.

Undeterred by this setback, he immediately joined Born in developing Heisenberg's idea of transition amplitudes into the idea of matrix mechanics (1925; work which he was to claim in 1964 as substantially his own, due to Born's illness at the time); then, with Born and Heisenberg, the principal paper defining quantum mechanics and first introducing the idea of quantum field theory (1925); again with Heisenberg, the quantum theory of electron spin (1926); simultaneously with Dirac, the relation between the Heisenberg and the Schrödinger forms of quantum mechanics (1926); stated the indeterminacy principle simultaneously with Heisenberg's statement of the Uncertainty Principle (1927); with Klein, *second* quantization applied to the nonrelativistic Bose-Einstein gas (1927); first alone and then with Wigner, he invented the anticommutation relations for second quantization according to the Pauli Exclusion Principle (1928); and then with Pauli, relativistic quantum electrodynamics (1928). And then at twenty six, it was over.

He was *Privatdozent* at Göttingen (1927); Lenz's assistant at Hamburg (1928); professor at Rostock (1929-44); director of the Institute of Theoretical Physics, Berlin (1944-45). He was a Nazi sympathizer, and had an unfortunate flirtation with Nazi authorities as an informant (1936). He was rehabilitated as a visiting professor (1947) and then full professor at Hamburg (1953-71). As a member of the German Bundestag (1957-61), he was an advocate of nuclear deterrence. His later scientific work was in a variety of biological, psychological, physiological, philosophical, political, and cosmological areas but

174

gained little if any acceptance. His invention of nonassociative *Jordan* algebras is notable (1932), although its application to quantum mechanics in a paper with von Neumann and Wigner (1934) seems to have been ignored.

We are reminded of a brief joust (possibly apocryphal and certainly hearsay) between Sydney Coleman (about twenty years old at the time) and Richard Feynman (about forty):

Coleman: Dick, why is it that great theorists do their best work when they're young, and end up doing crap?

Feynman: Well, Sydney, you're doing it the other way around, aren't you.

So who can criticize Jordan.

(† - from Karl von Meyern in *Dictionary of Scientific Biography* (Scribners, New York, 1994), edited by Frederic L. Holmes, pp. 448-454.)

Bibliography and References.

7.1) P. Jordan, Zeits. f. Phys. **44**, 1 (1927).

7.2) P. Jordan, Zeits. f. Phys. **44**, 473 (1927); see our App.7A.

7.3) P. Jordan and E. Wigner, Zeits. f. Phys. **47**, 631 (1928); see our App.7B.

APPENDIX 7.A:

Excerpt from: Zeitschrift für Physik 44, 473 (1927)

On the Quantum Mechanics of the Degenerate Gas

by **P. Jordan**, while in Copenhagen

(Received on 7 July 1927)

It has been shown by Dirac [1], following the prescription of Einstein, that the ideal molecular gas, in analogy to the gas of photons, can be represented by quantized waves in ordinary three dimensional space. In a quantum mechanically exact way and with the formulation given by Dirac [2], a unified discussion can be constructed following Schrödinger's method starting from a representation of eigenfunctions in an abstract space of infinite dimensions. In this paper a corresponding theory for the ideal Fermi-Dirac gas rather than the Bose-Einstein gas will be developed.

§1. Quantization of Schrödinger's Equation.

The purpose of this work is illustrated in the following example. We will start the analysis in §3 with the original work of Dirac [1]. We assume with Dirac, that the quantities occurring there, $-i\hbar\omega b_r^{\dagger}, b_r$ are canonically conjugate or, expressed otherwise, that

$$q_r = \frac{1}{2}(b_r + b_r^{\dagger}),$$
$$p_r = \frac{\hbar\omega}{i}(b_r - b_r^{\dagger}) \tag{1}$$

can be defined, where the real quantities q_r, p_r are conjugate. Here ω is a (real) constant (classical) c-number, which is set to 1 by Dirac [2]; we will leave its value open at first. We further assume with Dirac, that $b_r, b_r^{\dagger}$ can be represented by two conjugate quantities Θ_r, N_r in the form

$$b_r = e^{-i\Theta_r} N_r^{\frac{1}{2}},$$
$$b_r^{\dagger} = N_r^{\frac{1}{2}} e^{i\Theta_r}. \tag{2}$$

As we have recently established [3], it is not necessary to copy

$$q_r p_r - p_r q_r = i\hbar. \tag{3}$$

We have, in fact, a further freedom in the selection of eigenvalues of Θ_r and N_r.

Hypothesis A: The eigenvalues are

$$N'_r = 0, 1, 2, \cdots; \qquad 0 \leq \Theta'_r \leq 2\pi. \tag{4}$$

Then it follows as in [3] that

$$N_r e^{i\Theta_r} - e^{i\Theta_r} N_r = e^{i\Theta_r}, \tag{5}$$

and that with

$$\omega = 1, \tag{6}$$

(3) is satisfied. This is the case of Bose-Einstein statistics discussed by Dirac.

Hypothesis B: The eigenvalues are

$$N'_r = 0, 1; \qquad \Theta'_r = \pm \frac{\pi}{2}. \tag{7}$$

We then set

$$\xi_r = 2N_r - 1 (= -\sigma_z \ \text{[note added]}); \quad \eta_r = \frac{2}{\pi}\Theta_r (= \sigma_x); \quad \zeta_r = i\xi_r \eta_r (= \sigma_y). \tag{8}$$

Θ_r, N_r can be represented for multiplication and addition by quantities β of the form

$$\beta = a_0 + a_1 \xi_r + a_2 \eta_r + a_3 \zeta_r \tag{9}$$

with real c-numbers $a_0, \cdots, a_3$.

The quantities

$$k_1 = i\xi_r, \quad k_2 = i\eta_r, \quad k_3 = i\zeta_r \tag{10}$$

multiply as quaternions with

$$k_1^2 = -1, \quad k_1 k_2 = k_3,$$

and so on.

In (2) we have used $0^2 = 0$, $1^2 = 1$ to replace $N^{\frac{1}{2}}$ by N and write

$$b_r = e^{-i\Theta_r} N_r,$$
$$b_r^{\dagger} = N_r e^{i\Theta_r}. \tag{2}$$

For the exponentials, (7), (8) and (10) give:

$$e^{i\Theta_r} = \cos\frac{\pi}{2} + k_2 \sin\frac{\pi}{2} = k_2, \tag{11}$$

and finally we get

$$b_r = -k_2 N_r = -i\eta_r N_r = -i\frac{2}{\pi}\Theta_r N_r = -\frac{ik_3 + k_2}{2} = -i\begin{pmatrix} 0 & 1 \\ 0 & 0 \end{pmatrix},$$
$$b_r^{\dagger} = N_r k_2 = iN_r \eta_r = i\frac{2}{\pi}N_r \Theta_r = -\frac{ik_3 - k_2}{2} = i\begin{pmatrix} 0 & 0 \\ 1 & 0 \end{pmatrix} \tag{12}$$

and

$$q_r = \zeta_r,$$
$$p_r = \omega \hbar \eta_r; \tag{13}$$

these equations show that q_r, p_r will actually be conjugate if we set $\omega = \pi$. (Note added: Conjugate in the sense developed in his paper [3], that for $q \sim \sigma_x$ then a "conjugate" momentum $p \sim \sigma_y$ is the only possibility, necessarily requiring that we abandon the canonical commutation relation as the only criterion.)

Following Dirac, we next write the Hamiltonian

$$H = \sum_{rs} H_{rs} b_r^\dagger b_s. \tag{14}$$

The meaning of the variables is the same as in Dirac's Eqn.11. For the Case B we obtain from (14):

$$H = \left(\frac{2}{\pi}\right)^2 \sum_{rs} H_{rs} N_r \Theta_r \Theta_s N_s. \tag{15}$$

Next we write the wave equation

$$\left\{ H - i\hbar \frac{\partial}{\partial t} \right\} \psi(N_1', N_2', \cdots) = 0. \tag{16}$$

The matrix representation of the operators obtained in [3] is

$$\xi_r = \begin{pmatrix} -1 & 0 \\ 0 & 1 \end{pmatrix}, \quad \eta_r = \begin{pmatrix} 0 & 1 \\ 1 & 0 \end{pmatrix};$$
$$N_r = \begin{pmatrix} 0 & 0 \\ 0 & 1 \end{pmatrix}, \quad \Theta_r = \frac{\pi}{2} \begin{pmatrix} 0 & 1 \\ 1 & 0 \end{pmatrix}. \tag{17}$$

Therefore we obtain (16) symbolically in the form

$$\left\{ \sum_{rs} H_{rs} \begin{pmatrix} 0 & 0 \\ 0 & 1 \end{pmatrix}_r \begin{pmatrix} 0 & 1 \\ 1 & 0 \end{pmatrix}_r \begin{pmatrix} 0 & 1 \\ 1 & 0 \end{pmatrix}_s \begin{pmatrix} 0 & 0 \\ 0 & 1 \end{pmatrix}_s - i\hbar \frac{\partial}{\partial t} \right\} \psi = 0. \tag{18}$$

We will now describe the Fermi-Dirac gas by eigenfunctions of the kind given by Dirac [2] and Heisenberg [4]. Following Dirac, the corresponding Schrödinger equation is

$$\sum_{s_1, s_2, \cdots} H(r_1, r_2, \cdots; s_1, s_2, \cdots) \phi(s_1, s_2, \cdots) - i\hbar \frac{\partial}{\partial t} \phi(r_1, r_2, \cdots) = 0. \tag{19}$$

or

$$\sum_m \sum_{s_m \neq r_m} H_{r_m s_m} \phi(r_1, r_2, \cdots, r_{m-1}, s_m, r_{m+1}, \cdots)$$
$$+ \sum_n H_{r_n r_n} \phi(r_1, r_2, \cdots) \quad - \quad i\hbar \frac{\partial}{\partial t} \phi(r_1, r_2, \cdots) = 0. \tag{20}$$

178

These are the equations (14) and (15) in §3 of the earlier work of Dirac [1]. We have used the variable ϕ in place of the b of Dirac; otherwise all symbols have the meaning stated by Dirac. The functions ϕ should be antisymmetric; therefore all s in $\phi(s_1, s_2, \cdots)$ are different from one another. We order the numbers $s_1, s_2 \cdots$, which can assume the values $1, 2, 3, \cdots$ to the other numbers $N_1', N_2', N_3', \cdots$ in such a way, that $N_k' = 0$ when no s_j is equal to k, $N_k' = 1$ when one s_j is equal to k. Then we confirm that

$$\pm \phi(s_1, s_2, \cdots) = \psi(N_1', N_2', \cdots), \tag{21}$$

where ψ is the function defined in (16). This prescription in the theory of the Fermi-Dirac gas is the analog of that for the Bose-Einstein gas in §3 of Dirac's work.

We designate that function of $N_1', N_2', \cdots$ which corresponds to $\pm\phi(s_1, s_2, \cdots)$ - which we above wrote as $\psi(N_1', N_2', \cdots)$ - as $\phi(N_1', N_2', \cdots)$. The term

$$\pm H_{r_m s_m} \phi(r_1 \cdots r_{m-1} s_m r_{m+1} \cdots)$$

in (20) is then the same as Dirac's term

$$H_{rs} \phi(N_1', N_2', \cdots, N_r' - 1, N_s' + 1, \cdots), \tag{22}$$

when we abbreviate r_m, s_m as r, s. Then necessarily

$$N_r' = 1, \quad N_s' = 0.$$

The summation $\sum_m \sum_{s_m \neq r_m}$ indicates that we should sum the expression (22) over all r different from s, for which $N_s' = 0$. The sum therefore gets contributions only from those r for which $N_r' = 1$. We can now see that this sum can be written as

$$\sum_{r \neq s} H_{rs} \begin{pmatrix} 0 & 0 \\ 1 & 0 \end{pmatrix}_r \begin{pmatrix} 0 & 1 \\ 0 & 0 \end{pmatrix}_s \phi(N_1', N_2', \cdots). \tag{23}$$

Further, the sum $\pm \sum_n H_{r_n r_n} \phi(r_1, r_2, \cdots)$ in (20) can be expressed as

$$\sum_r H_{rr} \begin{pmatrix} 0 & 0 \\ 0 & 1 \end{pmatrix}_r \phi(N_1', N_2', \cdots) = \sum_r H_{rr} \begin{pmatrix} 0 & 0 \\ 1 & 0 \end{pmatrix}_r \begin{pmatrix} 0 & 1 \\ 0 & 0 \end{pmatrix}_r \phi(N_1', N_2', \cdots). \tag{24}$$

These remarks show the equivalence of (18) and (20) to the corresponding expressions for the functions $\phi(N')$ and $\psi(N')$. Then (18) can also be written in the form

$$\left\{ \sum_{rs} H_{rs} \begin{pmatrix} 0 & 0 \\ 1 & 0 \end{pmatrix}_r \begin{pmatrix} 0 & 1 \\ 0 & 0 \end{pmatrix}_s - i\hbar \frac{\partial}{\partial t} \right\} \psi = 0. \tag{25}$$

§2. Density Fluctuations of the Ideal Gas.

We will apply our method to an investigation of the fluctuation properties of the Fermi-Dirac gas. For simplicity we will treat a one dimensional gas and give the gas

179

atoms zero rest mass. When we ultimately set the velocity of light equal to one, then we have a system which differs from the earlier considered fluctuation properties of a vibrating string [5] only through the application of Fermi-Dirac rather than Bose-Einstein statistics. We will calculate fluctuations for the Bose-Einstein as well as for the Fermi-Dirac system, in order to demonstrate the difference. The method known from the Bose-Einstein case is mathematically equivalent with that required, and can be transferred almost immediately to the Fermi-Dirac case.

According to Eqn.2 of §1, in both cases

$$N_r = b_r^\dagger b_r. \tag{1}$$

We now represent the displacement of the string of length l as

$$u(x,t) = \sum_{r=1}^{\infty} b_r \sin \frac{r\pi x}{l}. \tag{2}$$

Then we have (l, x are c-numbers)

$$\frac{2}{l} \int_0^l u^\dagger u \, dx = \sum_{r=1}^{\infty} N_r \tag{3}$$

which equals the total number of quanta of the string; correspondingly we denote

$$N(x_1, x_2) = \frac{2}{l} \int_{x_1}^{x_2} u^\dagger u \, dx \tag{4}$$

as the number of quanta between (x_1, x_2). The number $N(0, a)$ we will abbreviate N; we get

$$
\begin{aligned}
\Delta &= N(0, a) - \overline{N(0, a)} = N - \overline{N} \\
&= \frac{2}{l} \int_0^a \sum_{r \neq s, =1}^{\infty} b_r^\dagger b_s \sin \frac{r\pi x}{l} \cdot \sin \frac{s\pi x}{l} \cdot dx \\
&= \frac{1}{l} \sum_{r \neq s, =1}^{\infty} b_r^\dagger b_s K_{rs} = \frac{1}{l} \sum_{r \neq s, =1}^{\infty} b_s b_r^\dagger K_{rs};
\end{aligned}
\tag{5}
$$

where

$$K_{rs} = \frac{\sin(\omega_r - \omega_s)a}{\omega_r - \omega_s} - \frac{\sin(\omega_r + \omega_s)a}{\omega_r + \omega_s} \tag{6}$$

where $\omega_r = r\pi/l$ and so on. Furthermore

$$\Delta^2 = \frac{1}{l^2} \sum_{r \neq s, =1}^{\infty} \sum_{t \neq u, =1}^{\infty} b_r^\dagger b_s b_u b_t^\dagger K_{rs} K_{tu}, \tag{7}$$

and the mean value is

$$\overline{\Delta^2} = \frac{1}{l^2} \sum_{r \neq s, =1}^{\infty} \left\{ \overline{b_r^\dagger b_s^2 b_r^\dagger} + \overline{b_r^\dagger b_s b_r b_s^\dagger} \right\} K_{rs}^2. \tag{8}$$

180

For large string lengths l one can replace the sums by integrals and get:

$$\overline{\Delta^2} = \frac{1}{\pi^2} \int_0^\infty \int_0^\infty d\omega_r d\omega_s \left\{ \overline{b_r^\dagger b_s^2 b_r^\dagger} + \overline{b_r^\dagger b_s b_r b_s^\dagger} \right\} K_{rs}^2. \tag{9}$$

Here, because a is also assumed large, we get the formula

$$\lim_{a \to \infty} \int_{-\Omega}^{\Omega'} \frac{\sin^2 \omega a}{\omega^2} f(\omega) d\omega = \pi f(0) \tag{10}$$

for $\Omega, \Omega' > 0$. It gives

$$\overline{\Delta^2} = \frac{2a}{\pi} \int_0^\infty d\omega_r \overline{b_r^\dagger b_r^2 b_r^\dagger}. \tag{11}$$

In order to evaluate these expressions set

a) in the Bose-Einstein case

$$\begin{aligned}
b_r &= e^{-i\Theta_r} N_r^{\frac{1}{2}} = (1 + N_r)^{\frac{1}{2}} e^{-i\Theta_r}, \\
b_r^\dagger &= N_r^{\frac{1}{2}} e^{i\Theta_r} = e^{i\Theta_r} (1 + N_r)^{\frac{1}{2}};
\end{aligned} \tag{12}$$

b) in the Fermi-Dirac case

$$\begin{aligned}
b_r &= e^{-i\Theta_r} N_r^{\frac{1}{2}} = (1 - N_r)^{\frac{1}{2}} e^{-i\Theta_r}, \\
b_r^\dagger &= N_r^{\frac{1}{2}} e^{i\Theta_r} = e^{i\Theta_r} (1 - N_r)^{\frac{1}{2}}
\end{aligned} \tag{13}$$

which follow from §1. One has only to note that

$$N_r^{\frac{1}{2}} = N_r, \qquad (1 - N_r)^{\frac{1}{2}} = 1 - N_r \tag{14}$$

for the Fermi-Dirac case. If one places less emphasis on the analogy between (12) and (13), then instead of (13) one might write

$$\begin{aligned}
b_r &= -i\frac{2}{\pi} \Theta_r N_r = -i\frac{2}{\pi}(1 - N_r)\Theta_r, \\
b_r^\dagger &= i\frac{2}{\pi} N_r \Theta_r = i\frac{2}{\pi} \Theta_r (1 - N_r); \quad \Theta_r^2 = \frac{\pi^2}{4}.
\end{aligned} \tag{15}$$

Applying (12) and (13) to the integrands in (11) yields

a) for Bose-Einstein statistics

$$b_r^\dagger b_r^2 b_r^\dagger = N_r(1 + N_r); \tag{16}$$

b) for Fermi-Dirac statistics

$$b_r^\dagger b_r^2 b_r^\dagger = N_r(1 - N_r). \tag{17}$$

181

Eqn.16 in fact is just another expression of the known Einstein fluctuation equation for Bose-Einstein statistics. In (17), on the other hand, we have obtained the corresponding expression for Fermi-Dirac statistics - in agreement with the result which was obtained by Pauli [6] from a thermodynamical-statistical argument.

§3. Concluding Remarks.

The considerations of §1,2 apply to the ideal gas without interactions between the gas atoms. If one also wishes to include an interaction of the particles with the method developed by Dirac comparable to §1,2, then one must replace the free Hamiltonian H in the wave equation (16) of §1, in which $b, b^\dagger$ appear quadratically - corresponding to linear wave equations - by a general function; I hope soon to come back to this problem. The preceding discussion, however, already gives the proof that the probability distribution for all impulsive interactions in the Fermi-Dirac gas are treated correctly by our method. Then this probability distribution, whose form corresponds to the thermodynamical-statistical result [7], differs from the corresponding result for the Bose-Einstein gas by the replacement of factors $(1 + N_r)$ by $(1 - N_r)$. The same difference occurs in Eqns.16,17 of §2; its origin lies in the corresponding difference between the basic equations (12) and (13) of §2, of the Bose-Einstein and the Fermi-Dirac systems.

The end result is that - in spite of the applicability of Fermi-Dirac statistics rather than Bose-Einstein statistics for electrons - a quantum mechanical wave theory of matter can be constructed, in which the electrons are represented by quantized waves in ordinary three-dimensional space. The natural formulation of the quantum theoretic electron is such that light and matter are treated simultaneously as interacting waves in three dimensional space [8]. The basic fact of the electron theory, the existence of discrete electrical particles, manifests itself as a characteristic quantum phenomenon, and indeed, equally important, the matter waves appear only in discrete quantum states. The Schrödinger eigenfunctions of material particles play a full role in the realm of these ideas constructed by Dirac and Heisenberg, in which they appear not just as an analog of the electromagnetic waves. Rather they appear as a special case of the general probability amplitudes, which are to be used as a mathematical tool for the description of the statistical behavior of the quanta of light- and matter-waves.

I am heartily thankful to Professor Bohr for stimulating conversations about problems of quantum theory. I also wish to thank the International Education Board for the award of a fellowship in Copenhagen.

Footnotes and References:

1) P.A.M. Dirac, Proc. Roy. Soc. A**112**, 661 (192).

2) P.A.M. Dirac, ibid **114**, 243 (1927).

3) P. Jordan, Zeits. f. Phys. **44**, 1 (1927).

4) W. Heisenberg, Zeits. f. Phys. **38**, 411 (1926).

5) M. Born, W. Heisenberg and P. Jordan, Zeits. f. Phys. **35**, 557, (1925).

6) W. Pauli jr., Zeits. f. Phys. **41**, 81 (1927).

7) P. Jordan, Zeits. f. Phys. **41**, 711 (1927); L.S. Ornstein and H.A. Kramers, ibid **42**, 481 (1927).

8) That a theory of the system with many identical material particles could be derived by quantization of the Schrödinger equation is, to the author, already implied by the conclusions of the first Schrödinger paper. However it seems that at the time, the Pauli exclusion principle was a serious obstacle for this perception.

APPENDIX 7.B:

Excerpt from: Zeitschrift für Physik **47**, 631 (1928)

On the Pauli Exclusion Principle

by **P. Jordan** and **E. Wigner**, in Copenhagen

(Received on 26 January 1928)

This work is a continuation of the short note "On the Quantum Mechanics of the Degenerate Gas" [1], written by one of the authors, whose results will be essentially extended. It concerns an ideal or non-ideal gas subject to the Pauli Exclusion Principle, to be described by ideas which make no reference to the abstract coordinate space of all the atoms of the gas, but only use the ordinary three dimensional space. That is made possible by means of a three dimensional wave field, where the particular non-commutative properties of the wave amplitudes are responsible both for the existence of the quanta of the gas atoms and for the applicability of the Pauli Exclusion Principle. The details of the theory permit analogies to the corresponding theory for the Bose-Einstein gas, as developed by Dirac and by Klein and Jordan.

§1. Introduction.

During the first investigations into the matrix theory of quantum mechanics, there already appeared hints that the obstacles to radiation theory could be overcome by applying quantum mechanical methods [2] not only to the atoms, but also to the radiation field. There are a number of recent publications [3,4,5,6] which deal both with a quantum mechanical description of the electromagnetic field, and with a formulation of the quantum mechanics of material particles. Here the wave description in an abstract coordinate space is avoided in favor of a representation by quantum mechanical waves in ordinary three dimensional space. Also the existence of material particles is explained in a way similar to that by which the quantization of electromagnetic waves explains the existence of photons, which in turn are required to explain the physical situation.

For this description, one writes the quantity N_r, which in particle language is the number of atoms in the rth quantum state, as a product of two q-number quantum operators

$b, b^\dagger$ in the form

$$N_r = b_r^\dagger b_r, \tag{1}$$

with

$$
\begin{aligned}
b_r &= e^{-i\Theta} N_r^{\frac{1}{2}}, \\
b_r^\dagger &= N_r^{\frac{1}{2}} e^{i\Theta},
\end{aligned} \tag{2}
$$

where one requires that N_r and Θ_r are canonically conjugate quantities. If one takes the fundamental definition of canonically conjugate proposed by one of the authors [7], then one obtains a new possibility. It is possible to represent in this form not only Bose-Einstein statistics for which the eigenvalues N_r' of N_r are

$$N_r' = 0, 1, 2, 3, \cdots, \tag{3}$$

but also Fermi-Dirac statistics for which only

$$N_r' = 0, 1 \tag{4}$$

are possible.

Besides (2), one gets other equations. For the Bose-Einstein case

$$
\begin{aligned}
b_r &= (1 + N_r)^{\frac{1}{2}} e^{-i\Theta_r}, \\
b_r^\dagger &= e^{i\Theta_r} (1 + N_r)^{\frac{1}{2}};
\end{aligned} \tag{5}
$$

but, in place of these, in the Fermi-Dirac case

$$
\begin{aligned}
b_r &= (1 - N_r)^{\frac{1}{2}} e^{-i\Theta_r}, \\
b_r^\dagger &= e^{i\Theta_r} (1 - N_r)^{\frac{1}{2}},
\end{aligned} \tag{5}
$$

as was shown earlier [1].

These formulas are based on the conviction that this way of representing the Pauli Exclusion Principle is the essence of the subject, and its further pursuit will lead to correct results. Eqns.5 are closely related to both the problem of interactions of identical particles, and the problem of density fluctuations of a quantum mechanical gas.

§2. Effect on Interactions.

First, what concerns the effect of possible interactions is repeated from an earlier note [8]: In an isolated volume let there be different kinds of particles (matter-quanta or photons). The density of the lth particle type per cell in phase space is $n^l(E)$, where E is the energy corresponding to the particular cell. The total number N^l (integral of $n^l(E)$ over the phase space) must satisfy many $(j = 1, 2, \cdots)$ linear supplementary conditions

$$\sum_l C_{jl} N^l = C_j = \text{const.} \tag{6}$$

185

(which are described elsewhere), where the C_{jl} are positive or negative integers, or zero. In statistical equilibrium

$$n^l(E) = \frac{1}{e^{\sum_j C_{jl}\alpha_j(T) + E/kT} \pm 1},$$ (7)

where the $\pm$ sign depends on whether the lth particle type satisfies Fermi-Dirac statistics (+) or Bose-Einstein statistics (-).

Only those interaction processes are permitted which do not violate the restrictions (6). A particular form of the interactions is described by the indices l and the rates of the elementary processes which occur and which particles take part. The values $n_1^+, n_2^+, \cdots, n_r^+$ of the corresponding densities $n^l(E)$ of the Bose-Einstein particles which occur in each of the processes must be given; and also the $n_1^-, n_2^-, \cdots, n_s^-$ of the $n^l(E)$ for all the Fermi-Dirac particles which take part. Also the numbers $m_1^+, \cdots, m_\rho^+; m_1^-, \cdots, m_\sigma^-$ of all the particles produced by the interactions are required. Then on the basis of thermodynamical-statistical equilibrium the probability of the elementary interaction must be assumed proportional to

$$\prod_{i=1}^{r} n_i^+ \prod_{j=1}^{s} n_j^- \prod_{k=1}^{\rho} (1 + m_k^+) \prod_{h=1}^{\sigma} (1 - m_h^-);$$ (8)

and the inverse elementary interaction is proportional to

$$\prod_{i=1}^{\rho} m_i^+ \prod_{j=1}^{\sigma} m_j^- \prod_{k=1}^{r} (1 + n_k^+) \prod_{h=1}^{s} (1 - n_h^-).$$ (9)

The existence of the factors $(1 + m^+)$ etc. for the Bose-Einstein particles has been proven by Dirac by calculation of the absorption and emission of light by atoms. Their form necessarily follows from the rule for the corresponding factors in Eqn.5. In the same way, the factors $(1 - m^-)$ etc. for the Fermi-Dirac particles can also be traced back to (5).

On the other hand, as determined from the fluctuation phenomenon in [1], the mean square fluctuation of the particle density given by the Einstein formula is proportional to

$$n_r(1 + n_r).$$ (10)

(For classical waves it is proportional to n_r^2.) Pauli calculated an analogous quantity for the Fermi-Dirac gas as proportional to

$$n_r(1 - n_r);$$ (10)

Eqns.10 show the difference between the Bose-Einstein gas and the Fermi-Dirac gas again in the same form as in Eqns.5.

For the Bose-Einstein gas or the Bose-Einstein waves one already has a far-reaching insight [3,4]. In the following, we explore in a similar way the theory of the Fermi-Dirac gas which was begun in [1].

§3. Comments on the Fermi-Dirac Case.

For clarity, we repeat some formulas from [1]. The quantities $b_r, b_r^\dagger, N_r, \Theta_r$ are represented by the matrices

$$
\begin{aligned}
b_r &= \begin{pmatrix} 0 & 1 \\ 0 & 0 \end{pmatrix}_r ; \quad b_r^\dagger = \begin{pmatrix} 0 & 0 \\ 1 & 0 \end{pmatrix}_r ; \\
N_r &= \begin{pmatrix} 0 & 0 \\ 0 & 1 \end{pmatrix}_r ; \quad \Theta_r = \frac{\pi}{2} \begin{pmatrix} 0 & 1 \\ 1 & 0 \end{pmatrix}_r .
\end{aligned}
\tag{11}
$$

Here, for the time being,

$$
\begin{pmatrix} \alpha_{11} & \alpha_{12} \\ \alpha_{21} & \alpha_{22} \end{pmatrix}_r
\tag{12}
$$

is a matrix whose elements take the values 0 or 1; (12) is a unit matrix with respect to all indices except r. In addition to those equations already discussed in §1, we have

$$
\begin{aligned}
b_r^\dagger b_r &= N_r; \quad b_r b_r^\dagger = 1 - N_r; \\
N_r^2 &= N_r; \quad (b_r^\dagger)^2 = (b_r)^2 = 0; \\
e^{i\Theta_r} &= i\frac{2}{\pi}\Theta_r.
\end{aligned}
\tag{13}
$$

All these quantities can be re-expressed in terms of the three quantities $k_1^{(r)}, k_2^{(r)}, k_3^{(r)}$ which follow the multiplication rules of quaternions:

$$
\begin{aligned}
k_1^{(r)} k_2^{(r)} &= -k_2^{(r)} k_1^{(r)} = k_3^{(r)}, \quad \text{and cyclic 1,2,3;} \\
\left(k_1^{(r)}\right)^2 &= \left(k_2^{(r)}\right)^2 = \left(k_3^{(r)}\right)^2 = -1.
\end{aligned}
\tag{14}
$$

To be specific:

$$
\begin{aligned}
b_r &= \frac{k_3^{(r)} - ik_2^{(r)}}{2}, \quad b_r^\dagger = -\frac{k_3^{(r)} + ik_2^{(r)}}{2}; \\
N_r &= -\frac{ik_1^{(r)} - 1}{2}; \quad \Theta_r = -\frac{\pi}{2} ik_2^{(r)}.
\end{aligned}
\tag{15}
$$

The quaternions k_1, k_2, k_3 can be represented by the matrices

$$
k_1^{(r)} = \begin{pmatrix} -i & 0 \\ 0 & i \end{pmatrix}; \quad k_2^{(r)} = \begin{pmatrix} 0 & i \\ i & 0 \end{pmatrix}; \quad k_3^{(r)} = \begin{pmatrix} 0 & 1 \\ -1 & 0 \end{pmatrix}.
\tag{16}
$$

§4. The Heisenberg-Dirac Determinantal Formula.

Just as in [5] for the Bose-Einstein states, for the Fermi-Dirac states one can represent the Heisenberg-Dirac determinantal formula for the Schrödinger wavefunction - antisymmetric under the interchange of individual atoms - by the relevant probability amplitudes.

Such an amplitude for a single atom would be given by

$$\Phi^{\beta q}_{\alpha p} = \Phi_{\alpha p}(\beta', q'). \tag{17}$$

In order to make the sign (which is to be found from the determinant) unambiguous, we put the eigenvalues β' of β in an arbitrary but definite order. We indicate this ordering for two specific eigenvalues β', β'' of β by $\beta' < \beta''$ where here "$<$" means "comes before" (not, as usual, "less than"), and so on. In this way we proceed to order the eigenvalues of each observable quantity for each individual atom. Then one can order each amplitude for the individual atom in a definite way in an antisymmetrized amplitude $\Psi^{\beta q}_{\alpha p}$ for a system of N energetically uncoupled particles. We write

$$\Psi^{\beta q}_{\alpha p} = \Psi_{\alpha p}(\beta^{(1)}, \cdots, \beta^{(N)}; q^{(1)}, \cdots, q^{(N)}), \tag{18}$$

where

$$\begin{aligned}
\beta^{(1)} &< \beta^{(2)} < \cdots < \beta^{(N)}, \\
q^{(1)} &< q^{(2)} < \cdots < q^{(N)};
\end{aligned} \tag{19}$$

and then

$$\Psi^{\beta q}_{\alpha p} = \frac{1}{N!} \sum_n \epsilon_n \prod_{k=1}^{N} \Phi_{\alpha p}(\beta_k, q_k). \tag{20}$$

The sum is over the $N!$ permutations $n_1, \cdots, n_N$ of the numbers $1, \cdots, N$, and $\epsilon_n = +1$ for even permutations and -1 for odd.

According to (20) $\Psi^{\beta q}_{\alpha p}$ vanishes whenever two quantities $\beta^{(k)}$ are equal to one another. That says physically: A non-degenerate quantity β can never simultaneously have the same value for two indistinguishable Fermi-Dirac particles. We must understand β to be the system of quantum numbers which gives the Pauli Exclusion Principle in its original form. We shall consider in the following, the simultaneous existence of these states for all quantities β as the essential content of the Pauli Exclusion Principle.

We consider the case where each β has eigenvalues of finite range, say K. Occasionally we will let $K \rightarrow \infty$. The K eigenvalues of each quantity β are numbered as $\beta'_1, \beta'_2, \cdots, \beta'_K$ in such a way that the required ordering takes the form

$$\beta'_1 < \beta'_2 < \cdots < \beta'_K. \tag{21}$$

§5. The Antisymmetric Amplitudes.

The antisymmetric amplitudes defined in such a way can now be represented as functions of arguments

$$N'(\beta') \quad \text{or} \quad N'(q') \tag{22}$$

188

with the following meaning: $N'(\beta')$ is the number of atoms for which the operator β has the value β'; if β' is a discrete eigenvalue, then according to the Pauli Exclusion Principle

$$N'(\beta') = 0 \quad \text{or} \quad 1. \tag{23}$$

If β' lies in a continuum of eigenvalues, then we have to write:

$$N'(\beta') = \sum_{k=1}^{N'} \delta(\beta' - \beta'_k), \tag{24}$$

when N' particles are involved; The integral of $N'(\beta')$ over a part of the spectrum is then the number of atoms for which the value of β falls in this part of the spectrum.

However, we are not satisfied with the purely mathematical introduction of the new quantities $N'(\beta'), N'(q')$, but turn to a new physical theory in which we assume that the entire gas can be described by a canonical system of q-number operators

$$N(\beta') \quad \text{and} \quad \Theta(\beta'), \tag{25}$$

where the $N'(\beta')$ are the eigenvalues of $N(\beta')$. Then the $N(\beta'), \Theta(\beta')$ can be represented by matrices in the way explained in §3; the different eigenvalues β' correspond to the different values which the indices 'r' in §3 can assume. In particular, for discrete β' the equation

$$N(\beta') \left[1 - N(\beta')\right] = 0 \tag{26}$$

holds; for non-discrete β' one writes

$$\begin{aligned} N(\beta') \left[\delta(\beta' - \beta'') - N(\beta'')\right] &= -N(\beta')N(\beta''), \quad \beta' \neq \beta'', \\ &= 0, \quad \beta' = \beta''. \end{aligned} \tag{27}$$

Although the q-numbers $N(\beta')$ are completely defined by their physical significance, the same is not true for the $\Theta(\beta')$ which we only require to be canonically conjugate to the $N(\beta')$. Naturally we would like to remove or reduce this non-uniqueness by introducing definite relations for the operators $N(\beta'), \Theta(\beta')$ and $N(q'), \Theta(q')$. We will see that it is possible to define the Θ conjugate to the given N, and still have the distinct possibilities of representing, as before, a Bose-Einstein gas or, as here, a Fermi-Dirac gas.

These relations have a close analogy to the relations in the Bose-Einstein case derived earlier [6], in so far as one can expect an analogy in view of the profound difference between the two cases.

§6. Comparison of Bose-Einstein and Fermi-Dirac Cases.

The $2K$ quantities

$$N(\beta'), \Theta(\beta') \tag{28}$$

189

must certainly be q-number functions of the q-numbers

$$N(q'), \Theta(q'); \tag{29}$$

these functional relations will now be discussed. In the Bose-Einstein case one has simply

$$
\begin{aligned}
b_\alpha(\beta') &= \sum_{q',p} \Phi_{\alpha p}(\beta', q') b_p(q'), \\
b_\alpha^\dagger(\beta') &= \sum_{q',p} b_p^\dagger(q') \Phi_{p\alpha}(q', \beta'),
\end{aligned} \tag{30}
$$

but these formulas do not hold for the Fermi-Dirac gas. Instead one has the formulas

$$
\begin{aligned}
a_\alpha(\beta') &= \sum_{q',p} \Phi_{\alpha p}(\beta', q') a_p(q'), \\
a_\alpha^\dagger(\beta') &= \sum_{q',p} a_p^\dagger(q') \Phi_{p\alpha}(q', \beta')
\end{aligned} \tag{30}
$$

where we define the quantities $a, a^\dagger$ by

$$
\begin{aligned}
a_p(q') &= v(q') \cdot b_p(q'), \\
a_p^\dagger(q') &= b_p^\dagger(q') \cdot v(q');
\end{aligned} \tag{31}
$$

$$v(q') = \prod_{q'' \le q'} \left\{ 1 - 2N(q'') \right\}. \tag{32}$$

Here $v(q')$ is the product of all the quantities $\{1 - 2N(q'')\}$ for all q'' coming before q'. Therefore $v(q')$ is a diagonal matrix with matrix elements which are either $+1$ or -1; and one has

$$[v(q')]^2 = 1. \tag{33}$$

The complete mathematical proof of the validity of Eqn.30a will be given in §8 and §9. Here we just investigate the multiplication properties of a and $a^\dagger$, and the invariance of these properties under certain transformations (30a).

First we have

$$
\begin{aligned}
b_p(q') \cdot v(q'') &= -v(q'') \cdot b_p(q') \quad \text{for} \quad q' \le q'', \\
&= +v(q'') \cdot b_p(q') \quad \text{for} \quad q' > q''; \\
b_p^\dagger(q') \cdot v(q'') &= -v(q'') \cdot b_p^\dagger(q') \quad \text{for} \quad q' \le q'', \\
&= +v(q'') \cdot b_p^\dagger(q') \quad \text{for} \quad q' > q''.
\end{aligned} \tag{34}
$$

The proof follows easily since, for example,

$$
\begin{aligned}
b_p^\dagger(q') &= N(q') e^{i\Theta_p(q')} \left\{ 1 - 2N(q') \right\} \\
&= -\left\{ 1 - 2N(q') \right\} b_p^\dagger(q').
\end{aligned} \tag{35}
$$

190

Then further

$$
\begin{aligned}
a_p(q')a_p(q'') &= -a_p(q'')a_p(q'), \\
a_p^\dagger(q')a_p^\dagger(q'') &= -a_p^\dagger(q'')a_p^\dagger(q').
\end{aligned}
\tag{36}
$$

One proves for example

$$
\begin{aligned}
a_p(q')a_p(q'') &= v(q')b_p(q)v(q'')b_p(q'') \\
&= -v(q')v(q'')b_p(q')b_p(q'') \quad \text{for} \quad q' \le q'', \\
&= +v(q')v(q'')b_p(q')b_p(q'') \quad \text{for} \quad q' > q'';
\end{aligned}
\tag{37}
$$

now however, from (36), for $q'' = q'$:

$$
[a_p(q')]^2 = -[a_p(q')]^2 \equiv 0,
\tag{38}
$$

and from (37), for $q'' = q'$ also:

$$
[b_p(q')]^2 = 0.
\tag{39}
$$

One sees also from (37) that the product (36) in fact is antisymmetric in q', q''.

Furthermore, we get

$$
a_p^\dagger(q')a_p(q'') + a_p(q'')a_p^\dagger(q') = \delta(q' - q''),
\tag{40}
$$

because the left hand side will be

$$
\begin{aligned}
&b^\dagger(q')v(q')v(q'')b(q'') + v(q'')b(q'')b^\dagger(q')v(q') \\
&= +v(q')v(q'')\left[b^\dagger(q')b(q'') - b(q'')b^\dagger(q')\right] \quad \text{for} \quad q' \le q''; \\
&= -v(q')v(q'')\left[b^\dagger(q')b(q'') - b(q'')b^\dagger(q')\right] \quad \text{for} \quad q' > q'';
\end{aligned}
\tag{41}
$$

therefore it vanishes in the case $q' \ne q''$, and for $q' = q''$ is equal to

$$
[v(q')]^2 \cdot \left[b_p^\dagger(q')b_p(q') - b_p(q')b_p^\dagger(q')\right] = 1.
\tag{42}
$$

Next we show that Eqns.36 and 40 are invariant under transformations (30a). They give

$$
\begin{aligned}
&a_\alpha(\beta')a_\alpha(\beta'') + a_\alpha(\beta'')a_\alpha(\beta') \\
&= \sum_{q',q'',p} \Phi_{\alpha p}(\beta', q')\Phi_{\alpha p}(\beta'', q'') \cdot \left\{a_p(q')a_p(q'') + a_p(q'')a_p(q')\right\} = 0;
\end{aligned}
\tag{43}
$$

$$
\begin{aligned}
&a_\alpha^\dagger(\beta')a_\alpha(\beta'') + a_\alpha(\beta'')a_\alpha^\dagger(\beta') \\
&= \sum_{q',q'',p} \Phi_{p\alpha}(q', \beta')\Phi_{\alpha p}(\beta'', q'') \cdot \left\{a_p^\dagger(q')a_p(q'') + a_p(q'')a_p^\dagger(q')\right\} = \delta(\beta' - \beta'').
\end{aligned}
\tag{44}
$$

191

Therefore, the quantities $b, b^\dagger$ in the Fermi-Dirac case just as in the Bose-Einstein case have the property that $b(\beta')$ commutes with $b(\beta'')$ and with $b^\dagger(\beta'')$ for $\beta' \neq \beta''$. The $a, a^\dagger$ do not possess this property. In spite of this fact, the $a, a^\dagger$ of the Fermi-Dirac gas have a close analogy to the $b, b^\dagger$ of the Bose-Einstein gas (as indeed to the $b, b^\dagger$ of the Fermi-Dirac gas itself). One sees this analogy particularly clearly from the comparison between

a)the Bose-Einstein case:

$$b_\alpha^\dagger(\beta')b_\alpha(\beta'') - b_\alpha(\beta'')b_\alpha^\dagger(\beta') = \delta(\beta' - \beta'');$$
$$b_\alpha^\dagger(\beta')b_\alpha(\beta') = N_\alpha(\beta');$$
$$b_\alpha(\beta') = \sum_{q',p} \Phi_{\alpha p}(\beta', q')b_p(q');$$

and b) the Fermi-Dirac case:

$$a_\alpha^\dagger(\beta')a_\alpha(\beta'') + a_\alpha(\beta'')a_\alpha^\dagger(\beta') = \delta(\beta' - \beta'');$$
$$a_\alpha^\dagger(\beta')a_\alpha(\beta') = N_\alpha(\beta');$$
$$a_\alpha(\beta') = \sum_{q',p} \Phi_{\alpha p}(\beta', q')a_p(q').$$

We have derived these equations in order from the outset to lay down the foundation of the Pauli Exclusion Principle. It happens however that conversely, these multiplication properties of the $a, a^\dagger$ determine the possible eigenvalues of $N(\beta')$ and the commutability (simultaneous observability) of $N(\beta')$ and $N(\beta'')$ as well. We can say that the existence of material particles and the validity of the Pauli Exclusion Principle can be understood as a consequence of the quantum mechanical multiplication properties of the deBroglie wave amplitudes, since these facts are completely expressed in the two equations

$$N(\beta')N(\beta'') - N(\beta'')N(\beta') = 0, \qquad (45)$$
$$N'(\beta') = 0 \quad \text{or} \quad 1. \qquad (46)$$

Eqn.45 follows immediately. The proof that (46) also follows from the multiplication rules of $a, a^\dagger$ is given in the following way:

On the basis of

$$a_\alpha^\dagger(\beta')a_\alpha(\beta') + a_\alpha(\beta')a_\alpha^\dagger(\beta') = 1 \qquad (47)$$

and the result that

$$[a_\alpha(\beta')]^2 = 0,$$

we get (suppressing common indices)

$$a^\dagger aa^\dagger a = a^\dagger \left(1 - a^\dagger a\right) a = a^\dagger a; \qquad (48)$$

192

therefore

$$N(1-N) = a^\dagger a \left(1 - a^\dagger a\right) = 0. \tag{49}$$

It should be noted: Since $v(\beta')$ is defined from the $N(\beta')$ (by a rule following from the eigenvalues) then one can write

$$
\begin{aligned}
b_\alpha(\beta') &= v(\beta')a_\alpha(\beta'), \\
b_\alpha^\dagger(\beta') &= a_\alpha^\dagger(\beta')v(\beta')
\end{aligned}
\tag{50}
$$

which defines the $b, b^\dagger$ unambiguously in terms of the $a, a^\dagger$. In fact one can consider the $a, a^\dagger$ as the fundamental quantities of the theory and all other quantities as functions of them. Finally it should be emphasized, that the total number of particles remains invariant under the considered transformations

$$N = \sum_{\beta'} N(\beta') = \sum_{q'} N(q'). \tag{51}$$

This invariance is expressible in another way, that (30a) is a unitary transformation.

Statement of the Proof: It follows from general considerations, which are given in the appendix, that the multiplication rules of the $a, a^\dagger$ not only determine the eigenvalues of the $N(\beta)$, but also that they remain invariant under a canonical transformation of the matrix representation.

§7. Example of the One Dimensional Continuum.

For a one dimensional continuum with the wave equation

$$\frac{\partial^2 \psi}{\partial x^2} = \frac{\partial^2 \psi}{\partial t^2}; \quad \psi = \psi(x, t) \tag{52}$$

and the boundary condition

$$\psi(0, t) = \psi(l, t) = 0, \tag{53}$$

it was shown in [1] that the particle density should be defined by

$$
\begin{aligned}
N(x) &= \psi^\dagger \psi, \tag{54} \\
\psi &= \sum_{r=1}^{\infty} b_r \sin\left(r\frac{\pi}{l}x\right), \tag{55}
\end{aligned}
$$

where $N_r = b_r^\dagger b_r$ is the number of particles in the rth quantum state.

We still have to modify (55) by replacing b_r by the corresponding a_r:

$$\psi = \sum_{r=1}^{\infty} a_r \sin\left(r\frac{\pi}{l}x\right). \tag{56}$$

The calculation of the density fluctuations carried out in [1] can be taken over directly from (55) to (56) and shows immediately that (56), as it must, gives the correct formula. From (55), the required mean square $\overline{\Delta^2}$ should be proportional to

$$\overline{b_r^\dagger b_r} \cdot \overline{b_r b_r^\dagger} = \overline{N_r} \cdot \overline{(1 - N_r)}, \tag{57}$$

where the average is taken over an infinitesimal frequency interval ν_r, so that the formula in §2 becomes

$$\overline{\Delta^2} = \text{const} \cdot n_r(1 - n_r). \tag{58}$$

If one calculates instead with (56), then $\overline{\Delta^2}$ is proportional to

$$\overline{a_r^\dagger a_r} \cdot \overline{a_r a_r^\dagger} = \overline{N_r} \cdot \overline{(1 - N_r)}, \tag{59}$$

that is, the expression remains unchanged.

§8. Equivalence with the Many Dimensional Coordinate Space.

We now demonstrate the equivalence, already used in §6, of Eqn.30a with the usual result in the many dimensional coordinate space. We must restrict ourselves to operators which are symmetrical in the identical particles; moreover, we limit ourselves to quantities which consist of a sum of terms where only a single electron contributes to each term. The energy of an ideal gas is of this sort. These operators have the form

$$V = V_1 + V_2 + \cdots + V_N, \tag{60}$$

where the V_i always have the same magnitude, but belong to different $(1^{st}, 2^{nd}, \cdots, N^{th})$ particles.

Our wave function (§5, Eqns.22,23) depends upon $N'(\beta'_k)$. From the standpoint of quantum mechanics, in fact, this appears to be the natural description because a "maximal experiment" - given the identity of the particles - can only determine how many particles are in state $\beta_1, \beta_2, \cdots, \beta_K$, whereas the question in which state a particular electron is, cannot be answered. In the sense of the Pauli Exclusion Principle, we have limited the values of $N'(\beta'_k)$ allowed in (23) to 0 or 1.

For simplicity, we assume that an electron can only be in a finite number of states K, designated by $\beta_1, \beta_2, \cdots, \beta_K$. (As already noted in §4, this is never exactly true.) The wave function

$$\Psi\left(N'(\beta'_1), N'(\beta'_2), \cdots, N'(\beta'_K)\right)$$

has exactly K arguments and is defined for 2^K values of its argument. The limit $K \to \infty$ leads to no essential difficulty.

These considerations have a simple illustration when applied to the state of a single electron in the many dimensional coordinate space with a wave function whose argument

194

is β'. This result states that for a single electron the only measurement is to determine the quantity β, whose allowed values span the K numbers $\beta'_1, \beta'_2, \cdots, \beta'_K$.

Suppose we have, in the multi-dimensional coordinate space, an antisymmetric wave function of N' electrons

$$\psi\left(\beta'^1, \beta'^2, \cdots, \beta'^{N'}\right). \tag{61}$$

Then we will describe these states in our new N-space by the wave function

$$\Psi\left(N'(\beta'_1), N'(\beta'_2), \cdots, N'(\beta'_K)\right).$$

We have

$$\Psi\left(N'(\beta'_1), \cdots, N'(\beta'_K)\right) = \frac{1}{\sqrt{N'!}}\psi\left(\beta'^1, \cdots, \beta'^{N'}\right). \tag{62}$$

The interpretation of this equation is that Ψ is zero unless exactly N of the K $(K > N)$ numbers $N'(\beta'_1), \cdots, N'(\beta'_K)$ are 1, and the rest are zero. In order to determine the value in these positions, one inserts directly for the $\beta'^1, \beta'^2, \cdots, \beta'^{N'}$ those values for which

in the new N-space, for each function

$$\psi\left(\beta'^1, \beta'^2, \cdots, \beta'^{N'}\right)$$

in the many dimensional coordinate space (even when ψ is a null wave function). For the sign of Ψ to be uniquely determined, it is necessary to order $\beta'_1, \cdots, \beta'_K$ in ψ - once and for all - as

$$(\beta'_1 < \beta'_2 < \cdots < \beta'_{N'}).$$

Conversely, ψ is uniquely determined by Ψ [9]: in the positions for which $\beta'_1 < \cdots < \beta'_{N'}$ holds, by (62), it always satisfies the requirement of antisymmetry.

The individual terms (eg. V_1) of the operator V in Eqn.60, in the many dimensional coordinate space, are essentially Hermitian matrices of K rows and K columns. In the νth row and μ^{th} column stands $H_{\nu\mu}$. Then the whole operator V is identical with the matrix

$$H_{\nu_1 \cdots \nu_{N'}; \mu_1 \cdots \mu_{N'}} = H_{\nu_1 \mu_1}\delta_{\nu_2 \mu_2} \cdots \delta_{\nu_{N'} \mu_{N'}} +$$
$$\delta_{\nu_1 \mu_1} H_{\nu_2 \mu_2}\delta_{\nu_3 \mu_3} \cdots \delta_{\nu_{N'} \mu_{N'}} + \cdots + \delta_{\nu_1 \mu_1}\delta_{\nu_2 \mu_2} \cdots H_{\nu_{N'} \mu_{N'}}. \tag{63}$$

We write for short

$$V\psi(\beta'^1 \cdots \beta'^{N'}) = \overline{\psi}(\beta'^1 \cdots \beta'^{N'}), \tag{64}$$

195

and then

$$\bar{\psi}(\beta'_{\nu_1} \cdots \beta'_{\nu_{N'}}) = \sum_{\mu_1 \cdots \mu_{N'} = 1}^{K} H_{\nu_1 \cdots \nu_{N'}; \mu_1 \cdots \mu_{N'}} \psi(\beta'_{\mu_1} \cdots \beta'_{\mu_K}). \tag{65}$$

From these $\bar{\psi}(\beta'_1, \cdots, \beta'_{N'})$ we form - exactly as in (62) - a $\bar{\Psi}$ by

$$\bar{\Psi}\left(N'(\beta'_1), \cdots, N'(\beta'_K)\right) = \bar{\psi}(\beta'^{1}, \cdots, \beta'^{N'}) \frac{1}{\sqrt{N'!}}.$$

We now have that

$$\bar{\Psi}\left(N'(\beta'_1), \cdots, N'(\beta'_K)\right) = \Omega \Psi\left(N'(\beta'_1), \cdots, N'(\beta'_K)\right), \tag{66}$$

where the operator Ω is

$$\Omega = \sum_{\kappa, \lambda = 1}^{K} H_{\kappa \lambda} a_\kappa^\dagger a_\lambda, \tag{66}$$

with the a from (31).

Eqn.65 is certainly correct in all columns (for all values of the argument), if the number of 1's in the argument of $\bar{\Psi}$ is not exactly equal to N'. Then the left side vanishes according to (62); and also vanishes on the right side because $a^\dagger a$ does not change the number of 1's.

In the columns, however, if some of the

$$N'(\beta'_{i_1}), N'(\beta'_{i_2}), \cdots, N'(\beta'_{i_{N'}})$$

(exactly N') are equal to 1, then $\sqrt{N'!} \cdot \bar{\Psi}$ equals

$$\bar{\psi}(\beta'_{i_1}, \cdots, \beta'_{i_{N'}}),$$

and therefore equals

$$\sum_{\mu_1 \cdots \mu_{N'}}^{K} H_{i_1 \cdots i_{N'}; \mu_1 \cdots \mu_{N'}} \psi(\beta'_{\mu_1}, \cdots, \beta'_{i_{N'}})$$

$$= \sum_{\mu_1 = 1}^{K} H_{i_1 \mu_1} \psi(\beta'_{\mu_1}, \beta'_{i_2} \cdots, \beta'_{i_{N'}}) + \cdots + \sum_{\mu_{N'} = 1}^{K} H_{i_{N'} \mu_{N'}} \psi(\beta'_{i_1}, \cdots, \beta'_{\mu_{N'}})$$

$$= \sum_{r=1}^{N'} \sum_{\mu=1}^{K} H_{i_r \mu} \psi(\beta'_{i_1}, \cdots, \beta'_{i_{r-1}}, \beta'_\mu, \beta'_{i_{r+1}}, \cdots, \beta'_{i_{N'}}), \tag{67}$$

as one sees from (63). Our aim now is to express the right side of (67) in terms of Ψ. For this, we observe that in (67) the β'_i are already in the correct order, but some of the β'_μ are not. If it happens [10] that

$$\beta'_{i_{s-1}} < \beta'_\mu \leq \beta'_{i_s},$$

196

then we can make the order of the β' correct, by moving the β'_μ over the β'_i lying between $\beta'_{i_{r-1}}$ and $\beta'_{i_{r+1}}$ to the place between $\beta'_{i_{s-1}}$ and β'_{i_s}. In doing that, the antisymmetric wave function must be multiplied by $(-)^z$, where z is the number of β_i standing between the two given places.

We can also write (67) as

$$\overline{\psi}\left(\beta'_{i_1}, \cdots, \beta'_{i_{N'}}\right) = \sum_{r=1}^{N} \sum_{\mu=1}^{K} \pm H_{i_r \mu} \psi\left(\beta'_{i_1}, \cdots, \beta'_\mu, \cdots, \beta'_{N'}\right), \qquad (67)$$

where the sign $\pm$ depends on r and μ, but holds for all

$$\beta'_{i_1} < \cdots < \beta'_\mu < \cdots < \beta'_{N'}.$$

When we examine (62), we can also write this as [11]

$$\overline{\Psi}(x_1 \cdots x_K) = \sum_{x_j=1} \sum_{x_l=0} \pm \quad H_{jl} \Psi(x_1, \cdots, x_{j-1}, 0, x_{j+1}, \cdots, x_{l-1}, 1, x_{l+1}, \cdots, x_K)$$

$$+ \sum_{x_j=1} H_{jj} \Psi(x_1, \cdots, x_K) \qquad (68)$$

as one easily shows. In detail: in the summands in (67a) where $i_r \neq \mu$ [first term in (68)], the same arguments occur as on the left side, except for β'_{i_r}, which on the left was $(x_j = 1)$; in addition β'_μ is added, and we assume that it is not zero $(x_l \neq 0)$, since otherwise ψ would vanish. If $i_r = \mu$ [second term in (68)], then the same arguments occur on the right as on the left.

The sign in (68) is determined in the obvious way, that one gets $+$ or $-$ depending on whether - between the involved zero and 1 - there is an even or an odd number of 1's. (Just the number of β'_i across which one must move the corresponding β'_μ.) This is the number of the 1's, which stand on the left of x_l, minus the number of 1's which stand on the left of x_j.

Although now the correctness of the preceding formula is no doubt clear, we will express the idea once again in conclusion. We know, that the validity domain of the argument of Ψ embraces 2^K total possibilities, in that each of the $x_k = N'(\beta'_k)$ can be either $+1$ or 0.

The operator a_λ introduced in §3 and §6 is to be defined (for the sake of clarity, we write the indices as arguments) by

$$a_\lambda(x_1, x_2, \cdots, x_K; y_1, y_2, \cdots, y_K)$$
$$= (-1)^{x_1+x_2+\cdots+x_{\lambda-1}} \delta_{x_1 y_1} \cdots \delta_{x_{\lambda-1} y_{\lambda-1}} \delta_{x_\lambda 0} \delta_{y_\lambda 1} \delta_{x_{\lambda+1} y_{\lambda+1}} \cdots \delta_{x_K y_K}, \qquad (69)$$

and corresponding to $a_\kappa^\dagger$

$$a_\kappa^\dagger(x_1 \cdots x_K; y_1 \cdots y_K)$$
$$= (-1)^{x_1 + \cdots + x_{\kappa-1}} \delta_{x_1 y_1} \cdots \delta_{x_{\kappa-1} y_{\kappa-1}} \delta_{x_\kappa 1} \delta_{y_\kappa 0} \cdots \delta_{x_K y_K}. \qquad (69)$$

With the help of this formula, (68) can also be written as:

$$\overline{\Psi}(x_1 \cdots x_K) =$$
$$\sum_{\lambda,\kappa=1}^{K} H_{\kappa\lambda} \sum_{y's, z's=0}^{1} a^\dagger(x_1 \cdots x_K; y_1 \cdots y_K) a(y_1 \cdots y_K; z_1 \cdots z_K) \Psi(z_1 \cdots z_K). \quad (70)$$

Here one can see with a little trouble what can only be described with difficulty. In this way (66) is obtained.

§9. Operators in the N-Space.

In §8 we have seen the following: every antisymmetric function, which is defined in the coordinate spaces with all numbers of dimensions $N' < K$, is expressed through (62) as a function in a new space. Corresponding to the operator (60)

$$V = V_1 + V_2 \cdots + V_N$$

with the matrix in (64) $H_{r_1 \cdots r_{N'}; \mu_1 \cdots \mu_{N'}}$ in the coordinate space, is the operator Ω of (68) in the new N'-space. The operator Ω is

$$\Omega = \sum_{\kappa,\lambda=1}^{K} H_{\kappa\lambda} a_\kappa^\dagger a_\lambda \qquad (66)$$

with the $a_\kappa^\dagger, a_\lambda$ of (69),(69a)).

It follows from this that an eigenfunction of V corresponds to an eigenfunction of Ω. If we can also show that the inner product of two functions in coordinate space has the same value as that of the corresponding functions in the new N'-space, then we are finished with the demonstration of their equivalence. In coordinate space

$$(\psi\phi) = \sum_{\beta'^1, \cdots, \beta'^{N'} = \beta'_1}^{\beta'_K} \psi(\beta'^1 \cdots \beta'^{N'}) \tilde{\phi}(\beta'^1 \cdots \beta'^{N'}), \qquad (71)$$

which because of the antisymmetry gives

$$(\psi\phi) = N'! \sum_{\beta'^1 < \cdots < \beta'^{N'} = \beta'_1}^{\beta'_K} \psi(\beta'^1 \cdots \beta'^{N'}) \tilde{\phi}(\beta'^1 \cdots \beta'^{N'}). \qquad (72)$$

198

On the other hand in the new space

$$(\Psi\Phi) = \sum_{x_1\cdots x_K=0,1} \Psi(x_1\cdots x_K)\tilde{\Phi}(x_1\cdots x_K), \tag{73}$$

which with the help of (62) is the same as (72).

We make a further remark, that the q-number anticommutation relations (36),(40) follow naturally from the formula (62).

§10. Construction of Admissible Operators.

We must finally examine the operators which cannot be expressed in the form (60). The energy of a non-ideal gas is of such a form. We limit ourselves here, for the time being, to operators containing only two different types of particles. We follow the corresponding formulas for the Bose-Einstein statistics [4]. The operator V can then be written

$$V = \sum_{j<k=1}^{N'} V_{jk}, \tag{60}$$

where V corresponds to the matrix

$$H_{\nu_1\cdots\nu_{N'};\mu_1\cdots\mu_{N'}}$$
$$= \sum_{j<k=1}^{N'} H_{\nu_j\nu_k;\mu_j\mu_k}\delta_{\nu_1\mu_1}\cdots\delta_{\nu_{j-1}\mu_{j-1}}\delta_{\nu_{j+1}\mu_{j+1}}\cdots\delta_{\nu_{N'}\mu_{N'}}. \tag{63}$$

Then

$$V\psi(\beta'^1,\cdots,\beta'^{N'}) = \overline{\psi}(\beta'^1,\cdots,\beta'^{N'}) \tag{64}$$

needs to be calculated with the help of (63b) and (65). Furthermore, the ordering of the β on the right side is almost the "correct ordering". For that, the β'_i remain in place, but the β'_μ must be moved past a number of the β'_i, which can change the sign.

If we consider (62) again, then we can make the correspondence between $\overline{\psi}$ and ψ on the right and $\overline{\Psi}$ and Ψ on the left. From (69) one now finds that the operator (60b) corresponds to the operator

$$\frac{1}{2!}\sum_{\lambda_1,\lambda_2,\kappa_1,\kappa_2=1}^{K} H_{\kappa_1\kappa_2;\lambda_1\lambda_2} a^\dagger_{\kappa_1} a^\dagger_{\kappa_2} a_{\lambda_2} a_{\lambda_1}. \tag{66}$$

It is important to note that the particle statistics is automatically included through the noncommutative multiplication properties of the amplitudes in the three dimensional space. In the Bose-Einstein case these properties were defined by formula (40) of the paper of Jordan and Klein. The same formula also holds here, as follows from the easily proved formula

$$a_k^\dagger a_k a_l^\dagger a_l - a_k^\dagger a_l^\dagger a_l a_k = \delta_{kl} a_k^\dagger a_k.$$

We come finally to the case of operators which contain terms in which $n > 2$ particles are symmetrized, but in the sum all such terms are excluded for which the same particle occurs twice (which would mean an interaction of the particle with itself). A generalization of the above analysis leads to the following formula which is the analog of the generalization [5] of formula (34) given by Jordan and Klein [4]:

$$\frac{1}{n!} \sum_{\lambda_1 \cdots \lambda_n = 1}^{K} \sum_{\kappa_1 \cdots \kappa_n = 1}^{K} H_{\kappa_1 \cdots \kappa_n; \lambda_1 \cdots \lambda_n} a_{\kappa_1}^\dagger \cdots a_{\kappa_n}^\dagger a_{\lambda_n} \cdots a_{\lambda_1}. \tag{66}$$

§11. Construction of Amplitudes.

Finally, we can express the results in a somewhat different form: for the Pauli many body problem, instead of the probability amplitude above, it gives a probability that - if the measurable quantities $N(\beta_1'), N(\beta_2'), \cdots, N(\beta_K')$ have a certain set of values $N'(\beta_1'), \cdots, N'(\beta_K')$ - then another set of quantities $N(q_1'), \cdots, N(q_K')$ will be found to have the values $N'(q_1'), \cdots, N'(q_K')$

Such an amplitude is given by

$$\Phi_{-\Theta_\alpha(\beta'), -\Theta_p(q')}^{N(\beta'), N(q')} = 0 \quad \text{for} \quad \sum_{\beta'} N'(\beta') \neq \sum_{q'} N'(q')$$

and

$$\Phi_{-\Theta_\alpha(\beta'), -\Theta_p(q')}^{N(\beta'), N(q')} = N'! \Psi_{\alpha p}^{\beta q} \quad \text{for} \quad \sum_{\beta'} N'(\beta') = \sum_{q'} N'(q'), \tag{74}$$

where $\Phi_{\alpha p}^{\beta q}$, $\Psi_{\alpha p}^{\beta q}$ are the functions defined in §4.

The functional equation given in [1] in the form

$$\left\{ \sum_{rs} H_{rs} b_r^\dagger b_s - W \right\} \Phi = 0$$

for the amplitude Φ of the total system, now reads (including the determinant for the sign determination of Φ):

$$\left\{ \sum_{rs} H_{rs} a_r^\dagger a_s - W \right\} \Phi = 0;$$

This modification is necessary because in [1] the ambiguity of the sign of the determinant was not sufficiently allowed for; the matrices a_r differ from the b_r, only in the signs of

200

their different matrix elements. It is very satisfying that the introduction of the operators $a, a^\dagger$ - necessary for the energy expression - lead also to a simple multiplication rule, as we have seen in §6.

Finally, it should be emphasized that the multiplication rules discussed in §6 for the quantized amplitudes $a_p(q')$ can easily be generalized relativistically, in analogy to the relativistically invariant multiplication rules for the charge free electromagnetic field as introduced by Jordan and Pauli [6]. In this way, one gets the quantization - in a relativistically invariant form - of deBroglie waves which satisfy the Pauli Exclusion Principle. A detailed exposition will be presented soon.

Appendix on the Proof: Between the $2K$ operators $a_1, \cdots, a_K; a_1^\dagger, \cdots, a_K^\dagger$ there are the relations

$$a_\kappa a_\lambda + a_\lambda a_\kappa = 0$$
$$a_\kappa^\dagger a_\lambda^\dagger + a_\lambda^\dagger a_\kappa^\dagger = 0 \tag{36}$$

and

$$a_\kappa^\dagger a_\lambda + a_\lambda a_\kappa^\dagger = \delta_{\kappa\lambda}. \tag{40}$$

We will now show that these q-number relations uniquely determine the operators $a, a^\dagger$, provided one considers only irreducible matrix systems and regards matrix systems related to one another by a similarity transformation as not different [12].

To see this, we form the following operators

$$\alpha_\kappa = a_\kappa + a_\kappa^\dagger,$$
$$\alpha_{K+\kappa} = (a_\kappa - a_\kappa^\dagger)/i. \tag{1}$$

The $2K$ matrices α conversely determine the a uniquely. Now, generally, for the α

$$\alpha_\kappa \alpha_\lambda + \alpha_\lambda \alpha_\kappa = 2\delta_{\kappa\lambda}. \tag{2}$$

One also shows for example, when both $\kappa < K$ and $\lambda < K$, that

$$\alpha_\kappa \alpha_\lambda + \alpha_\lambda \alpha_\kappa = 2\delta_{\kappa\lambda}.$$

One can also write (2) as

$$\alpha_\kappa^2 = 1$$
$$\alpha_\kappa \alpha_\lambda = -\alpha_\lambda \alpha_\kappa \quad \text{for} \quad \kappa \neq \lambda,$$

201

which, in other words, states that the $2K$ matrices α together with the matrix -1 form a group. For example, if $K = 2$, the group has the following elements

$$1 \quad ; \quad \alpha_1, \alpha_2, \alpha_3, \alpha_4; \alpha_1\alpha_2, \alpha_1\alpha_3, \alpha_1\alpha_4, \alpha_2\alpha_3, \alpha_2\alpha_4, \alpha_3\alpha_4;$$

$$-1 \quad ; \quad -\alpha_1, -\alpha_2, -\alpha_3, -\alpha_4; -\alpha_1\alpha_2, -\alpha_1\alpha_3, -\alpha_1\alpha_4, -\alpha_2\alpha_3, -\alpha_2\alpha_4, -\alpha_3\alpha_4;$$

$$\alpha_1\alpha_2\alpha_3, \alpha_1\alpha_3\alpha_4, \alpha_1\alpha_2\alpha_4, \alpha_2\alpha_3\alpha_4, \alpha_1\alpha_2\alpha_3\alpha_4.$$

$$-\alpha_1\alpha_2\alpha_3, -\alpha_1\alpha_3\alpha_4, -\alpha_1\alpha_2\alpha_4, -\alpha_2\alpha_3\alpha_4, -\alpha_1\alpha_2\alpha_3\alpha_4. \tag{3}$$

There are 32 elements, or generally 2^{2K+1} elements. The irreducible matrix system which satisfies (2) is itself an irreducible representation of this group (the reverse need not be the case, because the isomorphy can be in several steps). We will only determine the irreducible representations.

Our group has the normal divisor $1, -1$ (the centrum), its factor group of degree 2^{2K} is abelian. However for us this does not come into consideration, because these are not used in Eqn.2 (because they are in fact commutative).

How many classes has our group? The two elements 1 and -1 always form a class by themselves, and also each element R is in a class with $-R$. If in particular R in (3) has an uneven number of factors, then $\alpha R \alpha^{-1} = -1 \cdot R$ when α is not contained in R; if R has an even number of factors, then $\alpha R \alpha^{-1} = -1 \cdot R$ when α is contained in R. The number of classes is therefore $2^{2K}+1$; this is also the number of irreducible representations different from one another. Since we already have identified 2^{2K} representations which do not enter our considerations, there can be only one uniquely, the last, which satisfies Eqn.2. All other solutions of (2) arise from this one by similarity transformations.

We determine next the number of rows and columns, the dimension of the representation. (It naturally must be 2^K.) In fact, the order of the group 2^{2K+1} must be equal to the sum of the squares of the dimensions of its representations. It has 2^{2K} of dimension 1, so the last must have dimension 2^K, giving

$$(2^K)^2 + 2^{2K} \cdot 1^2 = 2^{2K+1}.$$

This also agrees with our analysis of the matrix system (69) or §3 and 6.

Göttingen, Institute for Theoretical Physics.

Footnotes and References:

1) P. Jordan, Zeits. f. Phys. **44**, 473 (1927).

2) M. Born, W. Heisenberg and P. Jordan, Zeits. f. Phys. **35**, 557 (1926).

3) P.A.M. Dirac, Proc. Roy. Soc. A**114**, 243, 719 (1927).

4) P. Jordan and O. Klein, Zeits. f. Phys. **45**, 751 (1927).

5) P. Jordan, Zeits. f. Phys. **45**, 766 (1927).

6) P. Jordan and W. Pauli jr., Zeits. f. Phys. (in press).

7) P. Jordan, Zeits. f. Phys. **44**, 1 (1927).

8) P. Jordan, Zeits. f. Phys. **41**, 711 (1927). Note added in proof: the same formulas were recently discussed independently by Bothe, Zeits. f. Phys. **46**, 327 (1928).

9) We assume that Ψ vanishes in general unless $N'(\beta_1') + \cdots + N'(\beta_K') = N'$.

10) The equality does not come into question, because then the corresponding ψ vanishes.

11) For convenience, we replace $N'(\beta_k')$ by x_k.

12) This is the transformation of all matrices a by the same matrix S to $S^{-1}aS$, that is, in the language commonly used in quantum mechanics, a canonical transformation of the matrix representation.

Chapter 8

From Hole Theory to Positrons

Summary: Dirac's invention of the relativistic spin-$\frac{1}{2}$ equation with its successful predictions of the electron spin and magnetic moment as well as the correct hydrogen spectrum, led also to a brief period of confusion about the negative energy states. The resolution of this confusion led Dirac, with the prompting of Weyl, Oppenheimer, and others, to the prediction of both the positron and the antiproton.

§1. The Dirac Equation.

Dirac reached the highest point of his creativity inventing what was uniquely his own, the Dirac equation which encompasses in a point electron the properties postulated by Goudsmit and Uhlenbeck of a particle with spin-$\frac{1}{2}$

indexelectron spin and an anomalous gyromagnetic ratio of 2. In great contradistinction to previous attempts, all of which had been *ad hoc*, Dirac obtained these results as the unforeseen consequences of imposing the requirements of relativity on the Hamiltonian formulation of quantum mechanics [8.1]. "One would like to find some incompleteness $\cdots$ such that, when removed, the whole $\cdots$ follow without arbitrary assumptions. $\cdots\cdots$ the simplest Hamiltonian $\cdots$ satisfying $\cdots$ relativity and the general transformation theory leads to an explanation $\cdots$ without further assumptions."

Dirac dispensed with the Klein-Gordon equation, although he did recognize that its conserved current referred to a charge-current and not to a failure to possess a positive definite probability current. His conclusion that the interpretation of the Klein-Gordon equation was not "just as general as that of the non-relativistic theory" was flawed and eventually set right by Pauli and Weisskopf [8.2]. In spite of that mistaken motivation, Dirac made the fruitful postulate of the existence of a relativistic wave equation linear in the time derivative.

Dirac also noted what he considered a second difficulty - the Klein-Gordon equation "refers equally well to an electron with charge $-e$ as to one with charge e. $\cdots$ the electron suddenly changing its charge from e to $-e$ $\cdots$ we shall be concerned only with $\cdots$ the first of these two difficulties."

Then Dirac - in what has become standard textbook lore, but is worth seeing in his original words - postulates the linear equation

$$E\psi = (\vec{\alpha} \cdot \vec{p} + \beta m)\psi, \tag{1}$$

where the operators $(\vec{\alpha}, \beta)$ must be independent of $(\vec{p}, E)$ for linearity, and of $(\vec{x}, t)$ for homogeneity of space-time. "We are therefore obliged to have other dynamical variables $\cdots$ in order that $(\vec{\alpha}, \beta)$ may be functions of them. The wave function ψ must then involve more variables than merely $(\vec{x}, t)$." He then derives the anticommutation relations for the $(\vec{\alpha}, \beta)$ and finds the 4×4 standard matrix representation, and the final form of the Dirac equation (9). Including electromagnetic potentials in the usual way he obtains Eqn.14: "This wave equation $\cdots$ accounts for all duplication phenomena. $\cdots$ leads to the same energy levels as those obtained by Darwin, $\cdots$ in agreement with experiment."

"$\cdots$ it will have four times as many solutions as the non-relativistic wave equation, $\cdots$ twice as many as the previous relativistic wave equation $\cdots$ half must be rejected as referring to a positive charge on the electron $\cdots$ ".

§2. The Reinterpretation of Holes as Antiparticles.

Almost two years later, Dirac returned to these "rejected" solutions and - with help from Oppenheimer and Weyl - invented the concept of antiparticles as also contained in the solutions of the relativistic wave equations. For reasons that seem difficult to appreciate today, Dirac [8.3] seized on the false premise that the extra solutions described protons. Oppenheimer [8.4] soon pointed out that this would lead to proton-electron annihilation and the instability of matter. Weyl [8.5] proved from time-reversal invariance that the extra solutions describe particles with the same mass as the electron, but opposite charge. Immediately upon having this marvelous and mathematically sophisticated insight, Weyl lapsed into the same error as Dirac and he also characterized these particles as protons. Finally, more than a year after his first suggestion of protons as the culprit, as an aside in a paper devoted to a completely different subject (magnetic monopoles in quantum mechanics!), Dirac [8.6] acknowledged the arguments of Oppenheimer and Weyl. "⋯ we must abandon the identification of holes with protons ⋯. A hole ⋯ would be a new kind of particle, unknown to experimental physics, having the same mass and opposite charge to an electron. We may call such a particle an anti-electron. ⋯ if they could be produced ⋯ in high vacuum they would be quite stable and amenable to observation. ⋯ protons will have their own negative energy states ⋯ an unoccupied one appearing as an antiproton."

Carl Anderson had the defining word [8.7] in the discussion when he discovered positrons pair produced in his Wilson cloud chamber.

It is fascinating to reread Dirac's second paper in this series, and to substitute 'positron' for 'proton' everywhere. The result is almost exactly physics as we now know it: " ⋯ the holes in the distribution of negative-energy electrons are the *positrons*. When an electron of positive energy drops into a hole and fills it up, we have an electron and a *positron* disappearing together with emission of radiation."

§3. Feynman's Interpretation of Dirac's Scenario.

The final paragraph of Dirac's realization of the positron interpretation of the negative energy states of the Dirac equation, is an especially prescient description of the Compton scattering processes. He came very close to describing the mechanism in the way familiar today, as time-ordered Feynman diagrams. These in turn will be reinterpreted in the Feynman Green's function using the concept of negative-energy electrons propagating backward

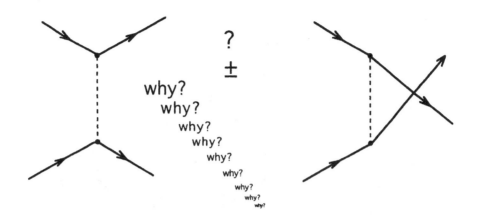

?
±
why?
 why?
  why?
   why?
    why?
     why?
      why?
       why?
        why?

Direct ± Exchange Amplitudes

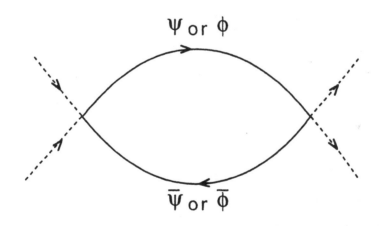

Ψ or ϕ

$\overline{\Psi}$ or $\overline{\phi}$

Feynman's Unitarity Requires S. S. T.

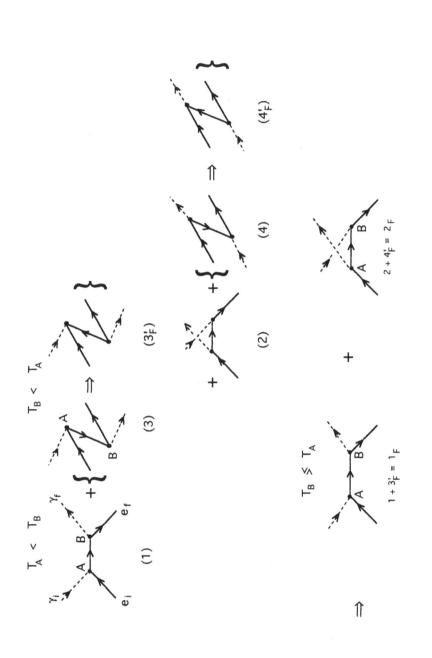

Feynman Diagrams for Compton Scattering.

in time as equivalent to positive-energy positrons propagating forward in time. It took eighteen years until Feynman [8.8], alone among all those who had pored over Dirac's writings, finally clarified Dirac's description of negative-energy electron states.

It is worth looking ahead to the Feynman diagram equivalent of Dirac's description of Compton Scattering in this last paragraph of his paper "A Theory of Electrons and Protons". The Feynman diagrams corresponding to the lowest order perturbative amplitudes for photon elastic scattering by an electron

$$\gamma_i + e_i \Rightarrow \gamma_f + e_f$$

consist of four types, in which electron e_i absorbs photon γ_i at space-time point A_1 in a transition to a virtual, energy non-conserving intermediate state of a positive energy electron e'_1, which propagates to space-time point B_1 where it emits photon γ_f in transition to electron e_f. This sequence of events is summarized

1)$\gamma_i + e_i \Rightarrow_{A_1} e'_1(E > 0) \Rightarrow_{B_1} \gamma_f + e_f$. In a self-explanatory notation we also get the possibilities:

2) $\gamma_i + e_i \Rightarrow_{A_2} \gamma_i + \gamma_f + e'_2(E > 0) \Rightarrow_{B_2} \gamma_f + e_f$,

3) $\gamma_i + e_i \Rightarrow_{A_3} e'_3(E < 0) \Rightarrow_{B_3} \gamma_f + e_f$,

4) $\gamma_i + e_i \Rightarrow_{A_4} \gamma_i + \gamma_f + e'_4(E < 0) \Rightarrow_{B_4} \gamma_f + e_f$.

These amplitudes correspond to the Feynman diagrams 1 − 4.

What Feynman made clear was that amplitudes (3) and (4) with negative energy electrons e'_3 and e'_4 in the intermediate states can be reinterpreted in a startling way, which leads finally to a great simplification. The Feynman propagator which precludes the creation of negative energy electrons propagating forward in time has only $t(A_3) > t(B_3)$ and $t(A_4) > t(B_4)$; that is, it has negative energy electrons propagating backward in time! The Feynman diagrams (3) and (4) can be redrawn showing this, with space-time point A_3 coming after B_3 (as read from left to right). The continuous electron line running from $i \Rightarrow A_3 \Rightarrow B_3 \Rightarrow f$ runs backward in time (with negative energy) from A_3 to B_3. This portion of the electron line should be reinterpreted as a positron $\bar{e}$ with positive energy (and charge) running forward in time from B_3 to A_3. Feynman goes on to show that amplitudes (1) and (3) combine to a single, simple, manifestly relativistic form which can be represented by a single diagram (1_F), and similarly (2) and (4) combine to (2_F), without regard to the time order of the points A, B, and without explicitly worrying about the "negative energy" states.

The net result of this tumult was the realization - owing much to Weyl's rigorous mathematical groundbreaking work on invariance properties - that a particle-antiparticle symmetry completed the prerequisites for an understanding of relativistic quantum field theory.

§4. Biographical Note on Dirac.(†,††)

Paul Adrien Maurice Dirac (1902-1984) is a mythic, godlike figure whose name and works permeate modern physics, and rightly so. His contributions to the genesis of quantum theory were accomplished in complete isolation compared to the community of Bohr, Heisenberg, Born, and Jordan. He worked more in the solitary mode of Schrödinger. His earlier contributions were certainly independent but not quite simultaneous by a matter of days and weeks with the parallel achievements of Heisenberg, Born, and Jordan, and of Fermi. However his insights were distinguished by such clarity, incisiveness, originality, and beyond everything, such elegance that he was immediately recognized as a founder of the quantum theory. His first completely unique contribution was his creation of the Dirac equation for the relativistic electron.

Dirac studied electrical engineering (1919-21) and applied mathematics (1921-23) at Bristol; was an Exhibitioner and research student at St John's College, Cambridge (1923-); made his reputation with his first paper establishing the fundamental principles of quantum mechanics, and demonstrating the equivalence of the Schrödinger and the Heisenberg formulations (1925); cofounder with Fermi of Fermi-Dirac statistics (1926); coinventer with Heisenberg of wave function symmetrization for identical particles (1926); coinventer, with Born, Heisenberg, and Jordan, of quantum field theory (1927); discovered the Dirac equation (1928); invented hole theory (1931); reinterpreted hole theory and predicted positron and antiproton (1933); wrote *The Principles of Quantum Mechanics* (1930); Fellow of the

Royal Society (1930); Lucasian Professor of Mathematics (1932-69), the chair once held by Newton; Nobel Prize (1933); Royal Medal of the Royal Society (1939); Copley Medal (1952); Max Planck Medal (1952), previously awarded only to Planck and Einstein; Order of Merit (1973); Professor of Physics, Florida State University (1971-1984).

His later work, although not popularly known, has had a deep impact on modern theoretical physics. Goddard and Taylor[tt] cite the quantum theory of magnetic monopoles, the dynamics of constrained systems, extended models of elementary particles, and steps toward a quantum theory of gravity as all having great importance fifty years after their publication. One of Dirac's speculative ideas on the role of the Lagrangian in quantum mechanics [8.9] was brought to fruition only after fifteen years, in Feynman's Path Integral formulation of quantum mechanics [8.10]. This in turn has become the most powerful formulation of quantum field theory, and in fact it is the only viable way to handle the complexities and constraints of modern theories [8.11].

Dirac's elegance frequently manifested itself in an economy of words which made his written work difficult. Painful years of study have been devoted to trying to discern what he meant by some enigmatic phrase. But his writing was practically effusive compared to his conversation, where he usually seemed content with his own thoughts and internalized those. One story has him on the occasion of leaving on a voyage to Japan, refusing the gift of a book with the words: "Reading precludes thought."

One of us (ID) had the task of hosting Dirac for an afternoon and explained that his interest was nuclear physics. Dirac responded: "That is a very difficult problem. I was more interested in the electron." End of conversation.

(† - The London Times, 23 Oct 1984, p10e; 25 Oct 1984, p20g; †† - P. Goddard and J.C. Taylor, The London Times, 15 Nov 1984, p18g.)

Bibliography and References.

8.1) P.A.M. Dirac, Proc. Roy. Soc. A117, 610 (1928); see our App.8A.

8.2) W. Pauli and V.F. Weisskopf, Helv. Phys. Acta 7, 709 (1934); see our App.9A.

8.3) P.A.M. Dirac, Proc. Roy. Soc. A126, 360 (1930); see our App.8B.

8.4) J.R. Oppenheimer, Phys. Rev. 35, 562 (1930); see our App.8C.

8.5) H. Weyl, *The Theory of Groups and Quantum Mechanics* (Dover, New York, 1930) translated by H.P. Robertson, p.225; see our App.8D.

8.6) P.A.M. Dirac, Proc. Roy. Soc. A**133**, 60 (1931); see our App.8E.

8.7) C.D. Anderson, Phys. Rev. **43**, 491 (1933).

8.8) R.P. Feynman, Phys. Rev. **76**, 749 (1949); see our App.15A.

8.9) P.A.M. Dirac, Phys. Zeits. der Sowjetunion **3**, 1' (1933) in *Selected Papers on Quantum Electrodynamics* (Dover, New York, 1958) edited by Julian Schwinger, pp.312-321; Rev. Mod. Phys. **17**, 195 (1945).

8.10) R.P. Feynman, Rev. Mod. Phys. **20**, 267 (1948).

8.11) M. Kaku, *Quantum Field Theory: A Modern Introduction* (Oxford, New York, 1993), Ch.8 *et seq.*

APPENDIX 8.A:

Excerpt from: Proc. Roy. Soc. A117, 610 (1928)

The Quantum Theory of the Electron

by P.A.M. Dirac

St. John's College, Cambridge

(Communicated by R.H. Fowler, F.R.S. - Received January 2, 1928)

The new quantum mechanics, when applied to the problem of the structure of the atom with point-charge electrons, does not give results in agreement with experiment. The discrepancies consist of a duplication phenomenon, the observed number of stationary states for an electron in an atom being twice the number given by the theory. To meet the difficulty, Goudsmit and Uhlenbeck have introduced the idea of an electron with a spin angular momentum of half a quantum and a magnetic moment of one Bohr magneton. This model for the electron has been fitted into the new mechanics by Pauli [1]; and Darwin [2], working with an equivalent theory, has shown that it gives results in agreement with experiment for hydrogen-like spectra to first order of accuracy.

The question remains as to why Nature should have chosen this particular model for the electron instead of being satisfied with the point-charge. One would like to find some incompleteness in the previous methods of applying quantum mechanics to the point-charge electron such that, when removed, the whole of the duplication phenomena follow without arbitrary assumptions. In the present paper it is shown that this is the case, the incompleteness of the previous theories lying in their disagreement with relativity, or, alternatively, with the general transformation theory of quantum mechanics. It appears that the simplest Hamiltonian for a point-charge electron satisfying the requirements of both relativity and the general transformation theory leads to an explanation of all duplication phenomena without further assumption. All the same there is a great deal of truth in the spinning electron model, at least as a first approximation. The most important failure of the model seems to be that the magnitude of the resultant orbital angular momentum of an electron moving in an orbit in a central field of force is not a constant, as the model leads one to expect.

211

§1. *Previous Relativistic Treatments.*

The relativistic Hamiltonian according to the classical theory for a point electron moving in an arbitrary electro-magnetic field with scalar potential A_0 and vector potential $\mathbf{A}$ is

$$F = -(W - eA_0)^2 + (\mathbf{p} - e\mathbf{A})^2 + m^2,$$

where $\mathbf{p}$ is the momentum vector. (*Note Added:* W is the total energy, m the rest mass, $\hbar$ and c have been set to 1. The charge of the electron has been taken as e, not $-e$ as in Dirac. Also, the metric sign $-W^2 + p^2 + m^2$ is made explicit, again, not as in Dirac.) It has been suggested by Gordon [3] that the operator of the wave equation of the quantum theory should be obtained from this F by the same procedure as in the non-relativistic theory, namely by putting

$$W = i\frac{\partial}{\partial t}, \quad p_r = -i\frac{\partial}{\partial x^r}, \quad r = 1, 2, 3,$$

in it.

This gives the wave equation

$$F\psi \equiv \left[-(W - eA_0)^2 + (\mathbf{p} - e\mathbf{A})^2 + m^2 \right] \psi = 0, \tag{1}$$

the wave function ψ being a function of x_1, x_2, x_3, t. This gives rise to two difficulties.

The first is in connection with the physical interpretation of ψ. Gordon, and also independently Klein [4], from the consideration of the conservation theorems, make the assumption that if ψ_m, ψ_n are two solutions

$$\rho_{nm} = \frac{e}{2m}[i(\bar{\psi}_n \frac{\partial \psi_m}{\partial t} - \frac{\partial \bar{\psi}_n}{\partial t}\psi_m) - 2eA_0\bar{\psi}_n\psi_m]$$

and

$$\mathbf{j}_{nm} = \frac{e}{2m}[i(\nabla\bar{\psi}_n\psi_m - \bar{\psi}_n\nabla\psi_m) - 2e\mathbf{A}\bar{\psi}_n\psi_m]$$

are to be interpreted as the charge and current associated with the transition $m \to n$. This appears to be satisfactory so far as emission and absorption of radiation are concerned, but is not so general as the interpretation of the non-relativistic quantum mechanics, which has been developed [5] sufficiently to enable one to answer the question: What is the probability of any dynamical variable at any specified time having a value lying between specified limits, when the system is represented by a given wave function ψ_n? The Gordon-Klein interpretation can answer such questions if they refer to the position of the electron (by the use of ρ_{nn}), but not if they refer to its momentum, or angular momentum or any other dynamical variable. We should expect the interpretation of the relativistic theory to be just as general as that of the non-relativistic theory.

The general interpretation of non-relativistic quantum mechanics is based on the transformation theory, and is made possible by the wave equation being of the form

$$W\psi = H\psi, \tag{2}$$

212

i.e., being linear in W or $\partial/\partial t$, so that the wave function at any time determines the wave function at any later time. The wave function of the relativistic theory must also be linear in W if the general interpretation is to be possible.

The second difficulty in Gordon's interpretation arises from the fact that if one takes the conjugate imaginary of Eqn.1, one gets

$$[-(-W - eA_0)^2 + (-\mathbf{p} - e\mathbf{A})^2 + m^2]\psi^* = 0,$$

which is the same as one would get if one put $-e$ for e. The wave equation (1) thus refers equally well to an electron with charge e as to one with charge $-e$. If one considers for definiteness the limiting case of large quantum numbers one would find that some of the solutions of the wave equation are wave packets moving in the way a particle of charge $-e$ would move on the classical theory, while others are wave packets moving in the way a particle of charge e would move classically. For this second class of solutions W has a negative value. One gets over the difficulty on the classical theory by arbitrarily excluding those solutions that have a negative W. One cannot do this on the quantum theory, since in general a perturbation will cause transitions from states with W positive to states with W negative. Such a transition would appear experimentally as the electron suddenly changing its charge from e to $-e$, a phenomenon which has not been observed. The true relativistic wave equation should thus be such that its solutions split up into two non-combining sets, referring respectively to the charge e and the charge $-e$.

In the present paper we shall be concerned only with the removal of the first of these two difficulties. The resulting theory is therefore still only an approximation, but it appears to be good enough to account for all the duplication phenomena without arbitrary assumptions.

§2. *The Hamiltonian for No Field.*

Our problem is to obtain a wave equation of the form (2) which shall be invariant under a Lorentz transformation and shall be equivalent to (1) in the limit of large quantum numbers. We shall consider first the case of no field, when Eqn.1 reduces to

$$(p_0{}^2 - \mathbf{p}^2 - m^2)\psi = 0 \tag{3}$$

if one puts

$$p_0 = i\frac{\partial}{\partial t}.$$

The symmetry between p_0 and p_1, p_2, p_3 required by relativity shows that, since the Hamiltonian we want is linear in p_0, it must also be linear in p_1, p_2 and p_3. Our wave equation is therefore of the form

$$(p_0 - \alpha_1 p_1 - \alpha_2 p_2 - \alpha_3 p_3 - \beta m)\psi = 0, \tag{4}$$

213

where for the present all that is known about the operators $\alpha_1, \alpha_2, \alpha_3, \beta$ is that they are independent of p_0, p_1, p_2, p_3, i.e., that they commute with t, x_1, x_2, x_3. Since we are considering the case of a particle moving in empty space, so that all points in space are equivalent, we should expect the Hamiltonian not to involve t, x_1, x_2, x_3. This means that $\alpha_1, \alpha_2, \alpha_3, \beta$ are independent of t, x_1, x_2, x_3, i.e., that they commute with p_0, p_1, p_2, p_3. We are therefore obliged to have other dynamical variables besides the coordinates and momenta of the electron, in order that $\alpha_1, \alpha_2, \alpha_3, \beta$ may be functions of them. The wave function ψ must then involve more variables than merely x_1, x_2, x_3, t.

Eqn.4 leads to

$$
\begin{aligned}
0 &= (p_0 + \alpha \cdot \mathbf{p} + \beta m)(p_0 - \alpha \cdot \mathbf{p} - \beta m)\psi \\
&= [p_0^2 - \sum_k \alpha_k^2 p_k^2 - \sum_{k>j}(\alpha_k \alpha_j + \alpha_j \alpha_k)p_k p_j - \beta^2 m^2 - \sum_k (\alpha_k \beta + \beta \alpha_k)m p_k]\psi. \quad (5)
\end{aligned}
$$

This agrees with Eqn.3 if

$$
\begin{aligned}
\alpha_k^2 &= 1, & \alpha_k \alpha_j + \alpha_j \alpha_k &= 0 \quad (k \neq j), \\
\beta^2 &= 1, & \alpha_k \beta + \beta \alpha_k &= 0,
\end{aligned}
$$

with $k, j = 1, 2, 3$. If we put $\beta = \alpha_4$, these conditions become

$$
\alpha_\mu^2 = 1 \qquad \alpha_\mu \alpha_\nu + \alpha_\nu \alpha_\mu = 0 \quad (\mu \neq \nu) \qquad \mu, \nu = 1, 2, 3, 4. \tag{6}
$$

We can suppose the α_μ's to be expressed as matrices in some matrix scheme, the matrix elements of α_μ being, say, $\alpha_\mu(\zeta, \zeta')$. The wave function ψ must now be a function of ζ as well as $\mathbf{x}, t$. The result of α_μ multiplied into ψ will be defined as

$$
(\alpha_\mu \psi)(\mathbf{x}, t, \zeta) = \sum_{\zeta'} \alpha_\mu(\zeta, \zeta')\psi(\mathbf{x}, t, \zeta').
$$

We must now find four matrices α_μ to satisfy the conditions (6). We make use of the matrices

$$
\sigma_1 = \begin{pmatrix} 0 & 1 \\ 1 & 0 \end{pmatrix} \qquad \sigma_2 = \begin{pmatrix} 0 & -i \\ i & 0 \end{pmatrix} \qquad \sigma_3 = \begin{pmatrix} 1 & 0 \\ 0 & -1 \end{pmatrix}
$$

which Pauli introduced [1] to describe the three components of spin angular momentum. These matrices have just the properties

$$
\sigma_k^2 = 1 \qquad \sigma_k \sigma_j + \sigma_j \sigma_k = 0 \quad (k \neq j), \tag{7}
$$

that we require for our α's. We cannot, however, just take the σ's to be three of our α's, because then it would not be possible to find the fourth. We must extend the σ's in a diagonal manner to bring in two more rows and columns, so that we can introduce three more matrices ρ_1, ρ_2, ρ_3 of the same form as $\sigma_1, \sigma_2, \sigma_3$, but referring to different rows and columns, thus:-

$$\sigma_1 = \left\{ \begin{matrix} 0 & 1 & 0 & 0 \\ 1 & 0 & 0 & 0 \\ 0 & 0 & 0 & 1 \\ 0 & 0 & 1 & 0 \end{matrix} \right\} \quad \sigma_2 = \left\{ \begin{matrix} 0 & -i & 0 & 0 \\ i & 0 & 0 & 0 \\ 0 & 0 & 0 & -i \\ 0 & 0 & i & 0 \end{matrix} \right\} \quad \sigma_3 = \left\{ \begin{matrix} 1 & 0 & 0 & 0 \\ 0 & -1 & 0 & 0 \\ 0 & 0 & 1 & 0 \\ 0 & 0 & 0 & -1 \end{matrix} \right\},$$

$$\rho_1 = \left\{ \begin{matrix} 0 & 0 & 1 & 0 \\ 0 & 0 & 0 & 1 \\ 1 & 0 & 0 & 0 \\ 0 & 1 & 0 & 0 \end{matrix} \right\} \quad \rho_2 = \left\{ \begin{matrix} 0 & 0 & -i & 0 \\ 0 & 0 & 0 & -i \\ i & 0 & 0 & 0 \\ 0 & i & 0 & 0 \end{matrix} \right\} \quad \rho_3 = \left\{ \begin{matrix} 1 & 0 & 0 & 0 \\ 0 & 1 & 0 & 0 \\ 0 & 0 & -1 & 0 \\ 0 & 0 & 0 & -1 \end{matrix} \right\}.$$

The ρ's are obtained from the σ's by interchanging the second and third rows, and the second and third columns. We now have, in addition to Eqns.7,

$$\rho_k^2 = 1 \qquad \rho_k \rho_j + \rho_j \rho_k = 0 \quad (k \neq j),$$

and also

$$\rho_k \sigma_n = \sigma_n \rho_k.$$

If we now take

$$\alpha_1 = \rho_1 \sigma_1, \quad \alpha_2 = \rho_1 \sigma_2, \quad \alpha_3 = \rho_1 \sigma_3, \quad \alpha_4 = \rho_3,$$

all the conditions (6) are satisfied, e.g.,

$$\alpha_1^2 = \rho_1 \sigma_1 \rho_1 \sigma_1 = \rho_1^2 \sigma_1^2 = 1$$
$$\alpha_1 \alpha_2 = \rho_1 \sigma_1 \rho_1 \sigma_2 = \rho_1^2 \sigma_1 \sigma_2 = -\rho_1^2 \sigma_2 \sigma_1 = -\alpha_2 \alpha_1.$$

The following equations are to be noted for later reference

$$\rho_1 \rho_2 = i\rho_3 = -\rho_2 \rho_1$$
$$\sigma_1 \sigma_2 = i\sigma_3 = -\sigma_2 \sigma_1, \tag{8}$$

together with the equations obtained by cyclic permutations of the suffixes.

The wave equation now takes the form

$$p_0 \psi = [\rho_1 \sigma \cdot \mathbf{p} + \rho_3 m]\psi, \tag{9}$$

where σ denotes the vector $(\sigma_1, \sigma_2, \sigma_3)$.

215

§3. *Proof of Invariance under a Lorentz Transformation.*

.... (*Note added:* We omit in Dirac's papers much material which has become standard textbook fare.)

§4. *The Hamiltonian for an Arbitrary Field.*

...... we adopt the usual procedure ···. From Eqn.9 we obtain

$$(p_0 - eA_0)\psi = [\rho_1 \sigma \cdot (\mathbf{p} - e\mathbf{A}) + \rho_3 m]\,\psi. \tag{14}$$

This wave equation appears to be sufficient to account for all duplication phenomena. On account of the matrices ρ and σ containing four rows and columns, it will have four times as many solutions as the non-relativistic wave equation, and twice as many as the previous relativistic wave equation [1]. Since half the solutions must be rejected as referring to the charge $-e$ on the electron (*Note added:* $e < 0$ for the electron, not as in the original), the correct number will be left to account for the duplication phenomena. ······

···The present theory will thus, in the first approximation, lead to the same energy levels as those obtained by Darwin, which are in agreement with experiment.

Footnotes and References.

1) W. Pauli, Zeits. f. Phys. **43**, 601 (1927).

2) C. Darwin, Proc. Roy. Soc. A**116**, 227 (1927).

3) W. Gordon, Zeits. f. Phys. **40**, 117 (1926).

4) O. Klein, Zeits. f. Phys. **41**, 407 (1927).

5) P. Jordan, Zeits. f. Phys. **40**, 809 (1927); P.A.M. Dirac, Proc. Roy. Soc. A**113**, 621 (1927).

APPENDIX 8.B:

Excerpt from: Proc. Roy. Soc. A126, 360 (1930)

A Theory of Electrons and Protons

by P.A.M. Dirac

St. John's College, Cambridge

(Communicated by R.H. Fowler, F.R.S. - Received December 6, 1929)

§1. *Nature of the Negative Energy Difficulty.*

The relativistic quantum theory of an electron $\cdots$ involves one serious difficulty $\cdots$. $\cdots$ the wave equation (1) (*Note:* Eqn.9 in the preceding Appendix 8A) $\cdots$ has, in addition to the wanted solutions for which the kinetic energy of the electron is positive, an equal number of unwanted solutions with negative kinetic energy $\cdots$. $\cdots$ take $\cdots$ periodic solutions of the form

$$\psi = ue^{-iEt}, \tag{2}$$

where u is independent of t, representing stationary states of total energy E $\cdots$. There will then exist solutions (2) with negative values for E as well as those with positive values; in fact, if we take a matrix representation of the operators $\rho_1\sigma_1, \rho_1\sigma_2, \rho_1\sigma_3, \rho_3$ with the matrix elements all real, then the complex conjugate of any solution of (1) will be a solution of the wave equation obtained from (1) by reversal of the sign of the potentials A, and either the original wave function or its complex conjugate must refer to a negative E.

The difficulty is not a special one connected with the quantum theory of the electron, but is a general one appearing in all relativistic theories. $\cdots\cdots$ The difficulty is not important in the classical theory, since here dynamical variables must always vary continuously, so that there will be a sharp distinction between those solutions for which $E \geq m$ and those for which $E \leq -m$, and we may simply ignore the latter.

We cannot, however, get over the difficulty so easily in the quantum theory. $\cdots$ if a perturbation is applied to the system it may cause transitions from one kind of state to the other. $\cdots$ we can make no hard-and-fast separation $\cdots$ transitions can take place in which the energy of the electron changes from positive to negative even in the absence of any external field $\cdots$. Thus we cannot ignore the negative energy states without giving rise to ambiguity in the interpretation of the theory.

217

Let us examine the wave functions representing states of negative energy a little more closely. If we superpose a number of these wave functions in such a way as to get a wave packet, the motion of this packet will be along a classical trajectory given by the relativistic Hamiltonian (*Note:* Eqn.1 of Appendix 8A.) with $W - eA_0$ negative. Such a trajectory $\cdots$ is a possible trajectory for an ordinary electron (with positive energy) moving in the electromagnetic field with reversed sign, or for an electron of opposite charge (and positive energy) moving in the original electromagnetic field. Thus *an electron with negative energy moves in an external field as though it carries a positive charge.*

This result has led people to suspect a connection between the negative-energy electron and the proton [1]. One cannot, however, simply assert that a negative-energy electron *is* a proton, as that would lead to the following paradoxes: -

(i) A transition of an electron from a state of positive to one of negative energy would be interpreted as a transition of an electron into a proton, which would violate the law of conservation of charge.

(ii) Although a negative-energy electron moves in an external field as though it has positive charge, yet, as one can easily see from a consideration of conservation of momentum, the field it produces must correspond to its having a negative charge, *e.g.*, the negative-energy electron will repel an ordinary positive-energy electron although it is itself attracted by the positive-energy electron.

(iii) A negative-energy electron will have less energy the faster it moves and will have to absorb energy in order to be brought to rest. No particles of this nature have never been observed.

A closer consideration of the conditions that we should expect to hold in the actual world suggests that the connection between protons and negative-energy electrons should be on a somewhat different basis and this will be found to remove all the above-mentioned difficulties.

§2. *Solution of the Negative Energy Difficulty.*

The most stable states for an electron (*i.e.*, the states of lowest energy) are those with negative energy and very high velocity. All the electrons in the world will tend to fall into these states with emission of radiation. The Pauli exclusion principle, however, will come into play and prevent more than one electron going into any one state. Let us assume that there are so many electrons in the world that all the most stable states are occupied, or, more accurately, that *all the states of negative energy are occupied except perhaps a few of small velocity.* Any electrons with positive energy will now have very little chance of

218

jumping into negative energy states and will therefore behave like electrons are observed to behave in the laboratory. We shall have an infinite number of electrons in negative energy states, and indeed an infinite number per unit volume all over the world, but if their distribution is exactly uniform we should expect them to be completely unobservable. *Only the small departures from exact uniformity, brought about by some of the negative energy states being unoccupied, can we hope to observe.*

Let us examine the properties of the vacant states or "holes". The problem is analogous to that of the X-ray levels in an atom with many electrons. $\cdots\cdots$ Thus the hole $\cdots$ is much the same thing as a single electron in a region that is otherwise devoid of them.

$\cdots\cdots$ the motion of one of these holes in an external electromagnetic field will be the same as that of the negative-energy electron that would fill it, and will thus correspond to its possessing a positive charge. We are therefore led to the assumption that *the holes in the distribution of negative-energy electrons are the protons*. When an electron of positive energy drops into a hole and fills it up, we have an electron and proton disappearing together with emission of radiation.

A difficulty occurs when we consider the field produced by the distribution of negative-energy electrons. $\cdots\cdots$ the field produced by a proton will correspond to its having a positive charge.

In this way we can get over the three difficulties $\cdots$ all the *distinctive* things in nature have positive energy.

Can the present theory account for the great dissymmetry between electrons and protons $\cdots\cdots$ Possibly some more perfect theory $\cdots$ is necessary $\cdots$.

§3. *Application to Scattering.*

As an elementary application of the foregoing ideas we may consider the problem of the scattering of radiation by an electron $\cdots$ consisting of first the absorption of a photon with the electron simultaneously jumping to any state, and then the emission with the electron jumping into its final state, or else first the emission and then the absorption. We therefore have to consider altogether three states of the whole system, the *initial state* with the incident photon and the electron in its initial state, the *intermediate state* with either two or no photons and the electron in any state, and the *final state* with the scattered photon and the electron in its final state. The initial and final states $\cdots$ must have the same total energy, but the intermediate state $\cdots$ may have a considerably different energy.

The question $\cdots$ is to interpret those scattering processes for which the intermediate state is one of negative energy for the electron. According to previous ideas these intermediate states had no real physical meaning, so it was doubtful whether scattering processes that arise through them should be included in the calculation. This gave rise to a serious difficulty, since in some important practical cases nearly all the scattering comes from intermediate states with negative energy for the electron [2]. In fact for $\cdots$ the classical formula $\cdots$, the whole of the scattering comes from such intermediate states.

According to the theory of the present paper it is absolutely forbidden, by the exclusion principle, for the electron to jump into a state of negative energy, so that the double transition processes with intermediate states of negative energy for the electron must be excluded. We now have, however, another kind of double transition taking place, namely, that in which first one of the negative-energy electrons jumps up into the required final state for the electron with absorption (or emission) of a photon, and then the original positive-energy electron drops down into the hole formed by the first transition, with the emission (or absorption) of a photon. Such processes result in a final state of the whole system indistinguishable from the final state with the more direct processes, $\cdots$. These new processes just make up for those of the more direct processes that are excluded on account of the intermediate state having negative energy for the electron, since the matrix elements that determine the transition probabilities are just the same in the two cases, though they come in the reverse order. In this way the old scattering formulas, in which no intermediate states are excluded, can be justified.

Footnotes and References.

1) H. Weyl, Zeits. f. Phys. **56**, 332 (1929).

2) I am indebted to I. Waller for calling my attention to this difficulty.

APPENDIX 8.C:

Excerpt from: Physical Review 35, 562 (1930)

On the Theory of Electrons and Protons

by J.R. Oppenheimer

The Norman Bridge Laboratory of Physics

California institute of Technology

Pasadena, California

February 14, 1930

In a recent paper, Dirac has suggested that the reason why the transitions of an electron to states of negative energy, which are predicted by his theory of the electron, do not in fact occur is that nearly all of the states of negative energy are already occupied. Dirac has further shown that the unoccupied states of negative energy have many of the properties of protons; that, for instance, they may be represented by wave functions which would be taken to correspond to a particle of positive charge and positive mass. He has further shown that the mass associated with these gaps is not necessarily the same as that of the electron, and he has suggested the assumption that the gaps are protons. In order to account for the fact that the divergence of the electric field is not, in spite of the infinite electron density, everywhere infinite, Dirac further assumes that only the departures from the normal state in which all the negative energy states are filled are to be counted in computing the charge density for Maxwell's fourth equation

$$\nabla \cdot E = 4\pi\rho. \tag{1}$$

Finally, Dirac is able to account for the validity of the Thomson formula for the scattering of soft light by a free electron, in spite of the fact that the derivation of this formula on his theory of the electron - a derivation which makes explicit use of the transitions to states of negative energy which are now forbidden - is invalid. According to Dirac, the scattering takes place by a double electron jump, in which a negative-energy electron jumps up to some state of positive energy, and the original positive-energy electron falls down into the gap which is left.

There are several grave difficulties which arise when one tries to maintain the suggestion that the protons are the gaps of negative energy, and that there are no distinctive particles of positive charge. In the first place, we can easily see that Dirac's theory requires an infinite density of positive electricity; and since we should expect the de Broglie

221

waves of this charge to be quantized, we should expect some corpuscular properties for the positive charges. The reason why the theory requires an infinite positive charge is this: If the explanation of the scattering of an electron is to be tenable, a negative-energy electron must interact with the electromagnetic field in the way predicted by Dirac's theory of the electron; for otherwise the scheme proposed would not give the Thomson formula. But this means that there must be a term involving the current and charge vector of the negative-energy electrons in the total energy-momentum tensor for matter and radiation. Thus by (1), the divergence of the electric field will be everywhere infinite unless there is an infinite density of positive electricity to compensate the negative-energy electrons.

A further difficulty appears when we try to compute the scattering of soft light by a proton. $\cdots\cdots$ is determined by precisely the same matrix elements as those which gave the electron scattering, and that the present theory gives equal scattering coefficients for electron and proton. $\cdots\cdots$ whereas the Thomson formula requires the latter to be smaller by a factor equal to the square of the ratio of the masses.

Finally, there is a numerical discrepancy to be noted. According to Dirac's suggestions, the filling of the proton gaps in the distribution of negative-energy electrons should correspond to the annihilation of an electron and a proton $\cdots\cdots$. The conservation laws require at least two photons be emitted $\cdots\cdots$ we obtain for the mean life time of the electron:

$$T = Gm^2c^3/(e^4 n_p) \tag{2}$$

where G is a numerical constant of the order of unity, e the charge, m the mass of the electron, c the speed of light, and n_p is the density of free protons. $\cdots\cdots$ the matrix elements are precisely the same type as those which give correctly the Thomson formula $\cdots$; and (2) gives a mean lifetime for ordinary matter of the order of 10^{-10} seconds.

Thus we should hardly expect any states of negative energy to remain empty. If we return to the assumption of two independent elementary particles, of opposite charge and dissimilar mass, we can resolve all the difficulties raised in this note, and retain the hypothesis that the reason why no transitions to states of negative energy occur, either for electrons or protons, is that all such states are filled. In this way, we may accept Dirac's reconciliation of the absence of these transitions with the validity of the scattering formulas.

Footnotes and References.

1) P.A.M. Dirac, Proc. Roy. Soc. A**126**, 360 (1930).

APPENDIX 8.D:

**Excerpt from: H. Weyl, 'The Theory of Groups and Quantum Mechanics',
p. 225, Dover (1930).**

. . . . . .

III. Interchange of Past and Future.

The action is so constructed that it is invariant under *interchange of right and left;*
the corresponding substitution is

$$t, \vec{x} \rightarrow t, -\vec{x};$$
$$\phi, \vec{A} \rightarrow \phi, -\vec{A};$$
$$\psi_1 \rightarrow \psi_3, \quad \psi_2 \rightarrow \psi_4; \quad \psi_3 \rightarrow \psi_1; \quad \psi_4 \rightarrow \psi_2. \tag{612}$$

Does a corresponding result hold for the interchange of past and future? The foundations of the theory lead to the hope that it will be able to take account of the essential difference between two time directions, so obvious in Nature. But *Dirac* has remarked that the action $S = \int d^4 x \mathcal{L}$ goes over into $-S$ under the substitution

$$t, \vec{x} \rightarrow -t, -\vec{x},$$
$$\phi, \vec{A} \rightarrow -\phi, -\vec{A};$$
$$\psi_1 \rightarrow \psi_1, \quad \psi_2 \rightarrow \psi_2; \quad \psi_3 \rightarrow -\psi_3, \quad \psi_4 \rightarrow -\psi_4. \tag{613}$$

Hence, when, in dealing with the motion of an electron in an external electromagnetic field, we obtain a solution ψ which contains the time in the factor $e^{-i\omega t}$, this substitution will lead to a new solution which contains the time in the factor $e^{+i\omega t}$; or, more precisely, a solution of the problem obtained by changing $(\phi, \vec{A})$ into $(-\phi, -\vec{A})$. But this can be done by retaining the same external field with potentials $(\phi, \vec{A})$ and replacing e by $-e$. We denote such a particle, whose mass is the same as that of the electron but whose charge is opposite, as a "positive-charged electron"; it is not observed in Nature! It follows from what has been said above that the energy levels of such a particle are $-\hbar\omega$, where those of the usual negative-charged electron are $+\hbar\omega$. Disregarding this difference in sign, the two particles behave the same. *The electron will possess, in addition to its positive energy levels, negative ones as well,* the latter arising from the positive energy levels of the positive-charged electron on changing signs as above. Obviously something is wrong here; we should be able to get rid of these negative energy levels of the electron. But that seems impossible, for under the influence of the radiation field transitions should occur between positive and negative terms. That we have twice as many terms as we should have is obviously related to the fact that our quantity ψ has *four* instead of *two* components (satisfying first order differential equations). The solution of this difficulty would seem to lie in the direction of interpreting our four differential equations as including the proton in addition to the electron.

⋯⋯ Our field equations as a whole ⋯ are *not* invariant under ⋯ (6.13). However, ⋯⋯ all remain invariant under the substitution

$$t, \vec{x} \rightarrow -t, \vec{x};$$
$$\phi, \vec{A} \rightarrow \phi, -\vec{A};$$
$$\psi_1 \rightarrow \bar{\psi}_2, \quad \psi_2 \rightarrow -\bar{\psi}_1; \quad \psi_3 \rightarrow \bar{\psi}_4, \quad \psi_4 \rightarrow -\bar{\psi}_4. \tag{614}$$

This shows that the past and the future enter into our field theory in precisely the same manner ⋯⋯.

APPENDIX 8.E:

Excerpt from: Proc. Roy. Soc. A133, 60 (1931)

Quantized Singularities in the Electromagnetic Field

by P.A.M. Dirac, F.R.S.
St. John's College, Cambridge

(Received May 29, 1931)

. . . . . .

A recent paper by the author [1] may possibly be regarded as a small $\cdots$ advance. The mathematical formalism $\cdots$ involved a serious difficulty through its prediction of negative kinetic energy values for an electron. It was proposed to get over this difficulty, making use of Pauli's Exclusion Principle which does not allow more than one electron in any state, by saying that in the physical world almost all the negative energy states are already occupied, so that our ordinary electrons of positive energy cannot fall into them. The question then arises as to the physical interpretation of the negative energy states, which on this view really exist. We should expect the uniformly filled distribution of negative energy states to be completely unobservable to us, but an unoccupied one of these states, being something exceptional, should make its presence felt as a kind of hole. It was shown that one of these holes would appear to us as a particle with a positive energy and a positive charge and it was suggested that this particle should be identified with a proton. Subsequent investigations, however, have shown that this particle necessarily has the same mass as an electron [2] and also that, if it collides with an electron, the two will have a chance of annihilating one another much too great to be consistent with the known stability of matter [3].

It thus appears that we must abandon the identification of the holes with protons and must find some other interpretation for them. Following Oppenheimer [4], we can assume that in the world as we know it, *all*, and not merely nearly all, of the negative energy states for electrons are occupied. A hole, if there were one, would be a new kind of particle, unknown to experimental physics, having the same mass and opposite charge to an electron. We may call such a particle an anti-electron. We should not expect to find any of them in nature, on account of their rapid rate of recombination with electrons, but if they could be produced experimentally in high vacuum they would be quite stable and

225

amenable to observation. An encounter between two hard γ-rays (of energy at least half a million volts) could lead to the creation simultaneously of an electron and anti-electron, the probability of occurrence of this process being the same order of magnitude as that of the collision of the two γ-rays on the assumption that they are spheres of the same size as classical electrons. This probability is negligible, however, with the intensities of γ-rays at present available.

The protons on the above view are quite unconnected with electrons. Presumably the protons will have their own negative energy states, all of which are normally occupied, an unoccupied one appearing as an antiproton. Theory at present is quite unable to suggest a reason why there should be any difference between electrons and protons.

······ (*Note added:* Dirac returns to the primary point of this paper, which was to include magnetic monopoles in quantum mechanics.)

Footnotes and References.

1) P.A.M. Dirac, Proc. Roy. Soc. A**126**, 360 (1930).

2) H. Weyl, 'Gruppentheorie und Quantenmechanik,' 2nd ed. p.234(1931).

3) I. Tamm, Zeits. f. Phys. **62**, 545 (1930); J.R. Oppenheimer, Phys. Rev. **35**, 939 (1930); P.A.M. Dirac, Proc. Camb. Philos. Soc. **26**, 361 (1930).

4) J.R. Oppenheimer, Phys. Rev. **35**, 562 (1930).

APPENDIX 8.F:

Excerpt from: Physical Review 43, 491, 1933

The Positive Electron

CARL D. ANDERSON

California Institute of Technology, Pasadena, California
(Received February 28, 1933)

Out of a group of 1300 photographs of cosmic-ray tracks in a vertical Wilson chamber 15 tracks were of positive particles which could not have a mass as great as that of the proton. From an examination of the energy-loss and ionization produced it is concluded that the charge is less than twice, and is probably exactly equal to, that of the proton. If these particles carry unit positive charge the curvatures and ionizations produced require the mass to be less than twenty times the electron mass. These particles will be called positrons. Because they occur in groups associated with other tracks it is concluded that they must be secondary particles ejected from atomic nuclei. *Editor*

On August 2, 1932, during the course of photographing cosmic-ray tracks produced in a vertical Wilson chamber (magnetic field 15,000 gauss) designed in the summer of 1930 by Professor R.A. Millikan and the writer, $\cdots$ tracks $\cdots$ were obtained, which seemed to be interpretable only on the basis of the existence in this case of a particle carrying a positive charge but having a mass of the same order of magnitude as that normally possessed by a free negative electron. $\cdots\cdots$ The reason that this interpretation seemed so inevitable is that the track $\cdots$ cannot possibly have a mass as large as that of a proton for as soon as the mass is fixed the energy is at once fixed by the curvature. The energy of a proton of that curvature comes out 300,000 volts, but a proton of that energy $\cdots$ has a total range of about 5 mm in air while that portion of the range actually visible in this case exceeds 5 cm without a noticeable change in curvature. $\cdots$ assume $\cdots$ two independent electrons $\cdots$ give the impression of a single particle shooting through the lead plate $\cdots$ assumption was dismissed $\cdots$. Also $\cdots$ completely untenable $\cdots$ assumption of an electron of 20 million volts entering the lead on one side and coming out with an energy of 60 million volts on the other side. $\cdots$ other photographs were obtained which could be interpreted logically only on the positive-electron basis $\cdots$

While this paper was in preparation press reports have announced that P.M.S. Blackett and G. Occhiliani in an extensive study of cosmic-ray tracks have also obtained evidence for the existence of light positive particles confirming our earlier report.

I wish to express my great indebtedness to Professor R.A. Millikan for suggesting this research and for many helpful discussions during its progress. The able assistance of Mr. Seth H. Neddermeyer is also appreciated.

Chapter 9

Pauli-Weisskopf Canonical Quantization
of the Klein-Gordon Field

Summary: The controversies over the interpretation of the relativistic scalar wave equation were not finally resolved until 1934 when Pauli and Weisskopf showed that the paradoxes were overcome by canonical relativistic quantum field theory and, necessarily, a many body interpretation of the wave equation.

§1. Introduction.

The quantization of the relativistic scalar wave equation was achieved in its present form by Pauli and Weisskopf only in 1934. The Klein-Gordon relativistic scalar wave equation

$$(\Box + m^2)\phi = 0 \tag{1}$$

with $E = i\partial/\partial t$, $\vec{p} = -i\vec{\nabla}$ had been introduced by Schrödinger in his first paper on wave mechanics, and the Lagrangian and conserved current for the unquantized field discussed by Gordon already in 1926. There remained a controversy over the interpretation as a probability of the non-positive definite "density"

$$\rho = i\left(\phi^*\frac{\partial\phi}{\partial t} - \frac{\partial\phi^*}{\partial t}\phi\right), \tag{2}$$

and of the meaning of the "negative energy states". These could not be "filled" and reinterpreted following Dirac's hole hypothesis, because the integral spin particles of the Klein-Gordon equation do not obey the Pauli Exclusion Principle. The Klein-Gordon equation was in further disrepute because of its failure to describe the hydrogen spectrum fine-structure. For these reasons, Pauli and Weisskopf felt it necessary to justify revisiting the scalar wave equation. They conclude that the criticisms of the scalar wave

229

equation can be refuted: first, by interpreting the conserved density as a charge density which can be of either sign; and second, by giving up the single particle interpretation of the relativistic wave equation in favor of a many particle interpretation admitting pair creation and annihilation processes. In this way, Pauli and Weisskopf achieve a reinterpretation of the theory consistent with the requirements of quantum mechanics and relativity. The energy is shown to be positive definite "without a new hole-like hypothesis $\cdots\cdots$ or strange cancelling limits and subtraction artifices $\cdots\cdots$ as a result of the quantization of the fields" in Eqn.4.

As a brief reminder, we review

1) the Lagrangian-Hamiltonian formulation of mechanics for classical point particles with generalized coordinates q_j;

2) as extended to classical fields with amplitudes $\phi(\vec{x}, t), \phi^*(\vec{x}, t)$; then

3) "first"-quantized to express the particle-wave duality for particles; and finally

4) "second"-quantized to express the wave-particle duality of the field quanta.

In a biblical quotation used by Cornelius Lanczos in a similar context: *"Put off thy shoes from off thy feet, for the place whereon thou standest is holy ground." [EXODUS III,5]*

§2. Review of Lagrangian and Hamiltonian Classical Mechanics.

The Lagrangian $L(q_j, \dot{q}_j; t)$ for point particles with coordinates q_j, velocities $\dot{q}_j$ is generalized to the Lagrangian density

$$\mathcal{L}(\phi(x, t), \phi^*(x, t), \partial_\mu \phi(x, t), \partial_\mu \phi^*(x, t)), \qquad (3)$$

which must be a Lorentz invariant function of the (Lorentz invariant) field

ϕ and its complex conjugate ϕ^* which play the roles of generalized coordinates, and their derivatives. Then the classical wave equations emerge as the corresponding Euler-Lagrange equations. The formalism for the fields follows in close analogy from the equations of classical mechanics:

$$\begin{aligned}
L(q_j, \dot{q}_j; t) &\Rightarrow \mathcal{L}(\phi, \phi^*, \partial_\mu \phi, \partial_\mu \phi^*) \\
q_j(t) &\Rightarrow \phi(x,t), \quad \phi^*(x,t) \\
\dot{q}_j &\Rightarrow \phi_{,t} \quad (\equiv \partial_t \phi) \\
p_j = \frac{\partial L}{\partial \dot{q}_j} &\Rightarrow \Pi(x,t) = \frac{\partial \mathcal{L}}{\partial \phi_{,t}} \\
\frac{d}{dt}\frac{\partial L}{\partial \dot{q}_j} = \frac{\partial L}{\partial q_j} &\Rightarrow \partial_\mu \frac{\partial \mathcal{L}}{\partial \phi_{,\mu}} = \frac{\partial \mathcal{L}}{\partial \phi}.
\end{aligned} \tag{4}$$

As a simplest example, consider the simple harmonic oscillator and the free complex scalar classical field:

$$\begin{aligned}
L = T - V = \frac{m}{2}\dot{x}^2 - \frac{k}{2}x^2 &\Rightarrow \mathcal{L} = \phi^*_{,t}\phi_{,t} - \phi^*_{,x}\phi_{,x} - m^2\phi^*\phi \\
p = \frac{\partial L}{\partial \dot{x}} = m\dot{x} &\Rightarrow \Pi = \frac{\partial \mathcal{L}}{\partial \phi_{,t}} = \phi^*_{,t}; \quad \Pi^* = \phi_{,t} \\
\frac{d}{dt}\frac{\partial L}{\partial \dot{x}} - \frac{\partial L}{\partial x} = \dot{p} + kx = 0 &\Rightarrow \partial_\mu \frac{\partial \mathcal{L}}{\partial \phi_{,\mu}} - \frac{\partial \mathcal{L}}{\partial \phi} = (\Box + m^2)\phi^* = 0.
\end{aligned}$$

The Hamiltonian

$$H(p,q) = p\dot{q} - L \Rightarrow \mathcal{H}(\Pi, \phi; \Pi^*, \phi^*) = \Pi \cdot \phi_{,t} + \phi^*_{,t} \cdot \Pi^* - \mathcal{L}.$$

In the example above

$$\begin{aligned}
p = m\dot{x} &\Rightarrow \Pi = \phi^*_{,t}, \quad \Pi^* = \phi_{,t} \\
H = p\dot{x} - \frac{m}{2}\dot{x}^2 + \frac{k}{2}x^2 &\Rightarrow \mathcal{H} = \phi^*_{,t}\phi_{,t} + \phi^*_{,x}\phi_{,x} + m^2\phi^*\phi \\
\rightarrow \frac{p^2}{2m} + \frac{k}{2}x^2 &\Rightarrow \rightarrow \Pi \cdot \Pi^* + \phi^*_{,x}\phi_{,x} + m^2\phi^*\phi.
\end{aligned}$$

Note that both H and $\mathcal{H}$ are intrinsically positive in these examples.

The Hamilton equations of motion follow as:

$$\dot{x} = \frac{\partial H}{\partial p} \equiv \frac{p}{m} \quad \Rightarrow \quad \phi_{,t} = \frac{\partial \mathcal{H}}{\partial \Pi} \equiv \Pi^*$$

$$\dot{p} = -\frac{\partial H}{\partial x} = -kx \quad \Rightarrow \quad \Pi_{,t} = -\frac{\partial \mathcal{H}}{\partial \phi} \rightarrow (\Box + m^2)\phi^* = 0.$$

(The last step is somewhat involved and requires a functional derivative under a differential, followed by an integration by parts.)

The Poisson bracket

$$[F, G]_{PB} = \sum_j \left(\frac{\partial F}{\partial q_j} \frac{\partial G}{\partial p_j} - \frac{\partial G}{\partial q_j} \frac{\partial F}{\partial p_j} \right), \tag{5}$$

is of central importance in constructing quantum mechanics from classical. Here, F and G are functions of the system point in the phase space spanned by the locally orthogonal p's and q's as phase space coordinates.

The canonical (meaning fundamental or standard, originally meaning 'prescribed by the high priests') Poisson brackets

$$[Q_j, Q_k]_{q,p} = [P_j, P_k]_{qp} = 0; \quad [Q_j, P_k]_{q,p} = \delta_{j,k} \tag{6}$$

are the test for generalized coordinates and their conjugate momenta (Q, P), with respect to a predetermined choice (q, p). The Poisson bracket reproduces the Hamilton equations of motion directly as

$$\dot{q}_j = [q_j, H]_{PB} = \frac{\partial H}{\partial p_j}$$

$$\dot{p}_j = [p_j, H]_{PB} = -\frac{\partial H}{\partial q_j}. \tag{7}$$

§3. Canonical Quantization, First and Second Quantization.

The prescription for canonical quantization is to replace the canonical Poisson brackets in the classical mechanics by the commutator brackets (divided by $i\hbar$) of the corresponding quantum operators:

$$[q_j, p_k]_{PB} = \delta_{jk} \Rightarrow [q_j, p_k]_- / i\hbar = [q_j p_k - p_k q_j] / i\hbar = \delta_{jk}. \tag{8}$$

The commutation relation can be realized by the identification of the momentum p as the differential operator

$$p = \frac{\hbar}{i} \frac{\partial}{\partial x},$$ (9)

which gives the desired result

$$[x, p]_- F(x) = i\hbar F(x),$$ (10)

which in turn is motivated by the Schrödinger momentum eigenfunction $\exp(ip' \cdot x/\hbar)$ involving the momentum eigenvalue p'. *First quantization* elevates the coordinates and their conjugate momenta to coordinates and operators in order to impose particle-wave duality. We obtain wave operators for deBroglie waves whose eigenvalues are the observable values of the dynamical variables.

For fields $\phi(x, t)$ labelled by $\vec{x}$ rather than j, the generalization for equal time commutation brackets is

$$[q, p]_{PB} \Rightarrow [\phi(x, t), \Pi(x', t)]_- = i\hbar \delta^3(x - x').$$ (11)

This step of *second quantization* elevates wave fields from classical c-number fields to quantum q-number operators, and completes the wave-particle duality by giving particle, that is quantum, properties to wave fields.

§4. Pauli-Weisskopf Quantization of the Klein-Gordon Field.

Pauli and Weisskopf follow the canonical quantization introduced by Heisenberg and Pauli [9.5] (Eqns.5-13 and 37) to deduce the conserved current (Eqns.14,16 and 42,43) and energy-momentum tensor (Eqns.21,22 and 45) in the field free case; and in the presence of an electromagnetic field. The fields $\phi(x), \phi^*(x)$ and their conjugate momenta $\pi(x), \pi^*(x)$ are made quantum field operators by requiring them to satisfy the canonical commutation relations (Eqn.10). They then introduce Fourier coefficients a_k, b_k and a_k^*, b_k^* (Eqns.27,28) which satisfy the commutation relations

$$[a_k, a_l^*]_- = \delta_{kl}, \quad [b_k, b_l^*]_- = \delta_{kl},$$ (12)

233

all others zero, familiar from the lowering and raising operators of the harmonic oscillator. These are interpreted as annihilation and creation operators of quanta of the complex field ϕ with their associated occupation number operators

$$N_k = a_k^* a_k, \qquad (13)$$

and so on, for the relevant energy, momentum and charge eigenstates. This interpretation is made evident (Eqns.29-33) for the energy, momentum, charge, and current. The fact that the energy is positive definite is clear (Eqns.4,7,21,29,37,and 45).

There is a residual element of confusion in the equations of Pauli and Weisskopf, connected with the question of factor ordering in various operators. This is commented on following Eqn.15 where the operator order is chosen to give zero vacuum expectation values "by requiring each term to be Hermitian $\cdots$".

The formulation of electrodynamics for the canonically quantized relativistic scalar field, whose quanta satisfy Bose-Einstein statistics, is complete in Eqn.45 and expressed in terms of annihilation and creation operators in Eqns.52,53. The various matrix elements required for computation (Eqn.54) are ready for the perturbative calculation of pair production and vacuum polarization following a similar calculation for the Dirac hole theory by Bethe and Heitler.

Pauli and Weisskopf, already in this paper, refer for the first time to the contradictions which would arise if one were to try to impose the Pauli Exclusion Principle on the spinless particles of the scalar field theory. "This is connected to the fact that the equation

$$\phi(x)\phi(x') + \phi(x')\phi(x) = 0$$

and its Hermitian conjugate, lead to $\phi(x) = 0$ and $\phi^*(x) = 0$." Pauli himself, then Belinfante, then deWet, and others tried to prove this statement. Finally after almost a quarter of a century, after the development of the

Hall-Wightman theorem, after the nearly complete proof of Lüders and Zumino, ultimately Burgoyne was able to give a definitive statement of this seemingly trivial fact, and of its counterpart for spinor fields, even including interactions which had never been done before, which established the spin-statistics connection in relativistic quantum field theory. But not to everyone's satisfaction.

In his autobiography "The Joy of Insight", Weisskopf reminisces at length about this interval of his life as Pauli's assistant in Zurich. He recalls being warned by Peierls before taking the appointment about Pauli's acerbic manner and did in fact run into some acid remarks $\cdots$ "I really wanted Bethe $\cdots$" $\cdots$ "I should have taken Bethe $\cdots$". But Weisskopf maintains that Pauli's real nature was that of childlike honesty which was sometimes hurtful but never malicious. If he could be perceived as bullying junior colleagues, he could just as well be seen contradicting Heisenberg or Bohr.

This is the last of the fundamental developments which we will present. In the following chapters we turn at last to the specific question of the Spin-Statistics Theorem as it was developed over the many years of its evolution from these - what have come to be seen as - mythically heroic origins of quantum mechanics.

§5. Biographical Note on Weisskopf.(†)

Victor Frederick Weisskopf (1908-) got his PhD at Gottingen with Wigner (1931); was visitor at Berlin (31-32); Rockefeller Fellow at Copenhagen and Cambridge (32-33); Pauli's Assistant at Zurich (33-36); Research Associate at Copenhagen (36-37); Assistant Professor at Rochester (37-43); Group Leader at Los Alamos (43-47); Professor at MIT (47-74); Emeritus Professor at MIT (1974-); Director General of CERN (61-65); awarded sixteen honorary doctorates; the Planck Medal (56); the Prix Mondial Cino Del Duca (72); President of the American Physical Society (1960); President of the American Academy of Arts and Sciences (1977-80).

Weisskopf was in a second wave of quantum mechanics with a very few other notables such as Bethe and Heitler, people who had the analytic power to explore the consequences of relativistic quantum field theory beyond the lowest order of perturbation theory. Their calculations were extremely cumbersome because the streamlined manifestly covariant techniques of Feynman and Schwinger were still ten years in the future (and not easy themselves). Weisskopf [9.14] by 1939 had shown that the electron self energy was only logarithmically divergent to all orders, a necessary precursor to the renormalization group invented fifteen years later by Gell-Mann and Low [9.15].

Weisskopf switched his research to nuclear theory where with Blatt he wrote the definitive textbook *Theoretical Nuclear Physics* (Wiley, New York, 1952). One of his great later accomplishments was as coinventer of the MIT Bag Model [9.16] of quark and gluon confinement in baryons and mesons with a direct (but not complete) connection to Quantum Chromodynamics.

Weisskopf has authored two autobiographical books [9.17], in both of which his kind, generous, gentle, and cultured nature are clearly seen.

(† - *American Men and Women of Science* (Bowker, New York, 1986) edited by Jaques Cattell Press, p.521; see also references [9.13,17]).

Bibliography and References.

9.1) W. Pauli and V.Weisskopf, Helv. Phys. Acta **7**, 109 (1934); see our App.9A.

9.2) E. Schrödinger, Ann. Physik **79**, 361, 489 (1926).

9.3) W. Gordon, Zeits. f. Phys. **40**, 117 (1926).

9.4) Cornelius Lanczos, *The Variational Principles of Mechanics* (Dover, New York, 1970), p.229.

9.5) W. Heisenberg and W. Pauli, Zeits. f. Phys. **56**, 1 (1929); **59**, 168 (1930).

9.6) H. Bethe and W. Heitler, Proc. Roy. Soc. A**134**, 83 (1934).

9.7) W. Pauli, Ann. Inst. H. Poincaré **6**, 137 (1936); see our App.10A.

9.8) F.J. Belinfante, Physica **VII**, 177 (1940); see our App.12B.

9.9) J.S. deWet, Phys. Rev. **57**, 646 (1940); see our App.13B.

9.10) A.S. Wightman, Phys. Rev. **101**, 860 (1956); see our App.18A,B.

9.11) G. Lüders and B. Zumino, Phys. Rev. **110**, 1450 (1958); see our App.17A.

9.12) N. Burgoyne, Nuov. Cim. **VIII**, 607 (1958); see our App.17B.

9.13) V.F. Weisskopf, *The Joy of Insight* (BasicBooks, Harper-Collins, New York, 1991), pp.75,76.

9.14) V.F. Weisskopf, Phys. Rev. **56**, 72 (1939).

9.15) M. Gell-Mann and F. Low, Phys. Rev. **95**, 1300 (1954).

9.16) A. Chodos, R.L. Jaffe, K. Johnson, C.B. Thorn, and V.F. Weisskopf, Phys. Rev. D**9**, 3471 (1974).

9.17) V.F. Weisskopf, *The Privilege of Being a Physicist* (W.H. Freeman, San Francisco, 1989); see also [9.13] above.

APPENDIX 9.A:

Excerpt from: Helvetica Physica Acta 7, 709 (1934)

On the Quantization of the Scalar Relativistic
Wave Equation

by **W. Pauli** and **V. Weisskopf** in Zurich

(27.VII.34)

Summary: In the following, the consequences of applying the Heisenberg-Pauli formalism for the quantization of fields are investigated for the scalar relativistic wave equation of matter fields satisfying Bose-Einstein statistics. The result, without further hypotheses, is the existence of particles of equal rest mass but of opposite charge, which - by absorption or emission of radiation - can be created or destroyed in pairs. The strength of these processes is of the same order of magnitude as for particles of the same charge and mass obeying Dirac's hole theory. The theory investigated here is the relativistic formulation of oppositely charged particles without spin satisfying a hole theory in the sense that the energy is always positive. As in the original hole theory, it turns out that the theory considered here involves infinite self energies and infinite polarizability of the vacuum.

§1. On the Connection of the Scalar Relativistic Wave Equation with the Existence of Oppositely Charged Particles.

The scalar relativistic wave equation is well known; with the introduction of the operators [1]

$$E = i\frac{\partial}{\partial t}, \quad p_k = -i\frac{\partial}{\partial x^k} \tag{1}$$

($k = 1, 2, 3$), in the force-free case we can write

$$E^2 - \sum_{k=1}^{3} p_k^2 - m^2 = 0. \tag{2}$$

Generally, it must be abandoned in favor of the Dirac four component wave equation since it does not yield spin for the particle and for the electron it represents an invalid approximation to reality. It therefore requires a separate justification to take up the discussion of the scalar wave equation once again. We believe such a justification can be given.

238

In particular, the empirical discovery of positrons and their theoretical prediction in the new interpretation by Dirac of the negative energy states appearing in his original theory, make necessary a revision in Dirac's previous argument. He based his original rejection of the scalar wave equation in favor of the spinor wave equation on that argument. In the following it will be shown, from the prescription for the quantization of fields formulated by Heisenberg and Pauli [2], not only that no general objections can be held against the scalar wave equation from the standpoint of quantum mechanical transformation theory, but also that relativistic invariance and gauge invariance of the theory can be preserved. *All this is accomplished without any further hypotheses about the consequences of the existence of oppositely charged particles, or of the occurrence of processes which create and destroy such particle pairs. In addition, in spite of all, the energy of the material fields is always positive.* For the particles, the statistics of symmetrical states (Bose-Einstein statistics) must be assumed; but perhaps things only appear to be satisfactory, because without introducing spin the introduction of the Exclusion Principle itself is not permitted under the assumption of relativistic invariance of the theory.

We turn now to the *a priori* argument of Dirac against the scalar wave equation [3], which depends essentially on two assumptions:

1) In the relativistic quantum theory, it should be possible to formulate a contradiction free one body problem.

2) The spatial density $\rho(x)$ (to be interpreted statistically) is a meaningful idea. Upon integration over an arbitrary finite volume one gets from it an "observable" (in the sense of transformation theory) with the eigenvalues 0 and 1.

As soon as the first assumption is made, it is not necessary to use the formalism of the quantization of fields on the problem; it is then possible to deal with ordinary fields in three dimensional space. The second assumption has the consequence that the particle density not only should be the fourth component of a four vector and must satisfy a continuity equation, but also must have the property that it should never be negative. Moreover the resulting density matrix, after integration over an infinite volume, as Dirac showed, only has the correct eigenvalue when the particle density is of the form [4]:

$$\rho(x) = \sum_r \phi_r^* \phi_r.$$

In contrast, the particle density for the scalar relativistic wave equation has the form

$$\rho(x) = \phi^* \left(i \frac{\partial \phi}{\partial t} - e \Phi_0 \phi \right) - \left(i \frac{\partial \phi^*}{\partial t} + e \Phi_0 \phi^* \right) \phi, \tag{3}$$

where e is the charge of the particle and Φ_0 is an external scalar potential. Since this is not of the permitted form, it appears to produce a contradiction.

As is known, Dirac has now - based on the result that, for his wave equation, a wave packet of states of negative energy in an external field moves like a particle with opposite charge, equal mass, and positive energy - explained that the states of negative energy represent the positrons in the following way: Only the difference from the case where all states of negative energy are occupied, the "holes" amongst the filled states of negative energy, should be observable, and contribute to the "real" (field generating) charge density and to the "actual" (in fact positive) energy.

On the difficulty of a contradiction-free formulation of this Dirac hole theory of the electron and positron in the presence of an external field, which is certainly much discussed in the literature, and is to be met again soon, we can say the following:

1) Because of the process of pair production and the new interpretation of the states of negative energy, it is no longer possible to limit oneself to a one body problem.

2) The particle density no longer has a direct physical meaning [5]. In the force free case, it is of course the number of particles with given momentum (probability density in momentum space) and therefore the total number of particles present is a significant "observable".

3) On the other hand, not only the total charge but also the charge density is a meaningful observable. After integration over a finite volume it must - even in the presence of an external field - have the eigenvalues $0, \pm 1, \pm 2, \cdots, \pm N$, which can be both positive as well as negative. The charge density $\rho(x)$ and the total number of particles present are however not equivalent.

The requirements are now sufficiently modified from the original ones of a truly relativistic one body problem, *that there is no longer a restriction to the special form* $\sum_r \phi_r^* \phi_r$ *for the charge density.* We will also show that the finally formulated requirements of the scalar relativistic theory for spinless particles with Bose-Einstein statistics are also fulfilled in the Dirac hole theory. For this, it is natural to interpret the expression (3) not as a particle density, but as a charge density.

The main interest in the latter theory, it seems to us, lies in the fact that automatically - without a new hole-like hypothesis and without any changes in the quantum theory or strange cancelling limits and subtraction artifices [6] - the energy of the particles is always positive as a result of the quantization of the fields. It follows that the Hamiltonian for the matter waves in the scalar theory discussed here - in contrast to the corresponding

240

expression in Dirac's spinor theory - always takes a *positive definite* form:

$$H = \int dV \left\{ |i\frac{\partial \phi}{\partial t} - e\Phi_0\phi|^2 + \sum_{k=1}^{3} |i\frac{\partial \phi}{\partial x^k} + e\Phi_k\phi|^2 + m^2|\phi|^2 \right\}. \tag{4}$$

In view of the hypothesis free nature of this scalar relativistic theory, one might be surprised that "nature has made no use" [7] of this possibility. There apparently are no oppositely charged particles with no spin with Bose-Einstein statistics, which can be created and destroyed by radiation. One must presume that the question of the applicability of the theory discussed here, for example to α-particles which are created by the effects of nuclear structure, lies beyond the validity range of the present quantum theory. Also the theory discussed here leads, as will be shown in §4 for the question of the polarization of the vacuum, to similar infinities as did the original form of the hole theory [8]. These lead, among other things, to an infinite self-energy not only for the charged particles, but also for the photon [9]. Any further progress on these questions will probably first require a theoretical understanding of the value of Sommerfeld's fine structure constant.

§2. Derivation of the Quantization of Fields in the Force-free Case.

The Lagrange function of the scalar relativistic theory [10] is (with $\mu, \nu \cdots = 1, \cdots 4$):

$$L = \sum_{\nu=0}^{3} \frac{\partial \phi^*}{\partial x_\nu} \frac{\partial \phi}{\partial x^\nu} - m^2\phi^*\phi = \frac{\partial \phi^*}{\partial t} \frac{\partial \phi}{\partial t} - \sum_{k=1}^{3} \frac{\partial \phi^*}{\partial x^k} \frac{\partial \phi}{\partial x^k} - m^2\phi^*\phi. \tag{5}$$

The relativistic energy-momentum tensor is

$$T_{\mu\nu} = \left(\frac{\partial \phi^*}{\partial x^\mu} \frac{\partial \phi}{\partial x^\nu} + \mu \leftrightarrow \nu \right) - L\delta_{\mu\nu}, \tag{6}$$

and the energy (Hamiltonian)

$$H = \int T_{00} dV = \int dV \left\{ \frac{\partial \phi^*}{\partial t} \frac{\partial \phi}{\partial t} + \sum_{k=1}^{3} \frac{\partial \phi^*}{\partial x^k} \frac{\partial \phi}{\partial x^k} + m^2\phi^*\phi \right\}, \tag{7}$$

and the momentum

$$G^k = -\int T_{0k} dV = -\int dV \left\{ \frac{\partial \phi^*}{\partial t} \frac{\partial \phi}{\partial x^k} + \frac{\partial \phi^*}{\partial x^k} \frac{\partial \phi}{\partial t} \right\}. \tag{8}$$

We interpret ϕ^* and ϕ as q-number operators (acting on the Schrödinger functional), where ϕ^* is the Hermitian conjugate of ϕ. We also have the momenta π and π^* canonically conjugate to ϕ and ϕ^* according to the rule

$$\pi = \frac{\partial L}{\partial (\partial \phi/\partial t)} = \frac{\partial \phi^*}{\partial t}, \qquad \pi^* = \frac{\partial L}{\partial (\partial \phi^*/\partial t)} = \frac{\partial \phi}{\partial t}. \tag{9}$$

241

For particles with Bose-Einstein statistics the canonical commutation relations are

$$I) \qquad i[\pi(x, t), \phi(x', t)]_- = \delta(x - x'), \quad i[\pi^*(x, t), \phi^*(x', t)]_- = \delta(x - x'),$$

where on the right side $\delta(x - x')$ is the well known Dirac δ-function and we have set

$$[A, B]_- = AB - BA. \tag{10}$$

All other commutators among ϕ, ϕ^*, π, π^* vanish.

The application of the rule

$$\frac{\partial f}{\partial t} = i[H, f]_- \tag{11}$$

to ϕ, ϕ^* leads to the identity; to π, π^* with (9) it leads to the wave equation

$$\frac{\partial^2 \phi}{\partial t^2} = \nabla^2 \phi - m^2 \phi \tag{12}$$

and its Hermitian conjugate for ϕ^*.

Further, as it must be, the rule

$$\frac{\partial f}{\partial x^k} = -i[G_k, f]_- \tag{13}$$

is satisfied for all quantities f. Only in the expression for the momentum does an ambiguity occur in the order of the factors. It is resolved by requiring that the integrand of the expression (8), which is the momentum density, be a Hermitian operator.

We come now to the expression for the charge density ρ and the current density $\vec{j}$ (measured in units of the electron charge e). They satisfy the continuity equation

$$\frac{\partial \rho}{\partial t} + \vec{\nabla} \cdot \vec{j} = 0 \tag{14}$$

or, in four-vector form,

$$\frac{\partial j^\nu}{\partial x^\nu} = 0. \tag{14}$$

These are given by

$$\rho = i \left(\phi^* \frac{\partial \phi}{\partial t} - \phi \frac{\partial \phi^*}{\partial t} \right), \quad \vec{j} = -i \left(\phi^* \vec{\nabla} \phi - \phi \vec{\nabla} \phi^* \right), \tag{15}$$

or

$$j_\nu = i \left(\phi^* \frac{\partial \phi}{\partial x^\nu} - \phi \frac{\partial \phi^*}{\partial x^\nu} \right). \tag{15}$$

The ambiguity in the order of the factors on the right side is not resolved by requiring each term to be Hermitian. A determination of the factor ordering on the basis of the

242

Hermiticity of the density operator alone is not possible. The order used is suitable because it gives rise to no zero point density, and is in agreement with relativistic invariance and the continuity equation.

The density operator can also be written using (9) as:

$$\rho = -i(\pi\phi - \pi^*\phi^*) = -i(\phi\pi - \phi^*\pi^*),\qquad(16)$$

which is useful for the proof that ρ has the eigenvalue at a given x_0

$$\rho(x) = N\delta(x - x_0)$$

with N an integer. For this, it is easiest to use Hermitian operators u_1 and u_2 where

$$\phi = \frac{1}{\sqrt{2}}(u_1 + iu_2),\quad \phi^* = \frac{1}{\sqrt{2}}(u_1 - iu_2)$$

and correspondingly p_1 and p_2 defined by

$$\pi = \frac{1}{\sqrt{2}}(p_1 + ip_2),\quad \pi^* = \frac{1}{\sqrt{2}}(p_1 - ip_2),$$

from which it follows

$$u_1 = \frac{1}{\sqrt{2}}(\phi + \phi^*),\qquad u_2 = \frac{-i}{\sqrt{2}}(\phi - \phi^*)$$

$$p_1 = \frac{\partial u_1}{\partial t} = \frac{1}{\sqrt{2}}(\pi + \pi^*),\qquad p_2 = \frac{\partial u_2}{\partial t} = \frac{-i}{\sqrt{2}}(\pi^* - \pi).$$

Then

$$[u_1(x), p_1(x')]_- = i\delta(x - x'),\quad [u_2(x), p_2(x')]_- = i\delta(x - x'),$$

and all other commutators among u_1, u_2, p_1, p_2 vanish. Now energy, momentum and charge density split up into the sum of expressions which depend only on p_1 or u_1 and p_2 or u_2; for example

$$\rho = p_1 u_2 - p_2 u_1.\qquad(16)$$

By analogy with the expression for a component of the angular momentum, one can immediately conclude that $\rho(x)$ has the eigenvalues $N\delta(x - x')$ with $N = 0, \pm 1, \pm 2, \cdots$. Since the values of the density at different space points commute with one another, it follows that an arbitrary finite region v contains the charge

$$Q_v = \int_v \rho dV$$

(in units of e) with the eigenvalues $0, \pm 1, \cdots \pm N$.

We remark also, that in the proposed theory all relations including the commutation relations remain valid under the interchange of all operators with their Hermitian conjugates (that is ϕ with ϕ^*, π with π^*). Since under this interchange, the four-vector current

243

changes sign, the symmetry of the theory with respect to positive and negative charges is demonstrated.

In addition it should be noted, that a resolution of the density ρ into commuting parts with only positive and only negative eigenvalues is possible in infinitely many ways, but that none of these parts by itself satisfies a continuity equation and none is relativistically invariant [11].

Next we investigate the theory in momentum space, which is important for applications and also of physical interest in its own right. In order to obtain momentum space sums in place of integrals, we use the well known formal method of confining the fields in a cube of edge length L and volume $V = L^3$, with periodic boundary conditions, so that the components of the propagation vectors $\vec{k}$ of the waves must be integral multiples of $2\pi/L$. We use

$$u_k = \frac{1}{\sqrt{V}} e^{i\vec{k}\cdot\vec{x}} \tag{17}$$

as a complete orthonormal system of c-number eigenfunctions for which

$$\int_V dV u_k^*(x) u_l(x) = \delta_{kl}. \tag{18}$$

For the sake of simplicity, we write the index k as a single index instead of the three components of $\vec{k}$ as three indices, and similarly for the sum over k.

We expand the functions ϕ, ϕ^*, π, π^* in terms of u_k

$$\phi(x) = \frac{1}{\sqrt{V}} \sum_k q_k e^{i\vec{k}\cdot\vec{x}} \quad \pi(x) = \frac{1}{\sqrt{V}} \sum_k p_k e^{-i\vec{k}\cdot\vec{x}}, \tag{19}$$

with the inverse formulas

$$q_k = \frac{1}{\sqrt{V}} \int_V dV \phi(x) e^{-i\vec{k}\cdot\vec{x}}, \quad p_k = \frac{1}{\sqrt{V}} \int_V dV \pi(x) e^{i\vec{k}\cdot\vec{x}}, \tag{19}$$

and similarly for their Hermitian conjugates. The q-numbers p_k, q_k, p_k^*, q_k^* satisfy the commutation relations

$$II) \qquad [q_k, p_l]_- = i\delta_{kl}, \quad [q_k^*, p_l^*]_- = i\delta_{kl},$$

with all other pairs commuting with one another. From (9) it follows that

$$p_k = \dot{q}_k^*, \quad p_k^* = \dot{q}_k. \tag{20}$$

For the Hamiltonian and the momentum, (7) and (8) give

$$H = \sum_k \left(p_k^* p_k + E_k^2 q_k^* q_k \right), \tag{21}$$

244

$$\vec{G} = -i \sum_k \vec{k} \left(p_k q_k - q_k^* p_k^* \right). \tag{22}$$

Here the abbreviation (always with the positive square root) is

$$E_k = +\sqrt{k^2 + m^2}. \tag{23}$$

One easily establishes the validity of Eqn.11 for the p_k, q_k, p_k^*, q_k^*; in particular

$$\dot{p}_k = i[H, p_k]_- = -E_k^2 q_k^*,$$
$$\dot{p}_k^* = i[H, p_k^*]_- = -E_k^2 q_k. \tag{24}$$

From (15) and (16) we get the total charge and the total current

$$Q = \int_V dV \rho, \quad \vec{J} = \int_V dV \vec{j},$$

expanded in terms of different momentum eigenfunctions. We get

$$Q = -i \sum_k \left(p_k q_k - p_k^* q_k^* \right), \tag{25}$$

and

$$\vec{J} = 2 \sum_k \vec{k} q_k^* q_k. \tag{26}$$

We will see that $\vec{J}$ is not constant in time.

We next show that the contributions of the individual wave numbers k to the total charge, to the energy, and to the momentum can be simultaneously separated into two parts which have a simple physical interpretation. For this, we introduce the following variables a_k, a_k^*, b_k, b_k^*:

$$p_k = \frac{\sqrt{E_k}}{\sqrt{2}} \left(a_k^* + b_{-k} \right), \quad q_k^* = \frac{-i}{\sqrt{2E_k}} \left(a_k^* - b_{-k} \right), \tag{27}$$

and their Hermitian conjugates for p_k^*, q_k; with the inverse formulas

$$a_k^* = \frac{1}{\sqrt{2E_k}} \left(p_k + iE_k q_k^* \right), \quad b_{-k} = \frac{1}{\sqrt{2E_k}} \left(p_k - iE_k q_k \right), \tag{28}$$

and their Hermitian conjugates for a_k, b_{-k}^*. The commutation relations for the new variables are

$$III) \quad [a_k, a_l^*]_- = \delta_{kl}, \quad [b_k, b_l^*]_- = \delta_{kl},$$

with all others zero.

One further obtains

$$H = \sum_k E_k \left(a_k^* a_k + b_k^* b_k + 1 \right), \tag{29}$$

$$\vec{G} = \sum_k \vec{k} \left(a_k^* a_k + b_k^* b_k \right), \tag{30}$$

and for the total charge

$$Q = \sum_k \left(a_k^* a_k - b_k^* b_k \right). \tag{31}$$

Finally from (26) for the total current

$$\vec{J} = \sum_k \frac{\vec{k}}{E_k} \left(a_k^* a_k - b_k^* b_k - a_k^* b_{-k}^* - a_k b_{-k} + 1 \right). \tag{32}$$

The commutation relations for the a, b, a^*, b^* have the consequence that the operators

$$N_k^+ = a_k^* a_k, \quad N_k^- = b_k^* b_k \tag{33}$$

commute and both have the non-negative integral eigenvalues $0, 1, 2, \cdots$. The expressions for charge, energy, and momentum justify us in the following interpretation (in particular for the force-free case):

N_k^+ *is the number of particles with charge* $+1$ *and momentum* $\vec{k}$, *and* N_k^- *the number of particles with the charge* -1 *and momentum* $\vec{k}$ [12].

It should be pointed out, that the term $+1$ in the energy expression is a zero point energy (vacuum energy) of the matter wave, which is completely analogous to the zero point energy of the electromagnetic field, and can be ignored for all applications and without damage to the relativistic invariance of the theory. A similar statement holds for the term with $+1$ in the expression for the current. It is of decisive importance that even apart from this term, the energy is always positive by itself.

The terms with $a_k b_{-k}$ and $a_k^* b_{-k}^*$ in the current are important because these prevent it from being time independent even in the force free case. As one sees from the equations of motion

$$\dot{a}_k = -i E_k a_k, \quad \dot{b}_k = -i E_k b_k, \tag{34}$$

and their integrals

$$a_k = a_k(0) e^{-i E_k t}, \quad b_k = b_k(0) e^{-i E_k t}, \tag{35}$$

and the Hermitian conjugates of these for a_k^*, b_k^*, which follow from (11), (III), and (29). These terms are analogous to Schrödinger's *Zitterbewegung* (vacuum fluctuations) and give rise to the processes of pair creation and pair annihilation, as will be shown in the next section.

As already explained in the introduction, a contradiction free derivation of the scalar theory for particles obeying the Exclusion Principle is not possible. A detailed investigation of the Hamiltonian function in terms of the a, b shows that, with the assumption

246

of Fermi-Dirac statistics, the relativistic invariance of the four-vector current can not be maintained. This is connected to the fact that the equation

$$\phi(x)\phi(x') + \phi(x')\phi(x) = 0,$$

and its Hermitian conjugate, lead to $\phi(x) = 0$ and $\phi^*(x) = 0$.

§3. The Presence of an External Force.

For a particle of charge e one goes from the force free case to the case of an external electromagnetic field with the four-potential Φ_μ by replacing the operator

$$p_\mu \rightarrow p_\mu - e\Phi_\mu. \tag{36}$$

The Lagrangian of the matter field is then

$$L^m = \sum_{\nu=1}^{4} ([p_\nu - e\Phi_\nu]\phi)^* ([p^\nu - e\Phi^\nu]\phi) - m^2\phi^*\phi, \tag{37}$$

and the Hamiltonian is

$$H^m = \int dV[(\partial_t\phi + ie\Phi_0\phi)^*(\partial_t\phi + ie\Phi_0\phi) +$$

$$\sum_{k=1}^{3}(\partial_k\phi - ie\Phi_k\phi)^*(\partial_k\phi - ie\Phi_k\phi) + m^2\phi^*\phi]. \tag{37}$$

If one adds to this the Lagrangian for the electromagnetic field

$$L^{em} = \frac{1}{8\pi}(E^2 - B^2) \tag{38}$$

corresponding to the electromagnetic energy

$$H^{em} = \frac{1}{8\pi} \int \left(E^2 + B^2\right) dV \tag{38}$$

then one gets the total energy integral

$$H^m + H^{em} = \text{const.}$$

In addition, the variation of the action integral

$$S = \int (L^m + L^{em}) \, dV \, dt$$

with respect to the fields ϕ, ϕ^*, Φ_ν leads to the wave equations

$$\left(\frac{\partial}{\partial t} + ie\Phi_0\right)^2 \phi = \sum_{k=1}^{3} \left(\frac{\partial}{\partial x^k} - ie\Phi_k\right)^2 \phi - m^2\phi, \tag{39}$$

247

and the Hermitian conjugate for ϕ^*; and to Maxwell's equations

$$\vec{\nabla} \times \vec{B} - \frac{\partial \vec{E}}{\partial t} = 4\pi e \vec{j} \tag{40}$$

and

$$\vec{\nabla} \cdot \vec{E} = 4\pi e \rho, \tag{41}$$

with the following expressions for ρ and $\vec{j}$

$$\rho = i \left(\phi^* \frac{\partial \phi}{\partial t} - \phi \frac{\partial \phi^*}{\partial t} \right) - 2e\Phi_0 \phi^* \phi, \tag{42}$$

and

$$j_k = -j^k = i \left(\phi^* \frac{\partial \phi}{\partial x^k} - \phi \frac{\partial \phi^*}{\partial x^k} \right) - 2e\Phi_k \phi^* \phi, \tag{43}$$

which correspond to the expressions (15) of the force free case with the characteristic additions. Use of the wave equation (39) as well as Maxwell's equations is necessary to demonstrate the validity of the continuity equation (14) for the new forms of $(\rho, \vec{j})$. An immediate result of (36) is the invariance of the Lagrangian and Hamiltonian and also the expressions for current and charge density under the gauge transformations

$$\Phi'_\mu = \Phi_\mu + \frac{\partial \lambda}{\partial x^\mu}, \quad \phi' = \phi e^{ie\lambda}. \tag{36}$$

It is also important that the wave equation, Maxwell's equations, and the Hamiltonian remain unchanged when ϕ is replaced by ϕ^* and at the same time e is replaced by $-e$, which has the consequence that the theory is symmetric with respect to positive and negative charge. All these statements remain valid when ϕ, ϕ^*, Φ are interpreted as q-numbers.

It is important that the definition of π, π^* be changed from (9) to

$$\pi = \frac{\partial L}{\partial(\partial \phi/\partial t)} = \frac{\partial \phi^*}{\partial t} - ie\Phi_0 \phi^*, \tag{44}$$

and its Hermitian conjugate for π^*. This new π still satisfies the commutation relation

$$I) \qquad i[\pi(x,t), \phi(x',t)]_- = \delta(x - x'),$$

and its Hermitian conjugate for π^* with ϕ^*, and they commute with the electromagnetic field quantities. Using (42), the charge density is always formally identical to (16)

$$\rho = i \left(\pi^* \phi^* - \pi \phi \right) = i \left(\phi^* \pi^* - \phi \pi \right). \tag{16}$$

Therefore the eigenvalues of the charge density remain the same in the presence of external potentials as in the force free case.

The matter part of the Hamiltonian is now written

$$H^m = H_0 + H_I$$

248

with

$$H_0 = \int dV \left\{ \pi \pi^* + \sum_{k=1}^{3} \frac{\partial \phi^*}{\partial x^k} \frac{\partial \phi}{\partial x^k} + m^2 \phi^* \phi \right\},$$

$$H_I = \int dV \left\{ ie \sum_{k=1}^{3} \Phi_k \left(\phi^* \frac{\partial \phi}{\partial x^k} - \frac{\partial \phi^*}{\partial x^k} \phi \right) + e^2 \sum_{k=1}^{3} \Phi_k^2 \phi^* \phi \right\}. \tag{45}$$

With the well known commutation relations for the field strengths and the electromagnetic potentials, quantum electrodynamics can be formulated in the customary way. Maxwell's equations (40) can be obtained by use of the Heisenberg equation of motion (11)

$$\frac{\partial \vec{E}}{\partial t} = i[H, \vec{E}]_-.$$

We prefer to go no further here, but only to state the known complications which occur, that Eqn.41 should contain only the gauge invariant quantities $(\phi \pi - \phi^* \pi^*), \vec{E}, \vec{B}$; but not others such as $\pi, \pi^*, \phi, \phi^*, \Phi$. If the rule (11)

$$\frac{\partial f}{\partial t} = i[H, f]_-$$

is to be valid for these quantities also, then one must formally add an expression of the form

$$\int dV \left\{ \Phi_0 (4\pi e \rho - \vec{\nabla} \cdot \vec{E}) \right\} \tag{46}$$

to the sums of (38) and (45) [13].

For the partition of H' different from (37)

$$H' = H_0 + H_1 + H_2 \tag{47}$$

$$H_2 = e \int dV \Phi_0 \rho = ie \int dV \Phi_0 (\pi^* \phi^* - \pi \phi), \tag{45}$$

the Hamiltonian then gives

$$\dot{\pi} = i[H', \pi]_-, \quad \dot{\phi} = i[H', \phi]_- \tag{48}$$

with the corresponding relations for π^*, ϕ^*.

This (not gauge invariant) Hamiltonian H' also has the property of being a time-independent constant if the potentials Φ_0 and Φ_k depend only on the space coordinates but not on the time. For many purposes, it is useful to consider the four-potential as a given c-number function. In this case, one must calculate with the Hamiltonian given by

H'. This result can also be obtained from the canonical transformation of L^m according to the formula:

$$H' = \int dV \left(\pi \dot{\phi} + \pi^* \dot{\phi}^* \right) - L^m.$$

We will now write the Hamiltonian terms H_1 and H_2 in momentum space, which has already been done for H_0 in Eqn.21. For this, using the orthogonal functions (17), the matrix elements of a function f (for example, the potential Φ) are defined by

$$f_{kl} = \frac{1}{V} \int dV f(\vec{x}) e^{i(\vec{k}-\vec{l})\cdot\vec{x}}. \tag{49}$$

Thus we obtain directly from (19)

$$H_2 = ie \sum_{k,l} \Phi^0_{kl} \left(p_l^* q_k^* - p_k q_l \right) \tag{50}$$

and

$$H_1 = - \sum_{k,l} \left[e \left(\vec{k} + \vec{l} \right) \cdot \vec{\Phi}_{kl} - e^2 \left(\vec{\Phi}^2 \right)_{kl} \right] q_k^* q_l; \tag{51}$$

or with the introduction of the variables a_k, b_k as in (27,28), in which H_0 is given by (29):

$$H_2 = \frac{1}{2} e \sum_{k,l} \Phi^0_{kl} \left[\frac{E_k + E_l}{\sqrt{E_k E_l}} (a_k^* a_l - b_{-l}^* b_{-k}) + \frac{E_k - E_l}{\sqrt{E_k E_l}} (a_l b_{-k} - a_k^* b_{-l}^*) \right] \tag{52}$$

$$H_1 = \frac{1}{2} \sum_{k,l} \frac{1}{\sqrt{E_k E_l}} \left[e(\vec{k} + \vec{l}) \cdot \vec{\Phi}_{kl} - e^2 \left(\vec{\Phi}^2 \right)_{kl} \right] (a_k^* a_l + b_{-k} b_{-l}^* - a_k^* b_{-l}^* - b_{-k} a_l). \tag{53}$$

§4. Pair Production by Photons and the Polarization of the Vacuum.

From the commutation relations (III) for the (a_k^*, a_k) and (b_k^*, b_k), the properties of these operators follow for the occupation number (N_k^+, N_k^-) dependence of the Schrödinger functional

$$| \cdots N_k^+ \cdots ; \cdots N_j^- \cdots \rangle.$$

These are

$$a_k^* | \cdots N_k^+ \cdots \rangle = \sqrt{N_k^+ + 1} \ | \cdots N_k^+ + 1 \cdots \rangle$$

$$a_k | \cdots N_k^+ \cdots \rangle = \sqrt{N_k^+} \ | \cdots N_k^+ - 1 \cdots \rangle$$

$$b_k^* | \cdots N_k^- \cdots \rangle = \sqrt{N_k^- + 1} \ | \cdots N_k^- + 1 \cdots \rangle$$

$$b_k | \cdots N_k^- \cdots \rangle = \sqrt{N_k^-} \ | \cdots N_k^- - 1 \cdots \rangle. \tag{54}$$

One easily sees that H_1 and H_2, which arise in the Hamiltonian from the presence of the external fields, involve as factors the terms $a_k^* b_{-l}^*$ and $a_k b_{-l}$, which give rise to pair

250

creation and pair annihilation. These terms lead to matrix elements between states which differ by one positive and one negative particle. In contrast, the factors $a_k^* a_l$ and $b_k b_l^*$ only yield transitions of a positive or negative particle from one state to another.

In the following, we calculate the probability for pair production by a photon of energy $h\nu > 2mc^2$, on the basis of expressions (52) and (53). The results agree with the corresponding expressions from hole theory, as calculated by Bethe and Heitler [14].

Because of the energy-momentum relation, this probability vanishes in free space. We therefore assume the existence of a time independent field Φ_0 (perhaps the coulomb potential of a nucleus), which can absorb the excess momentum.

We consider the effect of the field only in first approximation, just as Bethe and Heitler did. Thus we proceed as in field free space and consider Φ_0 both as a perturbation and also as the potential of the photon.

We now ask for the probability per unit time W, that a positive particle of momentum $\vec{k}$, energy E_k, and a negative particle of momentum $\vec{l}$, energy E_l should be created by absorption of a photon of energy $h\nu = E_k + E_l$. We get in second order perturbation theory the first non-vanishing result:

$$W = \left| \sum_c \frac{H_1(AC)H_2(CB)}{E_B - E_C} + \frac{H_2(AC)H_1(CB)}{E_A - E_C} \right|^2. \tag{55}$$

A refers to the vacuum state (all $N = 0$), B to the final state ($N_k^+ = N_l^- = 1$, all other $N = 0$): C represents any intermediate state. $H_1(AC)$ is the matrix element of H_1 between the state A and the state C. In H_2 the scalar potential is the introduced time-independent Φ_0; in H_1 the vector potential $\vec{\Phi}$ is that of the photon with frequency ν. Because of the momentum relations, in the calculation of H_1 only the following four intermediate states come into consideration:

$$C_1 \rightarrow N_k^+ = 1, \quad N_{n-k}^- = 1$$
$$C_2 \rightarrow N_{n-l}^+ = 1, \quad N_l^- = 1$$
$$C_3 \rightarrow N_k^+ = 1, \quad N_{l-n}^- = 1$$
$$C_4 \rightarrow N_{k-n}^+ = 1, \quad N_l^- = 1,$$

all other N are zero. · The matrix elements required can be calculated from (52) and (53) and lead to an expression which we write as a differential cross section $d\sigma$, for an unpolarized photon of frequency ν to produce a positive particle of energy between E_+ and $E_+ + dE$, a negative particle with energy between E_- and $E_- - dE$ ($E_+ + E_- = h\nu$)

and whose momenta $\vec{p}_+$ and $\vec{p}_-$ make angles θ_+, θ_- with the direction of the photon:

$$d\sigma = \frac{1}{(2\pi)^8} \frac{e^2}{\omega^3} p_+ \sin\theta_+ p_- \sin\theta_- d\theta_+ d\theta_- d\phi dE |\Phi_0(\vec{q})|^2 \times$$

$$\left\{ \frac{E_-^2 p_+^2 \sin^2\theta_+}{(E_+ - p_+ \cos\theta_+)^2} + \frac{E_+^2 p_-^2 \sin^2\theta_-}{(E_- - p_- \cos\theta_-)^2} + \frac{2E_+ E_- p_+ p_- \sin\theta_+ \sin\theta_- \cos\phi}{(E_+ - p_+ \cos\theta_+)(E_- - p_- \cos\theta_-)} \right\}. \quad (56)$$

[Note added: $\hbar$ and c are set to one, in accord with modern usage, and the photon energy $\omega = 2\pi\nu$ appears instead of ν.]

ϕ is the angle between the planes formed from the direction of the photon and the directions of $\vec{p_+}$ and $\vec{p_-}$.

$\Phi_0(\vec{q})$ is the matrix element

$$\Phi_0(\vec{q}) = \int dV \Phi_0(\vec{x}) e^{i\vec{q}\cdot\vec{x}}.$$

The excess momentum $\vec{q}$ taken up by the electric field is

$$\vec{q} = (\vec{p}_+ + \vec{p}_- - \vec{n}),$$

with $\vec{n}$ the momentum of the photon.

For a coulomb field $\Phi_0 = Ze/r$ and

$$\Phi_0(\vec{q}) = \frac{4\pi Ze}{\vec{q}^2}.$$

The corresponding expression in hole theory, due to Bethe and Heitler, differs from the expression (56) by the appearance of a fourth term in the large brackets, and in the appearance of q^2-dependent terms in the first three. These latter are negligible at high energy since $q << \omega$ for $\omega >> m$.

If one sets Φ_0 equal to the coulomb potential, then the integration over the angle can be easily done for the limiting case $\omega >> m$ [15]. One gets:

$$d\sigma = \frac{Z^2 e^6}{2\pi m^2} \frac{32}{3} \frac{E_+ E_-}{\omega^3} \left(\log \frac{E_+ E_-}{\pi \omega m} - \frac{1}{2} \right) dE$$

and for the total cross section:

$$\sigma = \frac{Z^2 e^6}{2\pi m^2} \left(\frac{16}{9} \log \frac{2\omega}{m} - \frac{104}{27} \right).$$

The corresponding formula for hole theory from Bethe and Heitler is:

$$d\sigma = \frac{Z^2 e^6}{2\pi m^2} 4 \frac{E_+^2 + E_-^2 + \frac{2}{3} E_+ E_-}{\omega^3} \left(\log \frac{E_+ E_-}{\pi \omega m} - \frac{1}{2} \right) dE$$

252

and

$$\sigma = \frac{Z^2 e^6}{2\pi m^2} \left(\frac{28}{9} \log \frac{2\omega}{m} - \frac{218}{27} \right).$$

The cross section for pair production in the theory treated here is smaller by a factor 4/7 in the limit of $\omega >> m$.

In conclusion, the polarization of the vacuum by an electric field will be calculated. For this purpose, we calculate the additional charge density $\delta\rho(x)$ induced in the space V due to the field Φ_0 of an "external" charge density $\rho_0(x)$; $\delta\rho(x)$ is the charge density induced by the potential Φ_0 in the space which is *empty* of the positive and negative particles described by the wave equation.

It is useful to Fourier analyse the density. From (16) and (19), one gets:

$$\rho(\vec{\xi}) = \frac{1}{V} \int dV \rho(\vec{x}) e^{-i\vec{\xi}\cdot\vec{x}} = \sum_k (p_l^* q_k^* - p_k^* q_l),$$

where

$$\vec{l} = \vec{k} + \vec{\xi}.$$

Further, in terms of the operators a and b:

$$\rho(\vec{\xi}) = \frac{1}{2} \sum_k \left\{ \frac{E_k - E_l}{\sqrt{E_k E_l}} (a_k^* a_l - b_{-l}^* b_{-k}) + \frac{E_k + E_l}{\sqrt{E_k E_l}} (a_k^* b_{-l}^* - a_l b_{-k}) \right\}.$$

We now use this operator to calculate the distortion of the Schrödinger functional of the empty space $|\cdots 0 \cdots; \cdots 0 \cdots\rangle$. Using first order perturbation theory, (52) gives

$$|\cdots 0 \cdots; \cdots 0 \cdots\rangle = |\cdots 0 \cdots; \cdots 0 \cdots\rangle_0$$
$$- \frac{1}{2} \sum_{kl} \Phi_{kl}^0 \frac{E_k - E_l}{\sqrt{E_k E_l}(E_k + E_l)} |\cdots 1_k \cdots; \cdots 1_l \cdots\rangle_0, \tag{58}$$

where $|\cdots\rangle_0$ represent free states in the vacuum, and $1_k, 1_l$ indicate $N_k^+ = 1$, $N_l^- = 1$, all other $N = 0$. If one now takes the expectation value of the operator $\rho(\vec{\xi})$ for the distorted state $|\cdots 0 \cdots; \cdots 0 \cdots\rangle$, one gets

$$\delta\rho(\vec{\xi}) = -\frac{1}{2} \Phi_0(\vec{\xi}) \sum_k \frac{(E_k - E_l)^2}{E_k E_l(E_k + E_l)}.$$

The sum over k diverges logarithmically, as one can easily see. After integration over the directions of $\vec{k}$, one gets:

$$\delta\rho(\vec{\xi}) = -\frac{e}{12\pi^2 \xi^2} \Phi_0(\vec{\xi}) \int \frac{d|k|}{|k|}$$

plus finite terms. In coordinate space one gets

$$\delta\rho(x) = K\nabla^2 \Phi_0$$

plus finite terms, with

$$K = \frac{e}{12\pi^2} \int \frac{d|k|}{|k|}.$$

The induced charge density has the opposite sign to the external charge density

$$\rho_0 = -\frac{1}{4\pi} \nabla^2 \Phi_0$$

and is proportional to it, with the divergent proportionality factor $4\pi K$. The result is that each external charge is compensated by the induced charge. This result agrees completely with that of Dirac [8] calculated on the basis of hole theory. Even the factor K of the divergent integral is the same.

Zurich, Physikalisches Institut der E.T.H.

Footnotes and References.

1) (Note added: $\hbar$ and c are set to one almost everywhere, in accordance with modern usage.)

2) W. Heisenberg and W. Pauli, Zeits. f. Phys. **56**, 1 (1929).

3) This appears in detail in the Leipzig Report (1932), following p.85.

4) Initially, this form for ρ was rejected because the wave equation was required to be of first order in $\partial/\partial t$.

5) If ϕ_κ^+ is the "positive" (corresponding to states of positive energy), and ϕ_κ^- the "negative" part of the wave function in Dirac hole theory, then the charge density operator has the form

$$\rho(x) = \sum_{\kappa=1}^{4} \left\{ \phi_\kappa^{+*}\phi_\kappa^+ - \phi_\kappa^{-*}\phi_\kappa^- + \phi_\kappa^+\phi_\kappa^- + \phi_\kappa^{+*}\phi_\kappa^{-*} \right\}.$$

Because of the mixed terms, even in the absence of external forces, it cannot be separated into two parts each by itself satisfying a continuity equation and forming the 4th-component of a four-vector.

6) P.A.M. Dirac, Proc. Cambr. Phil. Soc. **30**, Pt.II, 150 (1934); R. Peierls, Proc. Roy. Soc. A**146**, 420 (1934); W. Heisenberg, Zeits. f. Phys. **90**, 209 (1934).

7) P.A.M. Dirac, Proc. Roy. Soc. A**133**, 60 (1931), especially p.71.

8) P.A.M. Dirac, Solvay-Report 1933.

9) Compare W. Heisenberg [6]. One is inclined to doubt the value of the arguments

254

in many formulations of hole theory which claim the polarization effect is finite but the self-energy still infinite.

10) For the Lagrangian, energy-momentum tensor and current vector in the scalar relativistic theory, see for example: W. Gordon, Zeits. f. Phys. **40**, 117 (1926).

11) One obtains such a separation, for example, by introducing an arbitrary constant a (with the dimension $\sqrt{E}$) by the prescription

$$\pi = \frac{a}{\sqrt{2}}(\phi_1 + \phi_2^*), \quad \phi = \frac{-i}{\sqrt{2}a}(\phi_1^* - \phi_2),$$

and their Hermitian conjugates for π^* and ϕ^*. Then

$$[\phi_1(x), \phi_1^*(x')]_- = \delta(x - x'), \quad [\phi_2(x), \phi_2^*(x')]_- = \delta(x - x'),$$

whereas quantities with index 1 commute with those with index 2. It gives

$$\rho = \phi_2^*\phi_2 - \phi_1^*\phi_1.$$

From this, we get a new proof for the eigenvalues of ρ.

12) Again, it should be noted that a definition for a spacelike density $\rho^+(x)$ and $\rho^-(x)$ corresponding to the two particle types, is not possible in a physically meaningful way. For example, if one forms from a_k and b_k

$$a(x) = \frac{1}{\sqrt{V}}\sum_k a_k e^{i\vec{k}\cdot\vec{x}}, \quad b(x) = \frac{1}{\sqrt{V}}\sum_k b_k e^{-i\vec{k}\cdot\vec{x}},$$

then it can be shown that the expression

$$a^*(x)a(x) - b^*(x)b(x)$$

does *not* agree with the charge density.

13) W. Heisenberg and W. Pauli, Zeits. f. Phys. **59**, 168 (1930), in particular p.179, Eqn.38.

14) H. Bethe and W. Heitler, Proc. Roy. Soc. A**146**, 83 (1934).

15) We owe H. Bethe many thanks for a copy of his work before its appearance in the Proc. Camb. Phil. Soc., in which similar integrations were carried out.

Chapter 10

Pauli's First Proof of the Spin-Statistics Theorem

Summary: Pauli made the first attempt to prove that the spin-statistics relation is a necessary result of the postulates of relativistic quantum field theory. Almost simultaneously, Iwanenko and Socolow showed that quantization of the Dirac equation using anticommutation relations led naturally to results analogous to those of Pauli and Weisskopf for the spin-0 equation. In their paper, Iwanenko and Socolow also point out that problems arise from the wrong commutation relations.

§1. Pauli's First Proof.

Pauli's 1936 paper [10.1] is the first attempt to demonstrate that an exception to the spin-statistics relation is incompatible with physically motivated requirements of relativistic quantum field theory. This one paper set the agenda for all those following which seek to prove the Spin-Statistics Theorem from relativistic quantum field theory but which, as a consequence, leave the overwhelming majority of physicists hungering for a less formal, more intuitive, way to understand the spin-statistics relation [10.2]. The first two sections of Pauli's paper have become standard textbook lore but it is interesting to see it in Pauli's own words. A translation of the original (from the French) is reproduced in Appendix 10.A for convenient reference.

The essence of Pauli's paper is his Section III where he examines the possibility of quantizing scalar field theory according to the Pauli Exclusion Principle, using anticommutation relations for the creation and annihilation operators.

Pauli then proceeds through a series of questionable operations whose validity have subsequently been denied, and which were questioned immedi-

256

ately by Pauli himself. First, he splits the field ϕ into positive and negative frequency pieces ϕ_1 and ϕ_2 (Eqn.26) at the expense not only of doubling the number of degrees of freedom, but also of having non-local commutation relations and non-local connections between the fields and their conjugate momenta (Eqns.27,28,31). He admits that this split cannot survive in the presence of interactions. Then he equivocates "One could, it would seem - perhaps not unequivocally $\cdots$". But he proceeds nonetheless to construct bilinear currents as products of the field operators ϕ_1 and ϕ_2 (32). "One sees that we have changed the order of certain factors $\cdots$". "This changing of the order of factors opens up the possibility of Fermi-Dirac statistics." He then gives independent charges to the two parts ϕ_1 and ϕ_2 of the field ϕ, resulting in a current S_ν containing three parameters c_1, c_2, c_3 for the two charges and the pair production amplitude. Finally he is able to show that only the choice $c_1 = c_2 = \pm c_3$, for which one recovers the original local field theory, can have currents which commute for spacelike separation. All is well, and he has closed the circle.

Next he tries to quantize the freely reordered theory according to anti-commutation relations. Because of the reordering, he obtains the same expression as before for the current involving the three charges c_1, c_2, c_3. But now, when he tries to impose causal commutation relations on this "theory", he finds that only the choice $c_1^2 = c_2^2 = -c_3^2$ makes this possible. His conclusion that only the Exclusion Principle is to blame for the failure of the theory to have causal commutation relations, after having made a non-local split in the theory, followed by a free reordering, is not compelling, $\cdots$"but we ourselves are satisfied to state that the case of Bose-Einstein statistics is distinguished by its simplicity in the case of the relativistic scalar theory."

His final section introducing the electromagnetic interaction is of no concern to us here.

Summing up, it might be thought that this first paper, although not conclusive, at least did no harm with its introduction of mathematically nonlocal operations, as well as its arbitrary, unjustified and *ad hoc* reordering of operators. But in fact Pauli was to wield these invalid ideas as a club over following authors, especially Belinfante [10.3], with the result that their significant contributions were almost uniformly ignored. He was to return to these ideas over and over in the next decade before finally conceding the problem to others.

To his credit, his own classic and frequently cited proof [10.4], although not fully credited today, is free of such arguments. We will discuss it in its turn in a later chapter.

§2. Iwanenko and Socolow's Second Quantization of the Dirac Equation.

Next after Pauli, Iwanenko and Socolow, in a short "Note on the Second Quantization of the Dirac Equation", derived expressions in terms of occupation numbers for the energy, charge, current and momentum for the Dirac field. A translation of their paper [10.5] (from the German) is reproduced in Appendix 10.B. They used anticommutation relations for the creation and annihilation operators and obtained expressions of the same form as those obtained by Pauli and Weisskopf [10.6] for the Klein-Gordon equation, using commutation relations. They point out that the Pauli-Weisskopf forms for the scalar field can be obtained only with "wrong sign" anticommutation relations for the b coefficients (in their Eqn. 11), if one attempts quantization according to the Pauli Exclusion Principle. Alternatively, ordinary commutation relations and quantization according to Bose-Einstein statistics leads smoothly to the approved result in the paragraph following their Eqn.12. They conclude "One can obtain the correct Fermi-Dirac anticommutation relations only through special ordering $\cdots$ the scalar relativistic equation on the basis of other criteria. $\cdots$ Conversely, $\cdots$ application of Bose-Einstein statistics to the Dirac equation $\cdots$ leads always to negative energies for one of the two particle types."

Iwanenko and Socolow have written an elegant paper without pretensions to rigor or priority. Surely such manipulations were well known to Pauli, Weisskopf, Dirac and others. In this regard we appeal to a statement of van der Waerden [10.7] that we should only judge by *what* people *wrote* and *when*.

§3. Biographical Note on Iwanenko and Socolow.(†)

Dmitrij Dmitrievic Iwanenko (1904-), graduate Leningrad; professor at Leningrad, Kharkov, Tomsk, Sverdlovsk, Kiev (1930-); professor at Moscow University (1942-); USSR Academy of Science (1949-). From Pais we learn that Iwanenko deserves credit for the suggestion that Chadwick's neutron should be considered as an equal nuclear partner with the proton. Iwanenko resolved the statistical problems of nuclei constructed from protons and nuclear electrons by considering the neutron "$\cdots$not$\cdots$ as built up of an electron and a proton but as an *elementary particle* $\cdots$." Pais gives a day by day chronology of events of the spring and summer of 1932 taking place in Europe (Perrin, Auger, and Dirac still advocating electrons in the nucleus), in the US (Oppenheimer and Carlson somewhat equivocating about the neutron), and in Russia (Iwanenko with an explicit resolution of the statistics problems). Iwanenko also floated the explanation of beta decay as "the expulsion of an electron is similar to the birth of a new particle." For the first time we meet Socolow, in collaboration with Iwanenko trying to apply a Fermi-type interaction to understand charge independent nuclear forces by exchange of $e\bar{e}, e\bar{\nu}, \nu\bar{\nu}$ pairs [10.9]. Iwanenko, Socolow and Pomeranchuk (1944-48) developed the theory of synchrotron radiation. From Augenstein we learn of an even earlier achievement by Iwanenko, with Landau, in which [10.10] they - almost simultaneously, but independently - deduced the equivalent of the Dirac equation but were completely ignored.

After a protracted time, we find Iwanenko and Socolow at Moscow State University as authors of the text *Klassische Feldtheorie* (Akademie-Verlag, Berlin, 1953).

(† - A. Pais, *Inward Bound* (Oxford, New York, 1986), pp.409,410,411,418,426; B.W. Augenstein, Physics Today, May 1995, p.86; *World Who's Who in Science* (Marquis, Chicago, 1994), p.862.)

Bibliography and References.

10.1) W. Pauli, Annals de Institut Henri Poincaré **6**, 137 (1936); see our App.10A.

10.2) D.E. Neuenschwander, Am. J. Phys. **62**, 972 (1994); R.P.Feynman, R.B. Leighton, and M. Sands, *The Feynman Lectures on Physics*, *Vol.3* (Addison-Wesley, Reading, MA 1963), Ch.4, Sec.1.

10.3) W. Pauli and F.J. Belinfante, Physica **VII 3**, 177 (1960); see our App.12B.

10.4) W. Pauli, Phys. Rev. **58**, 716 (1940); see our App.14A.

10.5) D. Iwanenko and A. Socolow, Phys. Zeits. der Sowjetunion **11**, 590 (1937); see our App.10B.

10.6) W. Pauli and V. Weisskopf, Helv. Phys. Acta **7**, 709 (1934); see our App.9A.

10.7) B.L. van der Waerden, in *Theoretical Physics in the Twentieth Century: A Memorial Volume to Wolfgang Pauli* (Interscience, New York, 1960), edited by M. Fierz and V.F. Weisskopf, p.200.

10.8) D. Iwanenko, Nature **129**, 798 (1932); Comptes Rendus **195**, 439 (1932).

10.9) D. Iwanenko, Nature **133**, 981 (1934); D. Iwanenko and A. Socolow, Zeits. f. Phys. **102**, 119 (1936); also Nature **138**, 246, 684 (1936).

10.10) D. Iwanenko and L. Landau, Zeits. f. Phys. **48**, 340 (1928).

APPENDIX 10.A:

Excerpt from: Annals de Institut Henri Poincaré 6, 137 (1936)

Relativistic Quantized Theory of Particles
Obeying Bose-Einstein Statistics
by
W. PAULI, Zurich

§1. Introduction.

Dirac used a first order wave equation for a four component spinor ψ, and the postulate that the particle density ρ should be positive definite and have the form $\rho = \psi^\dagger \psi$. Dirac thought that the postulate should be imposed *a priori* on the basis of the general theory of transformations of wave mechanics and independent of the empirical fact that the electron has spin-$\frac{1}{2}$. This argument is correct as far as the single particle problem is concerned, or, more exactly, as long as only a single particle is involved. The subsequent development of Dirac theory led to the realization that the creation and annihilation of pairs of particles having opposite charges constitutes an inseparable part of the relativistic theory of material particles. In this case, the situation changes radically and the argument used by Dirac as a premise is no longer applicable *a priori*; in fact it is not a question of finding an expression for the density of the particles (which in general will no longer be measurable at distances of order $1/m$), but of finding an expression for the density of electric charge, with appropriate positive and negative values, and another expression for the density of energy with the appropriate uniquely positive values.

I do not write this paper as an attempt to repair the latest problem in the hole theory of Dirac, but I must insist on the fact that starting with the second order Schrödinger-Gordon wave equation, it is possible to develop a relativitistic theory of particles without spin and with Bose-Einstein statistics, a theory which implies the production of pairs and which may be more satisfactory from the point of view of logic and pedagogy than the hole theory of Dirac. In effect, in the theory in question, it is no longer necessary to introduce the artificial methods of passage to the limit to give a specific value to the difference of two infinite sums, which is the case in the theory of holes. In addition, in the scalar theory, the energy density is positive definite just as in the case where one considers the wave function ϕ and its complex conjugate $\phi^\dagger$ as ordinary numbers. Second quantization following the general formalism of Heisenberg-Pauli is necessary to get creation and annihilation of pairs by quantum transitions among the electric charge and energy eigenstates.

We give in Section 2 application of this formalism to the second quantization of the relativistic scalar equation with no forces. Section 3 contains a detailed analysis of the possibility of the formal development of a relativistic scalar theory for particles without spin, but obeying the Exclusion Principle. One arrives at the satisfying result that the properties of measurability of the charge density would be sufficiently more complicated in the case of the Exclusion Principle than in the case of Bose-Einstein statistics, and that the latter case is distinguished from the former hypothetical one even from the purely formal point of view. On the other hand, one must admit that it is not absolutely certain that the theory in question is applicable to reality, because those particles without spin - such as the alpha particle - are all complex. It is impossible to know the importance of the role which the effect of structure plays in the relativistic domain. Nevertheless, we give in Section 4 a brief estimate of the magnitude of certain effects which are produced by external forces and we compare the results with the analogous results from the hole theory.

§2. The Quantization of a Field Without Forces.

One can derive the relativistic scalar wave equation

$$\partial_t^2 \phi - \nabla^2 \phi + m^2 \phi = 0 \tag{1}$$

from the Lagrangian

$$L = \sum_{\nu=0}^{3} \partial^\nu \phi^\dagger \partial_\nu \phi - m^2 \phi^\dagger \phi$$

$$= \partial_t \phi^\dagger \partial_t \phi - \sum_{k=1}^{3} \partial_k \phi^\dagger \partial_k \phi - m^2 \phi^\dagger \phi$$

by using the variational principle

$$\delta \int L d^3 x dt = 0.$$

This gives the relativistic tensor for the energy-momentum density

$$T_{\mu\nu} = (\partial_\mu \phi^\dagger \partial_\nu \phi + \mu \Leftrightarrow \nu) - L\delta_{\mu\nu}$$

from which one derives for the energy (the Hamiltonian function)

$$H = \int T_{00} d^3 x = \int (\partial_t \phi^\dagger \partial_t \phi + \sum_{k=1}^{3} \partial_k \phi^\dagger \partial_k \phi + m^2 \phi^\dagger \phi) d^3 x \tag{2}$$

and for the momentum

$$P_k = -P^k = \int T_{0k} d^3 x = \int (\partial_t \phi^\dagger \partial_k \phi + \partial_k \phi^\dagger \partial_t \phi) d^3 x. \tag{3}$$

Note that the energy density is positive definite. Next, elevate $\phi^\dagger$ and ϕ to be Hilbert space operators with $\phi^\dagger$ the Hermitian conjugate of ϕ. According to the formalism of canonical quantization, it is necessary to introduce the generalized momenta Π and $\Pi^\dagger$ canonically conjugate to ϕ and $\phi^\dagger$ as

$$\Pi = \frac{\delta L}{\delta(\partial_t \phi)} = \partial_t \phi^\dagger, \tag{4}$$

and similarly for $\Pi^\dagger$, and to postulate the commutation relations

$$I.) \qquad i[\Pi(\vec{x}, t), \phi(\vec{x}', t)]_- = \delta^3(\vec{x} - \vec{x}'),$$

$$i[\Pi^\dagger(\vec{x}, t), \phi^\dagger(\vec{x}', t)]_- = \delta^3(\vec{x} - \vec{x}'),$$

with $[A, B]_- = AB - BA$ and $\delta(x)$ the Dirac delta function. All other commutators, $[\Pi, \phi^\dagger]_-$, $[\Pi, \Pi^\dagger]_-$ and $[\phi, \phi^\dagger]_-$ are chosen to be zero. One can easily verify the Heisenberg equations of motion

$$i\partial_t f = [f, H]_- \tag{5}$$

for $\phi, \phi^\dagger, \Pi, \Pi^\dagger$, and also

$$i\partial_k f = [f, P_k]_-.$$

The order of factors in the momentum is chosen to give a Hermitian operator.

The electrical current-density four vector $j_k = S_k$ for the current and $\rho = S_0$ for the density, which satisfies the continuity equation

$$\sum_{\nu=0}^{3} \partial_\nu S^\nu = 0 \tag{6}$$

or

$$\partial_t \rho + \vec{\nabla} \cdot \vec{j} = 0,$$

has the operator form

$$S_\nu = -ie(\partial_\nu \phi^\dagger \phi - \partial_\nu \phi \phi^\dagger), \tag{7}$$

or

$$\rho = -ie(\partial_t \phi^\dagger \phi - \partial_t \phi \phi^\dagger), \quad j_k = -j^k = -ie(\partial_k \phi^\dagger \phi - \partial_k \phi \phi^\dagger).$$

The order of factors in ρ is made - without altering its Hermitian character - so that there is no zero point density (as we shall see in more detail later).

With the aid of (7) we can write

$$\rho = -ie(\Pi \phi - \Pi^\dagger \phi^\dagger). \tag{8}$$

From the basic commutation relations we can deduce a fundamental property of ρ: the charge density at two spatially separated points (x, x') commute

$$[\rho(x), \rho(x')]_- = 0. \tag{9}$$

263

It is this property of ρ which allows us to speak of the charge density in space as measurable, even in the case that the spatial dimensions are of order $1/m$.

Later we want the total charge

$$Q = \int \rho d^3 x$$

to take the values $Q = e(0, \pm 1, \pm 2 \cdots)$. Since the eigenvalues of $\rho(x)$ are independent of x, this requires them to be $(0, \pm 1, \pm 2, \cdots \pm N \cdots) e \delta(x - x')$, with x' being arbitrary.

We next construct expressions for the various physical quantities in momentum space. In order to have sums instead of integrals in this space, we use wave functions periodic in a cube of side L, volume $V = L^3$. In this case, components of the momentum vector $\vec{k}$ in the phase of the wave function $e^{i\vec{k}\cdot\vec{x}}$ must be integer multiples of $2\pi/L$.

We can next write

$$\phi = \frac{1}{\sqrt{V}} \sum_k q_k e^{i\vec{k}\cdot\vec{x}}, \quad \Pi^\dagger = \frac{1}{\sqrt{V}} \sum_k p_k^\dagger e^{i\vec{k}\cdot\vec{x}}, \tag{10}$$

and their Hermitian conjugates for $\phi^\dagger$ and Π. The inversion formulas are

$$q_k = \frac{1}{\sqrt{V}} \int \phi e^{-i\vec{k}\cdot\vec{x}} d^3 x, \quad p_k^\dagger = \frac{1}{\sqrt{V}} \int \Pi^\dagger e^{-i\vec{k}\cdot\vec{x}} d^3 x,$$

and similarly for $q_k^\dagger$ and p_k.

The operators q_k, p_k, $q_k^\dagger$, $p_k^\dagger$ satisfy the commutation relations

$$II.) \qquad i[p_k, q_j]_- = \delta_{k,j}, \quad i[p_k^\dagger, q_j^\dagger]_- = \delta_{k,j},$$

with all others zero. With

$$\omega_k = +\sqrt{k^2 + m^2} \tag{11}$$

we get for the total energy and momentum

$$H = \sum_k (p_k^\dagger p_k + \omega_k^2 q_k^\dagger q_k) \tag{12}$$

and

$$\vec{P} = -i \sum_k \vec{k}(p_k q_k - q_k^\dagger p_k^\dagger). \tag{13}$$

The Heisenberg equations of motion yield

$$\begin{aligned} p_k &= \dot{q}_k^\dagger, & p_k^\dagger &= \dot{q}_k \\ \dot{p}_k &= -\omega_k^2 q_k^\dagger, & \dot{p}_k^\dagger &= -\omega_k^2 q_k. \end{aligned} \tag{14}$$

264

The total electric charge becomes

$$Q = \int \rho d^3 x = -ie \sum_k (p_k q_k - p_k^\dagger q_k^\dagger) \tag{15}$$

and the total current

$$\vec{J} = \int \vec{j} d^3 x = 2 \sum_k \vec{k} q_k^\dagger q_k. \tag{16}$$

Next we show that the energy, momentum and total charge, corresponding to a given momentum $\vec{k}$, can be separated into two parts which have a simple interpretation. For this purpose we introduce, instead of p_k, q_k, $p_k^\dagger$, $q_k^\dagger$, the variables a_k, $a_k^\dagger$, b_k, and $b_k^\dagger$ defined by

$$p_k = \frac{\sqrt{\omega_k}}{\sqrt{2}}(a_k^\dagger + b_{-k}), \quad q_k = \frac{-i}{\sqrt{2\omega_k}}(-a_k + b_{-k}^\dagger), \tag{17}$$

and their Hermitian conjugates, with the inverse formulas

$$a_k = \frac{1}{\sqrt{2\omega_k}}(p_k^\dagger - i\omega_k q_k), \quad b_{-k} = \frac{1}{\sqrt{2\omega_k}}(p_k - i\omega_k q_k^\dagger), \tag{18}$$

and their Hermitian conjugates. The new variables satisfy very simple commutation relations

$$III.) \qquad [a_k, a_j^\dagger]_- = \delta_{k,j}, \quad [b_k, b_j^\dagger]_- = \delta_{k,j},$$

all others zero.

We get finally

$$H = \sum_k \omega_k(a_k^\dagger a_k + b_k^\dagger b_k + 1), \tag{19}$$

$$\vec{P} = \sum_k \vec{k}(a_k^\dagger a_k + b_k^\dagger b_k), \tag{20}$$

$$Q = e \sum_k (a_k^\dagger a_k - b_k^\dagger b_k), \tag{21}$$

$$\vec{J} = e \sum_k \frac{\vec{k}}{\omega_k}(a_k^\dagger a_k - b_k^\dagger b_k - a_k^\dagger b_{-k}^\dagger - a_k b_{-k} + 1). \tag{22}$$

The Heisenberg equations of motion are particularly simple

$$\dot{a}_k = -i\omega_k a_k, \quad \dot{b}_k = -i\omega_k b_k, \tag{23}$$

or

$$a_k = a_k(0)e^{-i\omega_k t}, \quad b_k = b_k(0)e^{-i\omega_k t}, \tag{24}$$

and their Hermitian conjugates.

265

Consequently, we can say that a_k, b_k correspond to positive energy, $a_k^\dagger$, $b_k^\dagger$ to negative energy. Or, comparing (17), that ia_k is the positive energy part of q_k and $-ib_{-k}^\dagger$ is the negative energy part, and analogously $a_k^\dagger$ and b_{-k} for p_k. This result plays an important role in what follows.

The physical interpretation is made by noticing that $a^\dagger a$ and $b^\dagger b$ have the eigenvalues $0,1,2\cdots$. From (20,21), we can say that

$$N_{\vec{k}}^a = a_{\vec{k}}^\dagger a_{\vec{k}}, \qquad N_{\vec{k}}^b = b_{\vec{k}}^\dagger b_{\vec{k}} \tag{25}$$

are the number $N_{\vec{k}}^a$ of particles of charge e and momentum $\vec{k}$, and $N_{\vec{k}}^b$ of antiparticles of charge $-e$ and momentum $\vec{k}$.

The terms with coefficient one in (19) can be interpreted as zero point energy and current, that is, as unobservable quantities. The terms with $a_k b_{-k}$ and $a_k^\dagger b_{-k}^\dagger$ in $\vec{J}$ are very important. They prevent the current from being constant in time and represent Schrödinger quantum fluctuations, since they vary with the time as $e^{\pm 2i\omega_k t}$. As we shall see in Section 4, analogous terms give rise to creation and annihilation of pairs.

§3. Possible Generalizations of the Theory:
The Problem of a Scalar Theory with Exclusion Principle.

In order to know if a theory of particles with spin zero obeying an Exclusion Principle is possible, it is necessary to discuss in more detail the physical significance of the a, $a^\dagger$, b, $b^\dagger$. We have already remarked that a and $b^\dagger$ are the terms in q corresponding to the positive and negative energy parts.

Let us separate the terms in ϕ as

$$\phi_1 = \frac{1}{\sqrt{V}} \sum_k \frac{ia_k}{\sqrt{2\omega_k}} e^{i\vec{k}\cdot\vec{x}}, \quad \phi_2 = \frac{1}{\sqrt{V}} \sum_k \frac{-ib_k^\dagger}{\sqrt{2\omega_k}} e^{-i\vec{k}\cdot\vec{x}}, \tag{26}$$

and their Hermitian conjugates.

We find their commutation relations

$$[\phi_1(\vec{x},t), \phi_1^\dagger(\vec{x}',t)]_- = +\frac{1}{2} g(\vec{x}-\vec{x}'),$$

and

$$[\phi_2(\vec{x},t), \phi_2^\dagger(\vec{x}',t)]_- = -\frac{1}{2} g(\vec{x}-\vec{x}').$$

The function g is defined by the relations

$$g(x) \quad = \quad \frac{1}{V} \sum_k \frac{1}{\omega_k} e^{i\vec{k}\cdot\vec{x}},$$

$$= \int \frac{d^3k}{(2\pi)^3} \frac{e^{i\vec{k}\cdot\vec{x}}}{\sqrt{k^2+m^2}},$$

$$= \frac{1}{\sqrt{m^2-\nabla^2}} \int \frac{d^3k}{(2\pi)^3} e^{i\vec{k}\cdot\vec{x}},$$

$$= \frac{1}{\sqrt{m^2-\nabla^2}} \delta^3(x), \tag{27}$$

finally expressed symbolically in terms of the operator $1/\sqrt{m^2-\nabla^2}$ and the Dirac delta function. All other commutators $[\phi_1, \phi_2{}^\dagger]_-$ etc are zero. One also gets, from (23), again symbolically,

$$\Pi_1{}^\dagger = \partial_t \phi_1 = -i\sqrt{m^2-\nabla^2}\,\phi_1, \tag{28}$$

and

$$\Pi_2{}^\dagger = \partial_t \phi_2 = +i\sqrt{m^2-\nabla^2}\,\phi_2, \tag{29}$$

and similarly for Π_1 and Π_2.

The functions ϕ_1 and ϕ_2 are scalar representations of the Lorentz group. In fact, the property of a wave function of having frequencies of only one sign is conserved by the Lorentz transformations.

Noting that the commutation relations are also invariant representations of the relativistic group, one gets the generalization for $t \neq t'$ in the following way. Define

$$g_{\mp}(\vec{x}, t) = \int \frac{d^3k}{(2\pi)^3} \frac{e^{\pm i(\vec{k}\cdot\vec{x}-\omega_k t)}}{\omega_k}. \tag{30}$$

Then

$$[\phi_1(\vec{x}, t), \phi_1{}^\dagger(\vec{x}', t')]_- = +\frac{1}{2} g_-(\vec{x}-\vec{x}', t-t'),$$

and

$$[\phi_2(\vec{x}, t), \phi_2{}^\dagger(\vec{x}', t')]_- = -\frac{1}{2} g_+(\vec{x}-\vec{x}', t-t'), \tag{31}$$

which are seen to be relativistic scalars.

It is true that the definitions given for ϕ_1, ϕ_2 and their commutation relations are not invariant except for free particles and that it is necessary to change them in the presence of the electromagnetic field. One could, it would seem - perhaps not unequivocally - reformulate this problem following Dirac's method [2] for the analogous problem in the hole theory. The principle of this method is the characterization of the function g by its singularity on the light cone rather than by (30).

We do not discuss this last problem and proceed to the discussion of the expression for the four-vector of the density-current in the absence of forces. On substituting $\phi = \phi_1 + \phi_2$ in (7) one obtains three groups of terms:

$$S_\nu = -i[(\partial_\nu \phi_1{}^\dagger \phi_1 - \phi_1{}^\dagger \partial_\nu \phi_1) + (\phi_2 \partial_\nu \phi_2{}^\dagger - \partial_\nu \phi_2 \phi_2{}^\dagger)$$

$$+(\partial_\nu\phi_1^\dagger\phi_2 - \phi_2^\dagger\partial_\nu\phi_1 + \partial_\nu\phi_2^\dagger\,\phi_1 - \phi_1^\dagger\partial_\nu\phi_2)]. \tag{32}$$

One sees that we have changed the order of certain factors. In fact we have replaced:

$$\phi_2^\dagger\partial\phi_1 \Leftarrow \partial\phi_1\phi_2^\dagger, \quad \phi_1^\dagger\partial\phi_2 \Leftarrow \partial\phi_2\phi_1^\dagger$$

and also

$$-\phi_1^\dagger\partial\phi_1 + \phi_2\partial\phi_2^\dagger \Leftarrow -\partial\phi_1\phi_1^\dagger + \partial\phi_2^\dagger\phi_2,$$

in Eqn.7. This changing of the order of factors creates the opportunity for Fermi-Dirac statistics.

Note that each one of the three groups of terms in S_ν is covariant under transformations of the relativistic group independent of the other two. As a consequence, one can try the generalization of the theory by supposing

$$\begin{aligned} S_\nu &= -i[c_1(\partial_\nu\phi_1^\dagger\phi_1 - \phi_1^\dagger\partial_\nu\phi_1) + c_2(\phi_2\partial_\nu\phi_2^\dagger - \partial_\nu\phi_2\phi_2^\dagger) \\ &\quad + c_3(\partial_\nu\phi_1^\dagger\phi_2 - \phi_2^\dagger\partial_\nu\phi_1 + \partial_\nu\phi_2^\dagger\phi_1 - \phi_1^\dagger\partial_\nu\phi_2)] \end{aligned} \tag{32}$$

with the real coefficients c_1, c_2, c_3 undetermined. One easily verifies that this expression is 1) Hermitian, 2) relativistically covariant, and 3) satisfies the continuity equation. Also, the total charge is

$$Q = \int \rho d^3x = \sum_k (c_1 a_k^\dagger a_k - c_2 b_k^\dagger b_k)$$

instead of (21). The constants c_1 and $-c_2$ are the charge of the particles 1 and 2, whereas c_3 determines the amplitude for pair production.

The following question suggests itself: in what way is the case $c_1 = c_2 = c_3 = 1$ of our original theory of Section 1 distinguishable from the general case? The answer is given by the permutability of $\rho(x)$ and $\rho(x')$. If one calculates $[\rho(x), \rho(x')]_-$ using (31), one concludes that this expression is not equal to zero, unless $c_1^2 = c_2^2 = c_3^2$ and $c_1 c_3 = c_2 c_3$, that is $c_1 = c_2 = \pm c_3$. Indeed the sign of c_3 is arbitrary because one can substitute $\phi_2 \Rightarrow -\phi_2$ without changing the commutation relations. The theory of Section 1 is consequently the only one for which $\rho(x)$ and $\rho(x')$ commute.

One can now investigate the theory for the case of the Exclusion Principle. For N^a and N^b to have the eigenvalues 0,1 it is necessary to suppose

$$III'.) \qquad [a, a^\dagger]_+ = 1, \quad [b, b^\dagger]_+ = 1,$$

with the anticommutator $[A, B]_+ = AB + BA$, and all other anticommutators are zero. Furthermore

$$[\phi_1(\vec{x}, t), \phi_1^\dagger(\vec{x}', t]_+ = \frac{1}{2} g(\vec{x} - \vec{x}'),$$

and the same for $[\phi_2, \phi_2^\dagger]_+$.

The relations (28,29) remain valid for Fermi-Dirac statistics. Eqn.32 follows in the same way from the conditions of relativistic covariance, and the continuity equation is still satisfied. It is important to point out that the coefficients c_1, c_2, c_3 should be real in order for ρ to be Hermitian. If one looks at the conditions for which $[\rho(x), \rho(x')]_- = 0$ in the case of the Exclusion Principle, one finds $c_1^2 = c_2^2 = -c_3^2$ and $c_1 c_3 = c_2 c_3$. The minus sign in front of c_3^2 has the effect that the only solution possible with real coefficients is the trivial one with $c_1 = c_2 = c_3 = 0$. One can conclude: In the case of the Exclusion Principle it is impossible to satisfy simultaneously relativistic invariance of the theory, and the condition that $\rho(x)$ and $\rho(x')$ should commute for spacelike separations.

We have not tried to establish the possibility of a theory with the $\rho(x)$ not commuting at different points of space - a theory which would certainly give rise to difficulties for the commutation relations of the electromagnetic fields - but we are satisfied that the case of Bose-Einstein statistics is distinguished by its simplicity in the case of the relativistic scalar theory.

§4. Situation in the Presence of an Electromagnetic Field
Comparison of Results with the Hole Theory.

On returning to the theory developed in Section 2, we move on to the case where an electromagnetic field is present. In order to simplify things the potential A_ν and the field $f_{\mu\nu}$ are taken to be classical. One has the Lagrangian

$$L = \sum_{\nu=0}^{3} (\partial^\nu \phi^\dagger + ieA^\nu \phi^\dagger)(\partial_\nu \phi - ieA_\nu \phi) - m^2 \phi^\dagger \phi. \tag{33}$$

The momentum conjugate to $\phi, \phi^\dagger$ become

$$\Pi = \frac{\delta L}{\delta(\partial_t \phi)} = \{(\partial_t - ieA_0)\phi\}^\dagger \tag{34}$$

and its Hermitian conjugate $\Pi^\dagger$, replacing the earlier result. The Hamiltonian becomes

$$H = \int (\Pi \partial_t \phi + \Pi^\dagger \partial_t \phi^\dagger - L) d^3 x = H_0 + H_1, \tag{35}$$

with

$$H_0 = \int (\Pi^\dagger \Pi + \sum_{k=1}^{3} \partial_k \phi^\dagger \partial_k \phi + m^2 \phi^\dagger \phi) d^3 x$$

and

$$
\begin{aligned}
H_1 = {}& i \int eA_0 [\Pi^\dagger \phi^\dagger - \Pi \phi] d^3 x \\
& + \int \sum_{k=1}^{3} [eA_k(\phi^\dagger \partial_k \phi - \partial_k \phi^\dagger \phi) + e^2 A_k^2 \phi^\dagger \phi] d^3 x. \tag{36}
\end{aligned}
$$

269

The new expressions for Π and $\Pi^\dagger$ are to be used in the commutation relations (I).

The electric charge density-current four vector which satisfy the continuity equation becomes

$$S_\nu = -ie[(\partial_\nu \phi^\dagger + ieA_\nu \phi^\dagger)\phi - (\partial_\nu \phi - ieA_\nu \phi)\phi^\dagger],$$

which leaves

$$\rho = -ie(\Pi\phi - \Pi^\dagger \phi^\dagger)$$

and gives

$$j_k = -j^k = -ie(\partial_k \phi^\dagger \phi - \partial_k \phi \phi^\dagger) + 2e^2 A_k \phi^\dagger \phi.$$

The expression for ρ demonstrates that it retains the same commutation properties and eigenvalues as in the case of no forces. The wave equation, consistent with the Heisenberg equation of motion is

$$(\partial_t + ieA_0)^2 \phi = \sum_{k=1}^{3} (\partial_k - ieA_k)^2 \phi - m^2 \phi. \tag{37}$$

One can introduce into the Hamiltonian the a, $a^\dagger$, b, $b^\dagger$ defined in Sec 2. H_0 is given by (19) and, on introducing matrix elements f_{kj} of a function f defined by

$$f_{kj} = \frac{1}{V} \int f(x) e^{-i(\vec{k}-\vec{j})\cdot\vec{x}} d^3 x, \tag{38}$$

one gets

$$
\begin{aligned}
H_1 &= \frac{1}{2}\sum_{k,j} eA_{0;kj}\left[\frac{\omega_k + \omega_j}{\sqrt{\omega_k \omega_j}}(a_k^\dagger a_j - b_{-j}^\dagger b_{-k}) + \frac{\omega_k - \omega_j}{\sqrt{\omega_k \omega_j}}(a_j b_k - a_k^\dagger b_{-j}^\dagger)\right] \\
&\quad + \frac{1}{2}\sum_{k,j}\frac{1}{\sqrt{\omega_k \omega_j}}[e\vec{A}_{kj}\cdot(\vec{k}+\vec{j}) - e^2(\vec{A})_{kj}^2] \\
&\quad \times (a_k^\dagger a_j + b_{-k} b_{-j}^\dagger - a_k^\dagger b_{-j}^\dagger - b_{-k} a_j).
\end{aligned} \tag{39}
$$

One easily sees that the matrix elements coming from $a_k b_{-j}$ and $a_k^\dagger b_{-j}^\dagger$ correspond to the annihilation and production of pairs.

The amplitude of this process from the absorption of a single photon $\hbar\omega$ in the coulomb field of a nucleus of charge Ze is the same order of magnitude as in the hole theory. For $\hbar\omega \gg mc^2$ the total cross section is

$$\sigma \sim \frac{Z^2 e^2}{\hbar c}\left(\frac{e^2}{mc^2}\right)^2 \log\frac{2\hbar\omega}{mc^2},$$

the numerical factors being different in the two theories. In general, one can say that every process connected with pair production which occurs in the hole theory, also occurs in this theory. These include, for example, the scattering of light by light or the coherent

270

scattering of a photon by the field of a nucleus. Unfortunately the conclusions concerning the existence of infinite vacuum polarization and an infinite self energy are the same in the two theories. The two effects diverge like the integral $\int dk/k$ in momentum space just as in the original hole theory of Dirac [3].

I believe that these difficulties can not be removed from a theory which does not explain the numerical value of $e^2/\hbar c$ and that for this objective an entirely new point of view will be necessary.

References.

1) All the following considerations result from the work communicated with V. Weisskopf, of which a part has been published in Helvetica Physica Acta **7**, 709, 1934.

2) P.A.M. Dirac, Proc. Cambr. Phil. Soc. **30**, 150, 1934.

3) For the polarization see: Rapport du Congres Solvay 1933, by Dirac. For the particle energy see: V. Weisskopf, Zeits. f. Phys. **89**, 27 and **90**, 817, 1934.

APPENDIX 10.B:

Excerpt from: Physikalische Zeitschrift der Sowjetunion 11, 590 (1937)

NOTE ON THE SECOND QUANTIZATION
OF THE DIRAC EQUATION

by D. Iwanenko and A. Socolow
(received on 14 April 1937)

We investigate the Dirac Equation which is first order in the sense of Pauli and Weisskopf.

The problems of a single electron in a state of close confinement or of high energy are similar, since we have here to deal with pair production. The critical wavelength is $1/m$ and the critical energy is m, the electron mass. It is necessary at such energies to treat the Dirac equation by the method of second quantization. This property of electrons and positrons of great kinetic energy - in particular that they can be created and annhilated - is the exact analog of the well known property of photons (which always have the required energy) that they can be trivially created because the critical energy is zero. It should be remarked here, that it is in general impossible to isolate a single photon and even to set up the equation of the one photon problem; which also makes comprehensible the failure of many previous attempts in this direction.

In the following we obtain a simple formula for the Dirac hole theory by applying the notion of quantized plane waves. The relationship with the analogous development of QED and the theory of the relativistic scalar equation is made clear in this way.

§1. The Quantization of Wavefields in the Force Free Case.

We express the solutions of the free field Dirac Equation in the form of a Fourier series

$$\psi = \frac{1}{\sqrt{V}} \sum_{\vec{k},\sigma} \left(a_{\vec{k},\sigma} u(\vec{k},\sigma) e^{i(\vec{k}\cdot\vec{r}-\omega t)} + b_{\vec{k},\sigma}^{\dagger} v(\vec{k},\sigma) e^{-i(\vec{k}\cdot\vec{r}-\omega t)} \right), \qquad (1)$$

and its Hermitian conjugate $\psi^{\dagger}$, with $\omega_k = \sqrt{\vec{k}^2 + m^2}$, and

$$u(\vec{k},\sigma)^{\dagger} u(\vec{k},\sigma) = v(\vec{k},\sigma)^{\dagger} v(\vec{k},\sigma) = 1.$$

To determine the (anti)commutation relations among the amplitudes a and b we use

272

the Heisenberg equation of motion

$$i\dot{F} = [F, H]_- = FH - HF \qquad (2)$$

with the Hamiltonian

$$H = \sum_{\vec{k},\sigma} \omega_k (a^\dagger_{\vec{k},\sigma} a_{\vec{k},\sigma} - b_{\vec{k},\sigma} b^\dagger_{\vec{k},\sigma}), \qquad (3)$$

where we have used

$$H = \int d^3x \psi^\dagger i\partial_t \psi = \int d^3x \psi^\dagger (\vec{\alpha} \cdot \vec{p} + \beta m)\psi. \qquad (4)$$

From (2) and (3) we get for Fermi statistics

$$[a_{\vec{k},\sigma}, a^\dagger_{\vec{k}',\sigma'}]_+ = \delta_{\vec{k},\vec{k}'}\delta_{\sigma,\sigma'}, \quad [b_{\vec{k},\sigma}, b^\dagger_{\vec{k}',\sigma'}]_+ = \delta_{\vec{k},\vec{k}'}\delta_{\sigma,\sigma'}, \qquad (5)$$

and all other (anti)commutators zero. In the well known way, we can obtain a matrix representation of these and identify number operators

$$N^a_{\vec{k},\sigma} = a^\dagger_{\vec{k},\sigma} a_{\vec{k},\sigma}, \quad N^b_{\vec{k},\sigma} = b^\dagger_{\vec{k},\sigma} b_{\vec{k},\sigma} \qquad (6)$$

with possible eigenvalues 0 and 1, corresponding to the allowed number of electrons and positrons in the (momentum,spin)-state $(\vec{k}, \sigma)$.

The charge, current, energy and momentum of the system of Dirac particles are

$$
\begin{aligned}
Q &= e\int d^3x \psi^\dagger \psi = e\sum_{\vec{k},\sigma}(N^a_{\vec{k},\sigma} - N^b_{\vec{k},\sigma} + 1),\\
\vec{J} &= e\int d^3x \psi^\dagger \vec{\alpha}\psi = e\sum_{\vec{k},\sigma}\frac{\vec{k}}{\omega_k}(N^a_{\vec{k},\sigma} - N^b_{\vec{k},\sigma} + 1)\\
&\quad + \sim (a^\dagger b^\dagger)e^{2i\omega t} + \sim (ab)e^{-2i\omega t},\\
H &= \int d^3x \psi^\dagger i\partial_t \psi = \sum_{\vec{k},\sigma}\omega_k(N^a_{\vec{k},\sigma} + N^b_{\vec{k},\sigma} - 1),\\
\vec{P} &= \int d^3x \psi^\dagger \frac{1}{i}\vec{\nabla}\psi = \sum_{\vec{k},\sigma}\vec{k}(N^a_{\vec{k},\sigma} + N^b_{\vec{k},\sigma} - 1).
\end{aligned} \qquad (7)
$$

These formulas permit the usual interpretation of N^a, N^b as the number of electrons and positrons in the corresponding state. The infinite constants appear in energy, charge, current, and momentum of the electrons, because the states of negative energy are full. Even in the field free case, the total current is not constant, but involves fluctuations with frequencies $\pm 2\omega$.

It is interesting that the formulas obtained for the Dirac Equation agree with the work of Pauli and Weisskopf which second quantized the relativistic equations of second order.

273

The Hamiltonian function of the scalar equation

$$H = \int d^3x \left(\frac{\partial \phi^\dagger}{\partial t} \frac{\partial \phi}{\partial t} + \sum_{s=1}^{3} \frac{\partial \phi^\dagger}{\partial x_s} \frac{\partial \phi}{\partial x_s} + m^2 \phi^\dagger \phi \right) \tag{8}$$

with the Fourier decomposition

$$\phi = \frac{1}{\sqrt{V}} \sum_{\vec{k}} \frac{1}{\sqrt{2\omega_k}} \left(a_{\vec{k}} e^{i(\vec{k}\cdot\vec{r}-\omega t)} + b_{\vec{k}}^\dagger e^{-i(\vec{k}\cdot\vec{r}-\omega t)} \right) \tag{9}$$

gives the following expression for the total energy in terms of the Fourier coefficients

$$H = \sum_{\vec{k}} \omega_k (a_{\vec{k}}^\dagger a_{\vec{k}} + b_{\vec{k}} b_{\vec{k}}^\dagger). \tag{10}$$

This agrees with the case of Fermi statistics but with the wrong-sign anticommutation relations for the b coefficients

$$[a_{\vec{k}'}, a_{\vec{k}}^\dagger]_+ = +\delta_{\vec{k},\vec{k}'}, \quad [b_{\vec{k}'}, b_{\vec{k}}^\dagger]_+ = -\delta_{\vec{k},\vec{k}'}. \tag{11}$$

On the other hand, the case of Bose statistics leads to the right-sign commutation relations

$$[a_{\vec{k}'}, a_{\vec{k}}^\dagger]_- = \delta_{\vec{k},\vec{k}'}, \quad [b_{\vec{k}'}, b_{\vec{k}}^\dagger]_- = \delta_{\vec{k},\vec{k}'}, \tag{12}$$

with the interpretation of $N^a = a^\dagger a$ and $N^b = b^\dagger b$ as number operators of scalar electrons and positrons.

Then we get the total energy as the sum of the individual positive energies of the two particle types, up to the infinite constant. One can eliminate this infinite constant, which occurs also in the expressions for the energy, momentum, etc, only by special ordering of the factors in the expression for the energy, etc, exactly as in the analogous case of the radiation field. We take the matter waves as the starting point of quantization, and make such conversions as seem appropriate. The same observation holds for the application of Fermi statistics to the scalar relativistic equations (or to the d'Alembertian equations for the potential in the case of the photon fields). One can obtain the correct Fermi anticommutation relations only through special ordering, still following Pauli and Weisskopf, leading to the inadmissibility of the antisymmetric statistics for the scalar relativistic equation on the basis of other criteria.

Conversely, if we ask about the application of Bose statistics to the Dirac equation, then we come to the conclusion that - inevitably, without regard to further criteria - one can get the correct commutation relations, but these lead always to negative energies for one of the two particle types. All these characteristic complications of the application of commutation relations for the case of the two relativistic equations arise obviously because of the introduction of a number of independent free Fourier coefficients. We

274

have, for example, only one series of coefficients for the nonrelativistic equation, so both statistics appear completely equivalent. From the phenomenological standpoint, the Bose (or Fermi) statistics are the most natural for the scalar (or Dirac) relativistic equation.

§2. The Case of the Presence of External Forces and the Polarization of the Vacuum.

In the presence of external fields, the free Hamiltonian requires the addition of vector and scalar potential terms:

$$H_1 = -e \int d^3 x \psi^\dagger \vec{\alpha} \cdot \vec{A} \psi, \quad H_2 = e \int d^3 x \psi^\dagger A_0 \psi. \tag{13}$$

From these, substituting (2), one obtains

$$H_1 = -e \sum_{\vec{k}_f, \vec{k}_i} \vec{A}_{\vec{k}_f, \vec{k}_i} \cdot \langle \vec{k}_f | \vec{J} | \vec{k}_i \rangle, \quad H_2 = e \sum_{\vec{k}_f, \vec{k}_i} A_{0;\vec{k}_f, \vec{k}_i} \langle \vec{k}_f | \rho | \vec{k}_i \rangle \tag{14}$$

where $\langle \vec{J} \rangle$ contains four terms (expressed here in an abbreviated but obvious notation)

$$\begin{aligned}
\langle \vec{J} \rangle &= u_f^\dagger \vec{\alpha} u_i e^{i(\omega_f - \omega_i)t} + v_f^\dagger \vec{\alpha} v_i e^{-i(\omega_f - \omega_i)t} \\
&\quad + u_f^\dagger \vec{\alpha} v_i e^{i(\omega_f + \omega_i)t} + v_f^\dagger \vec{\alpha} u_i e^{-i(\omega_f + \omega_i)t},
\end{aligned} \tag{15}$$

and similarly for the matrix element $\langle \rho \rangle$. The matrix elements of $(\vec{A}, A_0)$ are defined as

$$\vec{A}_{k_f, k_i} = \frac{1}{V} \int d^3 x \vec{A} e^{-i(\vec{k}_f - \vec{k}_i) \cdot \vec{r}} \tag{16}$$

and so on. The above formulas allow many applications, of which we here mention only the relatively simple case of the polarization of the vacuum. We calculate the additional charge density ρ' which arises due to the field of the potential A_0 in a space with a pre-existing "external" charge density ρ. The details of this calculation depend on the use of functions of the electrons and positrons, which always only give positive energies.

To calculate the additional density, we set

$$\psi(\vec{r}, t) = \frac{1}{\sqrt{V}} \sum_{k,q} [c_a^{q,k}(t) a_k u_k e^{i(\vec{k} \cdot \vec{r} - \omega t)} + c_b^{q,k}(t) b_k^\dagger v_k e^{-i(\vec{k} \cdot \vec{r} - \omega t)}] \tag{17}$$

with $c_{a/b}^{q,k}(t = 0) = \delta_{q,k}$, because the initial state has only undisturbed waves. The perturbation calculation gives

$$\psi(\vec{r}, t) = \frac{e}{\sqrt{V}} \sum_k [(a_k + \delta a_k) u_k e^{i(\vec{k} \cdot \vec{r} - \omega t)} + (b_k^\dagger + \delta b_k^\dagger) v_k e^{-i(\vec{k} \cdot \vec{r} - \omega t)}] \tag{18}$$

where

$$\delta a_k = -e \sum_q A_{0;q,k} e^{i(\vec{q} - \vec{k}) \cdot \vec{r}} \left[\frac{a_q u_q^\dagger u_k}{(\omega_k - \omega_q)} + \frac{b_q^\dagger v_q^\dagger u_k}{(\omega_k + \omega_q)} \right],$$

$$\delta b_k^{\dagger} \quad = \quad e \sum_q A_{0;q,k} e^{i(\vec{q}-\vec{k})\cdot\vec{r}} \Big[\frac{a_q u_q^{\dagger} v_k}{(\omega_k + \omega_q)} + \frac{b_q v_q^{\dagger} v_k}{(\omega_k - \omega_q)} \Big]. \tag{19}$$

From this we form the expression for the additional density

$$\rho' = e[\psi(\vec{r},t)^{\dagger}\psi(\vec{r},t) - \psi(\vec{r},0)^{\dagger}\psi(\vec{r},0)]. \tag{20}$$

Here we must take the usual values for a,b from 3) and keep only the products $aa^{\dagger}$ and $bb^{\dagger}$ because the initial number of electrons $N^a = a^{\dagger}a$, and of positrons $N^b = b^{\dagger}b$ is equal to zero. This gives

$$\rho' = \frac{-e^2}{V} \sum_{k,q} \frac{(v_k^{\dagger} u_q u_q^{\dagger} v_k)}{(\omega_k + \omega_q)} 2 Re[A_{0;q,k} e^{i(\vec{q}-\vec{k})\cdot\vec{r}}], \tag{21}$$

where one gets for

$$v_k^{\dagger} u_q u_q^{\dagger} v_k = (1 - \frac{\vec{k}\cdot\vec{q} + m^2}{\omega_k \omega_q}). \tag{22}$$

Here a beautiful average over the spins is automatically produced. The momentum summation, which we replace by an integration, yields finally the known Dirac-Heisenberg result

$$\rho' = -\frac{1}{15\pi}(\frac{e}{m})^2 \nabla^2 \rho_0. \tag{23}$$

Siberian Physical-Technical Institute.Tomsk

References.

1) For the calculations and formulas, see Verh. Sib. Phys.-Techn. Inst., Tomsk. **5**, No. 1, 1937 (in Russian).

2) See L. de Broglie, L'Electron Magnetique, page 212 in the Russian Edition.

3) W. Pauli and V. Weisskopf, Helv. Phys. Acta **7**, 709, 1934.

4) The often conjectured connection of Fermi-statistics with half-integral spin requires further discussion, as for example the case of the one dimensional Dirac equation. See also A. Socolow about the possibility of a neutrino theory of light Sow. Phys. (in press).

5) P.A.M. Dirac, Proc. Camb. Phil. Soc. **30**, 150, 1934; W. Heisenberg, Zeits f. Phys **90**, 220, 1934.

Chapter 11

Fierz's Proof of the Spin-Statistics Theorem

Summary: Fierz's proof of the Spin-Statistics Theorem is the first formal and in any sense rigorous proof. It is based initially on the premise that each elementary particle corresponds to an irreducible relativistic spinor. Starting from this premise, Fierz proves the Spin-Statistics Theorem from the basic requirements of relativistic covariance, and positive definite energy. He points out in retrospect that the initial premise of irreducibility is actually unessential, thereby paving the way for Pauli's more general proof.

§1. Introduction.

The 1938 research of Markus Fierz, directed by Pauli but published without his authorship [11.1], constituted an abrupt escalation in formality, generality, and rigor in the discussion of the Spin-Statistics Theorem. The results are still limited to non-interacting fields but now include arbitrary integral and half-integral spins, although difficulties are noted for spins greater than one. The results for half-integral spins are expressed in the formalism of the van der Waerden-Dirac spinor representations of the proper Lorentz group, and are very obscure without a substantial investment by the reader. A readily available source is Misner, Thorne, and Wheeler [11.2]; another modern source is Roman [11.3]. The best advice to the reader is to get what one can from Fierz and later from Pauli [11.4], if only at a heuristic level. To quote Fierz, "It is not possible to represent this truly clumsy calculus with any clarity." It is unfortunate that Fierz's paper, the following one by Belinfante [11.5], the subsequent one by Belinfante and Pauli [11.6], and Pauli's historic proof [11.7], all using the spinor techniques, are destined to be detours on the eventual path to the current proof of the Spin-Statistics Theorem. They interest us today primarily for historical reasons, but do not provide strong motivation to master the spinor calculus. Nonetheless, we owe it to these pioneering authors, especially to the younger ones whose work has for so long been ignored, to understand and evaluate their contributions.

Fierz states, clearly and unequivocally, the scope and limitations of his conclusions: "From the very general requirement that the commutation relations should be relativistically invariant and local, and that the energy should be positive, it follows that identical particles with integral spin must always obey Bose-Einstein statistics, identical particles with half-integral spin must always obey Fermi-Dirac statistics."

§2. Guide to Fierz's Proof: Integral Spin.

In Part I §1 and §2, Fierz presents a beautiful discussion of massive tensor fields in terms of potentials A, and field tensors $B^{(q)}$ constructed by taking (q) successive 4-curls of A. From these he arrives at (q) real-symmetric rank-2 tensors which satisfy the continuity equation and are candidate energy-momentum tensors $T_{kl}^{(q)}$ (Eqns.2.1, 2.2). All integrate to the same total energy E which, in the particle rest frame, is a positive sum over the absolute squares of a minimal set of Fourier coefficients (Eqn.2.5), and therefore is positive in any Lorentz frame. In a similar way he is able to construct a current vector with appropriate zero value for real fields, and finally, a charge (Eqn.2.6) with the appropriate form in terms of absolute squares of Fourier coefficients. The expressions E and Q for the classical tensor fields are in a form very compatible with quantization according to Bose-Einstein statistics. Fierz has avoided any reference to the Lagrangian formalism and deals just with the free field equations and auxiliary conditions imposed to select massive fields of unique spin.

Fierz presents a valuable pedagogic discussion in §3 where he demonstrates the equivalence of the tensor formulation to the Dirac wave equations in van der Waerden spinor form. We will need these tools later when we get to his discussion of half-integral spin particles. Fierz elects to continue the treatment of integral spins in the more familiar tensor notation and proceeds to quantization of the tensor field theory.

He does this following Jordan and Pauli [11.8] by constructing manifestly Lorentz-covariant commutation relations for the tensor field Fourier coefficients. They must satisfy the field equations, the auxiliary conditions, the quantized energy requirements, and must produce the Heisenberg equations of motion for all field variables. This program is completed in Eqn.4.2 and following.

To summarize: Fierz presents the prescription for quantization of the classical massive tensor field in accordance with the requirements:

1) manifest Lorentz invariance;

2) local commutation relations compatible with the field equations and auxiliary conditions;

3) Heisenberg equations of motion for quantized fields;

4) quantized plane waves having integral energy (in units of ω) and integral charge (in units of e).

Quantization using commutation relations succeeds. Fierz returns to a discussion of the contradictions which arise using anticommutation relations after his discussion of the half-integral spin case.

§3. Guide to Fierz's Proof: Half-Integral Spin.

In Part II §5, Fierz extends the introductory discussion of spinors begun in §3 to the full treatment required for half-integral spin fields. The spinors are characterized by $2k$ (lower) undotted and $2l - 1$ (upper) dotted indices with $2k + 2l - 1 = 2f$. For fields with half integral spin $f = \frac{1}{2}, \frac{3}{2}, \cdots$, k and l are both integral or both half-integral. The spinor

$$a_{\delta\rho\cdots}^{\dot{\lambda}\dot{\mu}\cdots}$$

is separately symmetric in upper dotted indices and in lower undotted in-

279

dices. These can be converted one to the other using the equations of motion. The result is that there is a multiplicity of equivalent spinor representations, just as there is a multiplicity of representations for the tensor fields, which are labelled by the index q

$$a^{(q)} = a^{f-q-\frac{1}{2}}_{f+q+\frac{1}{2}} \quad \text{and} \quad b^{(q)} = b^{f+q+\frac{1}{2}}_{f-q-\frac{1}{2}} \tag{1}$$

with $f+q+\frac{1}{2}$ lower undotted indices and $f-q-\frac{1}{2}$ upper dotted indices and so on. q runs from 0 to $f-\frac{1}{2}$. In particular, for $q = f-\frac{1}{2}$, $a = a^0_{2f}$ has $2f$ lower undotted generally-symmetric indices and $2f+1$ linearly independent solutions as required (see the more detailed discussion following Eqn.5.3), thus identifying f as the spin.

Simple examples are helpful. For spin-0, $f = k = 0$ and $l = \frac{1}{2}$, the spinor is trivial with no indices. For spin-$\frac{1}{2}$, $2f = 1$, $k = 0$, $l = 1$, or $k = \frac{1}{2}$, $l = \frac{1}{2}$. The two possibilities correspond to $q = 0$ and $a^{(0)} = a_\lambda$ with one lower (undotted) index $\lambda = 1, 2$, and $b^{(0)} = b^{\dot{\lambda}}$ with one upper (dotted) index. For spin-$\frac{3}{2}$, $q = 0, 1$ leading to four possibilities $a^{(0)} = a^{\dot{1}}_2$ with one upper (dotted) index and two lower (undotted, symmetric) indices, and $b^{(0)} = b^2_{\dot{1}}$; also $a^{(1)} = a^0_3$ and $b^{(1)} = b^3_0$. For spin-1, $k = 1$, $l = \frac{1}{2}$ and $k = 0$, $l = 1$ lead to two choices a^0_2 and b^2_0, and $k = \frac{1}{2}$, $l = 1$ leads to one more, a^1_1. In this way a four vector is recognized as a rank-2 relativistic spinor. It can be returned to the explicit four vector form by contracting with the four (including the unit matrix) 2×2 Pauli spinors σ_μ

$$V_\mu = V_{\dot{\lambda}\beta}\sigma^{\dot{\lambda}\beta}_\mu. \tag{2}$$

The actual reduction of the spinor components to the physical degrees of freedom is deferred to Fierz's appendices, not reproduced here.

Fierz next constructs different but equivalent forms of a conserved current (Eqn.6.1)

$$S^{(q)}_{\dot{\lambda}\beta} = \frac{1}{2}(a^{(q)*}_{\dot{\lambda}}b^{(q-1)}_\beta + a^{(q)}_\beta b^{(q-1)*}_{\dot{\lambda}} + q \Leftrightarrow q - 1) \tag{3}$$

which all lead to the same total charge

$$Q = \int d^3x (S_{\overset{(0)}{11}} + S_{\overset{(0)}{22}}) \tag{4}$$

which he shows to be of the positive definite form $\sum(a^*a + bb^*)$ (Eqn.6.2).

Similarly he constructs rank-2 energy-momentum tensors from spinors of the form (Eqn.6.3)

$$t_{\dot\lambda\beta,\dot\tau\gamma}^{(q)} = \frac{1}{4}(a_{\dot\lambda}^* p_{\dot\tau\gamma} b_\beta \cdots) \tag{5}$$

which satisfy the continuity equations

$$p^{\dot\lambda\beta} t_{\dot\lambda\beta,\dot\tau\gamma}^{(q)} = p^{\dot\tau\gamma} t_{\dot\lambda\beta,\dot\tau\gamma}^{(q)} = 0. \tag{6}$$

The energy-momentum tensors

$$T_{kl}^{(q)} = \frac{1}{2}(t_{\dot\lambda\beta,\dot\tau\gamma}^{(q)} + t_{\dot\tau\gamma,\dot\lambda\beta}^{(q)})\sigma_k^{\dot\lambda\beta}\sigma_l^{\dot\tau\gamma} \tag{7}$$

are symmetric and satisfy the continuity equation. All lead to the same total energy, which, however, is now the non-positive form $\sum(a^*a - bb^*)$ with the "wrong-sign" for the "negative energy" coefficients bb^*.

Fierz concludes that the postulate of anticommutation relations for the quantized Fourier coefficients of the half-integral spin fields is required to make the total energy positive definite, and at the same time to permit the interpretation of negative energy states as antiparticles of the opposite charge.

Fierz constructs (anti)commutators for spinor fields of spin f, $a_{\alpha\beta\cdots\gamma}^{(f)}(x)$ and $a_{\dot\nu\dot\rho\cdots\dot\lambda}^{(f)*}(y)$, which are totally symmetric (see his Eqn.5.3 and following) in their $2f$ indices $\alpha\beta\cdots\gamma$ and $\dot\nu\dot\rho\cdots\dot\lambda$. He conjectures (8.1)

$$[a_{\alpha\beta\cdots\gamma}^{(f)}(x), a_{\dot\nu\dot\rho\cdots\dot\gamma}^{(f)*}(y)]_\pm$$
$$= \frac{1}{(m^{2f-1}(2f)!)}\sum\mathcal{P}(\alpha\cdots\gamma)p_{\alpha\dot\nu}\cdots p_{\gamma\dot\lambda}D(x-y). \tag{8}$$

Here D(x-y) is the invariant D function, and there are $2f$ differentiations $p_{\alpha\dot\nu}$. The symbol $\Sigma\mathcal{P}(\alpha\cdots\gamma)$ indicates a summation over all permutations

281

of the indices $\alpha \cdots \gamma$). The (anti)commutation relation in momentum space is

$$[a_{\alpha\beta\cdots\gamma}^{(f)}(k), a_{\dot\nu\dot\rho\cdots\dot\gamma}^{(f)*}(k')]_{\pm}$$
$$= \frac{\delta_{kk'}}{(m^{2f-1}(2f)!)} \sum \mathcal{P}(\alpha \cdots \gamma) k_{\alpha\dot\nu} \cdots k_{\gamma\dot\lambda} 1/k_0, \qquad (9)$$

using $D \sim exp^{-ik_0 t}/k_0$. In the rest-frame $k = k' = k_0 \delta_{\alpha\dot\nu}$ one obtains

$$(k_0/m)^{2f-1} \delta_{\alpha\beta\cdots\gamma} \delta_{\dot\nu\dot\rho\cdots\dot\lambda}.$$

For $k_0 = \pm m$, the righthand side is positive definite for half-integral f, in agreement with the choice of anticommutation on the left. In contrast, for integral f, the righthand side is $(\pm)^{2f-1}$ which can be positive or negative and is presumed to require the use of commutators on the lefthand side.

In this way Fierz concludes that he has proved the "long conjectured relationship between spin and statistics in a simple mathematical way." The requirements of the proof are:

1) the existence of a rest-frame for each plane-wave (ie $m \neq 0$).

2) the properties of the D-function that the commutation relations should be relativistically invariant and local.

3) the fact that the number of spinor indices $(2f)$ is even or odd for spin (f) integral or half-integral.

§4. Concluding Remarks.

Where then does Fierz's proof fall short? Pauli was silent on this subject although he was to criticize Belinfante's similar proof because it assumed irreducible spinors. Fierz addresses this question in the next-to-last paragraph of §8. He states unequivocally what is the foundation of Pauli's widely credited proof published almost two years later $\cdots$ "*For this it is moreover unessential that the spinors are irreducible.*" He goes on to provide the essential component to Pauli's proof, that " The demonstration uses only $\cdots$

3) the fact that the number of spinor indices is even or odd, depending on whether the spin is integral or half-integral." We appeal again to van der Waerden's edict that we should only judge by *what* people *wrote* and *when*. Fierz seems to have anticipated the essence of Pauli's proof.

Fierz ignored interactions, but so did Pauli in his subsequent proof. And finally, the assumption that all authors of this generation make, that $[\phi(x), \phi^\dagger(y)]_+$ must be positive for $(x - y)$ spacelike, could not be proved for some fifteen years until the Hall-Wightman theorem. We will see that Pauli's proof eliminates the initial premise of Fierz, that physical states should correspond to irreducible spinors, and builds a similar proof using instead only requirement (3) as a restriction on the spinor content of physical states. But be forewarned that almost all of these developments will eventually be replaced in the proofs due to Lüders and Zumino, and to Burgoyne.

§5. Biographical Note on Fierz.(†)

Markus Eduard Fierz (1912-) was Pauli's assistant (1936-39); to professor at Basel (1939-60); director of theory division at CERN (1959-60); and Pauli's successor at ETH, Zurich (1960-). His first research [11.10] was the application of Fermi's four-fermion interaction describing weak decays as a point interaction between fermion vector currents, modelled on electrodynamics. He applied the theory to the scattering of high energy neutrinos

$$\bar{\nu} + p \rightarrow n + e^+$$

and got the answer we know today, quadratically divergent in the neutrino momentum and requiring a unitarizing factor provided by the finite mass of the W-boson exchange. He was that close to leaping over the developments of the next twenty or more years in his very first research! His powerful ability in mathematical analysis is much in evidence in his proof of the Spin-Statistics Theorem . His name is prominent in the old four-fermion weak interaction theory, for his identification of interaction types which are invariant under permutation of the four fermions [11.11].

Fierz's later work extended to the statistical mechanics of fluctuation and condensation. He was ultimately recognized by his peers as a physicist-philosopher.

(†- in A. Pais, *Inward Bound* (Oxford, New York, 1986), pp.428-9; in V.F. Weisskopf, *The Joy of Insight* (BasicBooks, Harper-Collins, 1991); p.317; in *Theoretical Physics in the Twentieth Century* (Interscience, New York, 1960); in *World Who's Who in Science*

(Marquis, Chicago, 1994), p.564.)

Bibliography and References.

11.1) M. Fierz, Helv. Phys. Acta **12**, 3 (1939); see our App.11A.

11.2) C.W. Misner, K.S. Thorne, and J.A. Wheeler, *Gravitation* (Freeman, San Francisco, 1973), Ch.41, pp.1148-1155.

11.3) P. Roman, *Theory of Elementary Particles* (North Holland, Amsterdam, 1960), pp.80-85.

11.4) W. Pauli, Phys. Rev. **58**, 716 (1940); see our App.14A.

11.5) F.J. Belinfante, Physica **6**,870 (1939); see our App.12A.

11.6) W. Pauli and F.J. Belinfante, Physica **7**, 177 (1960); see our App.12B.

11.7) P. Jordan and W. Pauli, Zeits. f. Phys. **47**, 151 (1928).

11.8) A.S. Wightman, Phys. Rev. **101**, 860 (1956); see our App.18A.

11.9) G. Lüders and B. Zumino, Phys. Rev. **101**, 1450 (1958); N. Burgoyne, Nuov. Cim. **VIII**, 607 (1958); see our App.17A,B.

11.10) M. Fierz, Helv. Phys. Acta **9**, 245 (1936); see Pais above.

11.11) M. Fierz, Helv. Phys. Acta, **16**, 365 (1943); **22**, 489 (1949); see J.M. Blatt and V.F. Weisskopf, *Theoretical Nuclear Physics* (Wiley, New York, 1952), pp.639,718.

284

APPENDIX 11.A:

Excerpt from: Helvetica Physica Acta 12, 3 (1939)

ON THE RELATIVISTIC THEORY OF FORCE FREE
PARTICLES OF ARBITRARY SPIN

by MARKUS FIERZ
(3.IX.38.)

Abstract: In the force-free case it is possible to investigate quantized fields corresponding to particles with integral or half-integral spin greater than one. It is shown that particles with integral spin must always obey Bose-Einstein statistics, and particles with half-integral spin must always obey Fermi-Dirac statistics. In the force-free case, fields with spin less than or equal to one are special in that, for these, the charge density and energy density are well-defined, definite, and gauge-invariant quantities, whereas this is the case for higher spin only for the total charge and the total energy.

Introduction.

In the following, we investigate the relativistic theory of fields, quantized according to Jordan and Pauli, for particles with arbitrary integral or half-integral spin. We treat only the force-free case. By specification of the spin and mass of the particle, the corresponding field is uniquely determined, and moreover the statistics of the particles is fixed by the spin. From the very general requirement that the commutation relations should be relativistically invariant and local, and that the energy should be positive, it follows that identical particles with integral spin must always obey Bose-Einstein statistics, identical particles with half-integral spin must always obey Fermi-Dirac statistics.

Although it is possible to describe the fields for integral spin by tensors, to which one can associate van der Waerden's spinor calculus, for half-integral spin fields it is not possible to represent this truly clumsy calculus with any clarity. The difficulties with the representation techniques are perhaps due to the fact that a half-integral spin field, because of the Pauli Exclusion Principle, can never correspond to a classical field, and can never be understood in a classical sense, in contrast to a tensor field obeying Bose-Einstein statistics.

The $\cdots$ wave equations have been given $\cdots$ by Dirac. The physical interpretation $\cdots$

285

is not clear ⋯ . Sakata and Yukawa ⋯ proceed ⋯ much too formally ⋯ . Jauch has ⋯ the first correct energy-momentum tensors and current vectors for the Dirac fields. The special case of spin 1 ⋯ by Proca; ⋯ by Stueckelberg, ⋯ by Kemmer.

Our investigation shows that, at least for the force-free case, fields with arbitrary integral or half-integral spin are possible, and also that the small spin values 0, $\frac{1}{2}$, 1 are distinguished in many respects. In these three cases, both the energy density and the charge density are uniquely determined; and in the case of spin 0 and 1 the energy density, in the case of spin $\frac{1}{2}$ the charge density, of the classical c-number theory is of definite sign. This is no longer the case for spin larger than one, where only the total energy or the total charge is still well defined and of definite sign. Furthermore, in quantized theories for spins greater than 1, the charge densities at different places but for the same time are no longer commutable, but depend on derivatives of the D-function. Further complications occur during attempts to introduce interactions with other fields into the classical c-number theory, in the case of spin greater than 1. This last point therefore needs a more thorough investigation.

Part I. Integer Spin.

§1. Field Tensors and Wave Equations.

We consider the force-free classical field, from which one gets particles of mass m and spin f by means of the field quantization of Jordan and Pauli. f is a positive integer. Such a field can be represented by a complex, generally-symmetric four-tensor $A_{ij\cdots k}$ of rank f (with f indices running $i, j, \cdots k = 0, 1, 2, 3$ for $x_j = $t,x,y,z, and with metric (-+++) usually left implicit) which obeys the wave equation

$$(\Box - m^2)A_{ij\cdots k} = 0. \tag{11}$$

Furthermore, $A_{ij\cdots k}$ satisfies two supplementary conditions

$$A_{ii\cdots k} = 0, \tag{12}$$

$$\partial_i A_{ij\cdots k} = 0. \tag{13}$$

The physical interpretation ⋯ . The supplementary conditions guarantee that the tensor field contains only particles of spin f and not also particles of smaller spins. It follows from (1.2), and the general symmetry of $A_{ij\cdots k}$, that all traces vanish.

As shown in the appendix, these $A_{ij\cdots k}$ possess $(f+1)^2$ linearly independent components, which transform under Lorentz transformations according to the irreducible representation $\phi_{f/2,f/2}$ of the Lorentz group. Since the $A_{ij\cdots k}$ field corresponds to spin f, there must be $(2f+1)$ linearly independent plane waves, for each p satisfying $p_0^2 = (\vec{p})^2 + m^2$,

286

which differ in their spin orientations. To see this, one considers a plane wave in its rest system, which always exists for m≠0, for which $p_0 = \pm m$. Then the supplementary condition (1.3) says that all components of $A_{ij\cdots k}$ for which any of the indices is 0, must vanish; it follows that in the rest system the indices in fact run only from 1 to 3 and $A_{ij\cdots k}$ has the form

$$A_{ij\cdots k} = A^0{}_{ij\cdots k}e^{\pm imt}.$$

$A^0_{ij\cdots k}$ is a symmetric spacelike constant tensor of rank f in R_3, whose traces vanish. Such a tensor has $(2f+1)$ linearly independent components, which transform into one another, under rotation of the coordinate system, according to the irreducible representation ϕ_f of the rotation group. It follows that the particle states are identified with the $(2f+1)$ different orientations of the spin. Further, the supplementary condition (1.3) makes possible the construction of an energy-momentum tensor for the A-field for which the total energy is positive for every physical representation of the theory.···

One can replace the differential equations for the A-field by a system of first order differential equations, which are convenient. The equations are

$$B^{(1)}_{[ij]k\cdots r} = \partial_i A_{jk\cdots r} - (i \leftrightarrow j) \tag{14}$$

$$\partial_i B^{(1)}_{[ij]k\cdots r} = m^2 A_{jk\cdots r}. \tag{15}$$

These equations are the analog of Maxwell's equations. As follows from (1.2,4,5), $B^{(1)}$ also obeys the second order wave equation, as well as the further equations:

$$B^{(1)}_{[ij]j\cdots r} = 0, \qquad B^{(1)}_{[ij]k\cdots r} + ([ki]j) + ([jk]i) = 0, \tag{16}$$

$$\partial_k B^{(1)}_{[ij]k\cdots r} = 0, \qquad \partial_k B^{(1)}_{[ij]\cdots r} + (j[ki]) + (i[jk]) = 0. \tag{17}$$

In the unbracketted indices, $B^{(1)}_{[ij]k\cdots r}$ is symmetric; in the bracketted pair, antisymmetric. From (1.6) it follows, independent of the definition of $B^{(1)}$, that $B^{(1)}_{[ij]kk\cdots r} = 0$. Using this, one can count $2f^2+4f$ linearly independent components of $B^{(1)}$. From (1.6,7), all remaining equations can be derived, so that we can describe the field just as well with the $B^{(1)}$ as with the A. We can now construct a further quantity

$$B^{(2)}_{[ij][kr]\cdots t} = \partial_k B^{(1)}_{[ij]r\cdots t} - (k \leftrightarrow r).$$

$B^{(2)}$ is symmetric for the exchange of pairs of indices [ij] and [kr] as well as in the unbracketted indices and obeys analogous equations to $B^{(1)}$. Further it gives

$$B^{(2)}_{[ij][ir]\cdots t} = m^2 A_{jr\cdots t}.$$

Continuing to form the 4-curl with respect to the unbracketted indices, we get a sequence of $(f+1)$ field quantities

$$A_{ij\cdots k}, \quad B^{(1)}_{[ij]k\cdots r}, \quad \cdots B^{(f)}_{[ij][hr]\ [st]},$$

287

where $B^{(q)}_{[ij]\cdots[rs]\cdots t}$ contains q brackets which are symmetric with each other, and (f-q) individual indices symmetric among themselves. Within each bracketted pair, $B^{(q)}$ is antisymmetric. The $B^{(q)}$ satisfy the following equations:

$$(I) \qquad (\square - m^2)B^{(q)} = 0$$

$$(IIa) \qquad B^{(q)}_{\cdots[ij]j\cdots} = 0$$

$$(IIb) \qquad B^{(q)}_{\cdots[ij]k\cdots} + \quad ([ki]j) + \quad ([jk]i) = 0$$

$$B^{(q)}_{[ij][rs]\cdots} + \quad ([ri][js]) + \quad ([jr][is]) = 0$$

$$(IIIa) \qquad \partial_r B^{(q)}_{[ij]\cdots r} = 0$$

$$(IIIb) \qquad \partial_k B^{(q)}_{[ij]\cdots} + \quad (j[ki]) + \quad (i[jk]) = 0. \qquad (18)$$

These equations describe the field A exactly as do the equations (1.1,2,3). From above

$$\partial_i B^{(q)}_{\cdots[ik]\cdots} = m^2 B^{(q-1)}_{\cdots k \cdots}$$

$$\partial_i B^{(q-1)}_{\cdots k \cdots} - (i \leftrightarrow k) = B^{(q)}_{\cdots[ik]\cdots}$$

$$B^{(q)}_{[ki][kj]\cdots} = m^2 B^{(q-2)}_{\cdots ij \cdots}.$$

§2. Energy-Momentum Tensor and Current Vector.

For the field described by the $B^{(q)}$, it must be possible to construct from $B^{(q)}$ and its conjugate a real, symmetric, rank-2 tensor, which can be interpreted as the energy-momentum tensor of the field. $\cdots$ By means of the $B^{(q)}$ and the $B^{(q-1)}$ and their conjugates we can form the tensors $T^{(q)}_{kl}$:

$$T^{(1)}_{kl} = \frac{1}{2}[m^2 A^*_{ij\cdots k}A_{ij\cdots l} + B^{(1)*}_{ij\cdots[tk]}B^{(1)}_{ij\cdots[tl]} + \quad (k \leftrightarrow l)]$$

$$- \frac{1}{2}\delta_{kl}[m^2 A^*_{ij\cdots}A_{ij\cdots} + \frac{1}{2}B^{(1)*}_{ij\cdots[tr]}B^{(1)}_{ij\cdots[tr]}], \qquad (21)$$

and generally

$$T^{(q)}_{kl} = \frac{1}{2}[m^2 B^{(q-1)*}_{[rs]\cdots tk}B^{(q-1)}_{[rs]\cdots tl} + B^{(q)*}_{[rs]\cdots[tk]\cdots m}B^{(q)}_{[rs]\cdots[tl]\cdots m} + \quad (k \leftrightarrow l)]$$

$$- \frac{1}{2}\delta_{kl}[m^2 B^{(q-1)*}_{[rs]\cdots t}B^{(q-1)}_{[rs]\cdots t} + \frac{1}{2}B^{(q)*}_{[rs]\cdots[tm]\cdots n}B^{(q)}_{[rs]\cdots[tm]\cdots n}]. \qquad (22)$$

On the basis of the differential equations for the $B^{(q)}$, these tensors obey the continuity equation

$$\partial_k T^{(q)}_{kl} = 0.$$

It is now possible to require that the total energy of the field is positive. We first show that the total energy is to be identified with the integral over all space of $T^{(q)}_{00}$, starting

288

with $T_{00}^{(1)}$, which is constructed from A and $B^{(1)}$. To this end, we expand A in plane waves

$$A_{ij\cdots r}(\vec{x}, t) = \frac{1}{\sqrt{V}} \sum_k A_{ij\cdots r}^{(+)}(\vec{k}) e^{i(\vec{k}\cdot\vec{x} - \omega t)} + A_{ij\cdots r}^{(-)*}(\vec{k}) e^{-i(\vec{k}\cdot\vec{x} - \omega t)}, \tag{23}$$

where $\omega = +\sqrt{\vec{k}^2 + m^2}$. From (1.3), the $A^{(+)}$ and $A^{(-)}$ satisfy

$$\sum_{i=1}^3 k_i A_{i\cdots}^{(\pm)} - \omega A_{0\cdots}^{(\pm)} = 0. \tag{24}$$

$B^{(1)}$ can be expressed in terms of $A^{(\pm)}$ and we obtain

$$E = \int d^3 x T_{00}^{(1)} = \sum_k \omega^2 [A_{ij\cdots}^{(+)*}(\vec{k}) A_{ij\cdots}^{(+)}(\vec{k}) + A_{ij\cdots}^{(-)}(\vec{k}) A_{ij\cdots}^{(-)*}(\vec{k})]. \tag{25}$$

The energy is expressed as the sum of energies of individual Fourier components. To show that each individual contribution is positive, consider $A_{ij\cdots}^{(+)*} A_{ij\cdots}^{(+)}$ in the rest-system where k=(m,0,0,0). From (1.3), all terms with any of $(ij\cdots)$ equal zero must vanish. Consequently, all sums on $(i, j \cdots)$ run $(1 \cdots 3)$ and are positive, so that in the rest-system the energy of a plane wave is positive. In any other coordinate system the energy is multiplied by the positive factor $1/\sqrt{1 - v^2}$, so the total energy must be positive. $\cdots$

In the same way, one finds for the general tensor $T^{(q)}$

$$\int d^3 x T_{00}^{(q)} = (2m^2)^{(q-1)} \int d^3 x T_{00}^{(1)}.$$

All energy-momentum tensors give rise to the same total energy except for the overall factor $(2m^2)^{q-1}$. The energy density, however, is largely undetermined, in that it depends strongly upon q. Besides, $T_{00}^{(q)}$ is not positive definite for $f>1$. $\cdots$

In addition to the energy momentum tensor, a vector can be formed from $B^{(q)}$ and $B^{(q-1)}$, which obeys the continuity equation and can be used as the current vector:

$$S_v^{(q)} = \frac{1}{2i}[B_{[ik]\cdots ml}^{(q-1)*}\cdots B_{[ik]\cdots [mv]l}^{(q)}\cdots - B_{[ik]\cdots ml}^{(q-1)}\cdots B_{[ik]\cdots [mv]l}^{(q)*}\cdots].$$

This vector is "real". If the field tensors $B^{(q)}$ are "real", it vanishes identically. For a given field the current and charge density can again be defined in f different ways, which are equivalent in the force-free case. The total charge is the same in all cases (up to a factor)

$$Q = \sum_k \omega[A_{ij\cdots}^{(+)*}(k) A_{ij\cdots}^{(+)}(k) - A_{ij\cdots}^{(-)}(k) A_{ij\cdots}^{(-)*}(k)]. \tag{26}$$

These current-, charge-densities cannot be simply used as electric current-, charge-densities because they depend on the electric field. In general, the prescription $p \to (p - eA)$ is not possible $\cdots$ if $f>1$. Only for $f=1$, where the trace requirement is omitted, does one arrive at the contradiction-free theory of Proca.

§3. *Application to Dirac's Relativistic Wave Equation.*

Dirac has given differential wave-equations which have the form in spinor notation:

$$p^{\dot\rho\eta}a_{\eta\lambda\mu\cdots}^{\dot\tau\dot\nu\cdots} = m \quad b_{\lambda\mu\cdots}^{\dot\rho\dot\tau\dot\nu\cdots}, \tag{31}$$

$$p_{\dot\nu\rho}b_{\lambda\gamma\cdots}^{\dot\nu\dot\tau\mu\cdots} = m \quad a_{\rho\lambda\gamma\cdots}^{\dot\tau\mu\cdots}. \tag{32}$$

Here a has $2k$ undotted and $2l$-1 dotted indices which run 1,2; b has $2l$ dotted and $2k$-1 undotted indices running 1,2; k and l are integer or half-integer. a and b are symmetrical in dotted and in undotted indices, and therefore irreducible under Lorentz transformations, since these transform into one another like the representations $\phi_{k,l-\frac{1}{2}}$ or $\phi_{k-\frac{1}{2},l}$. If $k+l-\frac{1}{2}=f$ is an integer, then these equations are equivalent to the tensor equations we have used. All equations with the same f describe the same field, where the partition of $f+\frac{1}{2}$ into two summands corresponds to the different possibilities q in the tensor description.

In particular, the spinor $a_{\delta\rho\cdots}^{\lambda\mu\cdots}$, with the same number of dotted as undotted indices, that is $k = l - \frac{1}{2}$, is identical with the tensor $A_{rs\cdots}$. By means of the Pauli matrices $\sigma_{\dot\rho\rho}^{k}$, one transforms each dotted and each undotted index $\dot\rho, \rho$ into a vector index k. The tensor is then symmetric in all indices and its traces vanish, since the trace (defined as $a_\mu^\mu \equiv \epsilon^{\mu\nu}a_{\mu\nu}$ with $\epsilon^{AB} = -\epsilon^{BA}$, $\epsilon_{AB} = -\epsilon_{BA}$, and $\epsilon^{12} = \epsilon_{12} = 1$; $A, B = 1, 2$) of a symmetric spinor $a_{\mu\nu} = a_{\nu\mu}$ vanishes. Furthermore it also obeys (1.1,3). (1.3) follows from the spinor equation

$$p^{\dot\rho\eta}a_{\dot\rho\eta\lambda\mu\cdots}^{\dot\tau\cdots} = 0 \tag{33}$$

which follows by contraction from (3.1), using the fact just established that the trace of a symmetric spinor vanishes.

From $a_{\lambda\eta\cdots}^{\dot\rho\mu\cdots}$ one forms new spinors

$$mb_{\lambda\cdots}^{(1)\dot\rho\dot\tau\cdots} = p^{\dot\rho\eta}a_{\eta\lambda\cdots}^{\dot\tau\cdots}, \quad mb_{\mu\eta\cdots}^{(-1)\dot\nu\cdots} = p_{\mu\dot\rho}a_{\eta\lambda\cdots}^{\dot\rho\dot\nu\cdots},$$

and generally

$$mb_{\lambda\cdots}^{(q+1)\dot\rho\dot\tau\cdots} = p^{\dot\rho\eta}b_{\eta\lambda\cdots}^{(q)\dot\tau\cdots}, \quad mb_{\mu\eta\cdots}^{(q-1)\dot\nu\cdots} = p_{\mu\dot\rho}b_{\eta\lambda\cdots}^{(q)\dot\rho\dot\nu\cdots}, \tag{34}$$

starting with $b_{\eta\lambda}^{(0)\dot\mu\cdots} \equiv a_{\eta\lambda}^{\dot\mu\cdots}$. q is half the difference of the number of dotted and of undotted indices, and runs from -f to +f. $a_{\delta\rho}^{\lambda\dot\mu}$ has f dotted and f undotted indices and transforms as the $\phi_{f/2,f/2}$ irreducible representation of the Lorentz group. $b_{\delta\rho}^{(q)\lambda\dot\mu}$ has $f+q$ dotted and f-q undotted indices and transforms under proper Lorentz transformations as the irreducible representation $\phi_{(f-q)/2,(f+q)/2}$. Under reflections, $b^{(q)} \leftrightarrow b^{(-q)}$. The $b^{(q)}$ are irreducible quantities in the opposite sense to the $B^{(q)}$. The tensors with the index value q, which are antisymmetric and self-dual are of the form

$$\epsilon^{iklm}F_{\cdots[lm]}^{(q)} = F_{[ik]\cdots}^{(q)} \quad (q > 0), \qquad -\epsilon^{iklm}f_{\cdots[lm]}^{(q)} = F_{[ik]\cdots}^{(q)} \quad (q < 0).$$

Under reflections, $F^{(q)} \leftrightarrow F^{(-q)}$. Using the $b^{(q)}$, we can construct f energy-momentum tensors (just as with the $B^{(q)}$) which are symmetric and obey the continuity equation:

$$
\begin{aligned}
t^{(q-1)}_{\dot{\rho}\delta,\beta\dot{\nu}} = \quad & [b^{(q)*}_{\dot{\rho}\beta} b^{(-q)}_{\nu\delta} + \quad (q) \leftrightarrow (-q)] + [b^{(q)*}_{\nu\delta} b^{(-q)}_{\dot{\rho}\beta} + \quad (q) \leftrightarrow (-q)] \\
& + [b^{(q-1)*}_{\nu\dot{\rho}} b^{(-q-1)}_{\beta\delta} + \quad (q) \leftrightarrow (-q)] \\
& + [b^{(q+1)*}_{\beta\delta} b^{(-q+1)}_{\dot{\rho}\dot{\nu}} + (q) \leftrightarrow (-q)].
\end{aligned}
\tag{35}
$$

Here the unwritten indices should be contracted as

$$
a^* a \equiv a^{* \dot{\rho}\dot{\nu}\cdots}_{\alpha\delta\cdots} a^{\alpha\delta\cdots}_{\dot{\rho}\dot{\nu}\cdots}.
$$

From (3.4) it follows that

$$
p^{\dot{\rho}\delta} t^{(q-1)}_{\dot{\rho}\delta,\nu\gamma} = 0.
$$

Since the tensor formulation and the spinor notation are mathematically equivalent in this case, it follows that all tensors considered above are linear combinations of corresponding spinors. We forego any further discussion of (3.4) and refer to the tensor theory.

§4. Quantization of the Tensor Field Theory.

To represent particles by the above tensor fields, we construct Lorentz invariant commutation relations between the field quantities. It suffices to do this for the $A_{ij\cdots k}$, since those for other field quantities can be obtained from the differential equations. In the force-free case dealt with here, it is suitable to use commutation relations in four dimensional form, following Jordan and Pauli for the charge free electromagnetic field, which has the advantage of manifest Lorentz invariance. The commutation relations should have the result that the energy of a plane wave is an integer multiple of $\omega_k = +\sqrt{k^2 + m^2}$ and should give the Heisenberg equation of motion

$$
i\dot{F} = [F, H]_-
\tag{41}
$$

for all quantities F which do not depend explicitly on the time.

For $f=1$, one can introduce commutation relations between the 3 independent amplitudes corresponding to longitudinal and transverse waves for each momentum vector k_i. This approach seems unsuitable for higher spins because it destroys the symmetry and manifest Lorentz invariance of the commutation relations. One can instead define directly commutation relations between the $A_{ij\cdots k}$ and the $A^*_{ij\cdots k}$, which satisfy the same equations as the $A_{ij\cdots k}$. In this way the supplementary conditions (1.2,3) are satisfied, and one obtains the correct number of independent commutation relations.

We prescribe the commutation relations for the symmetric tensor $A_{i_1\cdots i_f}$ as

$$
[A_{i_1\cdots i_f}, A^*_{j_1\cdots j_f}]_- = iK[\sum P(j_k) R(i_1 j_1) \cdots R(i_f j_f)
$$

291

$$-\frac{1}{\Gamma} \sum_{n>m}^{f} R(i_n i_m) \sum P(j_k) R(i_1 j_1) \cdots R(i_n j_m) \cdots$$
$$R(i_f j_f)] D(x). \tag{42}$$

Here: $\Gamma = 2 + f(f-1)/2$; $A_{i_1 \cdots}$ is evaluated at $\xi + x/2$, $A^*_{j_1 \cdots}$ at $\xi - x/2$; $\sum P(j_k)$ is the sum over all permutations of the indices j_k; and

$$R(i_n i_m) = \delta_{i_n i_m} - \frac{1}{m^2} \partial_{i_n} \partial_{i_m} \tag{43}$$

operates on the invariant D-function of Jordan and Pauli, which is defined as

$$D(x) \equiv D(\vec{x}, t) = \frac{1}{V} \sum_{\vec{k}} e^{i \vec{k} \cdot \vec{x}} \frac{\sin \omega_k t}{\omega_k}.$$

D(x) obeys the wave equation

$$(\Box - m^2) D = 0$$

and the conditions

$$D(\vec{x}, 0) = 0, \quad \partial_t D|_{t=0} = \delta(\vec{x}).$$

Since $R(\alpha\beta)$ operates on D(x), which obeys the wave-equation, the following relations hold:

$$\begin{aligned} \partial_\alpha R(\alpha\beta) &= R(\alpha\beta)\partial_\alpha = 0, \\ R(\alpha\beta)R(\beta\gamma) &= R(\alpha\gamma), \\ R(\alpha\alpha) &= 3. \end{aligned} \tag{44}$$

The constant K in (4.2) must be chosen so that the eigenvalue of the charge is a whole number.

The commutation relations (4.2) satisfy equations (1.1) to (1.3) identically and are symmetric in primed and unprimed indices, since the same permutations of the j_k are summed over. Its divergence is zero from (4.4). $\cdots$ the traces vanish $\cdots$ to satisfy (1.2), the factor $1/\Gamma$ must precede the sum $\sum_{n>m}$.

We must now show that the equation $i\dot{F} = [F, H]_-$ holds, and that the energy has the correct eigenvalues. Recall the resolution of $A_{ij \cdots k}$ into plane waves in (2.3). It follows from the commutation relations, that $A(k)$ commutes with $A(k')$ for $k_i \neq k'_i$. Also $A^{(+)}$ commutes with $A^{(-)*}$. For given k_i there are $(2f+1)$ linearly independent linear combinations of the $A^{(+)}_{ij \cdots r}(k)$, which we call A_μ. The commutation relations of these can be written

$$[A_\mu, A^*_{\mu'}]_- = \mathcal{F}_{\mu\mu'}. \tag{45}$$

We now consider the A_μ in the rest system of k_i. There the A_μ and the A^*_μ transform like the irreducible representation ϕ_f of the rotation group which spans a space R_{2f+1}.

292

Since the right side of the commutation relations (4.2) satisfies the same relations as the A_μ, the matrix $\mathcal{F}_{\mu\mu'}$ has the property, that the vectors X_μ in R_{2f+1} generate $Y_{\mu'} = X_\mu \mathcal{F}_{\mu\mu'}$, which also span an irreducible space and, since all $Y_{\mu'}$ are non-zero, again has dimension $2f+1$. Therefore $\mathcal{F}_{\mu\mu'}$ has $2f+1$ eigenvalues different from zero. Also we choose $A_\mu, A_{\mu'}^*$ so that the representation formed from them is unitary.

Since the commutation relations are invariant under rotations, they have, for the proper normalization of the A_μ, the form

$$[A_\mu, A_{\mu'}^*]_- = \delta_{\mu\mu'}. \tag{46}$$

The energy is also rotation invariant and can be written in the standard form

$$E(k_i) = \sum_{\mu=1}^{(2f+1)} \omega(k) \cdot A_\mu^* A_\mu \cdot \mathcal{C}. \tag{47}$$

The constant K occurring in the commutation relations (4.2) is now determined so that $\mathcal{C} = 1$.

From (4.7)

$$\omega(k) A_{ij\cdots}^{(+)*} A_{ij\cdots}^{(+)} = \sum_\mu A_\mu^* A_\mu$$

for $\mathcal{C}=1$. Also,

$$\omega(k) [A_{ij\cdots}^{(+)}, A_{ij\cdots}^{(+)*}]_- = \sum_\mu [A_\mu, A_\mu^*]_- = 2f + 1.$$

Therefore it follows from comparison with (4.2)

$$\sum_{i,j} \delta_{ij} [\sum P(j_m) \cdots - \frac{1}{\Gamma} \sum_{n>m} \cdots] D(x) = \frac{2f+1}{K} D(x)$$

that the constant K is determined.

The commutation relations now have the property that the eigenvalue of the energy of a plane wave is an integer multiple of $\omega(k)$; the case for $A^{(-)}$ follows in exactly the same way. This can be represented by $2f+1$ quantities B_μ, which also have the commutation relations

$$[B_\mu, B_{\mu'}^*]_- = \delta_{\mu\mu'}. \tag{48}$$

And, finally, for every quantity F which can be constructed linearly from $A_\mu e^{i\omega t}$ and $B_\mu^* e^{-i\omega t}$ not explicitly containing the time, the Heisenberg equation of motion holds

$$i\dot{F} = [F, H]_-,$$

since it holds for A_μ and B_μ^*.

Therefore the commutation relations (4.2) constitute the solution to the problem set at the beginning, to quantize classical tensor fields to represent integral spin particles.

The commutation relations in spinor form are discussed in the second part, appropriate for the half-integral spin case, where we compare it with the above.

Part II. Half-Integral Spin.

§5. Fields and Wave-Equations.

The fields corresponding to particles of half-integral spin f $(f=\frac{1}{2}, \frac{3}{2} \cdots)$, cannot be represented by tensors but require spinors. In the spinor formulation the theories of whole- and half-integral spin are very much alike, although characteristic differences appear, which involve the structure of the energy-momentum tensor and the commutation relations. In most other respects however, the conclusions reached in the integral-spin case remain valid, so we can express ourselves here more briefly.

We start from a spinor $a_{\delta\rho\cdots}^{\lambda\mu\cdots}$, which has $2k$ undotted and $2l$-1 dotted indices, where $2k+2l\text{-}1=2f$ is an odd number. k and l are therefore both integral or both half-integral. The spinor a obeys the second order wave equation

$$(\Box - m^2)a = 0, \tag{51}$$

where m is the mass of the particle. Furthermore, $a_{\delta\rho\cdots}^{\dot\lambda\dot\mu\cdots}$ is symmetric in dotted and in undotted indices. This is equivalent to the requirement that all traces of a must vanish, where the trace is defined as

$$a_{\rho\cdots}^{\cdots} \equiv \epsilon^{\delta\rho}a_{\delta\rho\cdots}^{\cdots} = \epsilon_{\dot\lambda\dot\mu}a^{\dot\lambda\dot\mu\cdots}_{\cdots} = 0, \tag{52}$$

where $\epsilon^{1,1} = \epsilon^{2,2}=0$, $\epsilon^{1,2} = -\epsilon^{2,1}= 1$. In addition, a must satisfy the auxiliary conditions

$$\epsilon_{\dot\nu\dot\mu}p^{\dot\nu\rho}a_{\delta\rho\cdots}^{\dot\lambda\dot\mu\cdots} = 0, \tag{53}$$

which mean that $p^{\dot\nu\rho}a_{\delta\rho\cdots}^{\dot\lambda\dot\mu\cdots}$ and $p_{\dot\nu\tau}a_{\delta\rho\cdots}^{\dot\nu\dot\mu\cdots}$ are symmetric in dotted and in undotted indices.

The $a_{\delta\rho\cdots}^{\dot\lambda\dot\mu\cdots}$ is given when one knows how many dotted and how many undotted indices there are. Thus it has $(2k+1)(2l)$ linearly independent components, which transform among themselves under Lorentz transformations as the irreducible representation $\phi_{k,l-\frac{1}{2}}$. It remains to be shown, using the auxiliary conditions (5.3), that there are just $(2f+1)$ linearly independent plane waves for each momentum p_i. For this purpose we go to the rest system of p_i. There $p^{\dot\nu\rho}$ has the form $\omega(p)\delta^{\dot\nu\rho}$ which transforms under spacelike rotations as δ_ν^ρ. Eventually, all upper dotted indices can be made lower undotted indices at least as far as their rotation properties in the rest system are concerned. That is

$$a_{\delta\rho\cdots}^{\dot\lambda\cdots} \to a'_{\rho\cdots,\lambda\cdots}.$$

294

The auxiliary condition (5.3) yields

$$\epsilon^{\rho\lambda} a'_{\rho\cdots,\lambda\cdots},$$

and says that the spinor a' is symmetric in the indices ρ and λ and therefore in all indices. In the rest system, the spinor $a^{\dot\lambda\cdots}_{\dot\delta\rho\cdots}$ is equivalent to a spinor $a'_{\rho\cdots,\lambda\cdots}$ symmetric in all indices of rank $2k+2l-1=2f$ and has $2k+2l=2f+1$ linearly independent components which transform under rotations as the irreducible representation ϕ_f of the rotation group in R3. In this way, it is showed that the spin of the field is f. In the case m=0, there is no rest system and these considerations are no longer valid.

The differential equations (5.1) and (5.3) can be replaced by a system of first order equations:

$$p^{\dot\nu\rho} a^{\dot\lambda\mu\cdots}_{\dot\delta\rho\cdots} = m b^{\dot\nu\dot\lambda\mu\cdots}_{\dot\delta\cdots}, \qquad p_{\dot\nu\rho} b^{\dot\nu\dot\lambda\mu\cdots}_{\dot\delta\cdots} = m a^{\dot\lambda\mu\cdots}_{\rho\dot\delta\cdots}. \tag{54}$$

It follows from (5.3) that b is also a symmetric spinor which obeys the second order wave equation and the auxiliary conditions. One can replace all equations with just $(k+l)$ by differentiation. They all describe the same field. Let $a^{(0)\dot\lambda\cdots}_{\dot\delta\cdots}$ be that a for which $k=l$, which has $2k$ undotted and $2k-1$ dotted indices, and, with the corresponding $b^{(0)}$ having $2k$ dotted and $2k-1$ undotted indices, obeys equations of the form (5.4). From $a^{(0)}$ we generate the spinor $a^{(1)}$

$$p_{\dot\lambda\tau} a^{(0)\dot\lambda\dot\mu\cdots}_{\rho\dot\delta\cdots} = m a^{(1)\dot\mu\cdots}_{\tau\rho\dot\delta\cdots} \tag{55}$$

and in the same way $a^{(2)}$ and so on. These equations are again symmetric according to (5.3). In general we get

$$p_{\dot\nu\lambda} a^{(q)\dot\nu}_\lambda = m a^{(q+1)}_\lambda, \qquad p^{\dot\nu\lambda} a^{(q+1)}_\lambda = m a^{(q)\dot\nu}. \tag{56}$$

We generate the symmetric spinors $a^{(0)}\cdots a^{(f-\frac12)}$ which all obey the wave equation and, up to $a^{(f-\frac12)}$, also the auxiliary condition (5.3). $a^{(q)}$ has $(f+q+\frac12)$ undotted and $(f-q-\frac12)$ dotted indices. In the same way we generate the spinors $b^{(1)}\cdots b^{(f-\frac12)}$ from

$$p^{\dot\lambda\tau} b^{(q)\dot\mu\cdots}_{\tau\rho\cdots} = m b^{(q+1)\dot\lambda\dot\mu\cdots}_{\rho\cdots}, \qquad p_{\dot\lambda\tau} b^{(q+1)\dot\lambda\dot\mu\cdots}_{\rho\cdots} = m b^{(q)\dot\mu\cdots}_{\tau\rho\cdots}, \tag{57}$$

where $b^{(q)}$ has $(f+q+\frac12)$ dotted and $(f-q-\frac12)$ undotted indices. Under reflections $b^{(q)} \leftrightarrow a^{(q)}$, so the system of equations (5.6) and (5.7) together are reflection invariant. The case with $k=l$, $q=0$, which obeys separate equations like (5.4), is already reflection invariant by itself. In contrast, for integer spins there existed no spinor like $a^{(0)}$ which was reflection invariant.

§6. Energy-Momentum Tensor and Current Vector.

From the $a^{(q)}$ and $b^{(q)}$ we can construct rank-2 tensors and vectors, which can be interpreted as energy-momentum tensors and current vectors. It turns out, however, as known for the $f = \frac{1}{2}$ case, that the total energy is not positive definite, but the total charge is, as a result of the auxiliary condition (5.3). We first construct the vector

$$
\begin{aligned}
S^{(0)}_{\lambda\beta} &= a^{(0)*\rho\cdots}_{\lambda\dot\nu\cdots} a^{(0)\dot\nu\cdots}_{\beta\rho\cdots} + b^{(0)*\rho\cdots}_{\beta\dot\nu\cdots} b^{(0)\dot\nu\cdots}_{\lambda\rho\cdots} \\
S^{(q)}_{\lambda\beta} &= \frac{1}{2}(a^{(q)*}_{\dot\lambda} b^{(q-1)}_{\beta} + a^{(q)}_{\beta} b^{(q-1)*}_{\dot\lambda} + (q) \leftrightarrow (q-1)) \qquad (61)
\end{aligned}
$$

(one contracts over omitted indices as in $S^{(0)}_{\lambda\beta}$). $S^{(q)}_{\lambda\beta}$ obeys the continuity equation

$$
p^{\lambda\beta} S^{(q)}_{\lambda\beta} = 0
$$

as one shows from $a_\lambda b^\lambda = -a^\lambda b_\lambda$. This gives $f + \frac{1}{2}$ different but, in the force-free case, equivalent ways to define the current. All these possibilities also lead to the same value of the total charge $\int d^3x (S^{(q)}_{11} + S^{(q)}_{22})$, which is positive definite. To show this, consider first $\rho^{(0)}_{\lambda\beta}$: because ρ satisfies the continuity equation, the total charge is time independent and the integral breaks up into a sum over contributions of individual plane waves. Consider an individual plane wave in its rest system. There the auxiliary condition (5.3) states that

$$
a^{\dot 1\cdots}_{2\cdots} = a^{\dot 2\cdots}_{1\cdots}, \qquad b^{\dot 1\cdots}_{2\cdots} = b^{\dot 2\cdots}_{1\cdots}.
$$

Then in the rest system

$$
a^{(0)*\rho\cdots}_{\dot 1\mu\cdots} a^{(0)\dot\mu}_{1\rho\cdots} = a^{(0)*\rho\cdots}_{\dot 1\mu\cdots} a^{(0)\dot\rho\cdots}_{1\mu\cdots}. \qquad (62)
$$

But $a^{(0)*\rho\cdots}_{\dot 1\mu}$ is the complex conjugate of $a^{(0)\dot\rho\cdots}_{1\mu}$, so the right side of (6.2) is the positive form $\sum a^* a$. The charge density consists of just such terms as (6.2), therefore the total charge is positive.

In the general case

$$
\int d^3x (S^{(q)}_{11} + S^{(q)}_{22}),
$$

it is also sufficient to consider a single plane wave in its rest system. $a^{(q)}$ and $b^{(q-1)}$ are derived by q-fold application of the operator $p^{\lambda\beta}/m$ on $a^{(0)}$. In the rest system, this reduces to $\delta^{\lambda\beta}$. Therefore, we immediately conclude that

$$
\int d^3x (S^{(q)}_{11} + S^{(q)}_{22}) = \int d^3x (S^{(0)}_{11} + S^{(0)}_{22}),
$$

and that the vectors $S^{(q)}_{\lambda\beta}$ lead to the same total charge for all q, but not necessarily to the same local charge distribution.

We can also construct spinors corresponding to rank-2 tensors, which satisfy the continuity equation. These can be interpreted as energy-momentum tensors of our fields.

296

We consider the spinors

$$t^{(0)}_{\lambda\beta,\dot{r}\gamma} = \frac{1}{2}([a^{(0)*}_{\lambda} p_{\dot{r}\gamma} a^{(0)}_{\beta} - (a^*,a) \rightarrow (b,b^*)]$$

$$-[a^{(0)}_{\beta} p_{\gamma\dot{r}} a^{(0)*}_{\lambda} - (a,a^*) \rightarrow (b^*,b)]);$$

$$t^{(q)}_{\lambda\beta,\dot{r}\gamma} = \frac{1}{4}(a^*pb - bpa^* + b^*pa - apb^* +$$

$$b^*pa - apb^* + a^*pb - bpa^*) \qquad (63)$$

which obeys the equations

$$p^{\lambda\beta} t^{(q)}_{\lambda\beta,\dot{r}\gamma} = 0, \quad p^{\dot{r}\gamma} t^{(q)}_{\lambda\beta,\dot{r}\gamma} = 0.$$

One next constructs the spinor

$$\Theta^{(q)}_{\lambda\beta,\dot{r}\gamma} = \frac{1}{2}(t^{(q)}_{\lambda\beta,\dot{r}\gamma} + t^{(q)}_{\dot{r}\gamma,\lambda\beta}).$$

The tensor T_{kl} belonging to Θ

$$T_{kl} = \Theta^{(q)}_{\lambda\beta,\dot{r}\gamma} \sigma_k^{\lambda\beta} \sigma_l^{\dot{r}\gamma}$$

is symmetric in (k,l) and obeys the continuity equation

$$\partial_k T_{kl} = 0.$$

T_{kl} can be interpreted as the energy-momentum tensor of the field. One easily shows, with the aid of the Fourier coefficients of the fields, and with the assumption of a rest system, that all $\Theta^{(q)}$ lead to the same value of the total energy, but that the contribution to the energy of a plane wave with time dependence $\sim e^{i\omega t}$, has the opposite sign to that of a wave with time dependence $\sim e^{-i\omega t}$. The theory indicates states of positive and of negative energy, as is known from the Dirac theory of the electron. We must postulate, therefore, that for the field of particles which satisfy the Pauli Exclusion Principle, the energy can be made positive by a means analogous to Dirac's hole theory.

§7. The Matrices $u_\nu(p)$ and $v_\nu(p)$.

The spinors considered up to now were all symmetrical. These are given when the number of undotted and of dotted indices is known. One can represent a spinor $a^{\lambda\cdots}_{\dot{\delta}\cdots}$ with $2k$ undotted and $2l$ dotted indices which is symmetric in each, by a quantity A^r_s whose indices $\dot{r}, s$ give how many dotted and undotted indices there are, so s has $2k+1$ values, r has $2l+1$ values. $\cdots$ (This section serves no essential purpose for us and will be omitted.) $\cdots$

§8. Quantization of the Field Theory for Half-Integral Spin.

As in the case of integral spins, we set up the commutation relations which, together with the field equations, must be satisfied identically.

297

In order to construct such relations, one proceeds from those which are satisfied by the basic spinor $a^{(f)}_{\alpha\beta\ldots\gamma}$ which has only undotted indices. The number of indices is the odd number $2f$. $a^{(f)}_{\alpha\beta\ldots\gamma}$ is symmetric in all indices, however the auxiliary condition (5.3) brings a certain simplification.

As commutation relation between $a^{(f)}$ and $a^{(f)*}$ we try

$$[a^{(f)}_{\alpha\beta\ldots\gamma}, a^{(f)*}_{\dot\nu\dot\rho\ldots\dot\lambda}]_+ = \frac{1}{m^{2f-1}(2f)!}\sum Perm(\alpha\cdots\gamma)$$
$$p_{\alpha\dot\nu}\cdots p_{\gamma\dot\lambda}D(x). \qquad (81)$$

Here $[a,b]_+ = ab + ba$. $D(x)$ is again the invariant D-function defined after (4.3). On the right side of (8.1), there is an odd number (namely $2f$) of differentiations $p_{\alpha\dot\nu}$.

The relations (8.1) have the general consequence that the particles described satisfy the Pauli Exclusion Principle; a circumstance which makes it possible, through a "hole theory", to make the energy positive. That the relations are fulfilled, we show as follows:

Let $k_{\dot\lambda\beta}$ be the spinor $k_{\dot\lambda\beta} = k_i \sigma^i_{\dot\lambda\beta}$ constructed from the momentum vector k_i. Then the commutation relation in momentum space, using the definition of D(x), reads

$$[a^{(f)}_{\alpha\beta\ldots}(k), a^{(f)*}_{\dot\nu\dot\rho\ldots}(k')]_+$$
$$= \frac{\delta_{kk'}}{m^{2f-1}(2f)!}\sum Perm(\alpha\beta\cdots)k_{\alpha\dot\nu}k_{\beta\dot\rho}\cdots\frac{1}{2k_0}. \qquad (82)$$

One sees that spinors of different momentum have zero anticommutator. Consider now two spinors corresponding to the same 4-momentum, in their rest system where $k_{\alpha\dot\nu} = k_0\delta_{\alpha\dot\nu}$. Then the right-side of (8.2) is only nonzero when $a_{\alpha\beta\ldots\gamma}$ is the complex conjugate of $a^*_{\dot\nu\dot\rho\ldots\dot\gamma}$. So the left side of (8.2) is never negative, compatible with the right side which has the form $(k_0)^{(2f-1)}$ and is always positive for half-integral spin $\cdots$. The anticommutator has the desired form

$$[a_i, a^*_k]_+ = \delta_{ik}\cdot(constant). \qquad (83)$$

The charge has the correct eigenvalues, as follows from the anticommutation relation (8.2) if one calculates the charge of a plane wave in its rest-system. This has the positive definite form $\sum aa^*$ from (6.2), with the correct factor assured from the commutation relation (8.2).

If the spin is an integer, we can write the commutation relation between two quantities $a^{(f)}$ defined in §3, in the momentum representation as

$$[a^{(f)}_{\alpha\ldots\gamma}(k), a^{(f)*}_{\dot\nu\ldots}(k')]_-$$
$$= C\cdot\delta_{kk'}\sum Perm(\alpha\cdots\gamma)k_{\alpha\dot\nu}\cdots\frac{1}{k_0}. \qquad (84)$$

Now, however, there is an even number, $2f$, of factors $k_{\alpha\dot\nu}$. In the rest system the right side of (8.4) is

$$C(k_0)^{2f-1}.$$

This is positive or negative, depending on $k^0 = \pm m$. Therefore, one cannot write the left-side of (8.4) as an anticommutator which would have the positive definite form

$$aa^* + a^*a,$$

whereas the right side can be positive or negative. Consequently, one cannot quantize particles with integral spin according to the Pauli Exclusion Principle, without giving up the local character of the commutation relations, but only according to Bose-Einstein statistics [see: W. Pauli, Annales Poincare VI, 147 (1936)]. Particles with half-integral spin can and must be quantized according to the Pauli Exclusion Principle, so that the energy will be positive.

From the above analysis one seems to prove the long-conjectured relationship between spin and statistics in a simple mathematical way. *For this it is moreover unessential that the spinors are irreducible.* (Note: Italics added in translation.) The demonstration uses only 1) the existence of a rest system for each plane-wave; 2) the properties of the D-function; and 3) the fact that the number of spinor indices is even or odd, depending on whether the spin is integral or half-integral. The properties of the D-function determine that the commutation relations should be relativistically invariant and local.

From the commutation relations (8.1) for the $a^{(f)}_{\alpha\beta...}$, one can get those for arbitrary $a^{(q)}$ and $a^{(q)*}$, which have dotted and undotted indices $\cdots$ between $\cdots$ $a^{(q)}$ and $b^{(q)*}$ can also be obtained $\cdots$.

Appendices. (*Not Included*)
Determination of Coefficients of Equation (8.5).
Number of Independent Coefficients of Tensors.
Fields with Zero Rest Mass.
Special Cases as Example:$f = \frac{3}{2}, 2$.
Special Representation of the u^ν and v^ν.

Zurich, Physikalisches Institut der E.T.H.

Footnotes and References.

1) P. Jordan and W. Pauli, Zeits. f. Phys. **47**, 151 (1928).

2) van der Waerden, Die Gruppentheoret. Methode in der Quantenmechanik, Berlin 1932. III. Kapitel, § 20.

3) P.A.M. Dirac, Proc. Roy. Soc. **A155**, 447 (1936).

4) S. Sakata and H. Yukawa, Proc. Phys.-Nat. Soc. Japan **19**, 91 (1937).

5) Proca, C. R. **202**, 1490 (1936).

6) E.C.G. Stueckelberg, Helv. Ph. Acta XI, 299 (1938).

7) N. Kemmer, Proc. Roy. Soc. **A166**, 127 (1938).

Chapter 12

Belinfante's Proof of the Spin-Statistics Theorem

Summary: Belinfante based his proof of the Spin-Statistics Theorem on the requirement of invariance under a charge-conjugation transformation that had recently been invented by Kramers. The uniqueness of this requirement was criticized and, by using a series of his familiar but now recognizably invalid non-local transformations, seemingly refuted by Pauli, with Belinfante himself as an apparently futilely protesting coauthor.

§1. Introduction.

In a series of four papers [12.1] derived from his Leiden PhD thesis of 1939, Frederik Jozef Belinfante ultimately presented a proof of the Spin-Statistics Theorem based on charge-conjugation invariance. Kramers, Belinfante's PhD research professor, had proposed such an invariance and formulated the requirements for it [12.2]. Somewhat unfortunately, Belinfante's proof is expressed in the arcane and never viable language of "undors" which are an extension of the van der Waerden spinor formalism to representations of the improper Lorentz group which include space reflections. In Belinfante's words: "The interesting fact is that the statistical behavior of particles and quanta follows much more directly from the postulate of charge-conjugation invariance than from postulates concerning the positive character of the total energy of free particles or quanta." He cites Kramers for the charge-conjugation invariance and Fierz [12.3] for the proof based on the character of the total energy. In spite of some criticisms of the proof [12.4] (put forward by Pauli with Belinfante himself as coauthor, under apparent duress from Pauli), Belinfante deserves much credit for the germ of the connection between the Spin-Statistics Theorem and the yet undiscovered **TCP**-theorem.

301

Unfortunately, the undor language is at least as awkward as the spinor language characterized by Fierz as "···ugly···". Further compounding the felony, Belinfante was in the habit of inventing his own language (of neutrinors, neutrettos, hystatons, ··· and so on) and of using idiosyncratic mathematical notation. The result is that his papers are more difficult to read than even the undor formalism would dictate. We will give only a brief flavor of the character of the undor formalism and omit many details of their properties, and omit all reference to Belinfante's ancillary work on meson spins and magnetic moments. We have not reproduced his first publication in Nature, Dec 1938 (just three months after the submission of Fierz's paper) but do include selected directly relevant excerpts from his Physica paper of Oct 1939 in App.12A.

In brief, undors of the first rank Ψ_k are Dirac four-spinors in a representation which is irreducible with respect to proper Lorentz transformations. In this case $(\psi_1, \psi_2) \equiv U$ and $(\psi_3, \psi_4) \equiv W$ are van der Waerden spinors, U a covariant conjugated spinor $(u_{\dot{1}}, u_{\dot{2}})$ and W a contravariant regular spinor (w^1, w^2). One can define a metric tensor g for raising indices so that

$$\Psi_k = (\psi_1, \psi_2, \psi_3, \psi_4) = (u_{\dot{1}}, u_{\dot{2}}, w^1, w^2) \tag{1}$$

goes to

$$\Psi^k = (\psi^1, \psi^2, \psi^3, \psi^4) = (u^{\dot{1}}, u^{\dot{2}}, w_1, w_2). \tag{2}$$

The Dirac matrices are expressed as the product of Pauli matrices so $\vec{\alpha} = \rho_z \vec{\sigma}$, $\beta = \rho_x$, etc, and

$$\Psi_k = (\psi_{+\frac{1}{2}+\frac{1}{2}}, \psi_{-\frac{1}{2}+\frac{1}{2}}, \psi_{+\frac{1}{2}-\frac{1}{2}}, \psi_{-\frac{1}{2}-\frac{1}{2}}) \tag{3}$$

which can be labelled $\Psi_{s,r}$ with ρ operating on s and σ operating on r.

Space reflection is achieved by

$$\Psi'(-x, -y, -z, t) = j\beta\Psi(x, y, z, t) \tag{4}$$

with $j = \pm i$ chosen so that a double reflection reverses the sign.

Charge-conjugation is achieved by $\Psi^c = \mathcal{C}\Psi^*$ with $\mathcal{C} = \rho_y\sigma_y$ in Kramers' representation.

Undors of the second rank $\Psi_{k_1 k_2} = \Psi_{s_1 r_1, s_2 r_2}$ are sixteen component objects with $k_1, k_2 = 1, 2, 3, 4$; $s_1, r_1, s_2, r_2 = \pm\frac{1}{2}$, which transform like products $\psi_{k_1}\psi'_{k_2} = \psi_{s_1 r_1}\psi'_{s_2 r_2}$ with matrices $\rho^{(1)}$, $\sigma^{(1)}$ operating on (s_1, r_1), and so on. The sixteen components transform under Lorentz transformations as mixed tensor representations.

Reflections are achieved by successive reflections of ψ_{k_1} and ψ'_{k_2} using $\beta^{(1)}$ and $\beta^{(2)}$ as reflection operators. There are two classes of undors of the second rank, with $j_1 j_2 = \pm 1$. Further separation into symmetric and antisymmetric undors of the second rank simplifies their Lorentz content to

1) symmetric, $j_1 j_2 = -1$: 4-vector A_μ, 6-tensor $F_{\mu\nu} = -F_{\nu\mu}$.

2) symmetric, $j_1 j_2 = +1$: pseudo 4-vector B_μ, 6-tensor $G_{\mu\nu} = -G_{\nu\mu}$.

3) antisymmetric, $j_1 j_2 = -1$: pseudo 4-vector B'_μ, scalar S, pseudoscalar P.

4) antisymmetric, $j_1 j_2 = +1$: 4-vector A'_μ, scalar S', pseudoscalar P'.

Charge-conjugation is achieved in a similar way with successive operations $\rho_y^{(1)}\sigma_y^{(1)}$ and $\rho_y^{(2)}\sigma_y^{(2)}$. Conventions can be chosen so that the charge-conjugation operation on the fields $(A, F), (B, G), (B', S, P), (A', S', P')$ is simply the complex conjugate of these fields.

§2. Belinfante's Proof Using Charge-Conjugation.

The comments of §1 suffice to give at least the flavor of the undor calculus, and we now turn to the result of Belinfante's §5 of "The Undor Equations of the Meson Field". Here he obtains the spin-statistics relation as a consequence of charge-conjugation invariance, which had been postulated by Kramers as a general requirement on any physical theory.

Belinfante's demonstration of the spin-statistics relation is abstracted in App.12A. In an easily understood notation described above, the undor Lagrangian has a kinetic term

$$i\mathcal{K}\Psi^{\dagger}B\Gamma_{\mu}\nabla_{\mu}\Psi$$

with Ψ an undor of rank N, and $B = \prod_{n=1}^{N}\beta^{(n)}$. In the standard representation of Kramers $\beta^{(n)} = \rho_x^{(n)}$ so that

$$B = B^{\dagger} = B^{\star} = B^{\sim}. \tag{5}$$

Also

$$\Gamma_{\mu} = \sum_{n=1}^{N}\epsilon_n\gamma_{\mu}^{(n)}, \qquad (\epsilon_n = \pm 1). \tag{6}$$

By the prescription $p_{\mu}\to p_{\mu} - eA_{\mu}$ we recognize the charge density as

$$e\rho = e\mathcal{K}\Psi^{\dagger}B\Gamma_0\Psi. \tag{7}$$

The same interaction, in the charge-conjugate representation, is

$$e_c\rho_c = e\mathcal{K}\Psi_c^{\dagger}B\Gamma_0\Psi_c, \tag{8}$$

where $\Psi_c = \mathcal{C}\Psi^{\star}$ with $\mathcal{C} = \prod_{n=1}^{N}\rho_y^{(n)}\sigma_y^{(n)} = \mathcal{C}^{\star}$ and $\mathcal{C}^{\star}\mathcal{C} = 1$. By a simple calculation,

$$B\mathcal{C} = (-)^N\mathcal{C}B^{\star} \tag{9}$$

and with $\Gamma_0 = \sum_{n=1}^{N}\beta^{(n)} = \sum_{n=1}^{N}\rho_x^{(n)}$, also

$$\Gamma_0\mathcal{C} = -\mathcal{C}\Gamma_0^{\star} \tag{10}$$

and

$$\Gamma_0 B = B\Gamma_0. \tag{11}$$

Then

$$
\begin{aligned}
e_c\rho_c &= (-)e\mathcal{K}\Psi^{\sim}\mathcal{C}^{\dagger}B\Gamma_0\mathcal{C}\Psi^{\star} \\
&= (-)e\mathcal{K}\Psi^{\sim}\mathcal{C}^{\sim}B^{\sim}\Gamma_0^{\sim}\mathcal{C}^{\sim}\Psi^{\star} \\
&= -(-)^{N+1}e\mathcal{K}\Psi^{\sim}B^{\sim}\mathcal{C}^{\sim}\mathcal{C}^{\sim}\Gamma_0\Psi^{\star} \\
&= (-)^N e\mathcal{K}\Psi^{\sim}B\Gamma_0\Psi^{\star}.
\end{aligned} \tag{12}
$$

304

Changing the sign by $(-)^S$ upon changing the order of quantized fields gives

$$e_c \rho_c = (-)^{N+S} e \mathcal{K} \Psi^\star B \Gamma_0 \Psi = e\rho. \tag{13}$$

The last step, the requirement of charge-conjugation invariance, requires $S = 1$ for N odd, dictating Fermi-Dirac statistics for half-integral spin odd-rank undors. Conversely S=0 for N even dictates Bose-Einstein statistics for integral spin even-rank undors.

Belinfante goes on to include the infinite c-numbers which are ignored above but are present in principle. With them he obtains for an undor of the first rank, ie a Dirac particle, that

$$\rho_c = \psi_c^\dagger \psi_c - N^c = -\rho = -(\psi^\dagger \psi - N) \tag{14}$$

so the anticommutator

$$\psi^\dagger \psi + \psi_c^\dagger \psi_c = [\psi^\dagger, \psi]_+ = N + N_c \tag{15}$$

evaluated at the same space-time point is an infinite c-number, consistent with Fermi-Dirac statistics. A corresponding result is obtained for the commutator of integral spin particles.

Positive energy for half-integral spin particles is now obtained as a by-product of the anticommutation relations dictated by the requirement of charge-conjugation invariance.

The inevitable question must be asked: Where does Belinfante's proof fall short? Belinfante himself disclaimed any rigor because of the manipulations of field operators at the same space-time point, but that problem is common to all attempts. He did hint strongly in his concluding remarks at a proof including interactions: "Generally we can postulate that the *total Lagrangian itself* (integrated over all space and time) shall be charge-conjugation invariant on account of the commutation relations of the field components." A logical pursuit of this remark would have freed him from the burden of the undor formalism and led him as close as then possible to Schwinger's view of the spin-statistic relation being required for **TCP**-invariance. Why he did not simply address the example of the interacting spin-0, $\frac{1}{2}$, and 1 ϕ, ψ, and A_μ fields, in standard scalar, Dirac spinor and vector notation just boggles one's mind.

The germ of other ideas is also evident in Belinfante's undors, especially the idea of a constituent model similar to the quark model. Belinfante's work was at once too far ahead of its time and at the same time too concerned with the fashionable but illusory generality of including arbitrary spins. He was in any case immediately interrupted by WWII and apparently never resumed his interest in the problem of the Spin-Statistics Problem or the **TCP**-theorem, although both were to remain open questions for fifteen years.

§3. Pauli's Criticism.

Criticism of Belinfante's proof came almost immediately in a paper coauthored by Pauli and, remarkably, by Belinfante himself. One can only imagine the feelings of the twenty-six year old brand-new PhD compelled to defend his work against the determined assault of Pauli, and even to participate in its disparagement. One must also wonder why Kramers did not involve himself to defend ideas in which he certainly had a deep vested interest. So now we turn to Pauli's criticism of Belinfante's proof of the Spin-Statistics Theorem.

Five months after Belinfante's thesis papers, Belinfante and Pauli published the criticism of Belinfante's proof of the Spin-Statistics Theorem based on the postulate of charge-conjugation invariance and formulated in the undor notation. The assault was mounted on two fronts:

1) What are now considered unphysical, non-local splits of fields had been found earlier by Pauli, which do permit Fermi-Dirac quantization of spin-0 fields if only positive energy and charge-conjugation invariance are required. To rule out this possibility, Pauli required an additional postulate, commutation of observables at space-like separated points. The argument

here is a repeat of the arguments made originally in Pauli's 1936 paper, and conclude emphatically that only the postulate of causal commutativity of observables excludes Fermi-Dirac statistics for spin-0 fields.

2) Next, physical states are conjectured which are linear combinations of different undors. An example is presented with positive and negative mass Dirac particles, and physical states described by mixtures of irreducible undors. Depending on the phase of this mixture, charge-conjugation invariance can be achieved with either Fermi-Dirac or Bose-Einstein statistics. Bose-Einstein statistics can be ruled out only if the postulate of charge-conjugation invariance is supplemented with the requirement that the total energy is positive.

Let us state the conclusions of Pauli's criticism and then review the criticism itself and evaluate its validity.

The operative premise of the criticism is the introduction of an extra field where one would suffice. With this freedom, ambiguities are introduced and choices must be made so that the Lagrangian, the commutation relations and the charge-conjugation operation are not determined by the field equations. Charge-conjugation invariance is no longer sufficient to determine the spin-statistics relation and must indeed be supplemented or even replaced by the two postulates previously favored by Pauli, that the energy must be positive and the observables must be local causal operators commuting at spacelike separations.

The final paragraph, which comes across the years as an agonized protest from Belinfante, makes a statement that remains valid to this day (although one does have to update the postulate of charge-conjugation invariance to **TCP**-invariance): "The first order field equations of the *particles actually known until now (Note added: Italics in original.)* determine the law of charge-conjugation invariance unambiguously, where there is only one way of building up the (first order) Lagrangian without introducing superfluous quantities. Then, the commutation relations are determined either by (I)

$E \geq 0$ and (II) local commutativity of observables or by (III) **C**-invariance and (II) ····."

What then about the validity of the criticism of Belinfante's proof? The conclusion is that Belinfante's proof is not correct, but not for the reasons put forth by Pauli. Pauli's objections appear to us today as contrived and artificial. The tragedy is that - perhaps in the chaos of the early days of WWII - no one thought to impose charge-conjugation invariance on the various individual fields, ϕ, A_μ, ψ in the modern way. They could then have done away with the bothersome superstructure of undors. By postulating charge-conjugation invariance of the Lagrangian including interactions, they would have been led to the requirement of the spin-statistics relations. Belinfante was that close to vaulting the field more than a decade ahead of the actual course of events. It is perhaps not fair to blame anyone after the fact, but it is at least unfortunate that Pauli could not take a positive attitude and interpret Belinfante's contribution not as a threat to his own ideas (which it certainly was), but as a springboard to real progress. Again, the silence of Kramers is mystifying.

The counter-examples for spin-0 are of two types:

1) A redundant field is introduced, so that postulate (II) is required to restrict the choice of the charge-conjugation operation to

$$\Psi^c = \mathcal{C}\Psi^\star, \quad \Psi'^c = \mathcal{C}\Psi'^\star \tag{16}$$

and not

$$\Psi^c = \mathcal{C}\Psi'^\star, \quad \Psi'^c = \mathcal{C}\Psi^\star, \tag{17}$$

and thus to re-establish Bose-Einstein statistics.

2) A non-local split into positive and negative frequencies is made, followed by a re-ordering of the classical field, and then by quantization.

The first of these models would be considered today as a possible theory, but the Lagrangian would have to retain the appropriate invariance property by a choice of the charge-conjugation operation. A related contemporary situation is found in k-meson phenomena, where each of **C**, **P**, and **T** are violated, but **TCP**-invariance would require the spin-statistics relation in a way basically identical to Belinfante's development. The second model would be inadmissible and outside the bounds of local Lagrangian field theory. The objections based on both of these models would be considered not sufficiently compelling to abandon the proof, although repairs, extensions, and modifications might be in order.

The counter-example for spin-$\frac{1}{2}$ is likewise outside the limitations of Lagrangian field theory where the physical states - in agreement with experience then and now - are supposed to saturate the possible representations of the Lorentz group and whatever invariance groups exist.

The criticism includes statements like: "It was shown by Belinfante that, for a certain collection of hypothetical particles (which includes all known particles) described by one undor of an arbitrarily given rank, it follows from the postulate of charge-conjugation invariance that particles with integral spin must satisfy Bose-Einstein statistics and those with half-integral spin must satisfy Fermi-Dirac statistics." This statement remains true today, although, of course we must update **C**-invariance to **TCP**-invariance. And: "The first order field equations of the *particles actually known until now* determine the law of charge-conjugation unambiguously. $\cdots$" And right in the **Abstract** we find the explicit statement validating Belinfante's Thesis in the fundamental cases: "In the special case of scalar fields, vector fields and the Dirac electron, we only need a single particular type of undor in the theory, which follows by comparison to the unambiguous transformation law of charge-conjugation, so that in these cases, postulate (III) (charge-conjugation invariance) is enough to fix the statistics."

§4. **Concluding Remarks.**

We conclude that the counter-examples are not fundamental or conclusive, quite the contrary. And - in full agreement with the last paragraph

309

of Belinfante and Pauli - Belinfante's claim that invariance under charge-conjugation determines the choice of commutation or anticommutation relations for integral or half-integral spin particles stands. Positive energy (Pauli's postulate (I)) is a by-product. Local commutativity (Pauli's postulate (II)) is a concomitant of a local Lagrangian field theory and the Heisenberg equations of motion. Belinfante's charge-conjugation invariance proof of the spin-statistics relation (unfortunately obscured by the undor notation) is a natural precursor of Schwinger's proof [12.5] based on T-invariance, which in turn led to a proof based on the requirement of TCP-invariance. Belinfante's approach - because it can be extended easily to interacting fields - is more powerful than Fierz's already discussed, or the next one of deWet [12.6], or, following that, the much-acclaimed one of Pauli [12.7]. In all of these, the assumption of non-interacting fields is a seemingly necessary restriction.

Eventually, in the proofs of Lüders and Zumino [12.8] and of Burgoyne [12.9], the logic is reversed and the Spin-Statistics Theorem is proved first and serves as the foundation for the TCP-theorem, although Schwinger continued to advocate the converse of proving the Spin-Statistics Theorem based on the TCP-theorem.

§5. Biographical Note on F.J. Belinfante.(†)

Fredrik Jozef Belinfante (1913-91) was born in The Hague, Netherlands. His intellectual talent was recognized at an early age and he was educated at the Lyceum, a world-renowned high school for gifted students. He received his PhD from the University of Leiden under the direction of H.A. Kramers (1939) on the very eve of the outbreak of WWII in western Europe. He resumed his career at the University of British Columbia (46-48), before moving to Purdue (48-79). He was active in research on fundamental issues in quantum mechanics, quantum field theory, general relativity, and statistical mechanics, and authored the definitive text *A Survey of Hidden-Variable Theory*.

Surprising to us is the fact that he resumed collaboration with Pauli after coauthoring the paper in which Pauli criticized his thesis proof of the Spin-Statistics Theorem, in a way that we find to be unfair. They subsequently did important work together on molecular spectra. Belinfante's propensity for coining new nomenclature led to the word *nucleon*.

He showed that symmetrization of the energy-momentum tensor was necessary to define the field angular momentum and spin in special relativity.

Belinfante's many interests included photography, linguistics and Esperanto. 'He was a joy to converse with, willing to discuss any scientific question that might arise, and he never had a harsh word for anyone.'†

One might wish that he had, in particular where Pauli was concerned.

(† - A.N.Gerritsen, S. Rodriguez, A. Tubis, and J.C. Swihart, PHYSICS TODAY, July 1992, p.82.)

Bibliography and References.

12.1) F.J. Belinfante, Nature **143**, 241 (1939); Physica **6**, 848,870,887 (1939); see our App.12A for the second of these.

12.2) H.A. Kramers, Proc. Roy. Soc. Amsterdam **40**, 814 (1937).

12.3) M. Fierz, Helv. Phys. Acta **12**, 3 (1939); see our App.11A.

12.4) W. Pauli and F.J. Belinfante, Physica **6**, 177 (1940); see our App.12B.

12.5) J. Schwinger, Phys. Rev. **82**, 914 (1951); see our App.16A.

12.6) J.S. deWet, Phys. Rev. **57**, 646 (1940); see our App.13B.

12.7) W. Pauli, Phys. Rev. **58**, 716 (1940); see our App.14A.

12.8) G. Lüders and B. Zumino, Phys. Rev. **110**, 1450 (1958); see our App.17A.

12.9) N. Burgoyne, Nuov. Cim. **8**, 607 (1958); see our App.17B.

APPENDIX 12.A:

Excerpt from: Physica VI 9, 870 (1939)

THE UNDOR EQUATION OF THE MESON FIELD

by F.J. BELINFANTE

Instituut voor Theoretische Natuurkunde der Rijks-Universiteit, Leiden

Abstract: The meson equation of Proca, Kemmer and Bhabba is represented by means of an undor of second rank. A generalization of the equation leads to a new meson equation, which consists essentially of a combination of cases (b) and (d) of Kemmer. The neutretto equation is extended in a similar way. The magnetic moment of the mesons is derived.

The charge-conjugate wavefunction obeys an equation in which the signs of all charges are reversed. When one postulates, in the description of physical events with the charge-conjugate quantities, that the same values should be obtained for all physically significant quantities, then it can be concluded that particles with integer spin must obey Bose-Einstein statistics, and particles with half-integer spin must obey the Pauli Exclusion Principle.

(Note added: We skip to Belinfante's §5 where he substantiates the second paragraph of his abstract, the connection between charge-conjugation invariance and the requirement of the Spin-Statistics Theorem. A certain amount about undors must be taken on faith, because we hope to avoid too deep involvement in Belinfante's development, in favor of that of Fierz which has been of more lasting value, especially in Pauli's proof soon to come. A brief but hopefully sufficient explanation of undors is contained in the introduction to this chapter.)

§5. Charge-conjugation Invariance and Statistics.

It is well known that in the hole theory of electrons (the second-quantized theory of the Dirac electron) there is an infinite c-number difference between the q-number $e\rho_{electron} = e\psi^\dagger\psi$ (obtained by second-quantization of the wave function ψ with the expression for the charge density following in the usual way from the Lagrangian of unquantized wave-mechanics) and the q-number representing the *correct* (observable) electric charge density.

If the meson field is quantized, ρ_{meson} must also be corrected by addition of infinite c-numbers.

We have mentioned that to one description of Dirac particles, mesons, neutrettos and the electromagnetic field by undor wave-functions [4, *(see also note added)*], there is an equivalent *charge-conjugate description* in which some constants like e are replaced by e^C $\cdots$, and every quantized undor is replaced by its charge-conjugate. $\cdots$ This suggests a kind of symmetry between both ways of describing physical situations [10]. By way of hypothesis one might assume a *fundamental property of nature*. We shall call this property the "*charge-conjugation invariance*" of the physical world (not to be confused, however, with the principle of conservation of electric charge!).

Therefore we shall *postulate* that every physically significant quantity in quantum-mechanics (that is, every q-number *correctly* representing the value of an *observable*) is invariant under transformation from one description of the fields to the charge-conjugate description, or, in shorter terms, is *charge-conjugation invariant*.

This postulate can serve to distinguish between wave-mechanical expressions, which after quantization cannot have a physical meaning any longer, and other analogous expressions, which may represent observables. For the present we shall leave *this* question out of consideration, but we shall show here that the postulate of charge-conjugation invariance implies directly that photons and neutrettos *must be* neutral, that Dirac electrons *must* obey Fermi-Dirac statistics and that mesons *must* obey Bose-Einstein statistics. The interesting fact is that this statistical behavior of particles and quanta follows much more directly from the postulate of charge-conjugation invariance than from postulates concerning the positive character of the total energy of free particles or quanta [15,16].

From the Lagrangian of any kind of particle or quanta we can always deduce expressions for the electric charge density, the electric charge current, the total momentum and the total energy of these particles.

The terms of the Lagrangian depending on the derivatives of the field quantities always have [1,16] the form [9]

$$iK\Psi^\dagger B\Gamma_\mu\nabla^\mu\Psi. \tag{56}$$

If Ψ is an undor [4; Eqn.17] $\Psi_{k_1 k_2 \cdots k_N}$ of rank N, then [4; Eqn.12]:

$$B = B^\dagger = \prod_{n=1}^{N} \beta^{(n)};$$

313

$$\Gamma_\mu = \sum_{n=1}^{N} \epsilon_n \gamma^{(n)}, \quad (\epsilon_n = \pm 1);$$

$$B^*\Gamma^* = \Gamma^\sim B^\sim; \tag{57}$$

so that, if we put

$$\Psi^C = C\Psi^*, \quad C = C^\sim = \prod_{n=1}^{N} C^{(n)}; \quad C^*C = 1, \tag{58}$$

we have

$$BC = (-1)^N CB^*, \quad \Gamma C = -C\Gamma^*. \tag{59}$$

From (56) we find that the electric charge density, if it exists, is equal to

$$e\rho = eK\Psi^\dagger B\Gamma^0\Psi, \tag{60}$$

within an infinite c-number. In the charge-conjugate description this is turned into

$$e^C\rho^C = -eK\Psi^{C\dagger}B\Gamma^0\Psi^C, \tag{61}$$

again within an infinite c-number. Therefore, on account of (57)-(59):

$$e^C\rho^C = (-1)^N eK\Psi^\sim\Gamma^{0\sim}B^\sim\Psi^*. \tag{62}$$

If the expressions (60) and (62) for the electric charge density are postulated to be equal, the field operators Ψ and $\Psi^\dagger$ occurring in (60) and (62) must *commute* (apart from an infinite c-number) if N is *even*, and must *anticommute* if N is *odd*.

It is not true, of course, that the commutation rules follow *rigorously* from

$$e\rho = e^C\rho^C, \tag{63}$$

since in (63) the sum is taken over the undor indices, and only the operators Ψ and $\Psi^\dagger$ at *one and the same* point of space are multiplied with each other. In this case the δ-function appearing in the commutation relations becomes infinite; its value corresponds formally to the sum or difference of the two infinite c-numbers in (60) and (62) $\cdots$.

$\cdots$ For photons and neutrettos it follows from $\cdots$ the symmetry of the operator ρ_{op} with respect to both undor indices on which it operates, that

$$\rho \equiv \Psi^\dagger \rho_{op} \Psi = \Psi^{C\dagger}\rho_{op}^C\Psi^C = \Psi^{C\dagger}\rho_{op}\Psi^C \equiv \rho^C. \tag{64}$$

On the other hand we find from (63) $\cdots$ for any particles or quanta

$$\rho = -\rho^C. \tag{65}$$

314

Comparing (64) with (65) we conclude that the electric charge density of the fields of neutrettos and photons must vanish, if it is a charge-conjugation invariant expression. In a similar way we derive that neutrettors of the first rank can only describe neutral particles [17]. It does not follow from this, however, that neutral particles should necessarily be described by neutrettors!

For electrons we deduce from (65) that

$$\rho = \psi^\dagger \psi - \mathcal{N} \tag{66}$$

must be opposite but equal to

$$\rho^C = \psi^{C\dagger} \psi^C - \mathcal{N}_C, \tag{66}$$

where $\mathcal{N}_C$ takes the place of the infinite c-number $\mathcal{N}$ in the charge-conjugate description. From this we deduce

$$
\begin{aligned}
\psi^\dagger \psi + \psi^{C\dagger} \psi^C &= \psi^\dagger \psi + \psi^\sim \psi^* \\
&= \sum_k (\psi_k^\dagger \psi_k + \psi_k \psi_k^\dagger) \\
&= \mathcal{N} + \mathcal{N}_C.
\end{aligned}
\tag{67}
$$

Similar relations can be deduced by postulating the charge-conjugation invariance of the quantized expressions for electric charge current, total momentum and total energy. $\cdots$

It is obvious that the relation (67) is consistent with the anticommutation relations of Fermi-Dirac statistics

$$\psi_k(x)^* \psi_{k'}(x') + \psi_{k'}(x') \psi_k(x)^* = \delta_{kk'} \delta(x - x'), \tag{70}$$

but not with Bose-Einstein statistics.

In a similar way we find for mesons from

$$\rho = \Psi^\dagger \rho_{op} \Psi - \mathcal{N} \tag{71}$$

and

$$\rho^C = \Psi^{C\dagger} \rho_{op} \Psi^C - \mathcal{N}_C \tag{72}$$

that $\cdots$

$$\Psi^\dagger \rho_{op} \Psi - (\rho_{op} \Psi)^\sim \Psi^* = \mathcal{N} + \mathcal{N}_C. \tag{73}$$

It is obvious that (73) is consistent with Bose-Einstein commutation relations between the components of Ψ and $\Psi^\dagger$, and not with Fermi-Dirac anticommutation relations $\cdots$

315

For neutral particles, indication of the commutation relations can be derived in this way from the expressions for the total momentum and the total energy, which are also obtained directly from the Lagrangian. Generally we can postulate that the *total Lagrangian itself* (integrated over space and time) shall be charge-conjugation invariant on account of the commutation relations of the field components. It is therefore *not necessary to investigate the sign of the energy* in order to derive the statistical behavior of the particles concerned [15,16].

It is true, however, that charge-conjugation invariance of the quantized expression for the total energy *implies* that by quantization according to the scheme of Pauli and Weisskopf [18] the so-called "states of negative energy" of free particles (depending on the time by a factor $e^{+2i\omega t}$) can be interpreted, on account of the commutation relations (which do not need specification here!), as states of positive energy of particles with opposite electric charge. We can understand this in the following way. By charge-conjugation of the quantized wave-function these states pass into charge-conjugate states of positive energy. If, now, the expression for the total energy is charge-conjugation invariant on account of the (unspecified) commutation rules of the q-number operators **a** (Jordan-Wigner or Jordan-Klein matrices), the terms in this expression arising from the so-called states of negative energy are automatically equal to the terms in the charge-conjugate expression arising from states of *positive* energy of the *charge-conjugate* particles (which are described by the charge-conjugate operator $\mathbf{b}=\mathbf{a}^\dagger$). Using the latter (charge-conjugate) expression for the description of these terms in the total energy, the energy is given as a sum of only positive energies with operators $\mathbf{a}^\dagger\mathbf{a}$ or $\mathbf{b}^\dagger\mathbf{b}$.

We observe that both the statistical behavior of particles and the possibility of describing so-called states of negative energy (of free particles) as states of positive energy of charge-conjugate particles follow directly from the postulate of charge-conjugation invariance of quantum-mechanical theories. The relation between the positive character of the energy of free particles and the charge-conjugation invariance of energy seems to be still closer than that between charge-conjugation invariance and statistics.

I wish to thank Professor Kramers for his interest in this work.

Received July 15th 1939.
Footnotes and References

1) N. Kemmer, Proc. Roy. Soc. A**168**, 127, 1938.

2) H.J. Bhabba, Proc. Roy. Soc. A**166**, 501, 1938.

3) H. Yukawa, S. Sakata and M. Taketani, Proc. Phys. Math. Soc. Japan **20**, 319,

1938.

4) F.J. Belinfante, Physica **6**, 848, 1939. *(In the text referred to by [U.C.].)*

5) H. Frohlich, W. Heitler and N. Kemmer, Proc. Roy. Soc. **A166**, 154, 1938.

6) N. Kemmer, Proc. Camb. Phil. Soc. **34**, 354, 1938.

7) G. Breit, L.E. Hoisington, S.S. Share and H.M. Haxton, Phys. Rev. **55**, 1103, 1939.

8) C. Moller and L. Rosenfeld, Nature **143**, 241, 1939.

9) F.J. Belinfante, Physica **6**, 887, 1939.

10) H.A.Kramers, Proc. Roy. Acad. Amsterdam **40**, 814, 1937.

11) A. Proca, J. Phys. Radium **7**, 347, 1936; **8**, 23, 1937.

12) F.J. Belinfante, Nature **143**, 201, 1939.

13) L. de Broglie, Actual. Scient. Industr. 411, 1936.

14) N. Kemmer, Proc. Roy. Soc. A, 1939. *(In the press.)*

15) H.A. Kramers, "Les nouvelles theories de la physique", p. 164-167, 1938. (Inst. int. d. Coop. intell., Paris 1939).

16) M. Fierz, Helv. Phys. Acta **12**, 3, 1939.

17) E. Majorana, Nuov. Cim. **14**, 171, 1937.

18) W. Pauli and V. Weisskopf, Helv. Phys. Acta **7**, 709, 1934.

APPENDIX 12.B:

Excerpt from: Physica VII 3, 177 (1940)

ON THE STATISTICAL BEHAVIOR OF KNOWN
AND UNKNOWN ELEMENTARY PARTICLES

by W. PAULI and F.J. BELINFANTE

Physikalisches Institute der Eidgenossichen Technischen Hochschule, Zurich (Schweiz)
Instituut voor Theoretische Natuurkunde der Rijks-Universiteit, Leiden

Abstract: It is shown, as far as the statistical behavior is concerned, what hypothetical particles are possible in a relativistically invariant theory, when all or some of the following three postulates are assumed: I) the energy is positive; II) observables with spacelike separations commute; III) there are two equivalent descriptions of nature, in which the elementary charges are reversed and in which the corresponding field strengths are transformed in the same way under a Lorentz transformation.

Pauli has already shown, that integer spin particles by (II) always obey Bose-Einstein statistics, while half-integral spin particles by (I) always follow Fermi-Dirac statistics. Belinfante has shown from postulate (III), that for a certain class of particles (including all found in nature) which are described by an "undor" of a particular rank, Bose-Einstein statistics follows for integer spin, and Fermi-Dirac statistics follows for half-integer spin.

In the following note, however, we show from the example of the typical cases of spin-0 and spin-$\frac{1}{2}$, that in the general case of several undors of the same rank, the statistics of the particle can no longer be unequivocally determined from (III) alone. However, including (II) for integer and (I) for half-integer is always sufficient. In the special cases of scalar fields, vector fields and the Dirac electron, we only need a single particular type of undor in the theory, which follows by comparison to the unambiguous transformation law of charge-conjugation, so that in these cases, postulate (III) is enough to fix the statistics.

§1. Introduction

Belinfante [1] has postulated a general principle of *"charge-conjugation invariance"*, from which the statistical behavior of all known particles can be determined in a simple way. The principle is as follows. The field of the particles is described by a set of undors [2] $\Psi_{k_1 \cdots k_n}$ which satisfy first order partial differential equations (field equations), in which

318

some constants of the dimension of a charge may occur.

Under an infinitesimal Lorentz transformation where the components of a vector ξ transform according to

$$\delta\xi^\nu = \xi_\mu\delta\omega^{\mu\nu}, \tag{1}$$

the undors are transformed [3] according to

$$\delta\Psi = \frac{1}{4}\delta\omega^{\mu\nu}\Gamma_\mu\Gamma_\nu\Psi, \tag{2}$$

where

$$\Gamma_\mu = \sum_{j=1}^{n}\gamma_\mu^{(n)}. \tag{3}$$

Here the $\gamma_\mu^{(j)}$ are matrices operating on the index k_j of $\Psi_{k_1\cdots k_n}$ and satisfying the relations

$$\gamma_\mu^{(j)}\gamma_\nu^{(j)} + \gamma_\nu^{(j)}\gamma_\mu^{(j)} = 2g_{\mu\nu}. \tag{4}$$

Operating on the complex conjugate undor $\Psi^*{}_{k_1\cdots k_n}$ with a linear operator

$$C = \prod_{j=1}^{n}C^{(j)}, \tag{5}$$

where $C^{(j)}$ is a matrix operating on the index $\dot{k}_j$ of $\Psi^*{}_{k_1\cdots k_n}$ and satisfying the relations

$$\gamma_\mu^{(j)}C^{(j)} = C^{(j)}\gamma_\mu^{(j)*}, \tag{6}$$

we get back an undor

$$\Psi^C = C\Psi^*. \tag{7}$$

We can normalize C in such a way that

$$\Psi^{CC} = \Psi, \tag{8}$$

by putting

$$C^{(j)}C^{(j)*} = 1. \tag{9}$$

Instead of by (5)-(7), we might have constructed an undor $\Psi^{C'}$ from Ψ^* by

$$\Psi^{C'} = \mathcal{G}C\Psi^*, \tag{10}$$

where $\mathcal{G}$ is a scalar operator - for instance a product of an even number of factors $\gamma_5^{(j)}$.

In a similar way, we could form an undor from Ψ^* by means of

$$\Psi^Q = \mathcal{Q}\Psi^* \tag{11}$$

with

$$\mathcal{Q} = \gamma_5^{(1)}C, \tag{12}$$

319

so that

$$\Psi^{QQ} = -\Psi. \tag{13}$$

Now it can be shown [6,1] that, if the undors Ψ describing the known particles (such as electrons, nucleons, photons, mesons), are transformed according to (7), we obtain a set of undors Ψ^C, which again satisfy the field equations, except that all real constants of the dimension of a charge, in particular e, have reversed their signs:

$$e^C = -e. \tag{14}$$

This is true only [5] for the condition (7) and not for the alternative (12). Further, this invariance of the field equations and a similar *invariance of all physically significant formulae* (such as for instance the total energy or the total electric charge) exist *only on account of the commutation relations* holding between the undors. By the transformations

$$\Psi \to \Psi^L, \qquad e \to e^L, \tag{15}$$

therefore, where Ψ^L is given by

$$\Psi^L = \Psi^C, \tag{16}$$

we pass from one description of the fields of particles (by the undors Ψ) to another description (by Ψ^C), which is *completely equivalent* to the original one and was called - according to the terminology of Kramers [6]- the *charge-conjugate* description of the physical world.

It is clear that, by postulating that the transformation (15) specified by (16) should leave all physically significant quantities unaltered (which we shall call *specified* charge-conjugation invariance in the following), we obtain some information about the commutation relations holding between some of the undor components. This information is sufficient $\cdots$ to make a definite choice between commutators and anticommutators for the field components Ψ and Ψ^*, although the c-number $\cdots$ remains entirely undetermined. It was shown by Belinfante [1] that, for a certain collection of hypothetical particles (which includes all known particles) described by one undor of an arbitrarily given rank, it follows from this postulate: particles with integer spin must satisfy Bose-Einstein statistics and those with half-integer spin must satisfy Fermi-Dirac statistics.

At the same time, it was shown by Pauli [7] in a very general way, that a similar result for arbitrary particles described by a field satisfying a set of linear homogeneous differential equations, could be obtained from the following two postulates:

(I) for these particles *there is a finite number of states of negative energy;*

(II) *observables at spacelike separated points must commute.*

These two postulates can be understood from a physical point of view. The first postulate is necessary since otherwise particles would drop into states of lower and lower energy creating an infinite number of quanta or pairs of particles. According to the second postulate simultaneous measurements at spacelike separated points must always be possible; $\cdots$

Invariance under the transformation (15) *with the specification* (16) is a specialized case of invariance under some *unspecified* transformation (15) of the undors together with reversal of the signs of all charges (14). Invariance of all physical quantities under another transformation than (16), but still of the type (14)-(15), would also have meant that there is a description of the physical world, in which every "elementary charge" has the opposite sign. The existence of such a generalized charge-conjugate description may perhaps be inferred from some speculations on a possible symmetry between positive and negative charges, which - though not existing on earth - may still be a fundamental property of nature. (Anyhow, we know that it exists for all *known* particles [1]). Therefore, *without specialization* by (16), we shall postulate (what might be called the *unspecified charge-conjugation invariance* of the theory):

(III) *invariance of all physically significant formulae and quantities under some transformation* (15), where Ψ represents some set of undors $\psi^{(1)}, \psi^{(2)}, \cdots$ and Ψ^L a similar set of undors $\psi^{(1)L}, \psi^{(2)L}, \cdots$ transforming in the same way as Ψ. We do *not* postulate, however, that after the transformation from Ψ to Ψ^L - which we shall call "unspecified charge-conjugation" - the components of the undor $\psi^{(n)L}$ are linear combinations of the components of only $\psi^{(n)*}$ again.

In the following we shall discuss the bearing of the postulates (I), (II) and (III) on the $\cdots$ statistics of $\cdots$ particles with spin 0 or $\frac{1}{2}$ $\cdots$

§2. Particles with Spin 0.

In this section we discuss some possibilities of $\cdots$ particles $\cdots$ described by a $\cdots$ function s $\cdots$, which is a scalar $\cdots$ under spatial rotations and Lorentz transformations and which satisfies $\cdots$ a Klein-Gordon equation

$$(\Box - m^2)s = 0. \tag{17}$$

$\cdots$ For $m \neq 0$, putting

$$m\phi_\nu = \nabla_\nu s, \tag{18}$$

we find from (17):

$$\nabla^\nu \phi_\nu = ms, \tag{19}$$

where ν runs $0\cdots3$.

Equations (18)-(19) together form the first order field equations. They can be derived from the action integral

$$S = \int L_0 d^3x dt \tag{20}$$

where the Lagrangian $\cdots$ linear in the first derivatives of the fields $(s, s^*, \phi_\nu, \phi_\nu^*)$,

$$\begin{aligned}
L_0(s, s^*, \phi_\nu, \phi_\nu^*) &= K[m(\phi^{\nu*}\phi_\nu - s^*s) \\
&\quad -\frac{1}{2}(\phi^{\nu*}\nabla_\nu s + h.c.) + \frac{1}{2}(s^*\nabla_\nu \phi^\nu + h.c.)],
\end{aligned} \tag{21}$$

which is Hermitian for real K.

For the total energy we find [8]

$$E = mK \int d^3x[s^*s + \sum_{j=1}^{3} \phi_j^*\phi_j], \tag{22}$$

within a c-number.

As regards the commutation relations, we *assume* $\cdots$ the form

$$[s(x); s^*(x')] = c(x - x'); \tag{23}$$

$$[s(x); s(x')] = [s^*(x); s^*(x')] = 0; \tag{24}$$

$\cdots$ the bracket $\cdots$ denotes either the commutator $\cdots$ or the anticommutator $\cdots$ $c(x - x')$ is a c-number $\cdots$ scalar function, which $\cdots$ must satisfy the equations

$$(\Box - m^2)c(x - x') = (\Box' - m^2)c(x - x') = 0, \tag{26}$$

$\cdots$ postulate (II) of § 1 is satisfied, if we assume for all pairs of field components q_1, q_2:

$$[q_1(\vec{x}, x_0); q_2(\vec{x}', x_0)] = 0, \quad \vec{x} \neq \vec{x}'. \tag{27}$$

Since (27) must hold for s with s^* and for s with ϕ^*, both c and $\nabla_0 c$ must vanish for $x_0 = x_0'$, $\vec{x} \neq \vec{x}'$, so $\cdots$ [7,8]

$$c(x - x') = i\mathcal{C}\mathcal{D}(x - x'). \tag{28}$$

Here $\mathcal{C}$ is a constant and

$$\mathcal{D}(x) \equiv \mathcal{D}(\vec{x}, x_0) = \int \frac{d^3k}{(2\pi)^3} e^{i\vec{k}\cdot\vec{x}} \frac{\sin(\omega x_0)}{\omega}, \tag{29}$$

with $\omega = \sqrt{\vec{k}^2 + m^2}$, is a solution of (26) satisfying

$$\mathcal{D} = \mathcal{D}^*; \quad \mathcal{D}(\vec{x}, 0) = 0; \quad \frac{\partial \mathcal{D}}{\partial x_0}\Big|_{x_0=0} = \delta(\vec{x}). \tag{30}$$

All other solutions of (26) are excluded by postulate (II) $\cdots$ so$\cdots$

$$\mathcal{D}^*(x) = -\mathcal{D}(-x). \tag{31}$$

322

The commutation relation must be

$$[s(x), s^*(x')]_- = i\mathcal{C}\mathcal{D}(x - x'),\tag{32}$$

since Fermi-Dirac statistics are excluded by the fact that, for $x = x'$, from (32)

$$ss^* + s^*s = 0\tag{33}$$

it would follow that the field vanishes entirely [7]. $\cdots$ The total energy (22) is positive in accordance with (I).

If e is the charge $\cdots$ of the particles $\cdots$, we introduce the electromagnetic interaction by the replacement

$$\nabla_\nu s \rightarrow D_\nu s = (\nabla_\nu - ieA_\nu)s.\tag{36}$$

The only possibility of $\cdots$ charge-conjugation is $\cdots$ given by complex conjugation:

$$s^C = s^*, \quad \phi_\nu^C = \phi_\nu^*, \quad etc.\tag{37}$$

For Bose-Einstein statistics the Lagrangian is automatically invariant under this transformation, so that (III) is satisfied.

It must be emphasized that the fact that the law of charge-conjugation is determined by the Lagrangian is a consequence of the fact that, besides s^* and ϕ_ν^*, no other quantities transforming in the same way as s and ϕ_ν and describing particles with charge (-e) are considered as components of the field. If such field components *would* have occurred, there might also have been some ambiguity in the commutation relations. Still, any anticommutation of the form (23,24) can be excluded in this case *on account of the postulate* (II) [7], so that all laws of charge-conjugation which do not yield charge-conjugation invariance for Bose-Einstein statistics, must be excluded on the basis of (II).

The proof that *only (II)* - and no combination of (I) and (III) - necessitates Bose-Einstein statistics, for fields consisting of more than one scalar, is given by the following example, from [8], of a possible theory of particles with vanishing spin but obeying Fermi-Dirac statistics, in which the total energy is strictly positive and in which charge-conjugation invariance exists. (The postulate (II) will not be satisfied, of course.)

Define the operator $\sqrt{m^2 - \nabla^2}$ by

$$\begin{aligned}
f(\vec{x}, x_0) &= \int d^3k\, a(\vec{k}, x_0)e^{i\vec{k}\cdot\vec{x}}, \\
\sqrt{m^2 - \nabla^2}\, f(\vec{x}, x_0) &= \int d^3k\, a(\vec{k}, x_0)e^{i\vec{k}\cdot\vec{x}}\omega, \\
(\omega &= +\sqrt{m^2 + \vec{k}^2}\ ),
\end{aligned}\tag{38}$$

and the function

$$\mathcal{D}_1 = \int \frac{d^3 k}{(2\pi)^3} e^{i\vec{k}\cdot\vec{x}} \frac{\cos(\omega x_0)}{\omega},$$ (39)

so that $\mathcal{D}_1^*(x) = +\mathcal{D}_1(-x)$. Notice that

$$
\begin{aligned}
(\mathcal{D}_1 \pm i\mathcal{D})^*(x) &= (\mathcal{D}_1 \pm i\mathcal{D})(-x), \\
\nabla^0(\mathcal{D}_1 \pm i\mathcal{D}) &= \pm i\sqrt{m^2 - \nabla^2}(\mathcal{D}_1 \pm i\mathcal{D}).
\end{aligned}
$$ (40)

Now, we first consider the field given by (17, 19), but we shall split the fields (s, ϕ_ν) into two parts $\cdots$

$$s = s^{(+)} + s^{(-)}; \quad \nabla_\nu s^{(\pm)} = m\phi_\nu^{(\pm)},$$ (41)

$$\nabla^0 s^{(\pm)} = \pm i\sqrt{m^2 - \nabla^2} s^{(\pm)}.$$ (42)

From (42), the commutation relations for $s^{(\pm)}$ are

$$
\begin{aligned}
[s^{(\pm)}(x); s^{(\pm)*}(x')] &= \mathcal{C}^{(\pm)}(\mathcal{D}_1 \pm i\mathcal{D})(x - x'), \\
[s^{(\pm)}(x); s^{(\mp)*}(x')] &= 0, \\
[s; s] &= 0, \quad etc.
\end{aligned}
$$ (43)

Here $\mathcal{C}^{(\pm)}$ are real constants. The equations (17,19,42) remain valid under the charge-conjugation

$$s^{(\pm)C} = s^{(\mp)*}.$$ (44)

Now, instead of regarding (20,21) as the action and Lagrangian of the system - in which, after the substitution of (41), the integrals over time of cross products of (+) and (-) terms vanish - we shall interchange $s^{(-)}$ and $s^{(-)*}$ and regard

$$
\begin{aligned}
L &= L_0(s^{(+)}, s^{(+)*}, \phi_\nu^{(+)}, \phi_\nu^{(+)*}) \\
&\quad + L_0(s^{(-)*}, s^{(-)}, \phi_\nu^{(-)*}, \phi_\nu^{(-)})
\end{aligned}
$$ (45)

as the Lagrangian. From this Lagrangian the field equations (18,19) can be derived for $s^{(\pm)}$; the two equations (42) must then be added as an additional condition. The total energy is strictly positive [8]

$$E = mK \int d^3 x [s^{(+)*} s^{(+)} + s^{(-)} s^{(-)*} + \sum_{j=1}^{3} (\phi_j^{(+)*} \phi_j^{(+)} + \phi_j^{(-)} \phi_j^{(-)*})].$$ (46)

Charge-conjugation invariance (III) and positive energy (I) are now ensured by (44) and (46) for Fermi-Dirac as well as for Bose-Einstein statistics. Therefore, we may regard the bracket symbols of (43) as anticommutators; at least, if in (43) we take

$$\mathcal{C}^{(+)}/\mathcal{C}^{(-)} > 0,$$

to avoid situations like (33).

Thus we see that *the postulates (I) and (III) certainly do not exclude Fermi-Dirac statistics for Spin-0*, in this more general case that more than one scalar and its conjugate complex are regarded as field components; *the postulate (II), however, always suffices for this purpose in the case of integer spin [7].*

§3. Particles with Spin-$\frac{1}{2}$.

In this section we shall discuss some some possibilities of a quantum theory of particles, which are (at least partly) described by $\cdots \psi_k \cdots$, which transforms like an undor of the first rank [2] $\cdots$ and which satisfies $\cdots$ a Klein-Gordon equation

$$(\Box - m^2)\psi = 0. \tag{47}$$

$\cdots$ it can be shown that $\cdots$

$$\gamma^\nu \nabla_\nu \psi = -m\chi \tag{49}$$

transforms in exactly the same way as ψ. $\cdots$ (47) takes $\cdots$ the form

$$\gamma^\nu \nabla_\nu \chi = -m\psi. \tag{50}$$

The equations (49,50) together form the first order field equations. They take a particularly simple form, if [4]

$$\chi = \pm\psi. \tag{51}$$

$\cdots$ we can choose the $+$ sign $\cdots$ so (49,50) pass into the Dirac equation

$$(\gamma^\nu \nabla_\nu + m)\psi = 0. \tag{52}$$

Now we introduce $\gamma^5 = \gamma_5 = i\gamma_1\gamma_2\gamma_3\gamma_0$ and put

$$\alpha^\mu = \gamma_0\gamma^\mu = -\gamma^0\gamma^\mu, (\mu = 0, 1, 2, 3, 5), \tag{53}$$

so that [1]

$$\alpha^0 = 1; \quad \gamma_5 = -i\alpha_1\alpha_2\alpha_3. \tag{54}$$

Then it can be proved that a Hermitian matrix ϑ exists such that [2]

$$\vartheta\alpha^\mu = (\vartheta\alpha^\mu)^\dagger, \quad (\mu = 0, 1, 2, 3, 5), \tag{55}$$

are also Hermitian. The matrix ϑ shall be normalized so that $e\psi^\dagger\vartheta\psi$ is the charge density of the field. Further we put

$$\beta = i\vartheta\gamma^0 = \vartheta\alpha^1\alpha^2\alpha^3 = \beta^\dagger. \tag{56}$$

325

Now (52) can be derived from a Lagrangian

$$L = -K\psi^\dagger\beta(\gamma^\nu\nabla_\nu + m)\psi, \tag{57}$$

or, in covariant undor-notation [2],

$$L = iK\psi^{\#j}(\nabla_j^k + m\delta_j^k)\psi_k, \tag{58}$$

where we have put

$$\psi^{\#j} \equiv i\psi^\dagger\beta \tag{59}$$

and

$$\nabla_j^k \equiv (\gamma^\nu)_j^k\nabla_\nu,$$

so that

$$\nabla_j^k\nabla_k^m = \delta_j^m\Box.$$

The total energy is now given by

$$E = iK\int d^3x\,\psi^\dagger\vartheta\frac{\partial\psi}{\partial t} \tag{60}$$

within a c-number.

In analogy to the preceding section, we *assume* the commutation relations

$$[\psi_k(x); \psi^{\#j}(x')] = c_k^j(x - x'), \tag{61}$$

$$[\psi_k(x); \psi^j(x')] = [\psi_k^\#(x); \psi^{\#j}(x')] = 0. \tag{62}$$

Since $c_k^j(x - x')$ must be a mixed undor of second rank (61) and must satisfy the equations (26,27,47) and the postulate (II), it must be of the form $\cdots$

$$c_k^j(x - x') = C(\nabla_k^j - m\delta_k^j)\mathcal{D}(x - x'), \tag{63}$$

with C real. $\cdots$

Equation (52) is charge-conjugation invariant only for [5,6,2]

$$\psi^C = \psi^\# \equiv C\psi^*. \tag{67}$$

Charge-conjugation invariance of all physical quantities derivable from the Lagrangian (57) is obtained for (67) only in the case of *anticommutation* of the components of ψ^* and ψ [1], so that (III) requires Fermi-Dirac statistics. This $\cdots$ makes the total energy (60) strictly positive after quantization, so (I) and (III) are equivalent in this case.

Again everything is complicated if we admit that the field might consist of more undors than ψ and $\psi^\#$ (or ψ^*) only. *For instance* the *original* equations (49,50) *without*

326

the restriction (51) might then describe the field. In *this* case it is convenient to introduce the undors

$$\sqrt{2}\Psi_{(\pm)} = \psi \pm \chi, \tag{68}$$

which satisfy the first order field equations

$$\gamma^\nu \nabla_{(\pm)\nu} \Psi_{(\pm)} = 0, \tag{69}$$

so that we can regard this field as a mixture of Dirac particles of positive and negative mass. The equations (69) are invariant under the following transformations (where $\Psi^C = C\Psi^*$, $\Psi^{CC} = \Psi$; and $\Psi^Q = Q\Psi^*$, $\Psi^{QQ} = -\Psi$ for $Q = \gamma_5 C$, compare (6,7):

$$
\begin{aligned}
a) \qquad & \Psi_{(\pm)}^L = a_{(\pm)} \Psi_{(\pm)}^C, \quad |a_{(\pm)}|^2 = 1; \\
b) \qquad & \Psi_{(\pm)}^{L'} = a_{(\mp)} \Psi_{(\mp)}^Q, \quad (a_{(\pm)}^* a_{(\mp)})^2 = 1; \\
so \qquad & \Psi_{(\pm)}^{LL} = +(\text{a or b}) \text{ or } -(\text{b only}), \text{ times } \Psi_{(\pm)}.
\end{aligned}
\tag{70}
$$

In case (b) the particles with positive and negative mass are "interchanged" by charge-conjugation.

The field equations (69) can be derived from the Lagrangian

$$
\begin{aligned}
L = \quad & -K_{(+)} \Psi_{(+)}^\dagger \beta (\gamma^\nu \nabla_\nu + m) \Psi_{(+)} \\
& -K_{(-)} \Psi_{(-)}^\dagger \beta (\gamma^\nu \nabla_\nu - m) \Psi_{(-)}.
\end{aligned}
\tag{71}
$$

The total energy is

$$E = i \int d^3 x K_{(+)} \Psi_{(+)}^\dagger \vartheta \frac{\partial \Psi_{(+)}}{\partial t} \quad + (\Psi_{(+)} \to \Psi_{(-)}), \tag{72}$$

within a c-number. In analogy to (61, 62, 65), the fields can be quantized by

$$[\Psi_{(\pm)k}(x); \Psi_{(\pm)}^{Cj}(x')] = \mathcal{C}_{(\pm)} (\nabla_k^j \pm m\delta_k^j) \mathcal{D}(x - x'), \tag{73}$$

with $\mathcal{C}_{(\pm)}$ real, and all other brackets zero. For $x = x'$, (73) gives $\cdots$ in the usual representations $\cdots$

$$[\Psi_{(\pm)k}(x); (\Psi_{(\pm)k}(x))^*] = \mathcal{C}_{(\pm)} \delta(\vec{0}), \tag{75}$$

so that in the case of Fermi-Dirac statistics the $\mathcal{C}_{(\pm)}$ must be positive, whereas the fields vanish if the corresponding $\mathcal{C}$ is zero. Furthermore, $\cdots$ *canonical* commutation relations (for which $i\dot{F} = [F, H]_-$) require

$$\mathcal{C}_{(\pm)} K_{(\pm)} = 1, \tag{76}$$

so that *canonical commutation relations in the case of Fermi-Dirac statistics require* $K_{(\pm)} > 0$. On the other hand, *Bose-Einstein statistics are excluded entirely* by (I), since no choice of $K_{(\pm)}$ and $\mathcal{C}_{(\pm)}$ would make the energy (72) strictly positive in this case.

If we postulate (III) charge-conjugation invariance of all physical quantities derivable from the Lagrangian (71), the law (70b) of charge-conjugation gives Fermi-Dirac statistics only, if

$$|a_{(+)}/a_{(-)}|^2 = K_{(+)}/K_{(-)}, \tag{77}$$

but Bose-Einstein statistics, if

$$|a_{(+)}/a_{(-)}|^2 = -K_{(+)}/K_{(-)}. \tag{78}$$

Charge-conjugation invariance is not possible for either Fermi-Dirac or for Bose-Einstein statistics for any other choice of $K_{(\pm)}$ and $a_{(\pm)}$. Therefore, it is not possible to exclude Bose-Einstein statistics by (II) and (III) alone. If, however, we *combine* (III) and (I), we can exclude the law (70b) of charge-conjugation, if (77) is not respected.

On the other hand, for *any* $K(\pm)$, we find Fermi-Dirac statistics from the postulate of charge-conjugation invariance alone, if we define charge-conjugation in the usual way (70a) and not (70b). In the case of (70a) (I) and (III) are equivalent, again.

§4. Conclusion.

In the preceding sections we have seen that, as soon as two scalars (38), (40) or two undors of a given rank (68)-(71) are introduced, where one scalar (21) or one undor (51)-(57) would suffice, there are several possibilities of building up a quantized field theory of free elementary particles. Generally neither the Lagrangian, nor the commutation relations, nor the law of charge-conjugation (which was left unspecified by (15) without (16)) is then *a priori* unambiguously determined by the first order field equations.

Still the postulates of positive energy (I) and of space-like commutation of observables (II) are sufficient in these cases to decide between commutation or anticommutation of field quantities, although the c-numbers are not determined. The postulate of *unspecified* charge-conjugation invariance $\cdots$ (III) will $\cdots$ *not* be sufficient $\cdots$ (compare $\cdots$ (44)-(45) in §2 and (70) and (77) in §3), although in the last case in §3, *specified* charge-conjugation invariance with (16) is sufficient.

In the case of *integer spin* the postulates (I) and (III) are superfluous, since they are fulfilled automatically if (II) is satisfied. On the other hand, there exist cases with integer spin and Fermi-Dirac statistics, where (I) and (III) are fulfilled but not (II).

The first order field equations of *the particles actually known until now* determine the law of charge-conjugation unambiguously, where there is only one way of building up the (first order) Lagrangian without introducing superfluous quantities. Then, the commutation relations are determined by either (I) and (II), or (III) and (II), if we assume they are of the usual type (see (23)-(24) or (61)-(62)). In this case (III) determines the sign in the brackets and (II) determines the the c-numbers. On the other hand, if the c-numbers are known from (II), the sign in the brackets is determined *automatically* for integer spins, and can be deduced from (III) or from (I) for half-integer spins.

Received Dec. 23rd, 1939.

References

1) F.J. Belinfante, Physica **6**, 870, 1939.

2) F.J. Belinfante, Physica **6**, 848, 1939.

3) F.J. Belinfante, Physica **6**, 887, 1939.

4) F.J. Belinfante, "Theory of Heavy Quanta" (thesis Leiden 1939), **III**, §4. (M. Nijhoff, The Hague, 1939).

5) W. Pauli, Ann. Inst. H. Poincare **6**, 109, 1936.

6) H.A. Kramers, Proc. Roy. Acad. Amsterdam **40**, 813, 1937.

7) W. Pauli, "Bericht uber die allgemeinen Eigenschaften der Elementarteilchen", I, §3. (8meConseil d. Phys. Solvay, 1939). *In press.*

8) W. Pauli, Ann. Inst. H. Poincare **6**, 137, 1936.

Chapter 13

deWet's Proof Based on Canonical Field Theory

Summary: deWet proved that Fermi-Dirac quantization is not possible for tensor equations, leaving Bose-Einstein quantization as the only possibility for integral spin fields. Both are possible for the Dirac equation but Bose-Einstein "leads to difficulties".

§1. Introduction.

The next candidate proof of the Spin-Statistics Theorem appeared in the Princeton PhD thesis of Jacobus Stephanus deWet, submitted to the Physical Review on January 25, 1940 and published in the issue of April 1, 1940 [13.1]. Although the thesis was submitted to the Mathematics Department, every reference is to the physics literature, and the only acknowledgements are to two physicists, E.P. Wigner and H.P. Robertson.

One is immediately struck by the wonderful insouciance, bordering on impudence, of the author. The opening words of his Physical Review article : "The problem of the connection between the spin and statistics of particles was first tackled by Pauli. He attempted to show that Fermi-Dirac quantization was not admissible for the scalar wave equation. His work was not correct ⋯ "! Unheard of! No mention of why or how Pauli's work was not correct. Where were the referees? Where was the editor? Where was Pauli?

After generous credit to Iwanenko and Sokolow [13.2], and to Fierz [13.3], deWet presents his own case: "Here we will prove a very general result. We will show that on the basis of the Heisenberg-Pauli theory [13.4] of the quantization of wave fields it is not possible to carry out the Fermi-Dirac quantization of tensor equations ⋯ "; and "Sokolow and Iwanenko have shown ⋯ with spin-$\frac{1}{2}$ the Bose-Einstein quantization leads to serious difficulties with the negative energy values $\cdots$ "; and " $\cdots$ for integral spin $\cdots$, Fermi-Dirac quantization is not possible $\cdots$ half-integral spins $\cdots$ admit both $\cdots$, but Bose-Einstein leads to serious difficulties $\cdots$".

deWet puts forth as a theorem the fundamental result that Fermi-Dirac quantization is not possible for Hamiltonians derived from Lagrangians nonlinear in the derivatives. This is the case for the Heisenberg-Pauli Lagrangians for integral spin, in contrast to the Dirac Lagrangian - linear in the derivatives - for half-integral spins.

deWet's published work has the immediate attraction that it is stated in the familiar canonical formalism, in the standard variables, and only for the lowest spin fields. Both Fierz and Belinfante [13.5] obfuscated their work by the use of arcane spinor formulations made necessary by trying to be completely general before ever making a transparently clear case for spins 0, $\frac{1}{2}$, and 1. Pauli will do the same [13.6]. In his thesis, deWet does include a discussion of general spins almost equivalent to, but preempted by, the already published spinor formulation of Fierz. This is the basis for his published statements that the generalization to arbitrary spins is straightforward. He leaves all mention of this excursion out of his published work.

deWet is the first to clearly feature the crucial assumption on which the Spin-Statistics Theorem is based, even though he did not recognize it as something that needed justification. That step would require more than fifteen years and the work of Jost, Hall and Wightman, Lüders and Zumino, Burgoyne, and others.

§2. deWet's Proof of the Spin-Statistics Theorem.

The key statement in deWet's paper follows his Eqn.204 in the discussion of the Fermi-Dirac anticommutation relations. Given

$$[\pi_\alpha(x), \psi^\beta(x')]_+ = i\delta_\alpha^\beta \delta^3(x - x'), \tag{1}$$

331

and its Hermitian conjugate, with all other pairs anticommuting, he establishes the conditions required for the Heisenberg equations of motion to hold

$$i\frac{\delta H}{\delta \pi^\alpha(x)} = [\psi^\alpha(x), H]_-. \tag{2}$$

(Here, "$\delta H/\delta \pi(x)$" and so on, indicate a functional derivative which, for intuitive purposes, can usually be interpreted as the ordinary partial derivative "$\partial \mathcal{H}/\partial \pi(x)$" where $H = \int d^3x \mathcal{H}(\cdots, \pi(x), \cdots)$.) He then notes that included in the "all others anticommuting", one generally has

$$[\psi^\alpha(x), \psi^{\alpha\dagger}(x')]_+ = 0, \tag{3}$$

which requires

$$\psi^\alpha(x)\psi^{\alpha\dagger}(x) = -\psi^{\alpha\dagger}(x)\psi^\alpha(x) = 0 \tag{4}$$

and leads to the ostensibly trivial result that

$$\psi^\alpha(x) = 0.$$

This observation had been made before, by Pauli and Weisskopf [13.7], by Pauli [13.8], by Fierz, and by Pauli and Belinfante [13.9], and used by them to rule out anticommutation relations at least for scalar fields satisfying the Klein-Gordon equation. The only case where this requirement is avoided is where $\psi^{\alpha\dagger}(x')$ is proportional to $\pi^\alpha(x')$ and they satisfy one of the inhomogeneous anticommutation relations. This in turn requires the Dirac form of the Lagrangian

$$\mathcal{L} = \bar{\psi}\left(\beta i \frac{\partial}{\partial t} - \beta\vec{\alpha}\cdot\frac{\vec{\nabla}}{i} - m\right)\psi \equiv \bar{\psi}(\not{p} - m)\psi, \tag{5}$$

giving

$$\pi = \frac{\partial\mathcal{L}}{\partial\dot{\psi}} = i\bar{\psi}\beta \equiv i\psi^\dagger \tag{6}$$

so that

$$[\pi, \psi]_+ = i[\psi^\dagger, \psi]_+ = i\delta^3(x - x') \tag{7}$$

so

$$\psi^\dagger(x)\psi(x) \sim \delta^3(0).$$

332

The fact that one can manipulate highly singular products of field-operators and evaluate the fields at the same space-time point requires proof, and deWet's theorem is not conclusive. It has, however, withstood the intense scrutiny of the Hall-Wightman analysis of such questions [13.10], culminating in contemporary proofs of the Spin-Statistics Theorem.

deWet's further demonstration that no linear relation of the form

$$\pi_\alpha = A^4_{\alpha\beta}\psi^{\beta\dagger} = \frac{\partial\mathcal{L}}{\partial\dot{\psi}^\alpha} \tag{8}$$

can exist for tensor fields is extremely elegant, and consists of showing that $A^4_{\alpha\beta}$ must be an element of a numerically invariant tensor of the third rank $A^\sigma_{\alpha\beta}$. The only numerically invariant Lorentz tensors are $g_{\sigma\tau}$, δ^τ_σ, and $\epsilon_{\mu\nu\sigma\tau}$. No odd rank tensor can be constructed from these so, in deWet's words, "$\cdots$ it is not possible to find a Hamiltonian form for tensor equations which will be suitable for Fermi-Dirac quantization and will satisfy the requirements of relativistic invariance."

deWet appeals to the proof of Sokolow and Iwanenko to show that Fermi-Dirac statistics for spin-$\frac{1}{2}$ particles has the benefit over Bose-Einstein statistics that it makes the energy positive. Here again deWet is ahead of his time. Eventually Lüders and Zumino [13.11], and Burgoyne [13.12], will show that a proof also exists, very similar to the one used here to rule out anticommutators for integral spin fields, to rule out vanishing commutators for half-integral spin fields.

In summary, deWet made a key contribution which however lacked a necessary proof for more than fifteen years. His work had the merit of being based on canonical field theory, and of making use of familiar variables. He got to the heart of the matter for the simplest fields, without obscuring the results by using arcane formalisms. His work did not include the effect of interactions but is inferior in this respect only to the work of Belinfante [13.13], among his contemporaries. In common with so many others, he was to be unfairly criticized by Pauli, in his case for a lack of necessary generality.

Unlike Fierz or Belinfante, deWet's approach to the Spin-Statistics Theorem did not position him to make further progress. We will see that Fierz's spinor approach was a prelude to Pauli's more general proof. Belinfante's proof based on charge-conjugation invariance put him on the verge of a proof based on **TCP**-invariance which includes interactions. Ironically, however, deWet's path is closer to that finally used in conjunction the Hall-Wightman theorems in the eventual proofs of Lüders and Zumino, and of Burgoyne.

As was the case with Belinfante, deWet's career was interrupted by WWII and he apparently never resumed his interest in the Spin-Statistics Theorem.

§3. Biographical Note on J.S. deWet. (†, ††)

The same ebullient spirit which prompted deWet as a young graduate student to label work of Pauli's "wrong", propelled Jack deWet throughout his life (1913-1995).

Jacobus Stephanus deWet, grandson of the Boer general Christian deWet, was born in Rouxville, Orange Free State, South Africa. As a Rhodes scholar at Oxford (1935-37), graduate student at Cambridge (37-38), and PhD in Mathematics at Princeton (38-40), he attended lectures by such great physicists as Milne, Schrödinger, Dirac, Wigner, Robertson, and von Neumann, although his primary interest was mathematics. He served as a lecturer at Cape Town (40-42), Professor at Pretoria (42-46), and Mathematics Tutor (46-71) and Vice-Master (70-71) at Balliol College, Oxford; followed by a second career as Dean of Science at the University of Capetown (71-82), with outstanding services to science in South Africa (71-85); and a third career in retirement during which he taught at the Open University.

deWet was a legend at Balliol with his infectious enthusiasm for mathematics, and became a veritable cult figure, such was the great affection felt for him by the "family" of former students who met frequently as the "Balliol deWet Mathematicians", seventy of whom celebrated with him on his eightieth birthday.

deWet returned in 1948 to the work of his Princeton thesis in a series of papers aimed at extending the Heisenberg-Pauli quantization to Lagrangians with higher than first derivatives of the field. His interest in relativistic quantum field theory culminated in a landmark paper with Mandl on the asymptotic distribution of eigenvalues of Schrödinger operators.

(† - London Times, 8 Feb 1995. †† - private communications from Keith Hannabus, deWet's successor at Balliol; Frank Nabarro of the University of Witwatersrand, deWet's classmate in Milne's cosmological seminars at Oxford around 1935; and Johann Rafelski of the University of Arizona, deWet's colleague at Capetown in the 1980's.)

Bibliography and References.

13.1) J.S. deWet, Phys. Rev. **57**, 546 (1940); see our App.13B.

13.2) D. Iwanenko and A. Socolow, Phys. Zeits. d. Sowjetunion **11**, 590 (1936); see our App.10B.

13.3) M. Fierz, Helv. Phys. Acta **12**, 3 (1939); see our App.11A.

13.4) W. Heisenberg and W. Pauli, Zeits. f. Phys. **56**, 1 (1929).

13.5) F.J. Belinfante, Physica **6**, 870 (1939); see our App.12A.

13.6) W. Pauli, Phys. Rev. **58**, 716 (1940); see our App.14A.

13.7) W. Pauli and V. Weisskopf, Helv. Phys. Acta **7**, 709 (1934); see our App.9A.

13.8) W. Pauli, Annals de Institute Henri Poincaré **6**, 137 (1936); see our App.10A.

13.9) W. Pauli and F.J. Belinfante, Physica **7**, 177 (1940); see our App.12B.

13.10) A.S. Wightman, Phys. Rev. **101**, 860 (1956); see our App.18A.

13.11) G. Lüders and B. Zumino, Phys. Rev. **110**, 1450 (1958); see our App.17A.

13.12) N. Burgoyne, Nuov. Cim. **8**, 607 (1958); see our App.17B.

APPENDIX 13.A:

Excerpt from: The 1939 Princeton University PhD Thesis

ON THE CONNECTION BETWEEN THE
SPIN AND STATISTICS OF ELEMENTARY PARTICLES
by J.S. deWET

§1. Introduction.

The purpose of this dissertation is to study the connection between the spin and statistics of elementary particles. The study of this problem was started by Pauli, who attempted to show that the scalar wave equation did not admit Fermi-Dirac quantization. His work was incorrect, but the result he expected to find has turned out to be correct. His work was followed up by Sokolow and Iwanenko, who studied both the scalar wave equation and the Dirac equation for the electron. They showed that Bose-Einstein quantization of the scalar equation, which had been given by Pauli and Weisskopf, could not be extended to the Fermi-Dirac case. It was well known that both Fermi-Dirac and Bose-Einstein quantization could be carried out for the Dirac equation. These authors showed, however, that while the Fermi-Dirac quantization allowed the difficulties with the negative energy states to be overcome to some extent, by a theory of holes, this was not the case with Bose-Einstein quantization. This provides some argument in favor of the view that for the electron, Fermi-Dirac quantization is admissible, but not Bose-Einstein.

Sokolow and Iwanenko arrived at their result by considering the quantization in terms of the coefficients of the expansions of the operators in terms of a complete orthogonal set of functions. Before the present work was completed, Fierz gave a quantization of the general equations of Dirac for particles with integral and with half-integral spin. He showed that the quantization he gave for the equations for integral spin could be carried out for the Bose-Einstein case, but not for the Fermi-Dirac case. On the other hand, the quantization of the half-integral spin equations applied to both the Fermi-Dirac and Bose-Einstein case. His work thus showed that for a particular way of quantizing the integral spin equations, only Bose-Einstein quantization is possible.

Here we arrive at a much more general result. We will show that for the equations for integral spin, there does not exist the possibility of Fermi-Dirac quantization by the usual Heisenberg-Pauli procedure of going from classical equations to quantum equations. The way the argument goes is the following: ··· In considering the Fermi-Dirac anticommutation relations we find that they lead to difficulties. These difficulties can be avoided

336

if, and only if, in the classical Hamiltonian form of the equations, the pairs of conjugate variables are (one component of a set of functions, and a linear combination of the complex conjugate of the set of functions). It is shown that $\cdots$ *(for)* tensor equations, such a *(relativistic)* Hamiltonian form $\cdots$ does not exist. Thus for tensor equations a classical Hamiltonian form leading to the possibility of Fermi-Dirac quantization by the Heisenberg-Pauli procedure, does not exist. $\cdots$

In this way we have disproved the existence of a Hamiltonian form which leads to the *(possibility of)* Fermi-Dirac quantization $\cdots$ for integral spin particles. The existence of such a Hamiltonian form for half-integral spin is proved by actually giving it. From this follows the possibility of both Fermi-Dirac and Bose-Einstein quantization $\cdots$.

$\cdots$ In Section 4 we $\cdots$ derive a Hamiltonian form suitable for Fermi-Dirac quantization, and both Fermi-Dirac and Bose-Einstein quantization can be carried out. The work is presented in some detail and an account is given of the Veblen *(elsewhere called van der Waerden)* two component spinor formalism, which is useful $\cdots$.

In Section 5 the general spinor equations of Dirac are considered. The equations for half-integral spin are closely analogous to the electron equation $\cdots$. The integral spin equations $\cdots$ equivalent to tensor equations. The non-existence of $\cdots$ Fermi-Dirac quantization follows.

In Section 6 we consider the quantization of the electron equation further. We show essentially what has been pointed out by Sokolow and Iwanenko $\cdots$. The extension to general $\cdots$ half-integral spin is exactly analogous to the spin-$\frac{1}{2}$ case. $\cdots\cdots$

Most of the formal work in this thesis has been given by other authors. The work of Section 2, relating to the Fermi-Dirac commutation relations and the necessary form of the classical Hamiltonian, is new and forms the basis of the whole investigation. The rather trivial result of Section 4 relating to linear Lagrangians is also new. While trivial, it is useful in getting the Hamiltonian form of Dirac's equations for half-integral spins, which cannot be obtained from Lagrangians for which the usual theory holds.

In conclusion I must express my indebtedness to Professors E.P. Wigner and H.P. Robertson for useful criticism.

$\cdots$ *(followed by the body of the thesis for which we refer to the published account of de Wet's work).*

APPENDIX 13.B:

Excerpt from: Physical Review 57, 646 (1940)

ON THE CONNECTION BETWEEN THE
SPIN AND STATISTICS OF ELEMENTARY PARTICLES*

J.S. deWET[†]

Princeton University and Institute for Advanced Study, Princeton, New Jersey
(Received January 25, 1940)

It is shown that Fermi-Dirac quantization by the procedure of Heisenberg and Pauli cannot be carried out for tensor wave equations. Since the general wave equations for particles with integral spin are tensor equations, it follows that for these integral spin equations Fermi-Dirac quantization cannot be carried out. During the course of the discussion it appears that for equations derived from Lagrangians which are nonlinear in the derivatives of the functions, Fermi-Dirac quantization cannot be carried out. Since Heisenberg-Pauli theory applies only to nonlinear Lagrangians, a special discussion of linear Lagrangians (linear in the derivatives of the functions) is given. It is shown how equations derived from such Lagrangians can be put into Hamiltonian form. Lagrangians of this type occur for the equations for half-integral spin.

Introduction.

The problem of the connection between the spin and statistics of particles was first tackled by Pauli [1], ⋯ His work was not correct ⋯

(For the Introduction, we refer to deWet's PhD thesis in our App.13A)

In the course of the discussion of the Fermi-Dirac commutation relations it appears as an elementary result that Fermi-Dirac quantization is not possible for Hamiltonian forms of equations derived from Lagrangians which are nonlinear in the derivatives of the functions. Since the theory as given by Heisenberg and Pauli applies only to such Lagrangians, a discussion is given of Lagrangians linear in the derivatives of the functions. It is shown how equations derived from certain such Lagrangians can be put into Hamiltonian form. Such linear Lagrangians appear in the case of the equations for half-integral spin, and the discussion is of use in carrying out the quantization of these equations.

§1. Summary of the Classical Hamilton-Jacobi Theory for Continua.

⋯ see the Heisenberg-Pauli paper [3] ⋯ *(in our Appendix 9A.)*

338

The problem then is to choose commutation relations between the operators in such a way that

$$[\pi_\alpha(x), H]_- = -i\hbar \frac{\delta H}{\delta \psi^\alpha(x)}, \qquad [\psi^\alpha(x), H]_- = +i\hbar \frac{\delta H}{\delta \pi_\alpha(x)}. \qquad (110)$$

The choice of commutation relations such that Eqns.110 are satisfied is discussed in the next section. Two cases are discussed, the Bose-Einstein and the Fermi-Dirac commutation relations. It will be found that for the Fermi-Dirac anticommutation relations, there is a further restriction on the order of terms in H and hence in T_r^σ. When the relations of Eqn.110 are satisfied, the equations of motion can be written in the form

$$i\hbar \frac{\partial A}{\partial t} = [A, H]_-, \qquad (111)$$

where $A = \psi^\alpha, \pi_\alpha$, and so on.

These are the Heisenberg equations of motion. Thus if we choose commutation relations such that Eqns.110 are satisfied the Heisenberg equations of motion $\cdots$

§2. The Commutation Relations.

We consider in this section the commutation relations to be introduced between the operators $\pi, \psi, \cdots$ such that the relations of Eqn.110 hold $\cdots$.

a) The Bose–Einstein commutation relations.

$\cdots$ The proof is given by Heisenberg and Pauli. Given the commutation relations

$$[\psi^\alpha(x), \pi_\beta(x')]_- = i\hbar \delta_\beta^\alpha \delta^3(x - x'), \qquad (201)$$

and so on, with other pairs commuting. *Then* Eqns.110 hold, where

$$H = \int d^3 x \mathcal{H}(\psi^\alpha(x), \psi_{;r}^\alpha(x), \cdots), \qquad (202)$$

and $\mathcal{H}$ is quadratic in the operators $\psi^\alpha, \psi_{;r}^\alpha, \cdots \pi_{\alpha;r}$ and their Hermitian conjugates.

The result is not dependent on the order of terms in $\mathcal{H}$. We mention this because in the case of Fermi-Dirac anticommutation relations the order of terms is important.

b) The Fermi–Dirac commutation relations.

We establish a similar result with the Fermi-Dirac anticommmutation relations $\cdots$.

The result we prove is the following. Given the anticommutation relations:

$$[\pi_\alpha(x), \psi^\beta(x')]_+ = i\hbar \delta_\alpha^\beta \delta^3(x - x'), \qquad (204)$$

339

and its Hermitian conjugate, with all other pairs anticommuting $\cdots$ and

$$H = \int d^3 x \mathcal{H}(\pi_\alpha, \psi^\beta \cdots \psi^{\dagger\beta}_{;r}), \tag{205}$$

where $\mathcal{H}$ is a bilinear form of the type

$$\mathcal{H} = A_{\sigma\mu\alpha\beta} F^\sigma_\alpha G^\mu_\beta, \tag{206}$$

$A_{\sigma\mu\alpha\beta}$ being constants and $F^\sigma_\alpha, G^\mu_\alpha$ given by

$$F^\sigma_\alpha \equiv (\pi^\dagger_\alpha, \psi^\alpha, \psi^\alpha_{;r}, \text{ or } \pi^\dagger_{\alpha;r}), \quad G^\mu_\beta \equiv (\pi_\beta, \psi^\dagger_\beta, \psi^\dagger_{\beta;r}, \text{ or } \pi_{\beta;r}). \tag{207}$$

Then $\cdots$

$$[\psi^\alpha(x), H]_- = i\hbar \frac{\delta H}{\delta \pi_\alpha(x)}, \tag{208}$$

and so on $\cdots$

   Proof: We establish the result by straightforward verification $\cdots$ We note that the anticommutation relations of Eqn.204 include the following:

$$[\psi^\alpha(x), \psi^{\dagger\alpha}(x')]_+ = 0. \tag{210}$$

But $\psi^\alpha(x)\psi^{\dagger\alpha}(x)$ and $\psi^{\dagger\alpha}(x)\psi^\alpha(x)$ are essentially positive. Thus Eqn.210 leads to

$$\psi^\alpha(x)\psi^{\dagger\alpha}(x) = -\psi^{\dagger\alpha}(x)\psi^\alpha(x) = 0. \tag{211}$$

It is obvious that the relations of Eqn.211 while formally admissible, will lead to trivial results in our theory.

   The only case in which the difficulty *does not* arise is when

$$\pi_\alpha = A_{\alpha\beta}\psi^{\dagger\beta}, \quad \psi^{\dagger\beta} = a^{\beta\alpha}\pi_\alpha, \tag{212}$$

where

$$A_{\alpha\beta}a^{\beta\jmath} = \delta^\jmath_\alpha. \tag{213}$$

We are restricted to a linear relation between π_α and $\psi^{\dagger\beta}$ since our equations are linear. The $A_{\alpha\beta}$ or $a^{\alpha\beta}$ will not be perfectly general but will be subject to certain restrictions, which we will discuss as they arise. It is easily seen that the work of this section still applies, where $\mathcal{H}$ is now a function of $(\psi^\alpha, \pi_\alpha, \psi^\alpha_{;r}, \pi_{\alpha;r})$ only, and of the form

$$\mathcal{H} = A^{\alpha\beta}_{\sigma\mu} F^\sigma_\alpha G^\mu_\beta \tag{214}$$

with

$$F^\sigma_\alpha \equiv (\psi^\alpha, \psi^\alpha_{;r}), \quad G^\mu_\beta \equiv (\pi_\beta, \pi_{\beta;r}) \tag{215}$$

and the commutation relations $\cdots$ Eqn.204 $\cdots$ written in terms of $\psi^\alpha, \psi^{\dagger\alpha}$ become

$$[\psi^{\dagger\alpha}(x), \psi^\beta(x')]_+ = i\hbar a^{\alpha\beta}\delta^3(x - x'). \tag{217}$$

340

$\cdots$ it is clear that $ia^{\alpha\beta}$ must be Hermitian.

The additional condition on $ia^{\alpha\beta}$ is that it be positive definite; i.e., when the $\psi^{\dagger\alpha}, \psi^{\beta}$ undergo linear transformations which reduce $ia^{\alpha\beta}$ to a diagonal matrix, the diagonal elements must be positive. The *positive* definiteness is not essential. All that is essential is that $ia^{\alpha\beta}$ is definite. For if it is negative definite $\cdots$ change the sign of the anticommutation relations of Eqn.216 and $\cdots$ interchange all the terms in $\mathcal{H}$ $\cdots$.

Our general method has been to put the equations to be quantized into Hamiltonian form. Then we replace the functions by operators and introduce anticommutation relations $\cdots$ this can be done $\cdots$ if, and only if, the given equations can be put into a classical Hamiltonian form with ψ^{α} and $A_{\alpha\beta}\psi^{\beta}$ as conjugate variables (with $ia^{\alpha\beta}$ a definite Hermitian matrix). Our problem therefore reduces to the question of the existence of such classical Hamiltonian forms for given equations.

At this stage we note that when the Lagrangian $\cdots$ is nonlinear in the derivatives $\cdots$ the conjugate variables are not related $\cdots$ in the way necessary for the possibility of Fermi-Dirac quantization. $\cdots$ (We can) say at a glance that the Pauli-Weisskopf [3] Bose-Einstein quantization of the scalar wave equation and the Kemmer [7] Bose-Einstein quantization of Proca's vector-equations cannot be extended to the Fermi-Dirac case $\cdots$ since the Lagrangians $\cdots$ are nonlinear in the derivatives $\cdots$ further discussion is necessary $\cdots$ however for tensor equations a classical Hamiltonian form, suitable for Fermi-Dirac quantization by the anticommutation relations of this section, does not exist.

$\cdots$ It is by no means obvious that the commutation relations Eqns.201,216 are relativistically invariant. Heisenberg and Pauli have considered this in detail and have shown this to be the case if

(a) π_{α} transforms contragrediently to ψ^{α} and is in addition the 4th contravariant component of a vector; and

(b)
$$\frac{\partial}{\partial x^r} \frac{\partial^2 \mathcal{H}}{\partial \psi^{\alpha}_{;r} \partial \psi^{\alpha}_{;s}} = 0.$$

The second requirement is satisfied in all cases that arise since the equations we are dealing with are linear and the Lagrangians that occur are quadratic at most. The first is satisfied in the cases where the conjugate variables are obtained from an invariant Lagrangian by Eqn.106.

As far as our present considerations are concerned (a) requires that $A_{\alpha\beta}$ have the

341

transformation character $A_{\alpha\beta}^4$ as indicated by the indices, the 4 being a tensor index, and the α, β tensor or spinor indices as the case may be. A further requirement of the relativistic invariance is that $A_{\alpha\beta}^4$ shall be numerically invariant under Lorentz transformations (and the associated spin transformations if the α, β are spinor indices). This is necessary since we require that the form of the equations shall be the same in all Galilean frames.

When the α, β are tensor indices $A_{\alpha\beta}^\sigma$ must be a tensor of odd rank and must be numerically invariant under Lorentz transformations. The only such tensors are $g_{\sigma\tau}, \delta_\tau^\sigma, \epsilon_{\sigma\mu\tau\nu}$ and combinations of these [8]. It is not possible with these to construct a numerically invariant tensor $A_{\alpha\beta}^\sigma$ of odd rank. From this follows that it is not possible to find a Hamiltonian form for tensor equations, which will be suitable for Fermi-Dirac quantization and will satisfy the requirements of relativistic invariance.

Now the general wave equations for integral spin are tensor equations. For our present purposes the explicit form of the equations is not important and they are not given here. The tensor form of the equations is given in Fierz's paper. Since they are tensor equations it follows that their Fermi-Dirac quantization cannot be carried out by the above procedure.

§3. Lagrangians Linear in the Derivatives of the Functions and the Quantization of Half-Integral Spin Equations.

We have seen $\cdots$ that Lagrangians nonlinear in the derivatives $\cdots$ cannot yield Hamiltonian forms for which Fermi-Dirac quantization can be carried out. $\cdots$ we need some means of putting $\cdots$ linear Lagrangians into Hamiltonian form. $\cdots$ this is applied to the quantization of the equations for half-integral spin (which are not tensor equations, or equivalent to tensor equations). This will show that these equations admit Fermi-Dirac quantization, as well as Bose-Einstein, as has been shown by Fierz. $\cdots$

Consider a Lagrangian of the form

$$L = -\chi^\alpha A_{\alpha\beta}^\sigma \psi_{;\sigma}^\beta - F(\psi, \chi), \qquad (301)$$

where the $A_{\alpha\beta}^\sigma$ are constant matrices. The Euler equations are

$$
\begin{aligned}
A_{\alpha\beta}^\sigma \psi_{;\sigma}^\beta + \frac{\partial F}{\partial \chi^\alpha} &= 0, \\
-A_{\alpha\beta}^\sigma \chi_{;\sigma}^\alpha + \frac{\partial F}{\partial \psi^\beta} &= 0.
\end{aligned}
\qquad (302)
$$

$\cdots$ the equations can be put into Hamiltonian form, with $\mathcal{H}$ (and H) defined as before. The simplest way $\cdots$ is by verification.

We form the tensor T^σ_τ from L by

$$T^\sigma_\tau = (\partial L/\partial \psi^\beta_{;\sigma})\psi^\beta_{;\tau} - L\delta^\sigma_\tau, \cdots \qquad (303)$$

and $\partial_\sigma T^\sigma_\tau = 0$ from Eqn.302. As before $\cdots$

$$\mathcal{H} = T^t_t = A^r_{\alpha\beta}\chi^\alpha\psi^\beta_{;r} + F(\psi, \chi) \quad (r = 1, 2, 3). \qquad (304)$$

The conjugate variables are given by

$$\pi_\beta = \partial L/\partial \psi^\beta_{;t} = -A^t_{\alpha\beta}\chi^\alpha \qquad (305)$$

or

$$\chi^\alpha = -a^{\alpha\beta}\pi_\beta, \qquad (306)$$

where

$$A^t_{\alpha\beta}a^{\alpha\epsilon} = \delta^\epsilon_\beta. \qquad (307)$$

Substituting for χ^α in Eqn.304 in terms of π_α we have

$$\mathcal{H} = A^r_{\alpha\beta}a^{\alpha\epsilon}\pi_\epsilon\psi^\beta_{;r} + F(\psi, \chi(\pi)). \qquad (308)$$

Regarding π_β and ψ^β as conjugate variables and $\mathcal{H}$ as the energy density, the Hamiltonian equations are

$$\partial_t\pi_\beta = -\frac{\partial\mathcal{H}}{\partial\psi^\beta} + \partial_r\frac{\partial\mathcal{H}}{\partial\psi^\beta_{;r}} = -\partial F/\partial\psi^\beta + A^r_{\alpha\beta}a^{\alpha\beta}\pi_{\epsilon;r}, \qquad (309)$$

and

$$\partial_t\psi^\epsilon = \frac{\partial\mathcal{H}}{\partial\pi_\epsilon} = -a^{\alpha\epsilon}(A^r_{\alpha\beta}\psi^\beta_{;r} + \partial F/\partial\chi^\alpha). \qquad (310)$$

Multiplying Eqn.310 by $A^t_{\delta\epsilon}$ we get $\cdots$ the first of Eqns.302. Eqn.309 is the same as the second of Eqn.302 noting Eqn.305 and Eqn.306.

We have thus shown that the Hamiltonian equations with π_β and ψ^β as conjugate variables and $\mathcal{H}$ given by Eqn.308 are the same as the Euler equations (Eqn.302) of the Lagrangian (Eqn.301). $\cdots$ It is easily seen that the above holds also for Lagrangians

$$L = -\bar{\psi}^\alpha A^\sigma_{\alpha\beta}\psi^\beta_{;\sigma} - F(\psi, \bar{\psi}) \qquad (311)$$

and

$$L = -\chi^\alpha A^\sigma_{\alpha\beta}\psi^\beta_{;\sigma} - \bar{\chi}^\alpha B^\sigma_{\alpha\beta}\bar{\psi}^\beta_{;\sigma} - F(\chi, \psi, \bar{\chi}, \bar{\psi}). \qquad (312)$$

The general equations for particles with half-integral spin can be derived from a Lagrangian of the form Eqn.312 and can be put into the Hamiltonian form by the use of the theory of this section. The possibility of both Bose-Einstein and Fermi-Dirac quantization then follows $\cdots$. A complete discussion is given in my thesis [9] $\cdots$.

At this stage · · · we give a short account of Sokolow and Iwanenko's discussion · · ·

This completes the discussion of the connection between the spin and statistics of elementary particles. · · ·

The author wishes to thank Professors E.P. Wigner and H.P. Robertson for helpful criticism.

* Abstract of a PhD thesis submitted to Princeton University.

† Commonwealth Fund Fellow.

References and Footnotes:

1) W. Pauli, Ann. d. Inst. Henri Poincare **6**, 137 (1936).

2) A. Sokolow and D. Iwanenko, Physik. Zeits. Sowjetunion **11**, 590 (1937).

3) W. Pauli and V.F. Weisskopf, Helv. Phys. Acta **7**, 709 (1934).

4) M. Fierz, Helv. Phys. Acta **12**, 3 (1939).

5) P.A.M. Dirac, Proc. Roy. Soc. London **A155**, 47 (1936).

6) W. Heisenberg and W. Pauli, Zeits. f. Physik **56**, 1 (1929).

7) N. Kemmer, Proc. Roy. Soc. London **A166**, 127 (1938).

8) $g_{\sigma\tau}$ is the tensor which has the following components in all Galilean frames:

$$g_{11} = g_{22} = g_{33} = -g_{44} = 1, \quad g_{\sigma\tau} = 0(\sigma \neq \tau).$$

9) A copy is available at the Princeton University Library for inter-library loan.

Chapter 14

Pauli's Proof of the Spin-Statistics Theorem

Summary: Pauli's proof of the Spin-Statistics Theorem is based on a classification of the spinor representations of the proper Lorentz group into four classes different under the strong-reflection transformation $x_\mu \to -x_\mu$. From this very general beginning, Pauli obtains the spin-statistics connection for non-interacting particles of arbitrary spin which are not necessarily represented by irreducible spinors (as was the case with Fierz and Belinfante) or subject to canonical field theory (as was the case for Iwanenko and Socolow, and for deWet). He still had to assume (with all the above), without proof, the continuability of field products to zero separation.

§1. Introduction.

We come at last to the widely referred Pauli proof of the Spin-Statistics Theorem, finally submitted on August 19, 1940 and published in the Physical Review of October 15, 1940 [14.1]. Its publication had been promised for a year, and had been referred to as "In the press" in Pauli and Belinfante's joint paper of December, 1939 [14.2]. Pauli states in a footnote that "This paper is part of a report which was prepared by the author for the Solvay Congress 1939 and in which slight improvements have since been made. In view of the unfavorable times, the Congress did not take place, and the publication of the reports has been postponed for an indefinite length of time. The relation between the present discussion of the connection between spin and statistics, and the somewhat less general one of Belinfante, based on the concept of charge-conjugation invariance has been cleared up by W. Pauli and F.J. Belinfante, Physica **7**, 177 (1940)".

How widely circulated the manuscript was is not known. It seems certain that it had no influence on Belinfante's paper [14.3] or on deWet's [14.4]. Nor is it clear what were the "slight improvements" which had been made. One clearly was a begrudging acknowledgement, in a footnote, to deWet's paper: "On account of the existence of such conditions the canonical formalism is not applicable for spin > 1 and therefore the discussion about the connection between spin and statistics by J.S. deWet, Phys. Rev. **57**, 646 (1940), which is based on that formalism, is not general enough." There is no attempt here or elsewhere to rebut deWet's comment about Pauli's 1936 paper [14.5] being incorrect, but in another footnote he does "$\cdots$ exclude operations like $(m^2 - \nabla^2)^{\frac{1}{2}}$, which act at finite distances $\cdots$", which are present both in the 1936 paper and in his joint criticism of Belinfante's proof. Furthermore, his comment about the "$\cdots$ relation between the present discussion $\cdots$ and the somewhat less general one of Belinfante $\cdots$ has been cleared up $\cdots$", although mild enough, still misrepresents the situation, which had in no way been cleared up by the artificial counter-examples (including non-local ones which he now disavows) thrown at Belinfante's proof by Pauli.

It is interesting that deWet and Pauli both submitted their Physical Review papers from the Institute for Advanced Study, although some seven months apart. Whether they met or discussed their work remains to be determined.

There is no doubt that Pauli's personal behavior in this and other instances does invite criticism. There is also no doubt that Pauli's past scientific performance on the question of the Spin-Statistics Theorem showed serious lapses of taste and judgement, specifically on the points:

1) of abandoning canonical quantum field theory;

2) of insisting on solving the most general spin case without first understanding spins $0, \frac{1}{2}, 1$;

3) of introducing artificial counter-examples, including non-local ones;

4) of ignoring all interactions, which might have led him to appreciate

more the power of Belinfante's proof.

And ultimately, Pauli was to accept without proof the same unjustified argument as had Fierz [14.6], Belinfante, and deWet, in order to rule out anticommutation relations for integral spin particles. To this day, the fine judgement of deWet to confine himself to a minimal, fundamental, canonical set of fields is outstanding for its elegance.

§2. Outline of Pauli's Proof.

Having made all these critical remarks, we must now give Pauli due credit for his landmark paper "The Connection Between Spin and Statistics". He sought, and in the context of his time, achieved a proof of stunning economy for the Spin-Statistics Theorem for arbitrary spin.

Like Fierz (who was, after all, his assistant), Pauli used the spinor representations of the proper Lorentz group. Unlike Fierz, but presumably following a statement present in Fierz's paper that it was unessential to do so, Pauli did not assume that physical states correspond to an irreducible spinor representation. Pauli was able instead to identify a property of spinors which divides them into four classes depending on the number of dotted and of undotted indices. He labels the spinors (for reasons to be met soon) $U(+1), U(-1), U(+\epsilon), U(-\epsilon)$, where $U(\pm 1)$ describe integral spin particles, and $U(\pm\epsilon)$ describe half-integral spin particles. In the spinor representation, a Lorentz vector, x or p for example, is in the class $U(-1)$. From this very general classification, Pauli proceeds to construct permissible energy-momentum tensors and current-vectors for the unquantized theory. It becomes immediately apparent that relativistic particles of integral spin have no positive classical density which is part of a four-vector satisfying a continuity equation. Conversely, it is equally apparent that relativistic particles of half-integral spin have no positive classical energy-density.

After postulating causal commutation or anticommutation relations for the four classes of spinors that are available to represent physical fields, Pauli

347

is then able to establish the Spin-Statistics Theorem in ways that have been used already, especially by deWet. For the integral spin spinor fields $U(+1)$ the symmetrized commutator vanishes for an individual field and its Hermitian conjugate, even when evaluated at a single space-time point. For the choice of anticommutation relations, this vanishing was assumed - without proof - to require that the fields must vanish identically, and therefore that integral spin fields can only be quantized using commutation and not anticommutation relations. The same argument was used by Fierz, and by Belinfante, who identified irreducible spinor fields with physical states; and by deWet who used scalar and Dirac fields and limited his discussion to spin-0 and spin-$\frac{1}{2}$.

For half-integral spins, Pauli finds no such general restriction. Both commutation and anticommutation relations are permitted *a priori*. He then must have recourse to the requirement of positive energies to restrict the possibilities to anticommutation relations. In this respect, his proof is the same as that of Iwanenko and Socolow [14.7] for spin-$\frac{1}{2}$ (as noted and expanded on by deWet), and of Fierz. Pauli's proof has the merit of elegance and generality but does not differ in substance from these earlier proofs. His principal contribution was the classification of spinors into four classes which include any physical state.

Pauli was generous in his reference to Fierz: "For the positive proof that a theory with a positive total energy is possible by quantizing integral spin according to Bose-Einstein statistics, and half-integral spin according to Fermi-Dirac statistics, we must refer to the already mentioned paper by Fierz $\cdots$ "

§3. Guide to Pauli's Proof.

We now look at the details of Pauli's proof which requires only very simple and general properties of van der Waerden spinors [14.8]. Recall the

348

basic spinor-vector connection

$$U^{\alpha\dot{\beta}} = V^\mu \sigma_\mu^{\alpha\dot{\beta}} = (V^0 + \vec{\sigma} \cdot \vec{V})^{\alpha\dot{\beta}} \tag{1}$$

and

$$U_{\alpha\dot{\beta}} = V_\mu \sigma^\mu_{\alpha\dot{\beta}}, \tag{2}$$

where Lorentz indices $\mu = 0, 1, 2, 3$, spinor indices $\alpha, \dot{\beta}=1,2$, and the Pauli matrices $\sigma_\mu = 1, \vec{\sigma}$. Spinor indices are raised and lowered with the alternating symbol $\epsilon^{\alpha\beta} = -\epsilon^{\beta\alpha}$, $\epsilon_{\alpha\beta} = -\epsilon_{\beta\alpha}$ with $\epsilon^{12} = \epsilon_{12} = 1$, etc. The inverse relation for the vector is

$$V^\mu = -\frac{1}{2}\sigma^\mu_{\alpha\dot{\beta}}U^{\alpha\dot{\beta}} \quad \text{and} \quad V_\mu = -\frac{1}{2}\sigma_\mu^{\alpha\dot{\beta}}U_{\alpha\dot{\beta}}. \tag{3}$$

More generally, spinors $U^{\mu\dot{\nu}\dot{\sigma}\cdots}_{\alpha\beta\gamma\cdots}$ can be characterized by two positive integers (r, q) expressed in terms of "angular momentum quantum numbers" (j, k) with $r = 2j + 1$ and $q = 2k + 1$ and (j, k) integral or half-integral. The number of lower, undotted indices is $2j$, and of upper, dotted indices is $2k$. The number of independent components is $rq = (2j + 1)(2k + 1)$. For example $(j, k) = (0, 0)$ is a scalar, $(\frac{1}{2}, \frac{1}{2})$ is a 4-vector, $(1, 0)$ is self-dual antisymmetric tensor, $(1, 1)$ is a traceless symmetric tensor, and so on. A Dirac four-spinor is a combination of two irreducible spinors $(\frac{1}{2}, 0)$ and $(0, \frac{1}{2})$.

A product of two irreducible representations $U_1(j_1, k_1)U_2(j_2, k_2)$ decomposes into a sum of irreducible representations $U(j, k)$ with

$$j = j_1 + j_2, j_1 + j_2 - 1, \cdots, |j_1 - j_2|,$$
$$k = k_1 + k_2, k_1 + k_2 - 1, \cdots, |k_1 - k_2|. \tag{4}$$

Under 2π-space rotations, representations undergo a sign change if one of (j, k) is half-integral and the other integral, and no sign change if both of (j, k) are integral or both half-integral. Thus $(1, 0)$, $(1, 1)$, $(\frac{1}{2}, \frac{1}{2})$, etc are single-valued under rotations; $(\frac{1}{2}, 0)$, $(\frac{1}{2}, 1)$, etc are double-valued. The product of two single-valued or two double-valued representations is a single-valued representation; the product of one single-valued and one double-valued representation is double-valued.

Consider first single-valued representations which all have $j + k$ integral. We can characterize these as $U(+1)$ with j and k integral, $U(-1)$ with j and k half-integral. They satisfy a simple multiplication table

$$U(+1)U(+1) = U(+1), \qquad U(-1)U(-1) = U(+1),$$

$$U(+1)U(-1) = U(-1)U(+1) = U(-1). \tag{5}$$

Note that the complex conjugate $U(\pm 1)^* = U(\pm 1)$. Also the momentum vector $p \sim U(\frac{1}{2}, \frac{1}{2})$ is in the class $U(-1)$, so $pU(-) = U(+)$, $p^2 U(-) = pU(+) = U(-)$ and so on. The strong-reflection operation

$$p \to -p, \qquad U(\pm) \to \pm U(\pm), \qquad [U(\pm)]^* \to \pm [U(\pm)]^* \tag{6}$$

leaves these equations invariant. Observables bilinear in the fields are of importance. The traceless symmetric energy-momentum tensor

$$T \sim U(1,1) \sim U(+)$$

can contain terms of the form $U^*(+)U(+)$, $U^*(+)pU(-)$, and so on, which reduce to

$$T \sim U(+)U(+) + U(-)U(-) + U(+)pU(-) \tag{7}$$

Similarly the current vector

$$S \sim U(+)U(-) + U(+)pU(+) + U(-)pU(-). \tag{8}$$

Next, Pauli observes that the strong-reflection operation $p_\mu \to -p_\mu$, $x_\mu \to -x_\mu$ leaves the energy-momentum tensor unchanged but changes the sign of the current vector. Here Pauli comments how remarkable it is to discover this extra strong-reflection invariance from the properties of the spinor equations under proper Lorentz transformations. We can see how very close Pauli was to the full **TCP**-theorem, and to anticipating results which would lie undiscovered until work of Schwinger, of Lüders, and of Lüders and Zumino [14.9], more than ten years later. Pauli was able to use the **TP**-reflection (hereafter called strong-reflection) invariance of the

350

equations to prove that for every solution of the equations of motion, there is another obtained by strong-reflection with opposite sign charge-current four-vector. He is then able to state that no relativistic integral spin field can have a positive density satisfying a continuity equation.

A different classification of spinors is required for double-valued representations with one of (j, k) half-integral, one integral. call these $(+\epsilon)$ for j integral, k half-integral and $(-\epsilon)$ for k integral, j half-integral. Then we can complete the multiplication table with

$$U(\pm\epsilon)U(\pm\epsilon) = U(+), \qquad U(\pm\epsilon)U(\mp\epsilon) = U(-),$$

$$U(\pm\epsilon)U(+) = U(\pm\epsilon), \qquad U(\pm\epsilon)U(-) = U(\mp\epsilon). \tag{9}$$

Also

$$pU(\pm\epsilon) = U(\mp\epsilon), \qquad [U(\pm\epsilon)]^* = U(\mp\epsilon), \tag{10}$$

and so on. The strong-reflection transformation is

$$U(\pm\epsilon) \rightarrow \pm iU(\pm\epsilon). \tag{11}$$

Note that the "i" is required to give a sign change under double reflection.

The energy-momentum tensor bilinear in U^*, U must be of the form

$$T \sim U(\pm\epsilon)U(\pm\epsilon) + U(\pm\epsilon)pU(\mp\epsilon) \tag{12}$$

and the current four-vector is of the form

$$S \sim U(\pm\epsilon)pU(\pm\epsilon) + U(\pm\epsilon)U(\mp\epsilon). \tag{13}$$

In this case, strong-reflection gives $T \rightarrow -T$ but $S \rightarrow S$, so classically a definite sign of the energy-density is not possible for a half-integral spin field.

Finally, Pauli postulates causal commutation or anticommutation relations [14.10] for the four classes of spinor fields. He requires

$$[U^{(r)}(x'), U^{(r)*}(x'')] \sim D(\vec{x}' - \vec{x}'', x_0' - x_0'') \tag{14}$$

where

$$D(x) = \int \frac{d^3p}{(2\pi)^3} e^{i\vec{p}\cdot\vec{x}} \frac{sin(\omega x_0)}{\omega} \tag{15}$$

is the standard invariant D function satisfying

$$(\Box - m^2)D = 0, \qquad D(\vec{x}, 0) = 0, \qquad \partial_0 D(\vec{x}, 0) = \delta(\vec{x}).$$

The solution is of the form

$$D(\vec{x}, x_0) = \frac{-1}{4\pi r} \partial_r F(r, x_0)$$

with $r = |\vec{x}|$. $F(r, x_0) = 0$ for $r^2 > x_0^2$ is the essential property which makes the commutation or anticommutation relations causal and limits us to this choice. All physical quantities are required to commute for spacelike separations.

For half-integral spin fields, the left-side of the commutation relations transforms as

$$[U(\pm\epsilon), U(\pm\epsilon)^*] = [U(\pm\epsilon), U(\mp\epsilon)] = U(-), \tag{16}$$

corresponding to odd derivatives of D. Just the opposite occurs for integral spin fields where

$$[U(\pm), U(\pm)^*] = [U(\pm), U(\pm)] = U(+), \tag{17}$$

corresponding to even derivatives of D. Pauli then symmetrizes these relations in $x' \leftrightarrow x''$. The result on the left

$$[U(\pm; x'), U(\pm; x'')^*] + (x' \leftrightarrow x''), \tag{18}$$

(and similarly for half-integral spins) is even for $(\vec{x}' \leftrightarrow \vec{x}'', x_0' \leftrightarrow x_0'')$.

Now however, for integral spins the right side is originally even for $(\vec{x}' \leftrightarrow \vec{x}'')$ but odd for $(x_0' \leftrightarrow x_0'')$ so must vanish after symmetrization. Pauli can therefore rule out the possibility of anticommutation relations for integral spin particles because at $x' = x''$ the left side would be positive and

352

could vanish only for fields identically zero. This is essentially the same argument used originally by Pauli and Weisskopf [14.11], and subsequently by Fierz, by Belinfante, and by deWet. It would take some eighteen years for its justification by Lüders and Zumino [14.12], and by Burgoyne [14.13] using the Hall-Wightman theorem [14.14].

For half-integral spins, there is no requirement that the right side should vanish. The equations can hold for commutation or anticommutation relations. Pauli, like the others, had recourse to the positive energy requirement to select the anticommutation relations.

§4. Summary of Pauli's Proof.

To review and summarize, Pauli's achievements were:

1) the classification of van der Waerden spinors into four types - two for integral spin, two for half-integral spin - which permits him to make a very general discussion for particles of arbitrary integral or half-integral spin which was not possible in previous work. Physical states are not at any stage required to be irreducible spinor representations, a result already indicated by Fierz.

2) discovery of the remarkable strong-reflection symmetry of the spinor solutions to the equations of motion was very close to anticipating the **TCP**-invariance. If he had included minimal electromagnetic interactions and charge-conjugation invariance (following his recent experience with Belinfante), he, like Belinfante, would have been in a position to anticipate events by more than a decade.

Unfortunately, Pauli

1) stopped short and his proof - more general in some respects - was an incremental advance on those of Fierz, Belinfante, and deWet.

2) like everyone else of his era, fell understandably short by not recog-

nizing the necessity of proving the validity of field manipulations required to rule out anticommutation relations for integral spin particles. The same operations can even be extended to rule out, in a parallel way, commutation relations for half-integral spin fields, as we see eventually in Burgoyne's proof in 1958.

3) failed to address the problem of interactions, or to recognize the possibility of doing so in Belinfante's work and even in his own later work.

4) put undue emphasis on proving the theorem for arbitrary spin without first clarifying the situation for simpler and more fundamental situations. His rejection of canonical field theory impeded progress, as did his invocation of spurious counter-examples. The use of spinor representations - appearing now as an unnecessary detour which acted as a deterrent to understanding for generations of physicists - could have been avoided by using spin-0, spin-$\frac{1}{2}$, and spin-1 in a first discussion. Higher spins could have been treated by a constituent model, or, if one insisted, as an embellishment (perhaps in Pauli's style). In some sense, this is the program put forward by Iwanenko and Socolow, and by deWet, but deprecated and ignored because of Pauli.

§5. Comparison of Fierz and Pauli Proofs.

The question must be asked: Did Pauli do anything essentially more or different from Fierz?

Fierz identified the property of the commutator of spinors as an even number of momentum factors times an inverse energy, for integral spins; an odd number of momentum factors times an inverse energy, for half-integral spins. Compare his Eqn.8.4 in Appendix 11.A of our Chapter 11. Fierz was working in the momentum representation. In the coordinate representation these results can be recognized as an even or odd number of derivatives acting on the D function (whose Fourier transform is just the inverse energy). He then observed that this conclusion was independent of his initial premise that the physical fields should be irreducible spinors, and he removed this restriction in the paragraph following his Eqn.8.4.

Pauli identified the property of the commutator of spinors as transforming like $U(+)$ for integral spins, that is, an even derivative of D; or $U(-)$ for half-integral spins, that is, an odd derivative of D.

These results are the same as those obtained by Fierz and from this point their arguments for the Spin-Statistics Theorem are the same.

The difference between Fierz and Pauli is that Fierz *said* the premise of irreducible spinors could abandoned. With his discovery of strong-reflection invariance in the spinor representation of the proper Lorentz group, and by separating spinors into the four classes $U(\pm), U(\pm\epsilon)$, Pauli *showed how*.

This brings to an end the first generation of papers on the proof of the Spin-Statistics Theorem. The long interruption of WWII was to end the direct participation of Fierz, and of Belinfante and deWet - who made their contributions as their PhD theses - and only Pauli would return, and then only in the role of critic to the next generation, but even Pauli had little new to say on the subject.

§6. Biographical Note on Pauli.(†)

Wolfgang Pauli (1900-1958) was the son of a distinguished chemist at Vienna, and godson of Ernst Mach the physicist-cosmologist-philosopher whose Mach's Principle underlies general relativity. He mastered Einstein's new theory of general relativity while still in high school. Studying with Sommerfeld at Munich, he was entrusted at age 20 while in his fourth semester with the task of writing the general relativity article for the *Encyklopädie der Mathematischen Wissenschaften*. The result was praised by Sommerfeld to Einstein as "simply masterful" and has become a classic reference. Pauli was a PhD with Sommerfeld (1922); assistant to Born at Göttingen (1922-25); to Lenz at Hamburg (1926-28); professor at ETH Zurich (1928-58); and professor at the Institute for Advanced Studies (1940-46). He won the Nobel Prize (1945) for the discovery of the Exclusion Principle. His many discoveries and achievements include the neutrino hypothesis (1930); the development of field theory with Heisenberg and Jordan (1928-30); the canonical quantization of the Klein-Gordon field with Weisskopf (1934); the development of the **TCP**-theorem (1955).

From Fierz we learn that Pauli as a lecturer mumbled, was disorganized, wrote too small, was difficult to understand, but was fascinating and stimulating; had a caustic way of jumping at people which put them into disarray; was occasionally malicious, but not mean, and had a humorous, ironic side.

Surely the most outrageous - and costly - remark ever made by Pauli is one to Max Born, when Born asked Pauli's collaboration to understand Heisenberg's very first ideas of the new quantum mechanics. van der Waerden reports Born's recollection: "··· I asked him whether he would like to collaborate with me in this problem. But instead of the expected interest, I got a cold and sarcastic refusal. 'Yes, I know you are fond of tedious and complicated formalism. You are only going to spoil Heisenberg's physical ideas by your futile mathematics···'." And Pauli was Born's assistant! One can conjecture that Pauli was seriously traumatized by being replaced in the front rank of young geniuses by Heisenberg and was going through great psychological turmoil.

His association with the psychiatrist C.J. Jung evolved from therapy for depression (1932-), to friendship, to research collaborations on the nature of creativity.

(† - M. Fierz, in *Dictionary of Scientific Biography, Vol. X* (Scribner's, New York, 1988), edited by C.C. Gillespie, p.422; see also *Theoretical Physics in The Twentieth Century: A Memorial Volume to Wolfgang Pauli* (Interscience, New York, 1960), edited by M. Fierz and V.F. Weisskopf; and H. Atmanspacher and H.Primas, Journal of Consciousness Studies **3**(2), 112 (1996); and B.L. van der Waerden, *Sources of Quantum Mechanics* (North-Holland, Amsterdam, 1967), p.37.)

Bibliography and References.

14.1) W. Pauli, Phys. Rev. **58**, 717 (1940); see our App.14A.

14.2) W. Pauli and F.J. Belinfante, Physica **7**, 177 (1940); see our App.12.B.

14.3) F.J. Belinfante, Physica **6**, 870 (1939); see our App.12A.

14.4) J.S. deWet, Phys. Rev. **57**, 646 (1940); see our App.13B.

14.5) W. Pauli, Ann. d. Inst. H. Poincaré **6**, 137 (1936); see our App.10.A.

14.6) M. Fierz, Helv. Phys. Acta **12**, 3 (1940); see our App.11A.

14.7) D. Iwanenko and A. Socolow, Phys. Zeits. der Sowjetunion **11**, 590 (1937); see our App.10B.

356

14.8) C.W. Misner, K.S. Thorne, and J.A. Wheeler, *Gravitation* (Freeman, San Francisco, 1973), Ch.41, pp.1148-1155; P. Roman, *Theory of Elementary Particles* (North Holland, Amsterdam, 1960), pp.80-85.

14.9) J. Schwinger, Phys. Rev. **82**, 914 (1951); G. Lüders, Annals of Physics **2**, 1 (1957); G. Lüders and B. Zumino, Phys. Rev. **106**, 385 (1957).

14.10) W. Heisenberg and W. Pauli, Zeits. f. Phys. **59**, 168 (1930).

14.11) W. Pauli and V. Weisskopf, Helv. Phys. Acta **7**, 709 (1934); see our App.9A.

14.12) G. Lüders and B. Zumino, Phys. Rev. **110**, 1450 (1958); see our App.17A.

14.13) N. Burgoyne, Nuov. Cim. **8**, 607 (1958); see our App.17B.

14.14) A.S. Wightman, Phys. Rev. **101**, 860 (1956); see our App.18A.

APPENDIX 14.A:

Excerpt from: Physical Review **58**, 716 (1940)

The Connection Between Spin and Statistics[1]

W. Pauli

Physicalisches Institut, Eidg. Technischen Hochschule, Zurich, Switzerland
and Institute for Advanced Study, Princeton, New Jersey
(Received August 19, 1940)

Abstract: For the relativistically invariant free-particle wave equations, we conclude: I) for the energy to be positive, *Fermi-Dirac* statistics must be used for particles of half-integral spin; II) for observables at spacelike separated points to commute, *Bose-Einstein* statistics must be used for particles of integral spin.

[1]This paper is part of a report which was prepared by the author for the Solvay Congress 1939 and in which slight improvements have since been made. In view of the unfavorable times, the Congress did not take place, and the publication of reports has been postponed for an indefinite length of time. The relation of the present discussion between spin and statistics, and the somewhat less general one of Belinfante, based on the concept of charge-conjugation invariance, has been cleared up by W. Pauli and F.J. Belinfante, Physica **7**, 177 (1940).

. . . .

§2. Irreducible Tensors. Definition of Spins.

We use a few general properties of irreducible representations of the Lorentz group [1]. The proper Lorentz group is that continuous linear group the transformations of which leave invariant the form

$$\sum_{k=0}^{3} x^k x_k \equiv x^2 = \vec{x}^2 - x_0^2$$

and, in addition, satisfy the condition that they have determinant $+1$ and do not reverse the time.

The spinor representations of the Lorentz group are useful. A tensor or spinor

358

which transforms under this group can be characterized by two positive integers (r, q). (The corresponding "angular momentum quantum numbers" (j, k) are given by $r = 2j+1$, $q = 2k+1$, with integral or half-integral j, k. In the spinor calculus this is a spinor $U(j, k)$ with $2j$ undotted and $2k$ dotted indices.) The quantity $U(j, k)$ has $r \cdot q = (2j + 1)(2k + 1)$ independent components. Hence to $(0, 0)$ corresponds the scalar, to $(\frac{1}{2}, \frac{1}{2})$ the vector, to $(1, 0)$ the self-dual antisymmetric tensor, to $(1, 1)$ the traceless symmetric tensor, etc. Dirac's spinor u_ρ reduces to two irreducible quantities $(\frac{1}{2}, 0)$ and $(0, \frac{1}{2})$. If $U(j, k)$ transforms by Λ like

$$U'_r = \sum_{s=1}^{(2j+1)(2k+1)} \Lambda_{rs} U_s,$$

we represent with U^* the quantity which transforms by Λ^*.

The most important operation is the reduction of the product of two irreducible representations

$$U_1(j_1, k_1) U_2(j_2, k_2)$$

which decompose into a sum of irreducible representations $U(j, k)$ where j, k independently run through the values

$$j = j_1 + j_2, j_1 + j_2 - 1, \cdots, |j_1 - j_2|$$

$$k = k_1 + k_2, k_1 + k_2 - 1, \cdots, |k_1 - k_2|.$$

By limiting the transformations to the subgroup of space rotations alone, the distinction between the two indices j, k disappears and $U(j, k)$ behaves like the product of two irreducible representations $U(j)U(k)$ which in turn reduces to a sum of irreducible representations $U(l)$ each having $2l + 1$ components, with

$$l = j + k, j + k - 1, \cdots, |j - k|.$$

Under the space rotations the $U(l)$ with integer l transform as single-valued representations, whereas those with half-integral l transform as double-valued representations. Thus the unreduced quantities $T(j, k)$ with integral $j + k$ are single-valued representations, and those with half-integral $j+k$ are double-valued representations.

To determine the spin of a given representation $U(j, k)$, it might seem at first that these are given by $l = j + k$. Such a conclusion could not correspond to the physical facts, however, because then there would exist no relation between the spin value and the number of independent plane waves for a given value of the momentum $\vec{p}$ in the phase factor $e^{i\vec{p}\cdot\vec{x}}$. To correctly define the spin [2], we consider the case of rest-mass $m \neq 0$, in the rest-system of the particle where $\vec{p} = 0$, and the wave function depends only on the time. In this coordinate system we reduce the field components into parts irreducible under space rotations. To each part, with $r = 2s + 1$ components, belong r different eigenfunctions which transform among themselves under space rotations and correspond to a particle of spin s. For field equations describing particles with only one spin, there exists only one such irreducible group of components. From Lorentz invariance, it follows that r eigenfunctions always belong to a given momentum $\vec{p}$. The number of components $U(j, k)$ is generally more complicated, because these components together with the momentum $\vec{p}$ have to satisfy subsidiary conditions.

In the case of zero rest-mass $\cdots$.

In an arbitrary system of reference and for $m \neq 0$, the U which transform as double-valued representations with half-integral $j + k$ correspond to particles with half-integral spin. Conversely, single-valued U with integral $j + k$ correspond to particles with integral spin.

§3. Proof of the Indefinite Sign of a) Charge for Integral, and b) Energy for Half-Integral Cases.

We consider first $U(j, k)$ with $j + k$ integral. We divide the U into two classes: (1) the "+1 class" with j and k integral, and (2) the "-1 class" with j and k half-integral. The notation is justified by the multiplication rule derived from the rule for combining $j's$ and $k's$ to reduce a product of $U's$, that

$$U^+ U^+ \sim U^+, U^- U^- \sim U^+, U^+ U^- \sim U^-.$$

note that $U(j, k)^*$ belongs to the same ± 1 class as $U(j, k)$ and $U(k, j)$. The momentum vector belongs to the -1 class, because it corresponds to $U(\frac{1}{2}, \frac{1}{2})$.

Consider a homogeneous linear equation in the quantities U which, however, does not have to be first order. Assuming a plane wave, we may put p_a for $-i\partial/\partial x_a$. Solely because of invariance under *proper* Lorentz transformations it must have the typical form

$$pU^+ = U^-, \quad pU^- = U^+. \tag{1}$$

This typical form means that there may be many different terms of the same type

360

present. Furthermore, among the U^+ may occur also $(U^+)^*$, and so on. Finally we have omitted any *even* number of p factors. These may be present in arbitrary number on the left- or right-hand side of the equations. It is evident that these equations remain invariant under the substitution

$$p_a \to -p_a; \qquad U^+ \to U^+, \quad (U^+)^* \to (U^+)^*;$$
$$U^- \to -U^-, \quad (U^-)^* \to -(U^-)^*. \tag{2}$$

Let us now consider tensors T of even rank (scalars, antisymmetric or symmetric tensors of second rank, etc), which are bilinear in the $U's$. They are then composed solely of quantities with integer j and k and are of the typical form

$$T \sim U^+U^+ + U^-U^- + U^+pU^-, \tag{3}$$

where again a possible even number of p factors is omitted and no distinction is made between U and U^*. Under the substitution (2), $T \to T$.

The situation is different for tensors of odd rank S (vectors, etc) which consist of quantities of half-integral j and half-integral k. These are of the typical form

$$S \sim U^+pU^+ + U^-pU^- + U^+U^- \tag{4}$$

and change sign under the substitution (2), $S \to -S$. This is the case for the current vector J_k.

The transformation $p_a \to -p_a$ corresponds to the transformation $x_a \to -x_a$. It is remarkable that from the invariance of Eqn.1 under the proper Lorentz group alone follows an invariance property for the sign change of all coordinates. In particular, the indefinite sign of the current density and the total charge for integral spin particles follows, because to every solution to the field equations belongs another solution for which the components of J_k change sign. The definition, for particles of integral spin, of a definite sign particle density which transforms like the time-component of a four-vector is therefore impossible in the classical c-number theory.

We now proceed to the discussion of the half-integral spin case. Here we divide the quantities U with half-integral $j+k$ into the following classes:(3) the "$+\epsilon$ class" with j integral and k half-integral, and (4) the "$-\epsilon$ class" with j half-integral and k integral. The multiplication of the classes $(1) \cdots (4)$ follows with $\epsilon^2 = 1$. The multiplication table of the four classes is:

	+1	-1	$+\epsilon$	$-\epsilon$
+1	+1	-1	$+\epsilon$	$-\epsilon$
-1	-1	+1	$-\epsilon$	$+\epsilon$
$+\epsilon$	$+\epsilon$	$-\epsilon$	+1	-1
$-\epsilon$	$-\epsilon$	$+\epsilon$	-1	+1

It is important that the complex-conjugate quantities for which $j, k \rightarrow k, j$ do not belong to the same class so $U^{+\epsilon} \sim (U^{-\epsilon})^*$ belong to the $+\epsilon$ class, and so on. We shall therefore cite the complex conjugate quantities separately.

Instead of (1) we now obtain as a typical form

$$
\begin{aligned}
pU^{+\epsilon} + p(U^{-\epsilon})^* &= U^{-\epsilon} + (U^{+\epsilon})^* \\
pU^{-\epsilon} + p(U^{+\epsilon})^* &= U^{+\epsilon} + (U^{-\epsilon})^*,
\end{aligned}
\tag{5}
$$

because a factor p always changes class $\pm\epsilon$ to $\mp\epsilon$. As before any even number of p factors has been omitted.

Now we consider instead of (2), the substitution

$$
p \rightarrow -p; \qquad U^{+\epsilon} \rightarrow \pm iU^{+\epsilon}, \quad (U^{-\epsilon})^* \rightarrow \pm i(U^{-\epsilon})^*;
$$

$$
(U^{+\epsilon})^* \rightarrow -\pm i(U^{+\epsilon})^*, \quad U^{-\epsilon} \rightarrow -\pm iU^{-\epsilon}.
\tag{6}
$$

This transformation is in accord with the requirement that $U^{+\epsilon} \sim (U^{-\epsilon})^*$, etc, are members of the same class and transform in the same way. Furthermore, it does not interfere with possible reality conditions like $U^{+\epsilon} = (U^{-\epsilon})^*$, etc. Eqns.5 are invariant under the substitution (6).

We again consider tensors of even rank which are bilinear in the U and their complex conjugates. They must be of the form

$$
\begin{aligned}
T \quad \sim \quad &U^{\pm\epsilon}U^{\pm\epsilon} + U^{\pm\epsilon}pU^{\mp\epsilon} + U^{\pm\epsilon}(U^{\mp\epsilon})^* + (U^{\pm\epsilon})^*pU^{\pm\epsilon} + \\
&(U^{\pm\epsilon})^*p(U^{\mp\epsilon})^* + (U^{\pm\epsilon})^*(U^{\pm\epsilon})^*.
\end{aligned}
\tag{7}
$$

Furthermore, odd rank tensors must be of the form

$$
\begin{aligned}
S \quad \sim \quad &U^{\pm\epsilon}pU^{\pm\epsilon} + U^{\pm\epsilon}U^{\mp\epsilon} + U^{\pm\epsilon}p(U^{\mp\epsilon})^* + U^{\pm\epsilon}(U^{\pm\epsilon})^* + \\
&(U^{\pm\epsilon})^*p(U^{\pm\epsilon})^* + (U^{\pm\epsilon})^*(U^{\mp\epsilon})^*.
\end{aligned}
\tag{8}
$$

362

The result of the substitution (6) is now the opposite of the result of the sub-stitution (2): the tensors of even rank change their sign, the tensors of odd rank remain unchanged:

$$T \rightarrow -T; \quad S \rightarrow +S. \tag{9}$$

In case of half-integral spin, therefore, a definite sign for the energy density, as well as a definite sign for the total energy, is impossible in the classical c-number theory. This follows from the fact that under the above substitution (6), the field equations remain satisfied but the energy density at every space-time point changes its sign as a result of which the total energy also changes its sign.

It must be emphasized that it was not only unnecessary to assume that the wave equation is of the first order (we do exclude operations like $\sqrt{p^2 + \kappa^2}$ which are nonlocal in coordinate space), but also to assume that the theory is left invariant under space-reflections ($\vec{x} \rightarrow -\vec{x}, x^0 \rightarrow x^0$). This scheme therefore also covers Dirac's two component wave equations (with rest-mass zero).

These considerations do not prove that for integral spins there always exists a definite energy density, or for half-integral spins a definite charge density. In fact, it has been shown by Fierz [2] that this is not the case for spin > 1 for the density. There does exist, however, in the classical c-number theory, a definite total charge for half-integral spins, and a definite total energy for integral spins. The spin $\frac{1}{2}$ is distinguished by the possibility of a definite sign for the charge *density*, and the spins 0 and 1 are distinguished by the possibility of a definite sign for the energy *density*. Nevertheless, the present theory permits arbitrary values of the spin quantum numbers of elementary particles as well as arbitrary values of the rest-mass, the electric charge, and the magnetic moments of the particles.

§4. Quantization of the Free Fields. Connection Between Spin and Statistics.

The impossibility of defining physically acceptable particle density for integral spin and energy density for half-integral spin, in the classical c-number theory, is an indication that a satisfactory interpretation of the theory within the limits of the one-body problem is not possible. The fields must, therefore, be second-quantized. We do not apply the canonical formalism, in which time is distinguished from space, and which is only suitable if there are no supplementary conditions between the canonical variables [3]. Instead, we apply a generalization of this method due to Jordan and Pauli for the electromagnetic field [4]. This method is especially convenient in the absence of interaction, where all fields U satisfy the wave-equation

$$(\Box - m^2)U = 0. \tag{10}$$

An important tool for second quantization is the invariant D-function, which satisfies the above wave-equation,

$$D(\vec{x}, x_0) = \int \frac{d^3p}{(2\pi)^3} e^{i\vec{p}\cdot\vec{x}} \frac{sin\omega x_0}{\omega}, \tag{11}$$

with

$$\omega \equiv p_0 = +(\vec{p}^2 + m^2)^{\frac{1}{2}}. \tag{12}$$

The D-function is uniquely determined by the conditions:

$$(\Box - m^2)D = 0; \quad D(\vec{x}, 0) = 0; \quad \partial_0 D|_{(x_0=0)} = \delta(\vec{x}). \tag{13}$$

For $m = 0$ we have

$$D(\vec{x}, x_0) = [\delta(r - x_0) - \delta(r + x_0)]/4\pi r, \tag{14}$$

with $r = |\vec{x}|$. This expression also determines the light-cone singularity of D for $m \neq 0$, but in this case D is no longer zero inside the light-cone. One finds

$$D(\vec{x}, x_0) = -\frac{1}{4\pi r} \partial_r F(r, x_0)$$

with

$$
\begin{aligned}
F(r, x_0) &= +J_0[m(x_0^2 - r^2)^{\frac{1}{2}}], & x_0 > r \\
&= 0, & r > x_0 > -r \\
&= -J_0[m(x_0^2 - r^2)^{\frac{1}{2}}], & -r > x_0.
\end{aligned} \tag{15}
$$

For the following discussion it will be of decisive importance that D vanish outside the light-cone (ie. for $r^2 > x_0{}^2$).

The form of the factor d^3p/ω is determined by the requirement of Lorentz invariance. For this reason, in addition to D, there exists only one other invariant function which satisfies the wave-equation (10), namely

$$D_1(\vec{x}, x_0) = \int \frac{d^3p}{(2\pi)^3} e^{i\vec{p}\cdot\vec{x}} \frac{cos\omega x_0}{\omega}. \tag{16}$$

D_1 is given by

$$D_1(\vec{x}, x_0) = \frac{1}{4\pi r} \partial_r F_1(r, x_0) \tag{17}$$

with

$$
\begin{aligned}
F_1(r, x_0) &= N_0[m(x_0{}^2 - r^2)^{\frac{1}{2}}], & x_0 > r \\
&= -iH_0^1[im(r^2 - x_0{}^2)^{\frac{1}{2}}], & r > x_0 > -r \\
&= N_0[m(x_0{}^2 - r^2)^{\frac{1}{2}}], & -r > x_0,
\end{aligned} \tag{18}
$$

expressed in terms of Neumann and Hankel functions.

We shall postulate *that all physical quantities commute for spacelike separations* (this postulate is satisfied for the canonical quantization formalism but is much more general). It follows that the bracket expressions of all quantities which satisfy the free-field equation (9) can be expressed in terms of the D-function and (a finite number of) derivatives of it, without using the D_1 function. This is also true for brackets with the $+$ sign $\cdots$.

We can draw further conclusions about the number of derivatives of the D-function which can occur in the bracket expressions, if we take into account the invariance of the theories under the transformations of the restricted Lorentz group and if we use the results of the preceding section on the class division of the tensors. We assume the quantities $U^{(r)}$ to be ordered in such a way that each field component is composed only of quantities of the same class. We consider especially the bracket expression of a field component $U^{(r)}$ with its own complex conjugate

$$[U^{(r)}(x'), U^{*(r)}(x'')] \sim D[\vec{x}' - \vec{x}'', x_0' - x_0''].$$

We distinguish now the two cases of half-integral and integral spin. In the half-integral spin case, this bracket transforms according to (8) under Lorentz transformations as a tensor of odd rank. In the integral spin case, the bracket transforms as a tensor of even rank. Therefore we must have, for half-integral spin

$$a) \quad [U^{(r)}(x'), U^{*(r)}(x'')] = \quad odd \quad derivatives \quad of \quad D(x' - x''), \tag{19}$$

and similarly for integral spin

$$b) \quad [U^{(r)}(x'), U^{(r)}(x'')] = \quad even \quad derivatives \quad of \quad D(x' - x''). \tag{19}$$

$\cdots$ Consider now the bracket expression symmetrized in the two points

$$X = [U^{(r)}(x'), U^{*(r)}(x'')] + \quad (x' \leftrightarrow x''). \tag{20}$$

365

Because the D-function is even in the space coordinates and odd in time coordinate, which is clear from Eqns.11,15, it follows that X must have an even number of spacelike derivatives and an odd number of timelike derivatives of D. This is fully consistent with the result (19a) for half-integral spin, but inconsistent with (19b) for integral spin unless X vanishes. We have therefore for integral spin

$$[U^{(r)}(x'), U^{*(r)}(x'')] + [U^{(r)}(x''), U^{*(r)}(x')] = 0. \tag{21}$$

So far we have not distinguished between the two cases of Bose-Einstein and Fermi-Dirac statistics. In the Bose-Einstein case one has the commutator bracket with the minus sign; in the Fermi-Dirac case, the anticommutator bracket with the plus sign. *By inserting the brackets with the + sign into (21) we have an algebraic contradiction,* because the left-hand side is essentially positive for $x' = x''$ and cannot vanish unless both $U^{(r)}$ and $U^{*(r)}$ vanish [4].

Therefore we come to the result: *For integral spin, quantization according to the exclusion principle is not possible. For this result it is essential that use of the D_1-function instead of the D-function is ruled out; that is, that physical quantities commute at spacelike separations.*

On the other hand, it is formally possible to quantize the theory for half-integral spins according to Bose-Einstein statistics, *but according to the general result of the preceding section the energy of the system would not be positive.* Because it is necessary for physical reasons to postulate this, we must quantize half-integral spin according to Fermi-Dirac statistics.

For the positive proof that a theory with a positive total energy is possible by quantizing integral spin according to Bose-Einstein statistics, and half-integral spin according to Fermi-Dirac statistics, we refer to the paper by Fierz $\cdots$.

In conclusion we wish to state, that according to our opinion the connection between spin and statistics is one of the most important applications of relativity theory.

Footnotes and References.

1. B.L. van der Waerden, *Die Gruppentheoretische Methode in der Quantentheorie* (Berlin, 1932).

366

2. M. Fierz, Helv. Phys. Acta **12**, 3 (1939); also L. de Broglie, Comptes Rendus **208**, 1697 (1939); **209**, 265 (1939).

3. On account of the existence of such conditions the canonical formalism is not applicable for spin >1 and therefore the discussion about the connection between spin and statistics of J.S. de Wet, Phys. Rev. **57**, 646 (1940), which is based on that formalism, is not general enough.

4. This contradiction may be seen also by resolving $U^{(r)}$ into plane-wave states according to

$$U^{(r)} = \sum_p [U_+(p)e^{ip\cdot x} + U_-^*(p)e^{-ip\cdot x}]$$

and its Hermitian conjugate. Recall $p\cdot x = \vec{p}\cdot\vec{x} - \omega t$. The equation (21) leads to the relation

$$[U_+^*(p), U_+(p)] + [U_-(p), U_-^*(p)] = 0,$$

a relation which is not possible for anticommutator brackets unless $U_\pm(p)$ and $U_\pm^*(p)$ vanish.

Chapter 15

Feynman's Proof and Pauli's Criticism

Summary: In the very first paper introducing his novel contributions to relativistic quantum field theory, Feynman shows that the one-loop contribution to the vacuum-to-vacuum amplitude, calculated according to the new Feynman Rules, gives a vacuum survival probability greater than unity when calculated with the "wrong" spin-statistics relation. This impossible result is interpreted by Feynman as proving the necessity of the correct spin-statistics relation.

Pauli interprets the result as due to an indefinite metric in the Hilbert space of states when the "wrong" statistics is used, in violation of one of the basic postulates of his original proof.

§1. Introduction.

The second generation of contributions to the Spin-Statistics Theorem opened with an exchange between Feynman and Pauli, precipitated by a brief, tantalizing and marvelously original observation by Feynman in the first of his papers on quantum electrodynamics [15.1].

Using his propagator technique, Feynman calculates the vacuum to vacuum amplitude C_v, which differs from unity because of the possibility of virtual pair creation and annihilation, and more complicated processes of higher order. He defines a one-loop contribution $-L$ *(with an "extra" minus sign appropriate for Fermi-Dirac statistics)* which in lowest order is

$$L^{(1)} = -\frac{e^2}{2} \int d^4 x_1 d^4 x_2 Tr[\mathbf{A}(2) K_+(2,1)\mathbf{A}(1)K_+(1,2)]. \qquad (1)$$

He explains this result as

1) a minus sign from the perturbation $(-ie\mathbf{A})^2$ in second-order.

2) a factor of $\frac{1}{2}$ because we are counting twice pairs created at 1, annihilated at 2 and vice versa.

3) a sum on all indices for an overall Trace. Feynman's conventions are

$$\mathbf{A} = \gamma \cdot A = \gamma_\mu A_\mu \equiv \gamma_0 A_0 - \vec{\gamma} \cdot \vec{A}, \tag{2}$$

with the Dirac matrices

$$\gamma_\mu = \gamma_0, \vec{\gamma} \equiv \beta, \beta\vec{\alpha}. \tag{3}$$

Feynman remarks that the real part of L "appears" to be positive, and he argues that the one-loop contribution simply exponentiates to the vacuum survival amplitude C_v

$$C_v = 1 - L + L^2/2! + \cdots = e^{-L}, \tag{4}$$

from which the vacuum survival probability is

$$P_v = |C_v|^2 = e^{-2Re(L)}. \tag{5}$$

Clearly, we obtain an acceptable result only if $Re(L) \geq 0$, which raises the question of the "extra" minus sign in the definition of the one-loop vacuum polarization amplitude.

Feynman concludes that Fermi-Dirac statistics are required to give a survival probability of the vacuum $P_v \leq 1$, and that the Feynman propagator formulation requires the Pauli Exclusion Principle. He also shows that the opposite occurs for spinless particles obeying the Klein-Gordon equation which therefore require Bose-Einstein statistics for consistency.

§2. Feynman's Calculation of Vacuum Polarization.

It is interesting to examine Feynman's calculation in more detail because his conclusions, while characteristically persuasive, are neither obvious nor trivial. When he says that the vacuum to vacuum amplitude $C_v = e^{-L}$ for

369

Fermi-Dirac statistics, and that the real part of L appears to be positive as required, some points deserve closer scrutiny. For example:

1) How does this extra minus sign occur in the Feynman propagator formalism, as compared to the second-quantized formalism (or to the Feynman path-integral formalism, which we will not pursue here).

2) What does he mean that the real part "appears" to be positive? It turns out that the total amplitude is seriously divergent and not gauge invariant. Its evaluation is obscured by a variety of regularization and renormalization manipulations which, although they have become standard fare for some, continue to be viewed with suspicion by many non-experts. Fortunately, the real part of L is free of such complications and can be shown to be positive by a variety of arguments including old-fashioned perturbation theory, as had already been done by Iwanenko and Sokolow [15.2] and others before them; and shown again in the present context by Pauli, in the following commentary on Feynman's paper, using the unitarity of the S-matrix.

There is also a matter of taste involved. Can the Spin-Statistics Theorem *really* depend for its proof on such an obscure course of reasoning? Or, rather than being the result, can it better be viewed as the source of the consistency of quantum electrodynamics and have its own origin at a more fundamental point? This last view is developed in the proofs of Lüders and Zumino, and of Burgoyne [15.3], which will prevail over the Feynman proof as well as one due to Schwinger [15.4], based on time-reversal invariance.

We now return to the origin of the "extra" minus sign in the calculation of one-loop amplitudes. Using the formalism of second-quantization we require

$$- L \sim \langle 0 | (\bar{\psi} \gamma_\mu \psi)_2 (\bar{\psi} \gamma_\nu \psi)_1 | 0 \rangle, \tag{6}$$

which resolves itself, for $t_2 > t_1$, into

$$- L \sim \gamma_\mu \langle 0 | T(\psi_2 \bar{\psi}_1) | 0 \rangle \gamma_\nu \langle 0 | T(\psi_1 \bar{\psi}_2 | 0 \rangle \times (-1). \tag{7}$$

Here we express the vacuum expectation values as time-ordered products of $\psi\bar{\psi}$ with a characteristic (-1) from the anticommutation of the Dirac *field operators* $\bar{\psi}_2$ and ψ_1. The vacuum expectation values are just the Feynman propagators so

$$- L \sim (-1)\gamma_\mu K_+(2,1)\gamma_\nu K_+(1,2) \tag{8}$$

with the required "extra" minus sign appearing in the well known way. The same sign is built into the Feynman path-integral formalism where the Dirac fields are number-valued (but not yet second-quantized) anticommuting *Grassmann variables*.

In the Feynman propagator formalism the matrix element is

$$- L \sim (\bar{\psi}\gamma_\mu\psi)_2(\bar{\psi}\gamma_\nu\psi)_1 \tag{9}$$

which (still for $t_2 > t_1$,) is

$$- L \sim \gamma_\mu(\psi_2\bar{\psi}_1)\gamma_\nu(\psi_1\bar{\psi}_2), \tag{10}$$

where we have freely commuted the number-valued Dirac *wave functions*. Now impose the Feynman boundary conditions so

$$\psi_2\bar{\psi}_1 \rightarrow \sum_{E>0} \psi_2(E)\bar{\psi}_1(E) \equiv K_+(2,1) \tag{11}$$

summed over positive energy states propagating forward in time, and

$$\psi_1\bar{\psi}_2 \rightarrow \sum_{E<0} \psi_1(E)\bar{\psi}_2(E) \equiv (-1) \times K_+(1,2) \tag{12}$$

summed over "negative energy states propagating backwards in time". The "extra" minus sign appears from the construction of the Feynman propagator to satisfy the requirement that negative energy states should not propagate forward in time. (It is an interesting exercise - which Feynman does - to repeat the argument for Bose-Einstein particles satisfying the Klein-Gordon equation.)

Next, what about the calculation of C_v? We want the vacuum to vacuum S-matrix element. It is most easily obtained in the interaction representation in terms of the time-ordered exponential of the interaction Hamiltonian. The Schrödinger equation for the statevector

$$i\partial_t \Psi(t) = H'(t)\Psi(t) \tag{13}$$

with the formal solution

$$\Psi(t) = \mathrm{T} \exp\left[-i \int_{-\infty}^t H'(t)dt\right] \Psi(-\infty), \tag{14}$$

from which we recognize the S matrix as

$$
\begin{aligned}
S &= \mathrm{T} \exp\left[-i \int_{-\infty}^{+\infty} H'(t)dt\right], \\
&= 1 + (-ie)\int \mathcal{H}' d^4x + \frac{(-ie)^2}{2}\mathrm{T}\int \mathcal{H}_2' d^4x_2 \int \mathcal{H}_1' d^4x_1 + \cdots \tag{15}
\end{aligned}
$$

expressed in terms of the interaction Hamiltonian density $\mathcal{H}' = \bar{\psi}\mathbf{A}\psi$ integrated over all space-time $d^4x \equiv d\tau$.

We require

$$C_v = \langle 0|S|0\rangle = 1 - \frac{e^2}{2}\langle 0|\mathrm{T}\int\int \mathcal{H}_2'\mathcal{H}_1' d\tau_2 d\tau_1|0\rangle. \tag{16}$$

With $C_v = 1 - L$, Feynman defines $L^{(1)}$, the lowest order contribution to $-L$, as

$$
\begin{aligned}
L^{(1)} &= -\frac{e^2}{2}\langle 0|\mathrm{T}\int\int \mathcal{H}_2'\mathcal{H}_1' d\tau_2 d\tau_1|0\rangle, \\
&= -\frac{e^2}{2}\int Tr\left(\gamma_\mu K_+(2,1)\gamma_\nu K_+(1,2)\right) d\tau_2 d\tau_1 A_\mu^*(2)A_\nu(1). \tag{17}
\end{aligned}
$$

This is Feynman's Eqn.28 (in our App.15A), but it does not yet include the sign change from $K_+(1,2)$. Keep in mind that $L^{(1)}$ is the lowest order contribution to (minus L). In terms of momentum space propagators, using

$$K_+(2,1) = \frac{i}{4\pi^2}\int (\mathbf{p} - m)^{-1}\exp(-ip\cdot x_{21})d^4p, \tag{18}$$

and

$$A_\nu(1) = \epsilon_\nu \exp(iq\cdot x_1), \tag{19}$$

372

we get

$$L = -L^{(1)} = -\frac{e^2}{2} \int Tr \left(\gamma_\mu \frac{1}{\mathbf{p} + \mathbf{q} - m} \gamma_\nu \frac{1}{\mathbf{p} - m} \right) d^4p. \qquad (20)$$

For the evaluation of this integral, we compare to Feynman's Eqn.32 (in App.15B) where he evaluates the characteristic vacuum polarization amplitude $J_{\mu\nu}(q^2)$ which is just $-iL$ (within a positive real factor). So the real part of L has the opposite sign to the imaginary part of J.

Feynman obtains an explicit expression for J only after subtracting out a quadratically-divergent, gauge non-invariant piece which would have the effect of giving the photon a mass. The remaining regularized integral has a logarithmically divergent *but real* part, which is absorbed into the charge renormalization. Finally, the remaining piece - which is finite and independent of the cutoff parameters - is the piece of interest. It develops an imaginary part for $q^2 > 4m^2$. The continuation from $q^2 < 4m^2$ to $q^2 > 4m^2$ is determined by giving the mass a small negative imaginary part (required to produce the Feynman boundary conditions in the Green's function integral) which ultimately is taken to zero. The imaginary part of J is indeed negative, and the real part of L is positive as required. The vacuum survival probability is less than one as a consequence of, among many other things it would seem, the spin-statistics relation.

It is ironic that Feynman - revered as the intuitive master of theoretical physics - should put forth an argument that is at least as intricate as any other, although it is more dynamically motivated than the rather static arguments with which it is compared. One might make the apology for Feynman, that he was at the time monumentally engaged on other issues. But in fact he returned to just this same argument some thirty years later, when it was couched in hand-waving explanations for a general audience.

§3. Pauli's Analysis of Feynman's Proof.

Within a year, Pauli [15.5] made a detailed analysis of Feynman's calculation, using the techniques of second-quantization and canonical field theory much in the style of Schwinger. He showed that Feynman's "wrong sign" choice which leads to vacuum survival probabilities $P_v > 1$, can be interpreted as a violation of the third of Pauli's three basic postulates of relativistically invariant quantized field theories [15.6]:

1) The vacuum is the state of lowest energy.

2) Physical quantities at space-like separated points must commute.

3) The metric of the Hilbert space of the quantum states must be positive definite, guaranteeing positive probabilities.

Pauli shows that Feynman's calculation using Bose-Einstein statistics for spin-$\frac{1}{2}$ Dirac particles is his own "abnormal" case quantized with "wrong sign" commutation relations for the antiparticle creation operators (the Fourier coefficients of the negative energy part of Dirac field ψ). The theory requires an indefinite metric with a negative sign for states with an odd number of antiparticles $N_r^{(-)}$

$$\eta = (-)^{\sum_r N_r^{(-)}}. \tag{21}$$

In this theory, normally Hermitian operators satisfying $H^\dagger = H$, are required to be self-adjoint according to

$$\hat{H} \equiv \eta H^\dagger \eta = H \tag{22}$$

leading to the conservation of a probability defined as

$$\sum_{nm} \Psi_n^* \eta_{nm} \Psi_m = \sum_{N^{(-)}} (-)^{N^{(-)}} \sum_\alpha |\Psi_{N^{(-)},\alpha}|^2. \tag{23}$$

With these (unphysical) requirements Pauli obtains a positive energy

$$E = \sum \hbar\omega(N^{(+)} + N^{(-)} + 1) \tag{24}$$

where $N^{(+)}, N^{(-)} = 0, 1, 2, \cdots$ Similarly the charge is

$$Q = e \sum (N^{(+)} - N^{(-)}). \tag{25}$$

The pathology involved is the indefinite metric leading to the appearance of negative probability states. Pauli introduces a time-evolution operator which satisfies a generalized unitarity

$$\hat{U}U \equiv (\eta U^\dagger \eta)U = 1. \tag{26}$$

In the calculation of the survival probability of the vacuum in an external field A_μ one obtains

$$
\begin{aligned}
P_v &= |\langle 0|U|0\rangle|^2 = |\langle 0|T\exp\left(-i\int_{-\infty}^{+\infty} j\cdot A(\tau)d\tau\right)|0\rangle|^2 \\
&\simeq 1 - \frac{1}{2}\int d\tau d\tau' \langle 0|j\cdot A(\tau)|1,\bar{1}\rangle\eta\langle 1,\bar{1}|j\cdot A(\tau')|0\rangle \\
&\simeq 1 + \frac{1}{2}|\int d\tau\langle 0|j\cdot A(\tau)|1,\bar{1}\rangle|^2 > 1.
\end{aligned}
\tag{27}
$$

This agrees with Feynman's result that the wrong statistics gives a probability greater than one. The mechanism in the second-quantized formalism is identified as a violation of the requirement of a positive definite metric for the Hilbert space of quantum states, with the resulting negative probabilities. The negative probability necessarily ascribed to the $(1,\bar{1})$ states in order to satisfy the first postulate $(E \geq 0)$ (in spite of the "wrong statistics" commutation of the antiparticle operators) combines with the excess probability of the vacuum $(P_v > 1)$ to conserve the total probability at unity.

In this paper, Pauli [15.7] has refined an earlier discussion prompted by Dirac's suggestion of indefinite metric theories [15.8]. He makes clear that the indefinite metric device can be used to make positive energy, wrong statistics quantum field theories but at the price of negative probabilities. This device has in fact proven useful and makes its reappearance some twenty years later in the Faddeev-Popov "ghost" mechanism [15.9] for maintaining gauge invariance in non-Abelian gauge theories. Here the physical states are required to be ghost-free so Pauli's set of postulates remains valid and the ghosts are a mathematical artifice.

§4. Concluding Remarks.

Feynman returned to this argument in his 1986 Dirac Lecture "The Reason for Antiparticles" [15.10] where he resumed his search for an intuitive proof of the Spin-Statistics Theorem. He resurrects the argument presented here, apparently undeterred by Pauli's criticism which had shown that the "wrong statistics" implies an indefinite metric for a positive energy theory. He surely did not believe in either indefinite metrics or negative energy states. He also chose to ignore the - by that time - intervening proofs of Lüders and Zumino, and of Burgoyne, presumably in search of more easily understood alternatives. We postpone discussion of that lecture where we hope to do more justice to the charismatic if not compelling nature of Feynman's arguments.

§5. Biographical Note on Feynman.

Richard Phillips Feynman (1919-1988) because of his idiosyncratic but extremely attractive persona, and of course the sine qua non of his immense scientific talent and accomplishments, was a charismatic figure who became a cult idol inspiring popular biographies and even a movie. Some of the biographies are even scientifically informative. We have little to add to this part of the story.

For us the measure of the man resides in just three great accomplishments. The Feynman Green's function, the Feynman diagrams and, quite separately, the Feynman Path Integral formulation of quantum mechanics. Each of these ideas had a precursor in the work of others (notably Stückelberg, Bopp, and Dirac); each showed us how to solve problems already solved by others (most notably by Schwinger); but what Feynman did, was to make it possible for *everyone* to understand the great game at levels from the most qualitative to the most detailed. In the Path Integral formulation of quantum mechanics he created a necessary generalization of canonical field theory and later (in a bizarre incident at 2AM in a beer hall in Warsaw, pure Feynman) told us how to use it.

Feynman had an edge to his personality not often mentioned.

At a particularly boring theoretical seminar, he was soon almost supine in his front row seat. "Why do you bother with such crap?" The speaker - an old friend from Cornell days - was equal to the task, and replied without a quaver "Well, Dick, it's little steps for little feet." The man was well over six feet tall and his shoes had to be size fourteen.

At a colloquium, the speaker reported a delicate experiment measuring the force on an airfoil in a flowing superfluid which disagreed with Feynman's theory. He had the temerity to believe his own experiment and to suggest an alternative theory. Feynman stood up in the front row and raged at the man as an incompetent. A colleague sitting next to him pulled him back into his seat and calmed him with "Now, now, Dick, everyone has the right to think."

Feynman was not a good teacher. His lectures were wonderful but ordinary people could not operate in the same way. The undergraduate course which produced the three volume set of lectures was soon abandoned as a catastrophe for the freshman class. The books are still wonderful, but as leisure reading for the faculty.

Feynman did not produce many PhD students. His way of doing science seemed like someone creating a masterpiece from an empty canvas. Witness the derivation of the relativistic Green's function for the electron, or the evaluation of the path integral. Not for Feynman the systematic cranking of a machinery. His style left graduate students awe-struck and empty handed.

Far from being a carefree character just playing bongo drums or hanging out in bars, Feynman was very responsible. The wonderful lectures were meticulously prepared and rehearsed. He closed his office door for an hour to collect his thoughts even before a freshman lecture. Some of his brilliance was the result of a lot of deep thinking. He took a serious shot at every fundamental problem.

Feynman was not only a great man but he was also a great guy.

Bibliography and References.

15.1) R.P. Feynman, Phys. Rev. **76**, 749 (1949); see our App.15A,B.

15.2) D. Iwanenko and A. Socolow, Phys. Zeits. d. Sowejetunion **11**, 590 (1937); see our App.10B.

15.3) G. Lüders and B. Zumino, Phys. Rev. **110**, 1450 (1958); N. Burgoyne, Nuov. Cim. **8**, 607 (1958); see our App.17A,B.

15.4) J. Schwinger, Phys. Rev. **82**, 914 (1951); see our App.16A.

15.5) W. Pauli, Prog. Theor. Phys. **5**, 526 (1950); see our App.15C.

15.6) W. Pauli, Rev. Mod. Phys. **13**, 203 (1941).

15.7) W. Pauli, Rev. Mod. Phys. **15**, 145 (1943).

15.8) P.A.M. Dirac, Proc. Roy. Soc. **A180**, 1 (1942).

15.9) M.Kaku, *Quantum Field Theory* (Oxford, New York, 1993), pp.301-304.

15.10) R.P. Feynman, "The Reasons for Antiparticles," in *Elementary Particles and the Laws of Physics. The 1986 Dirac Memorial Lectures* (Cambridge, New York, 1987), pp.56-59.

APPENDIX 15.A:

Excerpt from: Physical Review 76, 749 (1949)

The Theory of Positrons

R. P. FEYNMAN

Department of Physics, Cornell University, Ithaca, New York

(Received April 8, 1949)

The problem of the behavior of positrons and electrons in given external potentials, neglecting their mutual interactions, is analyzed by replacing the theory of holes by a reinterpretation of the solutions of the Dirac equation. It is possible to write down a complete solution of the problem in terms of boundary conditions on the wave function, and this solution contains automatically all the possibilities of virtual (and real) pair formation and annihilation together with the ordinary scattering processes, *including the correct relative signs of the various terms.*

In this solution, the "negative energy states" appear in a form which may be pictured (as by Stückleberg) in space-time as waves travelling away from the external potential backwards in time. Experimentally, such a wave corresponds to a positron approaching the potential and annihilating the electron. A particle moving forward in time (electron) in a potential may be scattered forward in time (ordinary scattering) or backward (pair annihilation). When moving backward (positron) it may be scattered backward in time (positron scattering) or forward (pair production). For such a particle the amplitude for transition from an initial to a final state is analyzed to any order in the potential by considering it to undergo a sequence of such scatterings.

The amplitude for a process involving many such particles is the product of the transition amplitudes for each particle. *The Exclusion Principle requires that antisymmetric combinations of amplitudes be chosen for those complete processes which differ only by exchange of particles. It seems that a consistent interpretation is only possible if the Exclusion Principle is adopted.* The Exclusion Principle need not be taken into account in intermediate states. Vacuum problems do not arise for charges which do not interact with one another, but these are analyzed nevertheless in anticipation of application to quantum electrodynamics.

The results are also expressed in momentum-energy variables. Equivalence to the second quantization theory is proved in an appendix. (Note: *Emphases added.*)

1. INTRODUCTION

This is the first of a set of papers $\cdots\cdots$

$\cdots$ we deal $\cdots$ with the Dirac equation and show how the solutions may be interpreted to apply to positrons. The interpretation seems *not to be consistent unless the electrons obey the Exclusion Principle. (Emphasis added)* (Charges obeying the Klein-Gordon equations can be described in an analogous manner, but here consistency requires Bose-Einstein statistics.) $\cdots\cdots$.

2. GREEN'S FUNCTION TREATMENT OF SCHRÖDINGER'S EQUATION

$\cdots$ The Schrödinger equation

$$i\partial\psi/\partial t = H\psi, \tag{1}$$

$\cdots$ can always be written as

$$\psi(\mathbf{x}_2, t_2) = \int K(\mathbf{x}_2, t_2; \mathbf{x}_1, t_1)\psi(\mathbf{x}_1, t_1)d^3\mathbf{x}_1, \tag{2}$$

where K is the Green's function for the linear Eqn.1. (We have limited ourselves to a single particle of coordinate $\mathbf{x}$, but the equations are obviously of greater generality.) If H is a constant operator having eigenvalues E_n, eigenfunctions ϕ_n so that $\psi(\mathbf{x}, t_1)$ can be expanded as $\sum_n C_n\phi_n(\mathbf{x})\cdots$, one finds $\cdots$

$$K(2,1) = \sum_n \phi_n(\mathbf{x}_2)\phi_n^*(\mathbf{x}_1)\exp(-iE_n(t_2 - t_1)), \tag{3}$$

for $t_2 > t_1$. We shall find it convenient for $t_2 < t_1$ to define $K(2,1) = 0\cdots$ in general K can be defined by that solution of

$$(i\partial/\partial t_2 - H_2)K(2,1) = i\delta(2,1), \tag{4}$$

which is zero for $t_2 < t_1,\cdots$. When H is not constant, (2) and (4) are valid but K is less easy to evaluate than $(3)\cdots$.

3. TREATMENT OF THE DIRAC EQUATION

We shall now extend the method of the last section to apply to the Dirac equation. All that would seem to be necessary in the previous equations is to consider H as the Dirac Hamiltonian, ψ as a symbol with four indices (for each particle) $\cdots$. $\cdots$ we shall define a convenient relativistic notation $\cdots$. Call

$$\nabla \equiv \gamma_\mu \partial_\mu = \beta\partial_t + \beta\alpha\cdot\nabla.$$

$\cdots \phi_n^*$ in (3) is replaced by its adjoint $\bar{\phi}_n = \phi_n^* \beta$.

Thus the Dirac equation for a particle, mass m, in an external field $\mathbf{A} \equiv A_\mu \gamma_\mu$ is

$$(i\nabla - m)\psi = \mathbf{A}\psi, \tag{11}$$

and Eqn.4 determining the propagation of a free particle becomes

$$(i\nabla_2 - m)K_+(2,1) = i\delta(2,1), \tag{12}$$

$\cdots\cdots$ If a potential $\mathbf{A}$ is acting a similar function, say $K_+^{(A)}(2,1)$ can be defined. It differs from $K_+(2,1)$ by a first order correction $\cdots$

$$K_+^{(1)}(2,1) = -i \int d\tau_3 K_+(2,3)\mathbf{A}(3)K_+(3,1), \tag{13}$$

(where $d\tau_3 \equiv d^3 x_3 dt_3$) representing the amplitude to go from 1 to 3 as a free particle, get scattered there by the potential $\mathbf{A}$ and continue to 2 as free. The second order correction $\cdots$ is

$$K_+^{(2)}(2,1) = - \int d\tau_4 d\tau_3 K_+(2,4)\mathbf{A}(4)K_+(4,3)\mathbf{A}(3)K_+(3,1), \tag{14}$$

and so on. $\cdots$

We would now expect to choose, for the special solution of (12), $K_+ = K_0$ where $K_0(2,1)$ vanishes for $t_2 < t_1$ and for $t_2 > t_1$ is given by (3) where ϕ_n and E_n are the eigenfunctions and energy values of a particle satisfying Dirac's equation, and ϕ_n^* is replaced by $\bar{\phi}_n$.

The formulas arising from this choice, however, suffer from the drawback that they apply to the one electron theory of Dirac rather than to the hole theory of the positron. For example, consider $\cdots$ an electron after being scattered by a potential in a small region 3 of space-time. The one electron theory says (as does (3) with $K_+ = K_0$) that the scattered amplitude will proceed toward positive times with both positive and negative energies, that is with both positive and negative rates of change of phase. No wave is scattered to times previous to the time of scattering. These are just the properties of $K_0(2,3)$.

On the other hand, according to the positron theory negative energy states are not available to the electron after the scattering. But there are other solutions of (12). We shall chose the solution defining $K_+(2,1)$ so that $K_+(2,1)$ *for $t_2 > t_1$ is the sum of (3) over positive states only*. Now this new solution must satisfy (12) for all times in order that the representation be complete. It must therefore differ from the old solution K_0 by a solution of the homogeneous Dirac equation. It is clear from the definition that the difference $K_0 - K_+$ is the sum of (3) over all negative energy states, as long as $t_2 > t_1$.

381

But this difference must be a solution of the homogeneous Dirac equation for all times and must therefore be represented by the same sum over negative energy states also for $t_2 < t_1$. Since $K_0 = 0$ in this case, it follows that our new kernel, $K_+(2, 1)$, for $t_2 < t_1$ is the negative of the sum (3) over negative energy states. That is,

$$
\begin{aligned}
K_+(2, 1) &= \sum_{E_n > 0} \phi_n(2) \bar{\phi}_n(1) \exp(-i E_n(t_2 - t_1)) \quad for \quad t_2 > t_1 \\
&= -\sum_{E_n < 0} \phi_n(2) \bar{\phi}_n(1) \exp(-i E_n(t_2 - t_1)) \quad for \quad t_2 < t_1.
\end{aligned}
\tag{17}
$$

With this choice of K_+ our equations such as (13) and (14) will now give results equivalent to those of positron theory.

(COMMENT: The negative sign in front of the sum over negative energy states is the one which is crucial to the inclusion of the Pauli Exclusion Principle and the Fermi-Dirac statistics into the Feynman electron-positron propagator.)

······ The fact that in hole theory the hole theory proceeds in the manner of an electron of negative energy is reflected in the fact that $K_+(4, 3)$ for $t_4 < t_3$ is (minus) the sum of only negative energy components. In hole theory the real energy of these intermediate states is, of course, positive. This is true here too, since in the phases $\exp(-i E_n(t_4 - t_3))$ defining $K_+(4, 3)$ in (17), E_n is negative but so is $t_4 - t_3$. That is, the contributions vary with t_3 as $\exp(-i|E_n|(t_3 - t_4))$ as they would if the energy of the intermediate state were $|E_n|$. The fact that the entire sum is taken as negative in computing $K_+(4, 3)$ is reflected in the fact that in hole theory the amplitude has its sign reversed in accordance with the Pauli principle and the fact that the electron arriving at 2 has been exchanged with one in the sea.······

4. PROBLEMS INVOLVING SEVERAL CHARGES

··· No account need be taken of the Exclusion Principle in intermediate states. *(COMMENT: For the rather lengthy discussion of the cancellations involved, we must refer to the original literature.)*······

5. VACUUM PROBLEMS

···C_v, the vacuum to vacuum amplitude ··· is ··· the absolute amplitude that there be no particles both initially and finally. We can assume $C_v = 1$ if no potential is present during the interval, and otherwise we compute it as follows. It differs from unity because, for example, a pair could be created which eventually annihilates itself again. Such a path would appear as a closed loop on a space-time diagram. The sum of the amplitudes resulting from all such single closed loops we call L. To a first approximation L is

$$L^{(1)} = -\frac{1}{2} \int d\tau_1 \int d\tau_2 Tr[K_+(2,1)\mathbf{A}(1)K_+(1,2)\mathbf{A}(2)]. \tag{28}$$

For a pair could be created say at 1, the electron and positron could both go on to 2 and there annihilate. The trace Tr is taken since one has to sum over all possible spins of the pair. The factor $\frac{1}{2}$ arises from the fact that the same loop could be considered as starting at either potential, and the minus sign results since the interactions are each $-i\mathbf{A}$. $\cdots$ The sum of all such terms gives L [10].

In addition to these single loops we have the possibility that two independent pairs may be created and each pair may annihilate itself again. $\cdots\cdots$ The total contribution from all such pairs of loops (it is still consistent to disregard the Exclusion Principle for these virtual states) is $L^2/2$ for in L^2 we count every pair of loops twice. The total vacuum-vacuum amplitude is then

$$C_v = 1 - L + L^2/2 - L^3/6 + \cdots = e^{-L}, \tag{30}$$

the successive terms representing the amplitude from zero, one, two, etc., loops. The fact that the contribution to C_v of single loops is $-L$ is a consequence of the Pauli principle. For example, consider a situation in which two pairs of particles are created. Then these pairs later destroy themselves so that we have two loops. The electrons could, at a given time, be interchanged forming a kind of figure eight which is a single loop. The fact that the interchange must change the sign of the contribution requires that the terms in C_v appear with alternate signs. $\cdots$ the probability that a vacuum remain a vacuum is given by

$$P_v = |C_v|^2 = e^{-2Re(L)}$$

from (30) $\cdots$. The real part of L appears to be positive as a consequence of the Dirac equation and properties of K_+ so that P_v is less than one. Bose statistics gives $C_v = e^{+L}$ and consequently a value of P_v greater than unity which appears to be meaningless if the quantities are interpreted as we have done here. Our choice of K_+ apparently requires the Exclusion Principle.

Charges obeying the Klein-Gordon equation $\cdots$. The real part of L comes out negative $\cdots$ so that in this case Bose statistics appear to be required. $\cdots\cdots$

APPENDIX

a. Deduction from Second Quantization

(Derivation of some key results using second-quantization.)

b. Analysis of the Vacuum Problem

(Proof of the exponentiation of the series $C_v = 1 - L + L^2/2 + \cdots = e^{-L}.$)

References and Footnotes.

1) R.P. Feynman, Rev. Mod. Phys. **20**, 367 (1948).

2) The equivalence of the entire procedure (including photon interactions) with the work of Schwinger and Tomonaga has been demonstrated by F.J. Dyson, Phys. Rev. **75**, 486 (1949).

3) These are special examples of the general relation of spin and statistics deduced by W. Pauli, Phys. Rev. **58**, 716 (1940).

4) $\cdots$ 13) *(These have no direct bearing here.)*

APPENDIX 15.B:

Excerpt from: Physical Review 76, 769 (1949)

Space-Time Approach to Quantum Electrodynamics

R. P. FEYNMAN

Department of Physics, Cornell University, Ithaca, New York

(Received May 9, 1949)

In this paper two things are done. (1) It is shown that a considerable simplification can be attained in writing down matrix elements for complex processes in electrodynamics. Further, a physical point of view is available which permits them to be written down directly for any specific problem. Being simply a restatement of conventional electrodynamics, however, the matrix elements diverge for complex processes. (2) Electrodynamics is modified by altering the interaction of electrons at short distances. All matrix elements are now finite, *with the exception of those relating to problems of vacuum polarization. The latter are evaluated in a manner suggested by Pauli and Bethe, which gives finite results for these matrices also.* (Note: *Emphasis added.*) $\cdots\cdots$

7. THE PROBLEM OF VACUUM POLARIZATION

In the analysis of the radiative corrections to scattering one type of term was not considered. The potential which we can assume to vary as $a_\mu \exp(-iq\cdot x)$ creates a pair of electrons $\cdots$, momenta $p_a, -p_b$. This pair then reannihilates, emitting a quantum $q = p_b - p_a$, which quantum scatters the original electron from state 1 to state 2. The matrix element for this process (and the others which can be obtained by rearranging the order in time of various events) is

$$- (e^2/\pi i)\bar{u}_2 \gamma_\mu u_1 \int Tr[(p_a + q - m)^{-1}\gamma_\nu (p_a - m)^{-1}\gamma_\mu]d^4 p_a q^{-2} C(q^2)a_\nu. \tag{30}$$

This is because the potential produces the pair with amplitude proportional to $a_\nu \gamma_\nu$, the electrons of momenta p_a and $-(p_a + q)$ proceed from there to annihilate, producing a quantum (factor γ_μ) which propagates (factor $q^{-2}C(q^2)$) over to the other electron, by which it is absorbed (matrix element of γ_μ between states 1 and 2 of the original electron $(\bar{u}_2\gamma_\mu u_1)$). All momenta p_a and spin states of the virtual electron are admitted, which means the trace and the integral on $d^4 p_a$ are calculated.

One can imagine that the closed loop path of the positron-electron produces a current

$$4\pi j_\mu = J_{\mu\nu}a_\nu, \tag{31}$$

385

which is the source of the quanta which act on the second electron. The quantity

$$J_{\mu\nu} = -(e^2/\pi i) \int Tr[(\mathbf{p} + \mathbf{q} - m)^{-1} \gamma_\nu (\mathbf{p} - m)^{-1} \gamma_\mu] d^4 p, \qquad (32)$$

is then characteristic for this problem of polarization of the vacuum.

One sees at once that $J_{\mu\nu}$ diverges badly. The modification of $\cdots$ the amplitude with which the current j_μ will affect the scattering electron $\cdots$ can do nothing to prevent the divergence of the integral (32) and its effects. $\cdots\cdots$

A method of making (32) convergent without spoiling gauge invariance has been found by Bethe and by Pauli. The convergence factor for light can be looked upon as the result of superposition of the effects of quanta of various masses (some contributing negatively). Likewise if we take the factor

$$C(\mathbf{p}^2 - m^2) = -\lambda^2 (\mathbf{p}^2 - m^2 - \lambda^2)^{-1}$$

so that

$$(\mathbf{p}^2 - m^2)^{-1} C(\mathbf{p}^2 - m^2) = (\mathbf{p}^2 - m^2)^{-1} - (\mathbf{p}^2 - m^2 - \lambda^2)^{-1}$$

we are taking the difference of the result for electrons of mass m and mass $(\lambda^2 + m^2)^{\frac{1}{2}}$. But we have taken this difference for *each* propagation between interactions with photons. They suggest instead that once created with a certain mass the electron should continue to propagate with this mass through all the potential interactions until it closes its loop. That is if the quantity (32), integrated over some finite range of $\mathbf{p}$, is called $J_{\mu\nu}(m^2)$ and the corresponding quantity over the same range of $\mathbf{p}$, but with m replaced by $(\lambda^2 + m^2)^{\frac{1}{2}}$ is $J_{\mu\nu}(m^2 + \lambda^2)$ we should calculate

$$J_{\mu\nu}^P = \int_0^\infty [J_{\mu\nu}(m^2) - J_{\mu\nu}(m^2 + \lambda^2)] G(\lambda) d\lambda,$$

the function $G(\lambda)$ satisfying

$$\int_0^\infty G(\lambda) d\lambda = 1, \qquad \int_0^\infty G(\lambda) \lambda^2 d\lambda = 0.$$

Then in the expression for $J_{\mu\nu}^P$ the range of $\mathbf{p}$ can be extended to infinity as the integral now converges. The result of the integration using this method is the integral on $d\lambda$ over $G(\lambda)$ of (see Appendix C *(Comment: Not included.)*)

$$J_{\mu\nu}^P = -\frac{e^2}{\pi}(q_\mu q_\nu - \delta_{\mu\nu}\mathbf{q}^2)\left(-\frac{1}{3}\ln\frac{\lambda^2}{m^2} - \left[\frac{4m^2 + 2\mathbf{q}^2}{3\mathbf{q}^2}\left(1 - \frac{\theta}{\tan\theta}\right) - \frac{1}{9}\right]\right), \qquad (33)$$

with $\mathbf{q}^2 = 4m^2 \sin^2\theta$.

The gauge invariance is clear, since $q_\mu(q_\mu q_\nu - \mathbf{q}^2 \delta_{\mu\nu}) = 0$. Operating (as it always will) on a potential of zero divergence the $(q_\mu q_\nu - \delta_{\mu\nu}\mathbf{q}^2)a_\nu$ is simply $-q^2 a_\mu$, the

D'Alembertian of the potential, that is, the current producing the potential. The term $-\frac{1}{3}(\ln(\lambda^2/m^2))(q_\mu q_\nu - \mathbf{q}^2 \delta_{\mu\nu})$ therefore gives a current proportional to the current producing the potential. This would have the same effect as a change in charge, so that we would have a difference $\Delta(e^2)$ between e^2 and the experimentally observed charge, $e^2 + \Delta(e^2)$, analogous to the difference between m and the observed mass. This charge depends logarithmically on the cutoff, $\Delta(e^2)/e^2 = -(2e^2/3\pi)\ln(\lambda/m)$. After this renormalization of charge is made, no effects will be sensitive to the cutoff.

After this is done the final term remaining in (33), contains the usual effects [21] of polarization of the vacuum. It is zero for a free light quantum ($\mathbf{q}^2 = 0$). For small $\mathbf{q}^2$ it behaves as $(2/15)\mathbf{q}^2$ (adding $-\frac{1}{5}$ to the logarithm in the Lamb effect). For $\mathbf{q}^2 > (2m)^2$ it is complex, the imaginary part representing the loss in amplitude required by the fact that the probability that no quanta are produced by a potential able to produce pairs $((\mathbf{q}^2)^{\frac{1}{2}} > 2m)$ decreases with time. (To make the necessary analytic continuation, imagine m to have a small negative imaginary part, so that $(1 - \mathbf{q}^2/4m^2)^{\frac{1}{2}}$ becomes $-i(\mathbf{q}^2/4m^2 - 1)^{\frac{1}{2}}$ as $\mathbf{q}^2$ goes from below to above $4m^2$. Then $\theta = \pi/2 + iu$ where $\sinh(u) = +(\mathbf{q}^2/4m^2 - 1)^{\frac{1}{2}}$, and $-i/\tan\theta = i\tanh(u) = +i(\mathbf{q}^2 - 4m^2)^{\frac{1}{2}}(\mathbf{q}^2)^{-\frac{1}{2}}$.)

$\cdots\cdots$ Once the simple problem of a single closed loop is solved there are no further divergence difficulties for more complex processes [22]. $\cdots\cdots$

References and Footnotes.

$\cdots\cdots$

21) E.A. Uehling, Phys. Rev. **48**, 55 (1935); R. Serber, Phys. Rev. **48**, 49 (1935).

22) There are loops completely without external interactions. For example, a pair is created virtually, along with a photon. Next they annihilate, absorbing this photon. Such loops are disregarded on the grounds that they do not interact with anything and are thereby completely unobservable. Any indirect effects they may have via the Exclusion Principle have already been included.

$\cdots\cdots$

APPENDIX 15.C:

Excerpt from: Progress of Theoretical Physics, Vol.5, 526 (1950)

On the Connection between Spin and Statistics

W. PAULI

Swiss Federal Institute of Technology, Zurich.

(Received June 12, 1950)

§1. Introduction and Summary.

In relativistically invariant quantized field theories the following conditions are fulfilled in the normal cases of half-integral spin connected with the Exclusion Principle (fermions) and of integral spin connected with symmetrical statistics (bosons).

1) The vacuum is the state of lowest energy. $\cdots$

2) Physical quantities $\cdots$ commute with each other for two space-time points with a space-like separation. $\cdots$

3) The metric of the Hilbert-space of the quantum mechanical states is positive definite. This guarantees the positive sign of the values of physical probabilities.

There seems to be agreement now about the necessity of all three postulates in physical theories. In earlier investigations [1] I have shown that in the abnormal cases of half-integral spin with Bose-Einstein statistics and of integral spin with Fermi-Dirac statistics, which do not occur in nature, not all of the three postulates can be fulfilled in a relativistically invariant quantized field theory. $\cdots$ for the abnormal cases, postulate (1) was violated for half-integral spins and postulate (2) for integral spins, while postulate (3) was always fulfilled. Meanwhile *Dirac* [2] had directed attention to the possibility of mathematical theories in which postulate (3) is abandoned in favor of $\cdots$ indefinite metrics $\cdots$. In this theory the sum of all probabilities (which is conserved) contains in general also negative terms ("negative probabilities") and the square of "self-adjoint" operators (which replace Hermitian operators of the usual theory) can also have negative expectation values.

Recently *Feynman* in his "Theory of Positrons" [3], which does not use directly the concept of field quantization but more intuitive methods (which he proves to be equivalent $\cdots$) made the important remark, that the abnormal case of spin-$\frac{1}{2}$ bosons and also of spin-0 fermions could be treated in a way similar to the normal case. Considering the effect of an external electromagnetic field (which can produce and annihilate pairs of positive and negative particles) on the initial vacuum, he derived for the probability that the vacuum is left unchanged a value larger than unity in the abnormal case - in contrast to the expected value smaller than one for the normal case. In this paper it is shown for spin-$\frac{1}{2}$ (§2 and 3) and for spin-0 (§4) that Feynman's treatment of the abnormal case is equivalent to a different mathematical formulation of the field quantization than which I earlier took into consideration: The new form of the theory for the abnormal cases *preserves the postulates (1) and (2)* (and also the covariance of the theory with respect to charge-conjugation) but violates the postulate (3) introducing "negative probabilities" for states with an odd number of negative particles present. Feynman's result of a probability larger than one for the vacuum in such theories is then immediately understandable as the excess above unity this probability has to compensate the negative probability of states, where one pair is generated. The non-physical character of these negative probabilities for the abnormal cases is also stressed by the circumstance that the vacuum expectation value of the square of the integral of a component of the current over a finite space-time region becomes negative in these cases (compare §3, Eqns.A48 and B49. In the e^2-approximation these vacuum expectation values are indeed very closely connected with the value of the deviation from unity for the probability of the original vacuum in an external field.

······*(Pauli's paper will not be reproduced further. It is an interesting and challenging exercise which he carries out, if only to foreclose one possible mathematical path not chosen by the physical world.)*······

I am indebted to Dr. R. Jost for interesting discussions on this subject during my stay as visitor at the Institute for Advanced Study in Princeton.

References and Footnotes.

1) For the spin values 1/2 and 0: Ann.de l'Inst. Poincare **6**, 137 (1936).

For higher spin values: Phys. Rev. **58**, 716 (1940); Compare also my report: Rev. Mod. Phys. **13**, 203 (1941): in the following quoted as "I".

2) P.A.M. Dirac, Proc. Roy. Soc. A**180**, 1 (1942); Compare also my report: Rev. Mod. Phys. **15**, 145 (1943), which in the following is quoted as "II".

3) R.P. Feynman, Phys. Rev. **76**, 749 (1949), see especially p.756; compare also D. Rivier, Helv. Phys. Acta **22** (1949). ······

Chapter 16

Schwinger's Proof Using Time-Reversal Invariance

Summary: Schwinger's proof is based on the postulate of invariance under strong-reflection, following Pauli's original observation. Schwinger requires an anticommutation between Dirac field operators and an interchange of initial and final states, to maintain the invariance. This logic will eventually be reversed - most explicitly by Lüders and Zumino - and the full **TCP**-invariance will be based on the the Spin-Statistics Theorem, although Schwinger persistently defended his point of view.

§1. Introduction.

Schwinger's proof [16.1] of the Spin-Statistics Theorem is based on the requirement of invariance of relativistic quantum field theory under the time-reversal transformation $\mathbf{T}_s$. His ideas owe a lot to the earlier work of Weyl (time-reversal invariance) [16.2], Belinfante [16.3] (for a similar proof based on the requirement of invariance under Kramers' charge-conjugation transformation), and finally to Pauli's requirement of invariance under strong-reflection $\vec{x}, t \rightarrow -\vec{x}, -t$ [16.4]. In all cases the discussion remains valid today within the limited context of electrodynamics. Schwinger's proof differs in principle from the proofs soon to be constructed which prove the Spin-Statistics Theorem on the basis of other compelling postulates, and from this then prove the **TCP**-theorem.

The time-reversal operation $\mathbf{T}_s$ introduced by Schwinger is closely related to that of Weyl, and is called strong or Schwinger time-reversal. As explained in detail by Schwinger, $\mathbf{T}_s$ involves the transformation of a Hilbert space state vector $|\Psi_a\rangle$ into the the transpose of the Hermitian adjoint state vector $\langle\Psi_{\bar{a}}|$. This differs from the usual time-reversal operation $\mathbf{T}$ (also $\mathbf{T}_w$ for weak or Wigner time-reversal) which is defined to take $|\Psi_a\rangle \rightarrow |\Psi_{\bar{a}}\rangle$

but which has the somewhat artificial requirement that it changes c-numbers (classical numbers as distinct from q-numbers which are Hilbert space operators) to their complex conjugates ($\mathbf{T}_w i \mathbf{T}_w^{-1} = -i$) [16.5]. In modern notation Schwinger time-reversal is the succession of charge-conjugation $\mathbf{C}$ (particle $\leftrightarrow$ antiparticle transformation) and weak time-reversal, $\mathbf{T}_s = \mathbf{T}_w \mathbf{C}$ (or just $\mathbf{TC}$), and lacks the space reflection $\mathbf{P}$ for the full $\mathbf{TCP}$ transformation which leaves relativistic quantum field theory invariant. These requirements were all five years in the future when Schwinger wrote his first paper on the subject, and in fact are not necessary for quantum electrodynamics which satisfies $\mathbf{T}, \mathbf{C}$, and $\mathbf{P}$ invariances separately.

§2. Schwinger's Proof.

Schwinger requires a linear transformation (3.22) of the quantum fields

$$\psi' = \mathbf{R}\psi\mathbf{R}^{-1} = \mathcal{R}\psi \tag{1}$$

which leaves the equations of motion invariant under $t \to t' = -t$. The Dirac equation

$$i\frac{\partial}{\partial t'}\psi'(r',t') = (\alpha \cdot p' + \beta m)\psi'(r',t')$$
$$= -i\frac{\partial}{\partial t}\mathcal{R}\psi(r,t) = (\alpha \cdot p + \beta m)\mathcal{R}\psi(r,t). \tag{2}$$

Left multiplying by $\mathcal{R}^{-1}$ requires

$$i\frac{\partial}{\partial t}\psi(r,t) = -(\mathcal{R}^{-1}\alpha\mathcal{R} \cdot p + \mathcal{R}^{-1}\beta\mathcal{R}m)\psi(r,t). \tag{3}$$

So we require

$$\mathcal{R}^{-1}\alpha\mathcal{R} = -\alpha, \quad \mathcal{R}^{-1}\beta\mathcal{R} = -\beta. \tag{4}$$

The choice

$$\mathcal{R} = \gamma_0\gamma_5 \tag{5}$$

satisfies the requirements. It is helpful to keep in mind the convenient 'standard' representation [16.6]

$$\gamma_0 = \beta = \begin{pmatrix} 1 & 0 \\ 0 & -1 \end{pmatrix}, \quad \gamma_5 = \begin{pmatrix} 0 & 1 \\ 1 & 0 \end{pmatrix}, \quad \vec{\gamma} = \gamma_0\vec{\alpha} = \begin{pmatrix} 0 & \vec{\sigma} \\ -\vec{\sigma} & 0 \end{pmatrix}.$$

An alternative notation which is more intuitive is to express the time-reversal in terms of the "generator of rotations"

$$\sigma_{45} = -i\gamma_0\gamma_5 \tag{6}$$

so that a "rotation" through angle π in the (45) plane which turns $t \to -t$ is achieved by the operator

$$\exp(i\frac{\pi}{2}\sigma_{45}) = \gamma_0\gamma_5 \tag{7}$$

as before. Successive time-reversals are seen to reverse the sign of the Dirac field. Schwinger returned to this observation with Brown [16.7] in a paper in 1961, which is a predecessor of the "intuitive" proofs of the Spin-Statistics Theorem. In one form or another, this sign change under 2π rotations underlies all the responses to the original question raised by Neuenschwander about the existence of more transparent demonstrations of the spin-statistics relation.

Schwinger's crucial observation is that the generators of space-time translations - the integrated densities of the energy-momentum tensor, Eqn.302 - transform under time reflection different from the space-time translation itself. That is, the transformation $\vec{x}, t \to \vec{x}, -t$ (which defines the proper vector transformation) produces the transformation of the generators

$$\langle \vec{P}, P_0 \rangle \to \langle -\vec{P}, P_0 \rangle$$

as a (time-)pseudovector. The critical sign change occurs because the pseudoscalar volume element d^4x produces a pseudovector surface element $d\sigma_\mu$. This sign change makes it necessary to define the time-reversal as other than a unitary transformation. Schwinger's solution is to interchange initial and final states (see Eqn.317 and [6]). The alternative $\mathbf{T}_w$ would change $i \to -i$ as part of the weak time-reversal transformation (see [4]).

Even this does not suffice. The Dirac Lagrangian density $\mathcal{L}_D$ is a time-pseudoscalar whereas the Lagrangian densities of scalar fields and of the electromagnetic field are time-scalars (compare $m\bar{\psi}\psi$ with $\mu^2\phi^2$ or $E^2 - B^2$). The extra sign change necessary to make the full action (with $\mathcal{L}_D, \mathcal{L}_S, \mathcal{L}_V$ and $\mathcal{L}_{int}$) invariant under $\mathbf{T}_s$ comes from transposing the *anticommuting* Dirac operators, in accord with the requirement of the Spin-Statistics Theorem. The key relations are

1) Eqn.317 for the universal sign change of the action after the (non-unitary) time-reversal operation; and

2) the sign change from anticommutation of Dirac fields (Eqns.322-324) which brings the Dirac Lagrangian into conformity with the integral-spin Lagrangians, thanks to the spin-statistics relation.

Schwinger's argument is intricate, powerful and almost complete. The obvious shortcoming is to extend the proof to full **TCP** invariant theories. A further shortcoming - which is present in all proofs to this point - is to trivialize the extension of local (anti)commutativity to all spacelike separations including zero. This step will finally be taken - using the Hall-Wightman theorem [16.8] - by Lüders and Zumino and by Burgoyne [16.9], at the expense of some further subtlety in the proof but with the enormous logical benefit of giving priority to the Spin-Statistics Theorem. In this approach - most clearly stated in the introduction of Lüders and Zumino - the Spin-Statistics Theorem is proved on the basis of "common sense" postulates and the **TCP**-theorem follows as a result.

§3. Concluding Remarks.

Schwinger persevered with this problem in a series of papers [16.10] and in a book [16.11], in which he formulated his proof of the Spin-Statistics Theorem as a consequence of strong reflection invariance of relativistic field theory continued to a Euclidean space-time. His arguments seem contrived, especially in comparison to the work of Lüders and Zumino and of Burgoyne, to be discussed next. In this demonstration, the Spin-Statistics Theorem is proved on the basis of more "basic" postulates, and the **TCP**-theorem then

393

follows, as most clearly demonstrated in a pedagogical exposition by Lüders of his original proof [16.12].

Pauli wrote a lengthy commentary on Schwinger's proof of the Spin-Statistics Theorem for the 1955 Festschrift volume celebrating Bohr's seventieth birthday 16.13]. This paper contains Pauli's last published remarks on the subject. In it, he coincidentally withdrew one of his own most strongly held postulates on which he had based his proof of the Spin-Statistics Theorem, that the commutator of physical quantities must vanish for spacelike separation. This postulate had served the particular purpose to rule out the quantization of integral spin fields with anticommutation relations, which would have introduced the $\Delta^{(1)}$-function instead of the Δ-function for the free fields. He now stated that this anomalous quantization would be inconsistent with the field equations for any local interaction.

In this paper, Pauli also withdrew his objection [16.14] to Belinfante's proof of the Spin-Statistics Theorem. He had originally criticized Belinfante's proof as based on the unjustified assumption of a too simple prescription for the charge-conjugation transformation in which each species transforms separately. In his paper coauthored with Belinfante, criticizing Belinfante's thesis publication, he had maintained that such simplification was unjustified (although, as they mentioned there, it did hold true for all then known particles) and had to be replaced by the postulates of positive energies for half-integral spin fields, and commutation of physical observables for integral spin fields. Pauli states that he is now prepared to "*assume* the normal transformation for particle-antiparticle conjugation which connects every spinor-field with its own complex conjugate", which requires anticommutation relations.

Pauli then reverses Schwinger's argument and - following earlier work by Lüders - proves the **TCP**-theorem on the assumption of the spin-statistics relation. He characterizes as a remarkable gift of nature that invariance of quantum field theory under Schwinger's strong reflection (SR, now **TCP**) requires only the spin-statistics connection and invariance under the proper Lorentz group (as well as local interactions and the assumption of anticommutation of different half-integral spin fields).

We refer to the 1957 article by Lüders for a more transparent discussion than that put forward by Pauli in his critique of Schwinger.

In a later chapter we will make extensive reference to Schwinger's subsequent work on the Spin-Statistics Theorem, where it leads eventually [16.15] to a much simplified proof.

§4. Biographical Note on Schwinger.(†)

Julian Schwinger (1918-1994) played Heisenberg to Feynman's Dirac throughout their professional lives. In their Nobel Prize winning work on Quantum Electrodynamics (awarded in 1965, shared with Tomonaga), Schwinger's work was finished first and gained all the early recognition. But Schwinger's way of doing physics was so formal and abstract that it was and remains incomprehensible and unread by most physicists, including theoreticians, and conveys almost no intuitive grasp. Physics in Schwinger's writings seems like the struggles of Sisyphus to role a heavy stone up a hill. In contrast, Feynman is like a young Prometheus, stealing fire from the gods and showing ordinary mortals how to use it.

Schwinger was a child prodigy in New York City whose talents were recognized by Semat at CCNY who handed him on to Rabi at Columbia, who in turn saved him from a failing performance in English composition. Schwinger had his BA at 17 and his PhD at 21 from Columbia, working mainly on the theory of neutron-proton scattering and co-authoring papers with Rabi and with Teller by the age of 19; with Oppenheimer's group (1939-42) he collaborated with Corben, Rarita, Sachs, Gerjuoy, Hamermesh, Saxon, and Oppenheimer himself on groundbreaking work on few-nucleon problems, meson theory, theory of the nuclear force, and electron orbits in synchrotron acceleration; he spent (1942-45) with the MIT Radiation Laboratory where he developed the theory of microwave transmission using Green's function and variational techniques that later featured prominently in his development of quantum electrodynamics.

At Harvard (1945-72) Schwinger was an outstanding teacher with a large cadre of PhD students. He won the Nobel Prize (1965), the first Albert Einstein Prize (1951) and the National Medal of Science (1964). At UCLA (1972-), Schwinger made major contributions in his Source Theory as a phenomenological but field-theory informed treatment of high energy particle physics.

(† - *The London Times*, 16 Aug 1994, p.19a.)

Bibliography and References.

16.1) J. Schwinger, Phys. Rev. **82**, 914 (1951); see our App.16A.

16.2) H. Weyl, *The Theory of Groups and Quantum Mechanics* (Dover, New York, 1931), pp. 218-227.

16.3) F.J. Belinfante, Physica **6**, 870 (1939); see our App.12A.

16.4) W. Pauli, Phys. Rev. **58**, 716 (1940); see our App.14A.

16.5) S. Weinberg, *The Quantum Theory of Fields I* (Cambridge, New York, 1995), pp.74-81.

16.6) J.D. Bjorken and S.D. Drell, *Relativistic Quantum Mechanics* (McGraw-Hill, New York, 1964), p.282; F. Halzen and A.D. Martin, *Quarks and Leptons: An Introductory Course in Modern Particle Physics* (Wiley, New York, 1984), p.397.

16.7) L.S. Brown and J. Schwinger, Prog. Theor. Phys. **26**, 917 (1961).

16.8) A.S. Wightman, Phys. Rev. **101**, 860 (1956); D. Hall and A.S. Wightman, Matt. Fys. Medd. Dan. Vid. Selsk. **31**, 5 (1957); see our App.18A,B.

16.9) G. Lüders and B. Zumino, Phys. Rev. **110**, 1450 (1958); N. Burgoyne, Nuo. Cim. **8**, 607 (1958); see our App.17A,B.

16.10) J. Schwinger, Proc. Natl. Acad. Sci. USA **44**, 223, 617, 956 (1958).

16.11) J. Schwinger, *Particles, Sources, and Fields* (Addison-Wesley, Redwood City CA, 1989), p.99-114.

16.12) G. Lüders, Ann. Phys. **2**, 1 (1957).

16.13) W. Pauli, in *Niels Bohr and the Development of Physics, Essays Dedicated to Niels Bohr on the Occasion of His Seventieth Birthday* (Interscience, London and New York, 1955), edited by W. Pauli, pp.30-51.

16.14) W. Pauli and F.J. Belinfante, Physica **7**, 177 (1940); see our App.12B.

16.15) E.C.G. Sudarshan, "The Fundamental Theorem on the Connection Between Spin and Statistics," in: *Proc. Nobel Symposium 8*, ed. N. Svartholm (Almquist and Wiksell, Stockholm, 1968), pp.379-386; "Relation Between Spin and Statistics," Statistical Physics Supplement: Journal of the Indian Institute of Science, June 1975, pp.123-137.

APPENDIX 16.A:

Excerpt from: Physical Review 82, 914 (1951)

The Theory of Quantized Fields. I*

JULIAN SCHWINGER

Harvard University, Cambridge, Massachusetts

(Received March 2, 1951)

······ The fundamental dynamical principle is a variational principle for the transformation connecting state vectors associated with different spacelike surfaces,' which describes the time development of the system. The generator of the infinitesimal transformation is the variation of the action, the space-time volume integral of the invariant Lagrangian density. The invariance of the Lagrangian preserves the form of the dynamical principle under coordinate transformations, with the exception of those which include a reversal in the positive sense of time, where a separate discussion is necessary. It will be shown in Sec.III that *the requirement of invariance under time reflection imposes a restriction on the operator properties of the fields, which is simply the connection between the spin and statistics of particles.* (*Note:* Emphases added.) ········· In Sec.III, the exceptional nature of time reflection is indicated by the fact that the charge and the energy-momentum vector behave as a pseudoscalar and pseudovector, respectively, for time reflection transformations. ··· The contrast between the energy-momentum (pseudo-)vector and the displacement (proper-)vector then indicates, that time reflection cannot be described as a unitary transformation. ··· It is important to recognize here that the contributions to the Lagrangian of half-integral spin fields behave like pseudoscalars with respect to time reflection. The non-unitary transformation required to represent time reflection is the replacement of a state by its dual, or complex conjugate vector, together with the transposition of all operators. The fundamental dynamical principle is then invariant under time reflection if inverting the order of all operators in the Lagrangian leaves an integral spin contribution unaltered, and reverses the sign of a half-integral spin contribution. *This implies the essential commutativity, or anticommutativity, of integral and half-integral field components, respectively, which is the connection between spin and statistics.*

······(*Note:* We skip the **INTRODUCTION** and a monumental section **QUANTUM DYNAMICS OF LOCALIZABLE FIELDS** which has a daunting 139 equations.)

III. TIME REFLECTION

The general physical requirement of invariance with respect to coordinate transformations applies not only to translations and rotations of the coordinate system, but also to reflections of the coordinate axes. Among the latter transformations, time reflection has a singular position. Its special nature can be indicated by the transformation properties of some integrated physical quantities. Thus, the expectation value of the energy-momentum vector,

$$\langle P_\nu \rangle = \int_\sigma d\sigma_\mu \langle T_{\mu\nu} \rangle, \tag{301}$$

is actually a pseudovector with respect to time reflection. With the plane surface σ chosen perpendicular to the time axis, the components $\langle P_\nu \rangle$ are obtained as three-dimensional volume integrals,

$$\langle P_0 \rangle = \int d\sigma \langle T_{00} \rangle,$$

$$\langle P_k \rangle = \int d\sigma \langle T_{0k} \rangle, \quad k = 1, 2, 3, \tag{302}$$

and the time reflection $x_0 \to -x_0$, $x_k \to x_k$ induces $\langle P_0 \rangle \to \langle P_0 \rangle$, $\langle P_k \rangle \to -\langle P_k \rangle$, according to the transformation properties of tensors. This differs in sign from a proper vector transformation. In particular, the energy does not reverse sign under time reflection. More generally, this property of $\langle P_\nu \rangle$ is obtained from the pseudovector character of $d\sigma_\mu$, which expresses the pseudoscalar nature of a four-dimensional volume element with respect to time reflection. Similarly, the expectation value of the charge

$$\langle Q \rangle = \int d\sigma_\mu \langle j_\mu \rangle = \int d\sigma \langle j_0 \rangle \tag{303}$$

behaves as a pseudoscalar under time reflection. Hence, this transformation interchanges positive and negative charge, and both signs must occur symmetrically in a covariant theory. Indeed, for some purposes the requirement of charge symmetry can be substituted for the more incisive demand of invariance under time reflection.

The significant implication of these properties is that time reflection cannot be included within the general framework of unitary transformations. Thus, on referring to the Schrödinger equation for translations, or the operator equations (*Note*: Both are in Sec.II, and are not reproduced here.), we encounter a contradiction between the transformation properties of the proper vector translation operator δ_μ and of the pseudovector P_μ. This difficulty appears most fundamentally in our basic variational principle [4]. With the Lagrangian density $\mathcal{L}$ behaving as a scalar and $(d^4 x)$ as a pseudoscalar, reflection of the time axis introduces a minus sign on the right side of this equation. However, it is important to notice that the scalar nature of $\mathcal{L}$ cannot be maintained for that part of the Lagrangian which describes half-integral spin fields. Indeed, such contributions to $\mathcal{L}$

399

behave like pseudoscalars with respect to time reflection [5]. If we were to consider only such a half-integral spin field, the basic dynamical equation would preserve its structure under time reversal, but at the expense of violating the general transformation properties of all physical quantities. The charge would remain unaltered, and energy would reverse sign under time reflection. The latter difficulty simply indicates that, on inclusion of the contributions of integral spin fields, the various parts of $\mathcal{L}$ would transform differently, thus emphasizing again the general failure to admit time reflection as a unitary transformation.

To aid in investigating the extended class of transformations that is required to include time reflection, we shall introduce some notational developments. The scalar product of two vectors, Ψ_a and Ψ_b,

$$\langle a|b \rangle = \Psi_a^* \Psi_b = \Psi_b \Psi_a^*, \tag{304}$$

is the invariant combination of a vector Ψ_b with the dual, complex conjugate vector, Ψ_a^*. We allow operators to act both on the left and on the right of vectors, Ψ and Ψ^*. Thus, the transposed operator A^T, associated with an operator A, is defined by

$$A\Psi = \Psi A^T, \quad \Psi^* A = A^T \Psi^*, \tag{305}$$

or by

$$\langle a|A|b \rangle = \Psi_a^* A \Psi_b = \Psi_b A^T \Psi_a^*. \tag{306}$$

We also define the complex conjugate operator A^*,

$$(A\Psi)^* = A^* \Psi^*. \tag{307}$$

The Hermitian conjugate operator $A^\dagger$ is obtained from the definition,

$$(A\Psi)^* = \psi^* A^\dagger, \tag{308}$$

as

$$A^\dagger = A^{*T}. \tag{309}$$

Conventional quantum mechanics uses transformations only within the Ψ vector space, and contragredient transformations within the dual Ψ^* space. We shall now consider transformations that interchange the two spaces, as in

$$\Psi_a \rightarrow \Psi_{\bar{a}} = \Psi_a^*. \tag{310}$$

The effect of Eqn.310 is indicated by

$$\langle a|b \rangle = \Psi_a^* \Psi_b = \Psi_{\bar{a}} \Psi_{\bar{b}}^* = \langle \bar{b}|\bar{a} \rangle, \tag{311}$$

and

$$\langle a|A|b \rangle = \Psi_a^* A \Psi_b = \Psi_{\bar{a}} A \Psi_{\bar{b}}^* = \langle \bar{b}|A^T|\bar{a} \rangle. \tag{312}$$

400

More generally, if

$$\Psi_{\bar{a}}^* = \mathbf{R}\Psi_a, \tag{313}$$

where R is a unitary operator, we have

$$\langle a|b\rangle = \langle \bar{b}|\bar{a}\rangle, \quad \langle a|A|b\rangle = \langle \bar{b}|\bar{A}|\bar{a}\rangle, \tag{314}$$

in which

$$\bar{A} = (\mathbf{R}A\mathbf{R}^{-1})^T. \tag{315}$$

Now, we have

$$\overline{AB} = (\mathbf{R}AB\mathbf{R}^{-1})^T = (\mathbf{R}B\mathbf{R}^{-1})^T(\mathbf{R}A\mathbf{R}^{-1})^T = \bar{B}\bar{A}, \tag{316}$$

and therefore

$$\langle a|[A,B]_-|b\rangle = -\langle \bar{b}|[\bar{A},\bar{B}]_-|\bar{a}\rangle. \tag{317}$$

We have precisely the sign change that is required to preserve the structure of equations like Eqn.2110 (*Note:* reproduced in [6]) under time reflection.

We now examine whether it is possible to satisfy the requirement of invariance under time reflection by means of transformations of the type (3.13). When we introduce the coordinate transformation

$$\bar{x}_0 = -x_0, \qquad \bar{x}_k = x_k, \quad k = 1,2,3, \tag{318}$$

in connection with the state vector transformation

$$\Psi^*(\bar{\zeta}',\sigma) = \mathbf{R}\Psi(\zeta',\sigma), \tag{319}$$

the fundamental dynamical equation [7] becomes

$$\delta\langle \bar{\zeta}_2'',\sigma_2|\bar{\zeta}_1',\sigma_1\rangle = i\langle \bar{\zeta}_2'',\sigma_2|\delta \int_{\sigma_1}^{\sigma_2} (d\bar{x})\bar{\mathcal{L}}|\bar{\zeta}_1',\sigma_1\rangle, \tag{320}$$

where

$$\bar{\mathcal{L}} = (\mathbf{R}\mathcal{L}\mathbf{R}^{-1})^T = \mathcal{L}^T((\mathbf{R}\phi^\alpha\mathbf{R}^{-1})^T, \pm\bar{\partial}_\mu(\mathbf{R}\phi^\alpha\mathbf{R}^{-1})^T). \tag{321}$$

In the last statement, the $\pm$ sign indicates the effect of the coordinate transformation (3.18) on the components of the gradient vector, while the notation $\mathcal{L}^T(\ )$ symbolizes the reversal in the order of all factors induced by the operation of transposition. The operator $\mathbf{R}$ will now be chosen to produce that linear transformation of the ϕ^α,

$$\mathbf{R}\phi^\alpha\mathbf{R}^{-1} = \mathcal{R}^{\alpha\beta}\phi^\beta \tag{322}$$

which compensates the effect of the gradient vector transformation. Thus we have

$$\bar{\mathcal{L}} = (\pm)\mathcal{L}^T(\phi^{\alpha T}, \bar{\partial}_\mu \phi^{\alpha T}), \tag{323}$$

401

where the ($\pm$) sign refers to the fact that the structure of the Lagrangian, for half-integral spin fields, can be maintained only at the expense of a change of sign. We now see that if

$$\bar{\mathcal{L}} = \mathcal{L}(\phi^{\alpha T}, \bar{\partial}_{\mu}\phi^{\alpha T}), \tag{324}$$

the form of our fundamental dynamical equation will have been preserved under time reflection, since Eqn.320 will then differ from Eqn.214 only in the substitution of $\phi^{\alpha T}$ for ϕ^{α} as the appropriate field variable, and in the interchange of σ_1 and σ_2, which simply follows from the reversed temporal sense in which the dynamical development of the system is to be traced.

Invariance under time reflection thus requires that inverting the order of all factors in the Lagrangian leave a scalar term unchanged, and reverse the sign of a pseudoscalar term. This can be satisfied, of course, by an explicit symmetrization or antisymmetrization of the various terms in $\mathcal{L}$. When the Lagrangian, thus arranged, is employed in the principle of stationary action, the variations $\delta_0\phi^{\alpha}$ will likewise be disposed in a symmetrical or antisymmetrical manner. We must now recall that the equations of motion [8], which do not depend explicitly on the nature of the field commutation properties, have been obtained by postulating the equality of terms $\delta_0\mathcal{L}$ that differ basically only in the location of $\delta_0\phi^{\alpha}$. Since such terms appear with the same sign in scalar components of $\mathcal{L}$, and with opposite signs in pseudoscalar components, we deduce a corresponding commutativity, or anticommutativity, between $\delta_0\phi^{\alpha}$ and the other operators in the individual terms of $\delta_0\mathcal{L}$.

The information concerning commutation properties that has thus been obtained is restricted to operators at common space-time points, since this is the nature of $\mathcal{L}$. Commutation relations between field quantities located at distinct points of a spacelike surface are implied by the general compatibility requirement for physical quantities attached to points with a spacelike interval. Components of integral spin fields, and bilinear combinations of the components of half-integral spin fields, are the basic physical quantities to which this compatibility condition applies. By considering the general possibilities of coupling between the various fields, we may draw from these two expressions of relativistic invariance the consequence that the variations $\delta\phi^b(x')$, and therefore the conjugate variations $\delta\Pi^b(x')$, commute or anticommute with $\phi^a(x), \Pi^a(x)$ for all x and x' on a given σ, where the relation of anticommutativity holds when both a and b refer to components of half-integral spin fields. The consistency of this statement with the general commutation relations that have already been deduced from it is easily verified. By subjecting the canonical variables [9] to independent variations, we obtain

$$[\phi^a(x), \delta\phi^b(x')]_{\pm} = [\Pi^a(x), \delta\phi^b(x')]_{\pm} = 0,$$
$$[\phi^a(x), \delta\Pi^b(x')]_{\pm} = [\Pi^a(x), \delta\Pi^b(x')]_{\pm} = 0, \tag{325}$$

which is valid for all x, x' on σ. In addition, all physical quantities commute at distinct points of σ.

402

We conclude that the connection between the spin and statistics is implicit in the requirement of invariance under coordinate transformations [10].

Footnotes and References.

......

4) Eqn.214: $\delta\langle\zeta_1', \sigma_1|\zeta_2'', \sigma_2\rangle = i\langle\zeta_1', \sigma_1|\delta W_{12}|\zeta_2'', \sigma_2\rangle$ with $\delta W_{12} = \delta\int_{\sigma_2}^{\sigma_1}(dx)\mathcal{L}$.

5) The fundamental invariant of a spin-$\frac{1}{2}$ field is $\bar{\psi}\psi = \psi^\dagger\gamma_0\psi$. The transformation that represents time reflection, $\psi' = R\psi$, can be obtained from its equivalence with a rotation through an angle π in the (45) plane; $R = \exp[i\pi\frac{1}{2}\sigma_{45}] = i\sigma_{45}$ (*Note added*: $R = \gamma_0\gamma_5$). Accordingly,

$$\bar{\psi}'\psi' = \psi^\dagger R^{-1}\gamma_0 R\psi = -\bar{\psi}\psi,$$

which indicates the pseudoscalar character of the spin-$\frac{1}{2}$ field Lagrangian, with respect to time reflection. The corresponding behavior of fields with other spin values can be obtained from the observation that a spinor of rank n contains fields of spin-$\frac{1}{2}n, \frac{1}{2}n - 1, \cdots$. The basic invariant and time reflection operator for a spinor of rank n are

$$\bar{\psi}\psi = \psi^\dagger\prod_{k=1}^{n}\gamma_0^{(k)}\psi,$$

and

$$R = \exp[i\pi\frac{1}{2}\sum_{k=1}^{n}\sigma_{45}^{(k)}] = \prod_{k=1}^{n}i\sigma_{45}^{(k)}.$$

Therefore,

$$\bar{\psi}'\psi' = \psi^\dagger R^{-1}\prod_{k=1}^{n}\gamma_0^{(k)}R\psi = (-1)^n\bar{\psi}\psi,$$

which shows the pseudoscalar nature of the Lagrangian for half-integral spin fields.

6) Eqn.2110: $[G(\sigma), P_\mu]_- = -i\delta_\mu G(\sigma)$.

7) See [4], Eqn.214.

8) Eqn.218: $\partial_\mu\partial\mathcal{L}/\partial\phi_\mu^\alpha = \partial\mathcal{L}/\partial\phi^\alpha$.

9) Eqn.281:

$$[\phi^a(x), \Pi^b(x')]_\pm = i\delta_{ab}\delta_\sigma(x - x')$$
$$[\phi^a(x), \phi^b(x')]_\pm = [\Pi^a(x), \Pi^b(x')]_\pm = 0.$$

10) The discussion of the spin and statistics connection by Pauli [Phys. Rev. **58**, 716 (1940)] is somewhat more negative in character, although based on closely related physical requirements. Thus, Pauli remarks that Bose-Einstein quantization of a half-integral spin field implies an energy that possesses no lower bound and that Fermi-Dirac quantization of an integral spin field leads to an algebraic contradiction with the commutativity of physical quantities located at points with a spacelike interval. Another postulate which has been employed, that of charge symmetry [W. Pauli and F.J. Belinfante, Physica **7**, 177 (1940)], suffices to determine the commutation relations for sufficiently simple systems. As we have noticed, it is a consequence of time reflection invariance. The comments of Feynman on vacuum polarization and statistics [Phys. Rev. **76**, 749 (1949)] appear to be an illustration of the charge symmetry requirement, since contradiction is established when the charge symmetrical concept of the vacuum is applied to a Bose-Einstein spin-$\frac{1}{2}$ field, or Fermi-Dirac spin-0 field.

Chapter 17

The Proofs of Lüders and Zumino, and of Burgoyne

Summary: Lüders and Zumino, and independently Burgoyne, proved the Spin-Statistics Theorem from basic postulates on quantum field theory, including invariance under the proper Lorentz group with no reflections. The analytic continuation in the separation of two fields in a vacuum expectation value of their product, due to Hall and Wightman, as well as their ability to treat interacting fields, are the key ingredients which advance their proofs over those of Fierz, deWet, and Pauli.

§1. Introduction. The Lüders and Zumino Proof.

Lüders and Zumino [17.1] succeed in disentangling the Spin-Statistics Theorem from the **TCP**-theorem where Schwinger [17.2] had left it in a secondary role. Schwinger's work - which evolved from Belinfante's **C**-invariance postulate [17.3], and from Pauli's strong-reflection result [17.4]-postulated **TC**-invariance which required the "normal case" spin-statistics connection (commutation between Bose-Einstein fields and Bose-Einstein or Fermi-Dirac fields; anticommutation between Fermi-Dirac fields). The proof of **TCP** [17.5] postulated the validity of the normal case spin-statistics connection, and assumed the usual commutation relations for like fields. Lüders and Zumino state "The situation is rather unsatisfactory $\cdots$ no independent proof of either of these theorems has been given." They give a proof of the spin-statistics connection, which does not postulate **TCP**, for the operators of an individual field. They still have to postulate the normal case for unlike fields. Their proof is limited to spin-0 and spin-$\frac{1}{2}$ and first discusses only Hermitian (or Majorana) fields, which they then generalize to charged fields by assuming an underlying gauge invariance.

Lüders and Zumino start with a basic demonstration for a Hermitian

405

spin-0 field. They list five postulates which make anticommutation relations untenable for this case. These postulates are:

I) Invariance under the proper inhomogeneous Lorentz group (not containing any reflections).

II) Locality. Two operators of the same field at points separated by a spacelike interval either commute or anticommute.

III) The vacuum is the unique state of lowest energy.

IV) The metric of the Hilbert space of states is positive.

V) The vacuum state is not identically annihilated by a field.

From (I),
$$\langle 0|\phi(x)\phi(y)|0\rangle = f(\xi), \quad \xi = x - y. \tag{1}$$
For ξ spacelike $f(\xi)$ depends only on ξ^2, so

$$\langle 0| [\phi(x), \phi(y)]_- |0\rangle_{SL} = 0. \tag{2}$$

Lüders and Zumino next examine the possibility from (II) that

$$\langle 0| [\phi(x), \phi(y)]_+ |0\rangle = 0, \tag{3}$$

which turns out to be an untenable assumption in that it leads by postulates (III) and (IV) to a contradiction of postulate (V). Adding the commutator and anticommutator for spacelike ξ, the above assumption gives

$$\langle 0|\phi(x)\phi(y)|0\rangle_{SL} = 0. \tag{4}$$

By an analytic continuation justified by Hall and Wightman [17.6], this equality must hold for all ξ including zero. Inserting a complete set of states and using postulate (III) we obtain

$$\sum_n |\langle n|\phi(x)|0\rangle|^2 = 0 \tag{5}$$

406

and postulate (IV) requires

$$\phi(x)|0\rangle = 0 \tag{6}$$

which contradicts postulate (V), leading to the conclusion that the assumption of the anticommutator for spin-0 fields is untenable. Lüders and Zumino defend postulate (V) as being essential to the construction of the Hilbert space of states for the field $\phi(x)$. One has to get out of the vacuum state somehow.

We are reminded of the similar theorem put forward by deWet [17.7], and used before him by Fierz [17.8] and by Socolow and Iwanenko [17.9] and first by Pauli and Weisskopf [17.10], to rule out anticommutation relations for spin-0 fields. deWet simply stated that

$$\left[\phi^\dagger(x), \phi(y)\right]_+ = 0 \tag{7}$$

was not possible because it was positive definite at $x = y$. He used the same postulates (III), (IV), and (V) above, but he did not have the Hall-Wightman theorem to justify putting $x = y$.

For spin-$\frac{1}{2}$ fields - still Hermitian (Majorana) in the discussion of Lüders and Zumino - a key distinction from the spin-0 case occurs. *The vacuum expectation value transforms under Lorentz transformations as the fourth-component of a four vector.* The Majorana fields satisfy

$$\bar{\psi} = \tilde{\psi} C \tag{8}$$

and $\bar{\psi}(\not{p} - m) = (\not{p} - m)\psi$ requires $C\tilde{\gamma}\tilde{C} = \gamma$ and $C\tilde{C} = 1$. The vacuum expectation value involved is

$$\langle 0|\psi_\beta^\dagger(x)\psi_\beta(y)|0\rangle = \eta_{\alpha\beta}\langle 0|\psi_\alpha(x)\psi_\beta(y)|0\rangle, \tag{9}$$

with $\eta_{\alpha\beta} = C_{\alpha\delta}\gamma_{\delta\beta}^0 = \eta_{\beta\alpha}$. This can be seen easily in the standard representation [17.11] where $\eta = C\gamma^0 = i\gamma_0\gamma_2\gamma_0 = -i\gamma_2$, and $\tilde{\eta} = \widetilde{\gamma_0}\tilde{C} = -i\widetilde{\gamma_2} = \eta$. Then postulate (I) requires

$$\eta_{\alpha\beta}\langle 0|\psi_\alpha(x)\psi_\beta(y)|0\rangle = \xi_4 g(\xi), \quad \xi = x - y. \tag{10}$$

407

For ξ spacelike, $g(\xi)$ depends only on ξ^2 so

$$\eta_{\alpha\beta}\langle 0|\, [\psi_\alpha(x), \psi_\beta(y)]_+ \,|0\rangle_{SL} = 0. \tag{11}$$

In order to show that the assumption

$$\eta_{\alpha\beta}\langle 0|\, [\psi_\alpha(x), \psi_\beta(y)]_- \,|0\rangle = 0 \tag{12}$$

is untenable, add these to get

$$\eta_{\alpha\beta}\langle 0|\psi_\alpha(x)\psi_\beta(y)|0\rangle_{SL} = 0, \tag{13}$$

for ξ spacelike. Following Hall-Wightman, this can be continued to all ξ including zero, giving

$$\langle 0|\psi^\dagger(x)\psi(x)|0\rangle = 0 \tag{14}$$

and from postulate (IV), $\psi(x)|0\rangle = 0$ in violation of postulate (V). The conclusion is that spin-$\frac{1}{2}$ fields cannot satisfy commutation relations.

The proof for spin-$\frac{1}{2}$ develops in parallel to that for spin-0, in contrast to the Pauli proof which depended for spin-$\frac{1}{2}$ more directly on postulate (III), that the energy of every state should be positive.

It is remarkable how close deWet was to the same proof. If only he had recognized that the vacuum expectation value of the Dirac bilinear is a vector function of the separation ξ, he might have had - without the Hall-Wightman theorem to justify it, of course - the spin-$\frac{1}{2}$ analog of his theorem for spin-0.

The fact that

$$\langle 0|\bar{\psi}(x)\gamma_\mu\psi(0)|0\rangle \tag{15}$$

is a Lorentz vector - although a characteristically singular one - can be easily seen. First of all, we expect this to be the case because $\bar{\psi}(x)\gamma_\mu\psi(x)$ is the current vector. Furthermore, a simple calculation for free fields gives

$$\begin{aligned}
\langle 0|\bar{\psi}(x)\gamma_\mu\psi(0)|0\rangle &= \int \frac{d^3p}{(2\pi)^3 2p_0} e^{ip_\mu x_\mu} \frac{p_\mu}{m} \\
&= \frac{\partial_\mu}{im} \int \frac{d^4p}{(2\pi)^4} 2\pi\delta(p^2 - m^2)\theta(p_0)e^{ip_\mu x_\mu}. \tag{16}
\end{aligned}$$

Each factor under the integral sign is a Lorentz invariant, and the vector transformation property is contained in the four-vector gradient ∂_μ.

This is clearly a result which was available to deWet and others of that era, and the observation that the vector character of the product of spinors required an anticommutation relation was at his fingertips.

The integral can be expressed in terms of standard singular functions [17.12] as

$$\frac{\partial_\mu}{2m}\left(\Delta - i\Delta_1\right) \tag{17}$$

which takes a particularly simple form for $m \to 0$ in which case

$$\Delta - i\Delta_1 \to \frac{-i}{2\pi^2}\frac{1}{x_\mu^2 - i\epsilon}, \tag{18}$$

where finally $\epsilon \to 0^+$. In this simple case the vacuum expectation value of the product of the separated spin-$\frac{1}{2}$ field operators is

$$\langle 0|\psi^\dagger(x)\psi(0)|0\rangle \sim \frac{x_0}{m}\frac{i}{2\pi^2}\frac{1}{(x_\mu^2 - i\epsilon)^2}. \tag{19}$$

The vector character is clear, although made obscure and perhaps somewhat doubtful by the singular nature of the integral.

By choosing to start with Hermitian (Majorana) fields, Lüders and Zumino next have to resort to an artifice to extend their results to charged (complex) fields. They introduce a global gauge invariance under

$$\phi \to \phi' = e^{i\alpha}\phi, \tag{20}$$

which requires

$$\langle 0|\phi(x)\phi(y)|0\rangle = \langle 0|\phi^*(x)\phi^*(y)|0\rangle = 0. \tag{21}$$

Then the above proof for Hermitian fields is applied to the Hermitian fields

$$\phi_1 = \phi + \phi^*, \quad \phi_2 = i\left(\phi - \phi^*\right). \tag{22}$$

The assumption of anticommutation for ϕ_1 and for ϕ_2 leads to

$$\langle 0|\phi^*(x)\phi(y) + \phi(y)\phi^*(x)|0\rangle = 0. \tag{23}$$

Covariance gives, for spacelike separation,

$$\langle 0|\phi(x)\phi^*(y)|0\rangle_{SL} = f(x-y) = f(\xi^2)_{SL} = \langle 0|\phi(y)\phi^*(x)|0\rangle_{SL}, \tag{24}$$

so the assumed anticommutator becomes

$$\langle 0|\phi^*(x)\phi(y) + \phi(x)\phi^*(y)|0\rangle_{SL} = 0. \tag{25}$$

Gauge invariance returns this to

$$\langle 0| \left(\phi(x) \pm \phi^*(x)\right)\left(\phi(y) \pm \phi^*(y)\right)|0\rangle_{SL} = 0. \tag{26}$$

The Hall-Wightman theorem makes this generally true, even at $x = y$, and as for the Hermitian fields

$$\phi(x)|0\rangle \pm \phi^*(x)|0\rangle = 0 \tag{27}$$

so $\phi(x) = \phi^*(x) \equiv 0$, and the assumption of anticommutation is untenable.

The spin-$\frac{1}{2}$ development is similar. Here the key relation for spacelike separations is

$$\langle 0|\psi_\alpha(x)\psi_\alpha^*(y)|0\rangle_{SL} = [-]\langle 0|\psi_\alpha(y)\psi_\alpha^*(x)|0\rangle_{SL} \tag{28}$$

and the assumed commutation relation becomes

$$\langle 0|\psi_\alpha^*(x)\psi_\alpha(y) + \psi_\alpha(x)\psi_\alpha^*(y)|0\rangle_{SL} = 0. \tag{29}$$

This is the piece of the Majorana field bilinear $\langle 0|\psi_\alpha^M(x)\psi_\alpha^M(y)|0\rangle$ which does not vanish by the assumed gauge invariance. Now the Hall-Wightman continuation can be made to all $\xi = x - y$ including zero, and finally we arrive at the contradiction with postulate (V)

$$\psi_\alpha|0\rangle = \psi_\alpha^*|0\rangle = 0. \tag{30}$$

Before turning at last to the details of the Hall-Wightman theorem, we discuss the proof of the Spin-Statistics Theorem put forward by Burgoyne almost simultaneously with, but independent of Lüders and Zumino. They acknowledge each others papers, but Burgoyne is able to arrive at his proof without recourse to the postulate used by Lüders and Zumino of an underlying gauge invariance, which Burgoyne characterizes as superfluous.

At the same time, Lüders and Zumino have the merit, by limiting their explicit discussion to spin-0 and spin-$\frac{1}{2}$, that their arguments are substantially more transparent than Burgoyne's. This is especially the case where they establish the reflection properties of the vacuum expectation value of the product of two fields as

1) a function $f(\xi^2)$ depending only on the square of the separation $\xi^2 = (x - y)^2$ for spacelike separation of the two field points in the case of a spin-0 field, and

2) as a vector function $\xi_4 g(\xi^2)$ for spacelike separation in the case of a spin-$\frac{1}{2}$ field.

Neither paper is supported by a discussion of the Hall-Wightman justification of the continuation to zero separation. We will postpone this until the next chapter, after looking at Burgoyne's work.

§2. Burgoyne's Proof of the Spin-Statistics Theorem.

Burgoyne's proof [17.13] is essentially the same as that of Lüders and Zumino, but is accomplished in a more economical and elegant way, and is applicable to fields of arbitrary spin. In Burgoyne's words: "The argument uses recently developed techniques of quantum field theory $\cdots$ establishes the theorem in great generality $\cdots$ no assumptions $\cdots$ about the form of the field equations or interactions $\cdots$."

Burgoyne requires, as did Lüders and Zumino,

411

0) relativistically invariant field theory (compare Lüders and Zumino's postulate I) which has

a) no negative energy states ($\sim$III);

b) a positive definite Hilbert space metric (IV); and

c) distinct fields which either commute or anticommute for spacelike separations (II). Then he shows that the assumption of the wrong connection between spin and statistics leads to the unacceptable result (V) that the field must vanish.

Burgoyne considers fields $\Phi_\mu(x)$ which transform under some finite dimensional irreducible representation μ of the homogeneous Lorentz group (no inversions). He introduces a set of test functions f^μ and defines tempered field operators

$$\Phi(f) = \int d^4x f^\mu(x)\Phi_\mu(x), \quad \Phi^*(f) = \int d^4x f^{\mu*}(x)\Phi_\mu^*(x). \qquad (31)$$

The vacuum expectation values

$$F_{\mu\lambda}(\xi) \equiv \langle 0|\Phi_\mu(x)\Phi_\lambda^*(y)|0\rangle \quad \text{and} \quad G_{\mu\lambda}(\xi) \equiv \langle 0|\Phi_\mu^*(x)\Phi_\lambda(y)|0\rangle \qquad (32)$$

are functions only of $\xi = x - y$. Looking forward to the Hall-Wightman theorem, we accept Burgoyne's statement that F and G are functions of $z = \xi - i\eta$ for z^2 in the complex plane cut along the non-negative real axis. For $\xi^2 < 0$ (spacelike separation of (x,y))

$$G_{\mu\lambda}(-\xi) = \pm G_{\mu\lambda}(\xi), \qquad (33)$$

(+) for integral spin, (-) for half-integral spin. This is in agreement with the Lorentz scalar nature of $\langle 0|\phi^*(x)\phi(y)|0\rangle$ and the Lorentz vector (time-component) nature of $\langle 0|\psi^\dagger(x)\psi(y)|0\rangle$, as stated in Lüders and Zumino's proof, and as proved in an appendix by Burgoyne.

The "wrong-sign" commutation relation, at spacelike separation, reads

$$\langle 0|\Phi_\mu(x)\Phi_\lambda^*(y) \pm \Phi_\lambda^*(y)\Phi_\mu(x)|0\rangle_{SL} = 0. \qquad (34)$$

Now

$$
\begin{aligned}
\langle 0|\Phi_\lambda^*(y)\Phi_\mu(x)|0\rangle &\equiv G_{\lambda\mu}(-\xi) \\
&= \pm G_{\lambda\mu}(\xi) \\
&= \pm\langle 0|\Phi_\lambda^*(x)\Phi_\mu(y)|0\rangle \\
&= \pm\langle 0|\Phi_\lambda^*(-y)\Phi_\mu(-x)|0\rangle, \qquad (35)
\end{aligned}
$$

by various applications of Lorentz invariance at spacelike separations. By analyticity, the "wrong-sign" commutator

$$
F_{\mu\lambda}(\xi) + G_{\lambda\mu}(\xi) = 0 \qquad (36)
$$

vanishes everywhere in the cut ξ plane. Instead of going directly to $\xi = 0$, Burgoyne uses the tempered fields

$$
\langle 0|\int d^4x f^\mu(x)\left[\Phi_\mu(x)\Phi_\lambda^*(y) + \Phi_\lambda^*(-y)\Phi_\mu(-x)\right] f^{\lambda*}(y)d^4y|0\rangle \quad . \quad (37)
$$

which is just

$$
\langle 0|\Phi(f)\Phi^*(f) + \Phi^*(g)\Phi(g)|0\rangle \qquad (38)
$$

where we have introduced $g^\mu(-x) = f^\mu(x)$. Using Burgoyne's postulate (b) (Lüders and Zumino's (IV)) that the Hilbert space metric should be positive definite, we obtain

$$
|\Phi^*(f)|0\rangle|^2 + |\Phi(g)|0\rangle|^2 = 0, \qquad (39)
$$

and conclude that $\Phi^*(f) = \Phi(g) = 0$ for all f (and g), and the assumption of "wrong-sign" commutation relations is untenable.

§3. Concluding Remarks.

We still have a number of loose-ends in the Burgoyne and in the Lüders and Zumino proofs of the Spin-Statistics Theorem. In the notation of Burgoyne's proof:

1) In the proof that $G_{\mu\lambda}(\xi) = \pm G_{\mu\lambda}(-\xi)$, what if any use is made, even implicitly, of the strong-reflection invariance or, possibly, of **TCP**?

413

The Lüders and Zumino proof is independent of the assumption of **TCP**, and for spacelike $\xi = x - y$ depends only on rotational invariance to deduce the necessary properties of matrix elements of products of two fields, so involves only invariance under proper Lorentz transformations. We are returned very close to - but independent of - the Schwinger-Belinfante philosophy of postulating **TC(P)** or **C** and requiring the spin-statistics connection. In the Burgoyne and in the Lüders and Zumino proofs the Spin-Statistics Theorem is proved first, and the **TCP**-theorem follows. Their advantage lay in having the Hall-Wightman theorem and an understanding of the analyticity properties of the product of two fields.

2) Both of these proofs are done in the Heisenberg representation. What, if anything is lost by using the interaction representation and using the field equations to generate the interactions in a way that preserves the Spin-Statistics Theorem?

If this is possible, then we can resurrect the proofs of deWet, Fierz, and Pauli, among the pioneers who did not appeal to elements of **TCP**.

3) Similar questions arise about the necessity of all the extreme formalism. Is it really necessary? Where, in fact, do we go wrong if we proceed naively like deWet?

4) A whole different class of questions arises from the dissatisfaction of the vast majority of physicists who want an intuitive, heuristic, but convincing proof of the Spin-Statistics Theorem without all the dry formalism. It seems clear that such yearning and the resulting endeavors will never stop, much like the scrutinizing of quantum mechanics itself. This is, of course, as it should be.

§4. Biographical Notes.

a) Gerhart Claus Friedrich Lüders (1920-95) - diploma in Physics Hamburg (47); PhD (50); Max Planck Institute of Physics, Göttingen, Munich (50-60); Professor of Physics, Göttingen (60-); Physics prize of the Göttingen Acad. (59); member of the Göttingen

Acad. Scie.; Max Planck medal (66). Perhaps the most lasting impression of Lüders is the marvelous clarity he brought to the proof of the **TCP**-theorem (†). In all the papers of this period, even the ones he authored alone, he was most generous in his acknowledgements of his sometime collaborator Zumino.

(† - G. Lüders, Annals of Physics **2**, 1 (1957).)

b) Bruno Zumino (1923-) - PhD Rome (45) in Mathematics and Physics; to Professor at NYU (51-68); Senior Research Fellow CERN (68-81); Professor at Berkeley (82-); Member of the National Academy of Sciences; Fellow of the American Academy of Arts and Sciences; Fellow of the Italian Physical Society; Dirac Medal, Trieste (87); Heineman Prize, APS (88); Max Planck Medal, Germany (89).

In one great paper (††), Zumino with Wess constructed the theory of supergauge transformations which began Supersymmetry as an active field of research. Zumino remains a leader with many contributions in supergravity and string theory.

(†† - J. Wess and B. Zumino, 'Supergauge Transformations in Four Dimensions' in: Nucl. Phys. **B70**, 39 (1974); S. Deser and B. Zumino, 'Consistent Supergravity' in: Phys. Lett. **62B**, 335 (1976); B. Zumino, 'Relativistic Strings and Supergauges' in: *Renormalization and Invariance in Quantum Field Theory* (Plenum, New York, 1974), ed. E. Caianiello, p.367.)

c) Peter Nicholas Burgoyne (1932-) - BSc McGill (55); PhD Princeton (61); Asst. Professor and Sloan Fellow, Berkeley (61-66); Assoc. Professor, Illinois (Chicago) (66-68); to Mathematics Professor, UC Santa Cruz (68-95); research in elementary particle theory, finite groups and number theory. Burgoyne came under the influence of Wightman, whom he praises as an inspirational teacher and a fine person, while a visiting student at the Bohr Institute in 1956-57. It was then that he got his introduction to Wightman's new techniques which enabled him to construct his elegant proof of the Spin-Statistics Theorem when just beginning graduate studies.

Bibliography and References.

17.1) G. Lüders and B. Zumino, Phys. Rev. **110**, 1450 (1958); see our App.17A.

17.2) J. Schwinger, Phys. Rev. **82**, 914 (1951); see our App.16A.

17.3) F.J. Belinfante, Physica **6**, 870 (1939); see our App.12A.

17.4) W. Pauli, Phys. Rev. **58**, 716 (1940); see our App.14A.

17.5) G. Lüders, Ann. Phys. **2**, 1 (1957).

17.6) A.S. Wightman, Phys. Rev. **101**, 860 (1956); see our App.18A.

17.7) J.S. deWet, Phys. Rev. **57**, 646 (1940); see our App.13B.

17.8) M. Fierz, Helv. Phys. Acta **12**, 3 (1939); see our App.11A.

17.9) D. Iwanenko and A. Socolow, Phys. Zeits. d. Sowjetunion **11**, 590 (1937); see our App.10B.

17.10) W. Pauli and V. Weisskopf, Helv. Phys. Acta **7**, 709 (1934); see our App.9A.

17.11) J.D. Bjorken and S.D. Drell, *Relativistic Quantum Fields* (McGraw-Hill, New York, 1965), pp.113-118.

17.12) see Ref 17.11, p.388.

17.13) N. Burgoyne, Nuovo Cim. **8**, 607 (1958); see our App.17B.

APPENDIX 17.A:

Excerpt from: Physical Review 110, 1450 (1958)

Connection between Spin and Statistics

GERHART LÜDERS,

Max-Planck-Institute für Physik, Göttingen, Germany

and

BRUNO ZUMINO,

Department of Physics, New York University, Washington Square, New York, New York

(Received January 13, 1958)

Abstract: The proof of the connection between spin and statistics for interacting fields is divided into two parts: commutation relations involving components of a single field, and commutation relations between different fields. The first problem is treated in this paper: the connection between spin and statistics is shown to follow from a few simple postulates. The explicit discussion is limited to the cases of spin zero and of spin one-half.

1. INTRODUCTION

$\cdots\cdots$ The connection between spin and statistics was proved by Pauli [1] for noninteracting particles on the basis of a few simple postulates [2]. In the presence of interaction the theorem splits into two parts:

Commutation relations between two operators of the same field. - $\cdots\cdots$

Commutation relations between different fields. - $\cdots$ commutation $\cdots$ between different boson fields, and between one boson and one fermion field, anticommutation $\cdots$ between different fermion fields. $\cdots$ refer to $\cdots$ as the "normal case".

There is a close relation between the **TCP**-theorem [3] and the connection between spin and statistics. $\cdots$ Lüders, Bell, and Pauli use $\cdots$ the usual connection $\cdots$. On the other hand, Schwinger [5] postulated the **TCP**-theorem (or actually the validity of **TC** and **P** separately) and inferred the connection $\cdots$. Earlier work by Belinfante and Pauli [6] had shown that the connection $\cdots$ can be deduced for charged fields if one postulates **C** invariance for their interaction with the electromagnetic field; this interaction is invariant in any case under **T** and **P**. The $\cdots$ proof of $\cdots$ **TCP** $\cdots$ by Jost [7] still has to use the

417

usual connection $\cdots$

The situation is rather unsatisfactory: the $\cdots$ Spin-Statistics Theorem and the **TCP**-theorem support each other $\cdots$ but no independent proof of either has been given. It is not clear whether Pauli's arguments [1] do not lose their strength when interactions are present ($\cdots$ when **TCP** $\cdots$ is nontrivial $\cdots$). In the present paper a new derivation $\cdots$ for operators of the same field will be given. The proof is valid in the presence of interactions and does not postulate the **TCP**-theorem $\cdots$. We analyze fields which are Hermitian or $\cdots$ admit a gauge transformation of the first kind:

$$\phi(x) \rightarrow \phi(x)e^{i\alpha} \cdots \cdots . \tag{1}$$

The derivation will be given for spin zero and one-half $\cdots$

No detailed analysis appears to have been given $\cdots$ of the problem of $\cdots$ different fields. $\cdots \cdots$ The normal case is not always the only possible one $\cdots$ one of the authors (G.L.) has shown that the normal case provides a possible choice $\cdots$ other choices are possible $\cdots$ obtained by Klein transformations [10]. This freedom of choice $\cdots$ can be shown not to affect the validity of the **TCP**-theorem $\cdots$ by a simple redefinition of the **TCP** transformation.

2. HERMITIAN SPIN-ZERO FIELD

Let $\phi(x)$ be a Hermitian spin-zero field in the Heisenberg representation.

We first postulate:

I) *The theory is invariant with respect to the proper inhomogeneous Lorentz group* ($\cdots$ not including reflections).

$\cdots$ the expectation value $\langle 0|\phi(x)\phi(y)|0\rangle$ is an invariant function of the difference four-vector

$$\xi_\mu = x_\mu - y_\mu. \tag{2}$$

One then has

$$\langle 0|\phi(x)\phi(y)|0\rangle = f(\xi), \tag{3}$$

where $f(\xi)$, for spacelike ξ, depends only on the invariant $\xi_\mu\xi_\mu$ but for timelike ξ depends also upon whether this vector points into the future or past light cone [12]. $\cdots$ the physical vacuum is a Lorentz invariant (and nondegenerate) state. One concludes from Eqn.3 that

$$\langle 0|\,[\phi(x), \phi(y)]_-\,|0\rangle_{SL} = 0 \quad (\xi \text{ spacelike}). \tag{4}$$

We now postulate:

418

II) *Two operators of the same field* ⋯ *separated by a spacelike interval either commute or anticommute (Locality)* [13].

⋯⋯ We ⋯ have ⋯ to show that the assumption

$$\langle 0 | [\phi(x), \phi(y)]_+ | 0 \rangle_{SL} = 0 \tag{5}$$

leads to contradictions with postulates ⋯ given later. From Eqns.4,5 one concludes that

$$\langle 0 | \phi(x)\phi(y) | 0 \rangle_{SL} = 0. \tag{6}$$

If one further postulates:

III) *The vacuum is the state of lowest energy,*

one finds by the method of analytic continuation as used by Hall and Wightman [14] that Eqn.6 holds, not only for ξ spacelike but for all ξ. We assume as usual that

IV) *The metric of the Hilbert space is positive definite.*

This postulate allows one to conclude from Eqn.6 that

$$\phi(x)|0\rangle, \tag{7}$$

where $|0\rangle$ is the physical vacuum. We finally postulate:

V) *The vacuum is not identically annihilated by a field.*

Since Eqn.7 is in contradiction with this postulate, the assumption (5) is untenable.

⋯⋯ it seems unlikely that a Hilbert space can be constructed if Postulate V is not satisfied.

The whole analysis holds also in the absence of interactions, the case originally studied by Pauli [1]. His deduction rests ⋯ on Postulate I and ⋯ II ⋯ IV tacitly ⋯ V is evident for free fields. ⋯ III ⋯ does not seem to play any role in Pauli's proof ⋯.

3. HERMITIAN SPIN ONE-HALF FIELD

Let ⋯ $\psi_\alpha(x)$ be Hermitian, or rather be a Majorana field,

$$\bar{\psi}_\alpha(x) = \psi_\beta(x) C_{\beta\alpha}, \tag{8}$$

where C is the ⋯ charge conjugation matrix [16]. We shall analyze the vacuum expectation value

$$\langle 0 | \psi_\alpha^*(x)\psi_\beta(y) | 0 \rangle = C_{\alpha\delta} \gamma_{\delta\beta}^4 \langle 0 | \psi_\alpha \psi_\beta | 0 \rangle. \tag{9}$$

419

This $\cdots$ is positive definite $\cdots$ we write

$$C_{\alpha\delta}\gamma^4_{\delta\beta} = \eta_{\alpha\beta}. \tag{10}$$

Since expression (9) is the fourth component of a four-vector, it follows from *Postulate I* that

$$\eta_{\alpha\beta}\langle 0|\psi_\alpha(x)\psi_\beta(y)|0\rangle_{SL} = \xi_4 g(\xi), \tag{12}$$

where $g(\xi)$ is an invariant function of the difference vector ξ. Making use of the symmetry of η, one sees that

$$\eta_{\alpha\beta}\langle 0|\,[\psi_\alpha(x),\psi_\beta(y)]_+\,|0\rangle_{SL} = 0 \quad (\xi \text{ spacelike}). \tag{13}$$

Because of *Postulate II* we have only to show that the assumption

$$\eta_{\alpha\beta}\langle 0|\,[\psi_\alpha(x),\psi_\beta(y)]_-\,|0\rangle_{SL} = 0 \tag{14}$$

leads to contradictions with the other postulates. In analogy to Eqn.6, $\cdots$

$$\eta_{\alpha\beta}\langle 0|\psi_\alpha(x)\psi_\beta(y)|0\rangle_{SL} = 0. \tag{15}$$

From *Postulate III* it follows that Eqn.15 holds without restriction on ξ. From Eqn.9 and *Postulate IV*, one finds that

$$\psi_\alpha(x)|0\rangle = 0, \tag{16}$$

$\cdots$ in contradiction with *Postulate V*.

The case of no interaction is again a specialization$\cdots$ our analysis $\cdots$ parallel $\cdots$ for spin zero and spin one-half, Pauli's proof $\cdots$ different in the two cases. $\cdots$ his deduction rests very heavily on Postulate III. $\cdots$ Pauli's argument requires a modification in the case of Majorana fields $\cdots$.

4. NON-HERMITIAN FIELDS

(See the main text for a brief summary of this section.)$\cdots\cdots$

ACKNOWLEDGEMENTS

The authors would like to thank F.J. Dyson for encouragement, H. Lehman for a letter on the Hall-Wightman method, and K. Symanzik for discussions and critical reading of a preliminary version of the manuscript.

Note added in proof. We have been recently informed by Professor A.S. Wightman that results similar to those obtained in this paper have been independently obtained by N. Burgoyne (Copenhagen).

REFERENCES and FOOTNOTES

1) W. Pauli, Phys. Rev. **58**, 716 (1940).

2) ⋯ listed ⋯ where ⋯ needed.

3) G. Lüders, Kgl. Danske Videnskab. Selskab. Mat.-fys. Medd. **28**, No.5 (1954) and Ann. Phys. **2**, 1 (1957); J.S. Bell, Proc. Roy. Soc. **A231**, 79 (1955); W. Pauli, in *Niels Bohr and the Development of Physics* (McGraw-Hill, New York, and Pergamon, London, 1955),edited by W. Pauli.

4) ⋯

5) J. Schwinger, Phys. Rev. **82**, 914 (1951); **91**, 713 (1953).

6) F.J. Belinfante, Physica **6**, 870 (1939); W. Pauli and F.J. Belinfante, Physica **7**, 177 (1940).

7) R. Jost, Helv. Phys. Acta **30**, 409 (1957).

8) To derive the equality of masses (and lifetimes) of particles and antiparticles from ⋯ the **TCP**-theorem one also has to postulate the existence of gauge invariance ⋯ usually not stated ⋯.

9) ⋯

12) More explicit expressions for $f(\xi)$ under the additional assumption of Postulate III were given by G. Källén, Helv. Phys. Acta **25**, 417 (1952); H. Lehman, Nuovo Cimento **11**, 342 (1954); M. Gell-Mann and F.E. Low, Phys. Rev. **95**, 1300 (1954).

13) ⋯

14) D. Hall and A.S. Wightman, Kgl. Danske Videnskab. Selskab, Mat.-fys. Medd. **31**, No.5 (1957).

15) W. Pauli, Prog. Theoret. Phys. Japan **5**, 526 (1950).

16) We use the definition of **C** given by W. Pauli in [3]. ⋯.

APPENDIX 17.B:

Excerpt from: Il Nuovo Cimento, Vol. VIII, N0. 4, 607 (1958)

On the Connection of Spin and Statistics

N. BURGOYNE

Institute for Theoretical Physics-Copenhagen
(ricevuto il 20 Marzo 1958)

Summary. - The relation between spin and statistics is proved using only the most general assumptions.

The purpose of this note is to give a proof of Pauli's well known theorem on the connection of spin with statistics. The argument uses recently developed techniques of quantum field theory [1,2] and establishes the theorem in great generality. In particular, no assumptions are made about the form of the field equations or interactions [3]. We show that if

0) a relativistically invariant field theory has the properties:

a) No negative energy states;

b) The metric in Hilbert space is positive definite;

c) Distinct fields either commute or anticommute for spacelike separations,

then no field can have the ≪wrong≫ connection of spin with statistics.

Consider first a field Φ, transforming under some finite dimensional irreducible representation $\Lambda \rightarrow S(\Lambda)$ of the homogeneous Lorentz group (without inversions). Then

$$\mathbf{U}(0, \Lambda)\Phi(f)\mathbf{U}(0, \Lambda)^{-1} = \Phi(f_\Lambda), \tag{1}$$

where

$$\Phi(f) = \int d^4 x f^\mu(x)\Phi_\mu(x), \quad f_\Lambda^\mu(x) = f^\lambda(\Lambda^{-1}x)S_{\lambda\mu}(\Lambda^{-1}), \tag{2}$$

and the f belong to an invariant space of test functions. $\{a, \Lambda\} \rightarrow \mathbf{U}(a, \Lambda)$ is a unitary representation of the inhomogeneous Lorentz group. $\mathbf{U}$ operates on a Hilbert space of state vectors, containing a unique vacuum $|0\rangle$. We define $\Phi^*(f) = \left[\Phi(\bar{f})\right]^*$, where $\bar{f}$ is the complex conjugate of f. The metric is $x^2 = x_0^2 - \mathbf{x}^2$.

The twofold vacuum expectation values are

$$F_{\mu\lambda}(\xi) = \langle 0|\Phi_\mu(x)\Phi_\lambda^*(y)|0\rangle,$$
$$G_{\mu\lambda}(\xi) = \langle 0|\Phi_\mu^*(x)\Phi_\lambda(y)|0\rangle, \quad \xi = x - y. \tag{3}$$

F and G can be extended to functions of a complex 4-vector $z = \xi - i\eta$, analytic when z^2 varies in the complex plane cut along the non-negative real axis [4]. The region $\xi^2 < 0$ is within the cut plane, and in this region

$$G_{\mu\lambda}(-\xi) = \pm G_{\mu\lambda}(\xi) \tag{4}$$

the upper sign for integral spins (S single valued), the lower for half-integral spins (see Appendix for proof).

Using these results the $\ll$ *wrong* $\gg$ commutation relations lead to

$$F_{\mu\lambda}(\xi) + G_{\lambda\mu}(\xi) = 0 \tag{5}$$

and by analyticity

$$F_{\mu\lambda} + G_{\lambda\mu} = 0 \tag{6}$$

identically.

Now consider the state vectors $\Phi(f)|0\rangle$ and $\Phi^*(\widetilde{f})|0\rangle$, where $\widetilde{f}(x) = \bar{f}(-x)$. We compute,

$$|\Phi(f)|0\rangle|^2 + |\Phi^*(\widetilde{f})|0\rangle|^2 = \int d^4x\, \bar{f}^\mu(x)\left[G_{\mu\lambda}(x - y) + F_{\lambda\mu}(x - y)\right] f^\lambda(y) d^4y = 0. \tag{7}$$

Therefore $\Phi(f)|0\rangle = \Phi^*(\widetilde{f})|0\rangle = 0$ for all f. Using assumption (c) and the results of HALL and WIGHTMAN on the real analytic points we conclude that any vacuum expectation value containing Φ vanishes identically and consequently that such a field is zero [2,4].

***

The author is indebted to A.S. WIGHTMAN for his valuable comments and advice. He also thanks Professor NIELS BOHR for the hospitality extended to him at the Institute for Theoretical Physics.

APPENDIX

G satisfies the relation,

$$\bar{S}_{\mu\mu'}(\Lambda)S_{\lambda\lambda'}(\Lambda)G_{\mu'\lambda'}(\xi) = G_{\mu\lambda}(\Lambda\xi), \tag{1}$$

423

by standard invariance arguments [5] one finds that G is given uniquely as a sum of terms of the form

$$\Gamma_{\mu\lambda}(\alpha_1 \cdots \alpha_n)\xi_{\alpha_1} \cdots \xi_{\alpha_n} g(\xi), \quad (\alpha_i = 1 \cdots 4), \tag{2}$$

where

$$\bar{S}_{\mu\mu'}(\Lambda)S_{\lambda\lambda'}(\Lambda)\Gamma_{\mu'\lambda'}(\alpha_1 \cdots \alpha_n) = \Gamma_{\mu\lambda}(\alpha_1' \cdots \alpha_n')\Lambda_{\alpha_1'\alpha_1} \cdots \Lambda_{\alpha_n'\alpha_n}, \tag{3}$$

with $g(\Lambda\xi) = g(\xi)$, and n is always even for S single valued, and always odd for S double valued.

On extending G to complex values, it becomes

$$G_{\mu\lambda}(z) = \cdots + \Gamma_{\mu\lambda}(\alpha_1 \cdots \alpha_n)z_{\alpha_1} \cdots z_{\alpha_n} g(z) + \cdots, \tag{4}$$

where $g(z) = g(\Lambda_c z)$ for z in the cut plane, and Λ_c belongs to the complex Lorentz group (with determinant +1). In particular $g(\xi) = g(-\xi)$ for $\xi^2 < 0$. This proves the original statement.

REFERENCES and FOOTNOTES

1) A.S. WIGHTMAN: *Phys. Rev.* **101**, 860 (1956).

2) D. HALL and A.S. WIGHTMAN: *Dan. Vid. Selsk.* **31**, no. 5 (1957).

3) After the present work was complete, I received a preprint of a paper on the same subject by G. LÜDERS and B. ZUMINO which deals with the case of spin zero and spin one half fields. For the case of neutral fields their result coincides with mine, but for charged fields they introduce a spurious hypothesis of ≪gauge invariance≫. The reader is referred to their paper for a description of previous papers on the same subject.

4) The argument given in [1] is for scalar fields but is easily extended to the case of arbitrary fields. See, for example, A.S. WIGHTMAN hectographed notes on lectures at Institute Henri Poincaré (1957). The same remarks apply to the result that a theory is uniquely determined by its vacuum expectation values.

5) The property proved in this appendix follows from the analysis of W. PAULI: *Phys. Rev.* **58**, 716 (1940).

Chapter 18

The Hall-Wightman Theorem

Summary: The Hall-Wightman theorem establishes that the vacuum expectation value

$$\langle 0|\phi(x)\phi(y)|0\rangle = F^{(2)}(x-y)$$

of the product of two fields is analytically continuable to all separations $\xi = x - y$, which is crucial to the proof of the Spin-Statistics Theorem. Following Jost, we summarize the 'Pauli Era' status of the Spin-Statistics Theorem.

§1. Introduction. Analytic Continuation for Free Fields.

The essential feature required from the Hall-Wightman theorem [18.1] is previewed by A.S. Wightman [18.2]. Wightman states the theorem that the vacuum expectation value

$$\langle 0|\phi(x)\phi(y)|0\rangle \tag{1}$$

is continuable to all $\xi = x - y$ including zero, and must be positive definite at $\xi = 0$. This conclusion is in contradiction to the assumption of Fermi-Dirac statistics with anticommutation relations for spin-zero fields, as was originally assumed correctly but without proof, by deWet [18.3]. Wightman's demonstration, developed formally by Hall and Wightman, forms the basis of the Lüders and Zumino proof [18.4] of the Spin-Statistics Theorem. The extension to non-Hermitian fields is required in the Burgoyne proof [18.5].

It is instructive to specialize Wightman's general arguments for vacuum expectation values for n fields to the case of immediate interest $n = 2$, and discuss only the Wightman function

$$F^{(2)}(x, y) \equiv \langle 0|\phi(x)\phi(y)|0\rangle. \tag{2}$$

From Lorentz invariance, it is established that $F^{(2)}$ can depend only on the four-vector difference $\xi = x - y$. This is just $\Delta^{(+)}(\xi)$ in the standard notation [18.6].

For a non-interacting scalar Hermitian field of mass m

$$F^{(2)}(\xi) \to F_0^{(2)}(\xi) = \frac{m^2}{8\pi i} \frac{H_1^{(1)}(m(\xi^2)^{\frac{1}{2}})}{m(\xi^2)^{\frac{1}{2}}}, \tag{3}$$

(where the Hankel function $H_1^{(1)} = J_1 + iY_1$ in terms of the usual Bessel functions). deWet's argument would require the limit of $F_0^{(2)}(\xi)$ as ξ approaches zero from a spacelike direction so

$$(\xi^2)^{\frac{1}{2}} \to (-\vec{\xi}^2)^{\frac{1}{2}} \to +i|\vec{\xi}|, \tag{4}$$

and

$$H_1^{(1)} \equiv J_1(m(\xi^2)^{\frac{1}{2}}) + iY_1(m(\xi^2)^{\frac{1}{2}}) \to -\frac{2i}{\pi} \frac{1}{m(\xi^2)^{\frac{1}{2}}} \to -\frac{2i}{\pi} \frac{1}{mi|\vec{\xi}|}. \tag{5}$$

And finally

$$F_0^{(2)}(\xi)_{SL} \to \frac{m^2}{8\pi i} \frac{-2i}{\pi} \left(\frac{1}{mi|\vec{\xi}|} \right)^2 \to \frac{+1}{4\pi^2|\vec{\xi}|^2} \to +\infty. \tag{6}$$

The question not addressed by deWet was whether this same positive limit results if $\xi \to 0$ from inside the lightcone. The answer given by Wightman is that it does, but the way this happens is rather intricate. Wightman shows that $F^{(2)}(\xi)$ - even with interactions - is defined as the boundary value of an analytic function of a complex variable $z^2 = (\xi - i\eta)^2$ with $\eta = (\eta_0, 0, 0, 0)$ and $\eta_0 > 0$ (chosen for convenience in our discussion, but, in general, in the future lightcone). The function $F^{(2)}(z^2)$ is analytic in the z^2-plane cut along the positive real axis. z^2 real and negative corresponds to: $\eta_0 = 0$, $z = \xi$ and $z^2 = \xi^2 < 0$ spacelike. $Re z^2 > 0$, $Im z^2 \to 0^+$ corresponds to the inside of the past lightcone; $Re z^2 > 0$, $Im z^2 \to 0^-$ to the inside of the future lightcone. The point $z^2 = 0$ corresponds to the past and future lightcones and their intersection at $\xi = 0$. We have already approached the intersection on a real, spacelike path with $Rez^2 < 0$, $Imz^2 = 0$ and obtained $F_0^{(2)}(0)$ positive definite. To approach from ξ inside the future lightcone, say, we follow - in order - the path $z^2 = (t - i\eta_0)^2 - R^2 \to 0$ as $R \to 0, t \to 0^+, \eta_0 \to 0^+$. Then

$$
\begin{aligned}
F_0^{(2)} &= \frac{m^2}{8\pi i} \frac{H_1^{(1)}(m(z^2)^{\frac{1}{2}})}{m(z^2)^{\frac{1}{2}}} \\
&\to \frac{m^2}{8\pi i} \frac{H_1^{(1)}(m\,[(t - i\eta_0)^2]^{\frac{1}{2}})}{m\,[(t - i\eta_0)^2]^{\frac{1}{2}}}, \quad R \to 0 \\
&\to \frac{m^2}{8\pi i} \frac{H_1^{(1)}(m\,[(-i\eta_0)^2]^{\frac{1}{2}})}{m\,[(-i\eta_0)^2]^{\frac{1}{2}}}, \quad t \to 0 \\
&\to \frac{m^2}{8\pi i} \frac{-2i}{\pi} \left(\frac{1}{m\,[(-i\eta_0)^2]^{\frac{1}{2}}} \right)^2, \quad \eta_0 \to 0^+ \\
&= \frac{+1}{4\pi^2\eta_0^2} \to +\infty,
\end{aligned}
\tag{7}
$$

as before. The limit of $F_0^{(2)}$ is the same on spacelike and "timelike" paths to $\xi = 0$. Without Wightman's analytic continuation, we might have arrived at the naive result

$$
F_0^{(2)} \to \frac{-1}{4\pi^2 t^2} \to -\infty,
\tag{8}
$$

different from the limit on a spacelike path. The result would be that no firm conclusion could be deduced about $F^{(2)}(\xi)$ as $\xi \to 0$.

Wightman also points out [18.7] that in the presence of interactions we can always write the full Wightman function

$$
F^{(2)}(z^2) = \int dm^2 \rho(m^2) F_0^{(2)}(z^2; m^2)
\tag{9}
$$

as a superposition of Wightman functions for non-interacting fields of mass m, superposed with a positive, real mass distribution $\rho(m^2) \geq 0$. It is usually too naive to suppose that

$$
\frac{1}{Z} \equiv \int_0^\infty dm^2 \rho(m^2)
\tag{10}
$$

427

is finite, so we cannot simply assume that the singularity of $F^{(2)}(z^2)$ is the same order as in the noninteracting case. Nonetheless, the analytic continuation to $\xi = 0$ is valid and the deWet proof of the Spin-Statistics Theorem for spin-0 is finally justified. In addition, the technique can be neatly extended to the Fermi-Dirac case.

§2. The Essential Result of the Hall-Wightman Theorem.

With these heuristic results, we next turn to Wightman's pedagogical demonstration of the essential feature of the Hall-Wightman theorem. We again specialize his results from the vacuum expectation value of n fields to the case of immediate interest with $n = 2$,

$$F^{(2)}(x_1, x_2) = \langle 0|\phi(x_1)\phi(x_2)|0\rangle, \tag{11}$$

where $\phi(x)$ is a neutral scalar field. These singular functions are Schwartz distributions, which are defined with the help of well-behaved test functions $f(x_1, x_2)$ as

$$F^{(2)}(f) = \int d^4x_1 d^4x_2 f(x_1, x_2) F^{(2)}(x_1, x_2). \tag{12}$$

The structure of $F^{(2)}$ is constrained by the Lorentz invariance of the vacuum state and the Lorentz transformation property of the fields. For

$$x \rightarrow x' = \Lambda x + a \tag{13}$$

the field ϕ transforms as

$$\phi(x) \rightarrow \phi'(x') = \mathbf{U}(a, \Lambda)\phi(x)\mathbf{U}^{-1}(a, \Lambda) = \phi(\Lambda x + a), \tag{14}$$

the last step for scalar fields. $\mathbf{U}(a, \Lambda)$ is a unitary transformation for the Lorentz transformations of interest here, which do not include time-reversal. Then, using $\mathbf{U}^{-1}|0\rangle = |0\rangle$,

$$
\begin{aligned}
F^{(2)}(x_1, x_2) &\equiv \langle 0|\phi(x_1)\phi(x_2)|0\rangle \\
&= \left\{\mathbf{U}^{-1}|0\rangle\right\}^\dagger \phi(x_1)\mathbf{U}^{-1}\mathbf{U}\phi(x_2)\left\{\mathbf{U}^{-1}|0\rangle\right\}, \\
&= \langle 0|\left\{\mathbf{U}\phi(x_1)\mathbf{U}^{-1}\right\}\left\{\mathbf{U}\phi(x_2)\mathbf{U}^{-1}\right\}|0\rangle,
\end{aligned}
$$

$$
\begin{aligned}
&= \langle 0|\phi(\Lambda x_1 + a)\phi(\Lambda x_2 + a)|0\rangle \\
&= F^{(2)}(\Lambda x_1 + a, \Lambda x_2 + a).
\end{aligned}
\tag{15}
$$

For $\Lambda = 1$, and $a = -x_2$ this is $F^{(2)}((x_1 - x_2), 0)$, so $F^{(2)}$ is a function only of the difference $\xi = x - y$ and we write

$$
F^{(2)}(x_1, x_2) = F^{(2)}(\xi).
\tag{16}
$$

The Fourier transform of $F^{(2)}(\xi)$

$$
\begin{aligned}
G^{(2)}(p) &= \int d^4\xi\, e^{ip\cdot\xi}\langle 0|\phi(0)\phi(-\xi)|0\rangle \\
&= \int d^4\xi\, e^{ip\cdot\xi}\langle 0|\phi(0)e^{-i\mathcal{P}\cdot\xi}\phi(0)|0\rangle,
\end{aligned}
\tag{17}
$$

(where we write $\mathbf{U}(-\xi, 1) = \exp(-i\mathcal{P}\cdot\xi)$ in terms of the momentum operator $\mathcal{P}$). Inserting a complete set of momentum eigenstates $|\psi\rangle$, this becomes

$$
\begin{aligned}
G^{(2)}(p) &= \sum_\psi \int d^4\xi\, e^{i(p - p_\psi)\cdot\xi}|\langle\psi|\phi(0)|0\rangle|^2 \\
&= \sum_\psi (2\pi)^4\delta^4(p - p_\psi)|\langle\psi|\phi(0)|0\rangle|^2,
\end{aligned}
\tag{18}
$$

which is zero unless $p = p_\psi$, which are postulated to be in the future lightcone with $p_\psi^2 = p_{\psi 0}^2 - \vec{p}_\psi^2 \geq 0$ and $p_{\psi 0} \geq 0$ for all permissable states $|\psi\rangle$.

Then

$$
F^{(2)}(z) \equiv F^{(2)}(\xi - i\eta) = \int \frac{d^4p}{(2\pi)^4} e^{-ip\cdot(\xi - i\eta)}G^{(2)}(p)
\tag{19}
$$

is analytic for η in the forward lightcone where the convergence factor

$$
e^{-ip\cdot(\xi - i\eta)} \sim e^{-p_0\eta_0}
\tag{20}
$$

defines the integration, given the above properties of $G^{(2)}$. $F^{(2)}(\xi)$ is the boundary value of a function analytic in the 8-dimensional "future tube" of the 4-vectors (ξ, η). From Lorentz invariance (not including time-reversal)

$$
F^{(2)}(z) = F^{(2)}(\Lambda z), \quad G^{(2)}(p) = G^{(2)}(\Lambda p),
\tag{21}
$$

429

which leads to the theorem that $F^{(2)}$ is a function of z^2 in the future tube

$$- \infty < Re z_\mu < +\infty, \quad Im z_0 > 0. \tag{22}$$

One sees that for this set,

$$z^2 = \xi^2 - \eta^2 - 2i\xi\cdot\eta \tag{23}$$

fills the entire z^2 plane cut along the positive real axis. z^2 is real and negative for $\eta = 0$ and ξ^2 spacelike; z^2 is positive real with an infinitesimal positive imaginary part (above the cut) for ξ^2 timelike, $\xi_0 < 0$, ie in the past lightcone; and finally, z^2 positive real with infinitesimal negative imaginary part (below the cut) for ξ^2 timelike, $\xi_0 > 0$, in the future lightcone.

Wightman gives the example described above of $F_0^{(2)}(z^2)$ for free fields.

Finally Wightman establishes that $F^{(2)}$ has the required positivity. The vector

$$\int d^4x\, f_1(x)\phi(x)|0\rangle \equiv \phi(f_1)|0\rangle \tag{24}$$

has the positive definite (length)^2

$$\langle 0|\phi(f_1)\phi(f_1)|0\rangle = \int d^4x_1\, f_1(x_1)F^{(2)}(x_1 - x_2)f_1(x_2)d^4x_2 \geq 0. \tag{25}$$

This is the necessary and sufficient condition that $F^2(\xi)$ is the Fourier transform of a positive measure with the properties of $G^{(2)}(p)$. These are almost the properties used by Burgoyne in his proof. It still needs to be shown that $F^{(2)}(z) = F^{(2)}(\Lambda z) \to F^{(2)}(z^2)$, and that this can be extended to non-Hermitian fields and to Dirac fields.

The full Hall-Wightman analysis to understand the $8(n-1)$ dimensional complex manifold is complicated by the fact that the difference vectors are not all independent. The Hall-Wightman program was directed at the reconstruction of a field theory from all its vacuum expectation values and is vastly more ambitious and complicated than the result needed for the proofs of Lüders and Zumino, and of Burgoyne. For this reason, we include only the most brief excerpt from their much-referenced monumental work in order at least to get a flavor of their achievement.

§3. Jost's Summary of the Pauli Era.

Jost's tribute to Pauli [18.8] is a fitting summary of the understanding of the Spin-Statistics Theorem at the end of the "classical" or Pauli era from the perspective of that time. Jost clearly seems content with the classical proofs of Pauli and his contemporaries, based on the requirements of relativistic quantum field theory. In his expressed view, there is no need to pursue the problem further. This is the epitome of the state of affairs which generated the dissatisfaction of Neuenschwander, seconded by Feynman, and prompted our exploration of the Spin-Statistics Theorem. In the following chapters we will discuss suggested alternatives including finally, one of our own which we believe makes understandable in a new logical context all the preceding work.

§4. Biographical Note on A.S. Wightman.(†)

Arthur Strong Wightman (1922-) - BA Yale (42); PhD Princeton (49); from Instructor to Thomas D. Jones Professor of Mathematical Physics, Princeton (49-92); Emeritus (92-). DSc Swiss. Fed. Inst. Techno. (69); DSc Göttingen (86); Member of the National Academy of Science, AAAS, American Mathematical Society, Federation of American Scientists, Royal Society of Arts and Sciences; visitor, Copenhagen (51-52); Copenhagen and Naples (56-57); Inst. Adv. Study Sci., Bures-sur-Yvette (63-64, 68-69); Ecole Polytechnique (77-78); Adelaide (82). Author with R.F. Streater of the standard reference *PCT, Spin and Statistics, and All That* (Benjamin, New York, 1964).

Wightman, with Haag and Jost and a few others, is a pioneer who brought mathematical rigor and integrity to quantum field theory. His results were frequently difficult and sometimes unpopular among those whose idea of a sufficient proof of existence is a modification of the old aphorism to *"Cogito, ergo est."*

(† - *American Men and Women of Science* (Bowker, New York, 1986), 16th ed., Vol VII, p.615.)

§5. Biographical Note on Jost.(††)

Res Jost (1918-90)- studied at Bern, Zurich, and ETH; wrote his thesis with Wentzel on nuclear forces; Institute for Advanced Study (49-55); visiting appointments at Copenhagen, Princeton, Bures-sur-Yvette; Professor ETH Zurich (55-); awarded the Planck Medal of the German Physical Society (84). He invented the Jost function representation of the S-matrix to clearly distinguish bound state zeros in the analytic structure from "false"-zeros; with Kohn, he pioneered the inverse scattering problem in quantum mechanics; with Lehman, he pioneered the study of analytic properties of relativistic scattering amplitudes; he first recognized the **TCP**-theorem as a simple consequence of general principles of relativistic quantum field theory; author of *The General Theory of Quantized Fields* (AMS,1965). His later work on the history of modern physics was made notable by his deep understanding of physics, history and philosophy.

(†† - Physics Today, Feb 1992.)

Bibliography and References.

18.1) D. Hall and A.S. Wightman, Matt. Fys. Medd. Dan. Vid. Selskb. **39**, 5 (1957); see our App.18B.

18.2) A.S. Wightman, Phys. Rev. **101**, 860 (1956); see our App.18A.

18.3) J.S. deWet, Phys. Rev. **57**, 646 (1940); see our App.13B.

18.4) G. Lüders and B. Zumino, Phys. Rev. **110**, 1450 (1958); see our App.17A.

18.5) N. Burgoyne, Nuovo Cim. **8**, 607 (1958); see our App.17B.

18.6) J.D. Bjorken and S.D. Drell, *Relativistic Quantum Fields* (McGraw-Hill, New York, 1965), p.389.

18.7) H. Umezawa and S. Kamefuchi, Progr. Theor. Phys. **4**, 543 (1951); G. Källen, Helv. Phys. Acta **25**, 417 (1952); H. Lehman, Nuovo Cim. **11**, 342 (1954); see also Ref. 18.6, pp.139,227.

18.8) R. Jost, 'The Pauli Principle and the Lorentz Group' in *Theoretical Physics in the Twentieth Century: A Memorial Volume to Wolfgang Pauli* (Interscience, New York, 1960), edited by M. Fierz and V.F. Weisskopf, pp.107-137; see our App.18C.

APPENDIX 18.A:

Excerpt from: Physical Review 101, 860 (1956)

Quantum Field Theory in Terms of
Vacuum Expectation Values

A. S. WIGHTMAN

Palmer Physical Laboratory, Princeton University, Princeton, New Jersey
(Received July 18, 1955)

Vacuum expectations values of products of neutral scalar field operators are discussed. The properties of these distributions arising from Lorentz invariance, the absence of negative energy states and the positive definiteness of the scalar product are determined. The vacuum expectation values are shown to be boundary values of analytic functions. Local commutativity of the field is shown to be equivalent to a symmetry property of the analytic functions. The problem of determining a theory of a neutral scalar field given its vacuum expectation values is posed and solved.

1. INTRODUCTION

Recent work in relativistic quantum field theory has made heavy use of certain basic singular functions defined as vacuum expectation values of products of fields taken at various space-time points. In this paper, we present some results of a systematic study of relativistic field theory based on such vacuum expectation values. For simplicity, we treat the case of a neutral scalar field interacting with itself. The methods used have generalizations in any field theory.

The objects of our attentions are the singular functions:

$$F^{(n)}(x_1, \cdots x_n) = \langle 0|\phi(x_1)\phi(x_2)\cdots\phi(x_n)|0\rangle,$$

where $|\Psi_0\rangle = |0\rangle$ is the vacuum state, assumed to be the unique state of energy and momentum zero, and $\phi(x)$ is a neutral scalar field. As is well known, $F^{(n)}$ has to be understood as a distribution in the sense of L. Schwartz. It is a linear functional which gives a complex number for each infinitely differentiable function $f(x_1, \cdots x_n)$ which vanishes outside a bounded region of space-time:

$$F^{(n)} = \int d^4x_1 \cdots d^4x_n f(x_1 \cdots x_n) F^{(n)}(x_1 \cdots x_n).$$

(We call such f test functions.) Furthermore, $F^{(n)}(f_k) \to 0$ if a sequence of test functions $f_k(x_1 \cdots x_n)$ (vanishing outside a fixed bounded region) and all their derivatives converge to zero uniformly in space-time. We shall study the structure of $F^{(n)}$, exploiting systematically the Lorentz transformation properties of $\phi(x)$ which are given by

$$\mathbf{U}(a,\Lambda)\phi(x)\mathbf{U}(a,\Lambda)^{-1} = \phi(\Lambda x + a). \tag{1}$$

Here, $\{a,\Lambda\}$ is an element of the inhomogeneous Lorentz group $\cdots$ $\mathbf{U}(a,\Lambda)$ is the corresponding unitary $\cdots$ operator which yields the transformed wave functions. We shall also determine the consequences for $F^{(n)}$ of the assumptions that no negative energy states exist in the theory and that $\phi(x)$ is a local field. Finally, we shall show how, given a set $F^{(n)}, n = 1, 2 \cdots$, one can construct a theory of a neutral scalar field which has these $F^{(n)}$ as its vacuum expectation values. We do not show that any set of $F^{(n)}$ outside of those determined by the free field actually exists.

2. CONSEQUENCES OF LORENTZ INVARIANCE

For Lorentz transformations without time inversion, we know that $\mathbf{U}(a,\Lambda)$ is unitary. Thus,

$$
\begin{aligned}
F^{(n)}(x_1, \cdots x_n) &= \langle 0|\mathbf{U}(a,\Lambda)^{-1}\mathbf{U}(a,\Lambda)\phi(x_1)\cdots\phi(x_n)|0\rangle \\
&= \langle 0|\phi(\Lambda x_1 + a)\cdots\phi(\Lambda x_n + a)|0\rangle,
\end{aligned}
$$

and therefore

$$F^{(n)}(x_1, \cdots x_n) = F^{(n)}(\Lambda x_1 + a, \cdots \Lambda x_n + a) \tag{2}$$

for $\{a,\Lambda\}$ without time inversion. Here we have used Eqn.1 and the Lorentz invariance of the vacuum state:

$$\mathbf{U}(a,\Lambda)|0\rangle = |0\rangle. \tag{3}$$

......

3. CONSEQUENCES OF THE ABSENCE OF NEGATIVE ENERGY STATES

From Eqn.2, we see that $F^{(n)}$ is a function only of the differences of the $x_1, \cdots x_n$. We shall therefore write

$$F^{(n)} = F^{(n)}(\xi_1, \cdots, \xi_{n-1})$$

where

$$\xi_1 = x_1 - x_2, \cdots\cdots, \xi_{n-1} = x_{n-1} - x_n.$$

We shall assume that $F^{(n)}$ has a Fourier transform and shall show that if

$$F^{(n)} = \int \exp\left(-i\sum_{j=1}^{n-1} p_j \cdot \xi_j\right) G^{(n)}(p_1, \cdots p_{n-1}) d^4 p_1 \cdots d^4 p_{n-1},$$

434

then $G^{(n)}$ vanishes unless the p_j satisfy

$$p_j^2 = (p_j^0)^2 - (\vec{p})^2 \geq 0, \quad p_j^0 \geq 0,$$

by virtue of the assumption that no negative energy states exist.

If $|\Psi\rangle$ is an arbitrary state, its component of momentum p is

$$\int e^{-ip\cdot a}d^4a\, \mathbf{U}(a,1)|\Psi\rangle.$$

Thus, by our hypothesis that no negative energy states exist,

$$\int e^{ip\cdot a}d^4a\, F^{(n)}(\xi_1, \cdots \xi_j + a, \cdots \xi_{n-1})$$

$$= \int e^{ip\cdot a}d^4a\langle 0|\phi(x_1)\cdots\phi(x_j)\phi(x_{j+1}-a)\cdots\phi(x_n-a)|0\rangle$$

$$= \langle 0|\phi(x_1)\cdots\phi(x_j)\int e^{ip\cdot a}d^4a\, \mathbf{U}(-a,1)\phi(x_{j+1})\cdots\phi(x_n)|0\rangle$$

vanishes unless p is within or on the forward lightcone. Therefore, $G^{(n)}(p_1\cdots p_{n-1})$ vanishes if any of its arguments lie outside the forward lightcone. In the special case $n = 2$, this result is well known.

It is an important consequence of this property of the $G^{(n)}$ that the distributions $F^{(n)}$ are boundary values of analytic functions. This result is displayed in and simultaneously proved by the formula

$$F^{(n)}(\xi_1 - i\eta_1, \cdots \xi_{n-1} - i\eta_{n-1})$$

$$= \int \exp\left(-i\sum_{j=1}^{n-1} p_j\cdot(\xi_j - i\eta_j)\right)$$

$$\times G^{(n)}(p_1\cdots p_{n-1})d^4p_1\cdots d^4p_{n-1},$$

where the 4-vectors η_j are restricted to lie in the future lightcone. The $8(n-1)$-dimensional open region thus defined in the $8(n - 1)$-dimensional space of the components of the $\xi_1, \cdots \xi_{n-1}, \eta_1, \cdots \eta_{n-1}$ is called the future tube. Thus, $F^{(n)}(\xi_1, \cdots \xi_{n-1})$ is a boundary value of a function analytic in the future tube.

We introduce the notation

$$z_{j,j+1} = \xi_j - i\eta_j.$$

Eqn.2 implies the Lorentz invariance of $G^{(n)}$

$$G^{(n)}(p_1\cdots p_{n-1}) = G^{(n)}(\Lambda p_1, \cdots \Lambda p_{n-1}), \tag{4}$$

for Λ without time inversion $\cdots$. Consequently, we have throughout the tube

$$F^{(n)}(z_{12}, \cdots z_{n-1,n}) = F^{(n)}(\Lambda z_{12}, \cdots \Lambda z_{n-1,n}) \tag{5}$$

435

$\cdots$. We use the following theorem, whose proof will be published in another paper.

Theorem. - A function $f(z_1, \cdots z_n)$ of n 4-vectors $z_1 \cdots z_n$ analytic in the tube: $-\infty < Rez_{i\mu} < +\infty$, $Imz_{i\mu}$ in the future cone, and invariant under the homogeneous Lorentz group without time inversion:

$$f(z_1, \cdots z_n) = f(\Lambda z_1, \cdots \Lambda z_n)$$

is a function of the scalar products $z_j \cdot z_k$, $j, k = 1, 2 \cdots n$. It is analytic in the complex manifold over which the scalar products vary when the vectors $z_1, \cdots z_n$ vary over the future tube.

If we introduce the complex vectors

$$z_{ij} = \sum_{k=i}^{j-1} z_{k,k+1}, \quad i < j,$$

then the theorem tells us that the $F^{(n)}$ are analytic functions of the squares of the lengths of these vectors

$$F^{(n)} = F^{(n)}(z_{ij}^2). \tag{6}$$

As the variables $z_{j,j+1}$ vary over the tube each variable z_{ij}^2 varies in an open set of the complex plane. From the explicit form

$$z^2 = (\xi - i\eta)^2 = \xi^2 - \eta^2 - 2i\xi \cdot \eta,$$

one can easily see that this set fills the entire plane except for the positive real axis and the origin. The situation is is indicated in Fig. 1. $\cdots\cdots$ (*see Sec. 18.2*) $\cdots\cdots$.

4. CONSEQUENCES OF THE COMMUTATION RULES

The local commutation rules

$$[\phi(x), \phi(y)]_- = 0, \quad (x - y)^2 < 0,$$

imply that

$$F(n)(x_1, \cdots x_j, x_{j+1}, \cdots x_n) = F^{(n)}(x_1, \cdots x_{j+1}, x_j, \cdots x_n) \tag{11}$$

as long as x_j and x_{j+1} are space-like separated points. These relations can be extended by analytic continuation to relations of the analytic functions $F^{(n)}(z_{ik}^2)$:

$$F^{(n)}(z_{ik}^2) = F^{(n)}(Pz_{ik}^2), \tag{12}$$

where P stands for the operation of permuting the subscripts j and $j+1$, and, by definition, $z_{ik}^2 = z_{ki}^2$. The proof is simple. Eqn.11 coincides with Eqn.10 for all z_{ik}^2 on the negative real axis. Consequently, it holds everywhere by analytic continuation. Thus, the local property of a field $\phi(x)$ is characterized by a global symmetry relation on the analytic functions $F^{(n)}$.

(The rest of this discussion takes us too far afield and we include only a few isolated remarks.)······.

It is an immediate consequence of the Lorentz invariance of $G^{(2)}$, that the most general $F^{(2)}$ is of the form

$$F^2(z^2) = \int_0^\infty dg(m) \frac{m^2}{8\pi i} \frac{H_1^{(1)}(m(z^2)^{\frac{1}{2}})}{m(z^2)^{\frac{1}{2}}} \tag{16}$$

(apart from a constant), where dg is a weight function which will be shown to be positive in Sec. 5. It would be tempting to argue that the operation indicated in Eqn.14 can be carried out under the integral sign in Eqn.16 so that

$$1/Z = \int_0^\infty dg(m). \tag{17}$$

This is indeed correct if $\int dg < \infty$, but if $\int dg = \infty$, then Eqn.17 is somewhat misleading, as the following example shows ······.

5. POSITIVE DEFINITENESS CONDITIONS

Since the length of a vector is greater than or equal to zero, we have

$$\left| \left[\alpha_0 f_0 + \alpha_1 \int d^4 x_1 f_1(x_1)\phi(x_1) + \alpha_2 \int d^4 x_1 d^4 x_2 \phi(x_1)\phi(x_2) + \cdots \right] |0\rangle \right|^2 \geq 0$$

for all $\alpha_0, \alpha_1, \alpha_2, \cdots$ and all test functions $f_1, f_2, \cdots$. Therefore, ······.

The simplest consequences are obtained by setting all but one of the α_i equal to zero. For example, for $\alpha_1 \neq 0$, we find

$$\int \bar{f}_1(x_1) F^{(2)}(x_1 - x_2) f_1(x_2) d^4 x_1 d^4 x_2 \geq 0. \tag{22}$$

It can be shown that Eqn.22 is a necessary and sufficient condition that $F^{(2)}$ be a Fourier transform of a positive measure of not too fast increase [2] (generalized Bochner theorem). ······

APPENDIX 18.B:

Excerpt from: Mat. Fys. Medd. Dan. Vid. Selsk. 31, no. 5 (1957)

A THEOREM ON INVARIANT ANALYTIC FUNCTIONS WITH APPLICATIONS TO RELATIVISTIC QUANTUM FIELD THEORY

BY

D. HALL and A.S. WIGHTMAN

Synopsis: The paper is in three parts $\cdots$. In the first part, a detailed proof is given of a previously announced theorem: an analytic function of n 4-vector variables invariant under the orthochronous Lorentz group is an analytic function of their scalar products. The second part is devoted to a preliminary study of the domain of analyticity of such invariant analytic functions. The third part applies the preceding results to quantum field theory. It is shown that the vacuum expectation value

$$\langle 0|\phi(x_1)\cdots\phi(x_n)|0\rangle = F^{(n)}(x_1,\cdots x_n)$$

where $\phi(x)$ is a neutral scalar field, is an analytic function of the real variables $z_j - z_{j+1}$, $j = 1,\cdots n-1$ in a region where all these vectors are space-like. *It is shown that the values of $F^{(n)}$ for all values of its arguments are uniquely determined in terms of its values for space-like separations, and that, for $n = 2, 3, 4$, $F^{(n)}$ is determined from its values at points where all times are equal.* (*Note*: Emphasis added.) These results are applied to $\cdots$ two theorems of R. Haag $\cdots$.

Introduction.

In a preceding paper [1], the second-named author showed that the main content of a relativistic quantum theory of a scalar field, $\phi(x)$, is contained in the vacuum expectation values, $F^{(n)}$, $\cdots$. It was shown there that, as a consequence of the transformation law of the field under space-time translations and the absence of negative energy states, the distributions $F^{(n)}$ are boundary values of analytic functions. The analysis of the $\cdots$ $F^{(n)}$ was carried further, using a theorem quoted there without proof $\cdots$: an analytic function of n 4-vector variables invariant under the orthochronous Lorentz group is an analytic function of their scalar products.

The first part of the present paper is devoted to a proof of this theorem. $\cdots\cdots$ In the third part $\cdots$. It is further shown that the values of $F^{(n)}$ for all values of its arguments are uniquely determined in terms of its values for space-like separation. For $\cdots$ $n = 2, 3, 4$, $\cdots$: $F^{(n)}$ is determined everywhere from its values at points where all the times $\cdots$ are equal. $\cdots\cdots$

Outline of the Proof.

······(*Note added:* The proof, unfortunately, runs on for twenty-seven pages and is quite overwhelming, partly because it treats general values of n, whereas we are interested principally in $n = 2$, which turns out to be a special case, and, in fact, a subtle one. We therefore content ourselves with selected quotes which give the relevant results and, hopefully, some of the flavor of the work.)

··· To complete the proof of the theorem, it remains to show that f is analytic on $\mathcal{M}_n$. For $n \leq 4$, analyticity is a perfectly straightforward because $\mathcal{M}_n$ is an open set in complex Euclidean $n(n+1)/2$ space. However, for $n \geq 5$ ······ The reader may find it helpful to think of the example of the light cone. In that case, the point where the tips of the past and future light cones touch is singular and ··· not locally Euclidean. However, ··· the actual situation is much more complicated since singular points only appear on $\mathcal{M}_n$ for $n \geq 5$ ··· In the following, we prove analyticity at all points of $\mathcal{M}_n$ for $n \leq 4$ ···

3. Physical Applications.

Some physical applications of the theorem of Sec.(1) were already discussed in I ······ *the vacuum expectation value* $\langle \Psi_0 | \phi(x_1) \cdots \phi(x_n) | \Psi_0 \rangle$ *is uniquely determined from its values for space-like separated* $x_1, \cdots x_n$. ···

For $F^{(2)}(\xi_1)$, $F^{(3)}(\xi_1, \xi_2)$, and $F^{(4)}(\xi_1, \xi_2, \xi_3)$ an even more striking result holds: ··· (these) are uniquely determined from their values at equal times ······

Acknowledgements.

The authors wish to express their deep gratitude and indebtedness to V. BARGMANN. ··· Part of this paper is based on part of the Princeton thesis (1956) of the first-named author. The second named author is a National Science Foundation Fellow on sabbatical leave from Princeton University. He thanks Professor NIELS BOHR for the hospitality extended to him at the Institute of Theoretical Physics, University of Copenhagen.

Footnotes and References.

1) A.S. WIGHTMAN, Phys. Rev. **101**, 860 (1956) (referred to as I). ······
9) D. HALL, thesis, Princeton (1956), Chapter III, unpublished.
10) R. HAAG, Mat.Fys.Medd.Dan.Vid.Selsk. **29**, No. 12 (1955), pp.30-32. ······

APPENDIX 18.C:

Excerpt from: **Theoretical Physics in the Twentieth Century, a Memorial Volume to Wolfgang Pauli 1960, p.107-137**

The Pauli Principle and the Lorentz Group
Res Jost

Historical Part

§1.Introduction.

Ever since quantized field theory began, it was built on the dictates of Pauli. Not only has the theory progressed through his individual contributions, but also the work of others has profited from his studies and been influenced by his criticism. It is no exaggeration to see in Pauli the personification of the science of field theory.

It was Pauli's special goal to rationalize special relativity and quantum theory with one another. The resolution of this confrontation in a consistent theory is nontrivial even in a model. Nevertheless, embarking on the hard business of this problem, completely unexpectedly, Pauli produced a miracle, the discovery of the Exclusion Principle and the connection between spin and statistics: in a relativistic quantum theory, particles with half-integral spin satisfy the Exclusion Principle, those with integral spin satisfy Bose-Einstein statistics. ······ *(In the interest of brevity, we go to the portions of Jost's article directly concerned with the Spin-Statistics Theorem.)* ······

This ······ is the starting point of the work of Pauli and Weisskopf [11] on the quantization of the scalar relativistic equation. The wave function ϕ describing charged particles of spin-0 is complex but its phase - due to the existence of a gauge group - is fundamentally not observable. The localization argument used in the case of the Maxwell field can not be applied to ϕ without further examination. However, one can assume a local charge density. As for all other observables, this is bilinear in ϕ and ϕ^* and for the force free theory is

$$\rho = ie[(\partial_0\phi^*)\phi - \phi^*(\partial_0\phi)]. \tag{21}$$

Now obviously

$$[\rho(x), \rho(y)]_- = 0 \quad \text{for} \quad (x-y)^2 < 0, \tag{22}$$

in the case where $\phi(x)$ commutes with $\phi(x')$ and

$$[\phi^*(x), \phi(y)]_- = 0 \quad \text{for} \quad (x-y)^2 < 0. \tag{23}$$

The same is true when $\phi(x)$ and $\phi(y)$ anticommute and

$$[\phi^*(x), \phi(y)]_+ = 0 \quad \text{for} \quad (x-y)^2 < 0, \tag{24}$$

so that one cannot choose (2.3) on the basis of localizability. Eqn(2.3) corresponds to canonical quantization and Bose-Einstein statistics. Eqn(2.4) however does not correspond directly to quantization according to the Exclusion Principle, but involves a further puzzle. In fact, if we assume (as is correct for the force free theory) that the left hand side of (2.4) is a c-number, then from (2.4)

$$[\psi^*(x), \psi(y)]_+ = F(x-y), \tag{25}$$

using the wave equation

$$(\Box + m^2)F(\xi) = 0, \tag{26}$$

and Lorentz invariance, $F(\xi)$ is necessarily odd. Therefore, for arbitrary x and y,

$$\psi^*(x)\psi(y) + \psi^*(y)\psi(x) + \psi(y)\psi^*(x) + \psi(x)\psi^*(y) = 0, \tag{27}$$

which is obviously consistent only for $\psi \equiv 0$.

This is one of the arguments given against the quantization of the scalar wave equation according to the Exclusion Principle. In another treatment [12], it was shown that non-locality of the charge density resulted from certain non-local terms in the fields necessitated by "wrong" statistics for the scalar field.

Thus, in the important cases of spin-0, spin-$\frac{1}{2}$, and photons, one had a satisfactory explanation of the connection between spin and statistics. However, before the case of arbitrary spin could be handled, the theory of force free fields had to be extended to an arbitrary finite representation of the Lorentz group. This was done in the extension of Dirac theory [13] by Fierz [14], who showed with complete generality that for the (canonical) quantization of double valued representations (half-integral spin), the total energy is of indefinite sign but the total charge is positive definite. In order to obtain a stable vacuum, one must quantize according to the Exclusion Principle. Then one treats particles with both signs of charge in a symmetrical way. For single valued representations (integer spin) it is the total energy that is positive definite and the charge that is indefinite. Quantization according to Bose-Einstein statistics is therefore possible, but according to the Exclusion Principle is incompatible with a local charge density. The argument is therefore completely analogous to the one which led from (2.4) to the contradiction (2.7).

The argument must be modified somewhat for a real tensor field: Here, as for the Maxwell field, the observability of the field is assumed. An analogous remark holds for the double valued fields, which are subject to Majorana reality conditions. For these one must assume the localizability of a bilinear quantity.

In connection with Fierz's work we must mention also those based upon Wigner's analysis of the irreducible unitary representations of the inhomogeneous Lorentz group [15]. These show that the theory of Fierz produces a wave equation for a representation of the inhomogeneous Lorentz group with real non-vanishing masses. For zero mass one finds wave functions for representations which follow from non-zero masses in the limit. The other zero-mass representations, which contain a continuous instead of a discrete spin-variable, seem to play no physical role. We will not consider their quantization.

The famous work of Pauli [16] titled *The Connection Between Spin and Statistics* generalizes Fierz's result for irreducible spinor fields. Pauli gives a direct proof that, in a *c*-number theory of free fields respecting real Lorentz group invariance, the charge density for a single valued and the energy density for a double valued representation are of undetermined sign. This leads to the same conclusion which we had above from Fierz.

The new proof depends upon a classification of spinors into four classes, and on the observation that the free field wave equation is invariant under a mirror reflection $(x_\mu \rightarrow -x_\mu)$, even though one originally assumed invariance only under the proper Lorentz group.

Spinor fields are classified by the *character* $[(-1)^m, (-1)^n]$ where m and n are the number of dotted and undotted spinor indices. These are $[+, +]$ and $[-, -]$ for single valued representations, the first for tensors of even rank, the second for tensors of odd rank. For double valued representations the values are $[+, -]$ and $[-, +]$, with the definition

$$\psi^*(+, -) = \phi(-, +). \tag{28}$$

For single valued representations, the character does not change under complex conjugation.

The operator $i\partial_\nu = p_\nu$ has the character $[-, -]$ The first order wave equations are of the symbolic form

$$\sum p\psi(+, +) = \sum \psi(-, -), \quad \sum p\psi(-, -) = \sum \psi(+, +). \tag{29}$$

These are invariant under the transformation Θ

$$\psi(+, +) \rightarrow \psi(+, +), \quad \psi(-, -) \rightarrow -\psi(-, -), \quad p \rightarrow -p.$$

Thus a tensor field of odd rank constructed from $p, \psi(+, +), \psi(-, -)$ by repeated operations of addition and multiplication, changes sign under the transformation Θ. It can therefore have no definite charge density.

The double valued representations are of interest. Here the wave equations are of the

442

symbolic form

$$\sum p\psi(+,-) = \sum \psi(-,+), \quad \sum p\psi(-,+) = \sum \psi(+,-). \tag{210}$$

Eqn(2.10) is invariant under the transformation

$$p \to -p, \quad \psi(\sigma_1, \sigma_2) \to \sigma_1 \psi(\sigma_1, \sigma_2)$$

But this transformation is incompatible with (2.8). This completely decisive fact led Pauli to the transformation Θ':

$$\psi(+,-) \to i\psi(+,-), \quad \psi(-,+) \to (-i)\psi(-,+), \quad p \to -p.$$

So every even rank bilinear tensor constructed from $\psi(+,-), \psi(-,+)$ and their derivatives (of arbitrary order) must change sign under the transformation Θ'. Therefore for double valued fields the energy $\int T_{00} dV$ cannot be positive definite.

The existence of the transformations Θ and Θ' is of great significance, as will first become clear from the TCP theorem.

With this proof by Pauli the treatment of the subject came to a definite conclusion. For force free particles the connection between spin and statistics was explained by an important application of the theory of special relativity. (Note: Itallics added)

We list once more the assumptions essential for the proof of the connection between spin and statistics [17]:

1) Invariance under the proper Lorentz group.

2) The existence of a state of lowest energy, which is normalizable and non-degenerate and which we identify with the vacuum.

3) A positive definite scalar product in the linear state space.

4) Physical quantities such as the charge density should commute for spacelike separations. The fields themselves commute or anti-commute for spacelike separations (locality).

§3. The New Work on Spin and Statistics. The TCP Theorem.

New approaches to the subject of the already classical question of spin and statistics came from the development of QED by Schwinger, Feynman, and Dyson. *Note added: see our App.15A,B,C.······*

More noteworthy was the explicit way Schwinger [22] confirmed the spin-statistics connection. His demonstration required the invariance under simultaneous time reversal T and particle-antiparticle conjugation C transformations in theories separately invariant under space reflection P [23].

Independent of Schwinger, Lüders [24] came to the same conclusion, that under very broad assumptions a P invariant theory in which normal permutation relations hold (the spin-statistics connection), the theory is automatically invariant under CT.

The definitive formulation of this theorem comes from Pauli [25]: A field theory invariant under the proper Lorentz group, with normal permutation relations, is also TCP invariant.

The advance of the new wording is that (naturally before the discovery of parity violation) only invariance under the proper Lorentz group was required. Moreover the theorem was valid for arbitrary spin, whereas Lüders limited himself to spin-0, $\frac{1}{2}$ and 1.

It was now quite clear that the basis of the theorem was the fact that TP could be combined into an invariance of the complex Lorentz group. This observation clarifies the formulation of the problem and the essential result used in the proof.·········

To the question: "Commutator or anti-commutator for various fields?" a paper of Lüders [26] should be mentioned. Here the normal permutation relations ······ are assumed. The fundamentally important energy density can exist for no other possibility ······. In more recent research, field theory itself takes new forms ······. The Wightman scheme yields a satisfactory treatment and an appropriate understanding [30]. Less satisfactory, but probably more interesting is the problem of spin and statistics in this formalism. Since the Wightman fields seem to have nothing to do with particles, the only question that can be asked is whether the field commutes or anti-commutes with its complex conjugate. ······ One must provide these with the usual spin-statistics connection. For the treatment of these problems one must thank Burgoyne [31].······

In the following I give a most elementary but completely consistent example of the application of Wightman's theory, which is important for the investigation of TCP invariance and for the spin-statistics connection.

The first part summarizes facts about the real and the complex Lorentz groups, so that we can use them later. The second part gives a short reminder of the important properties of the W-functions. The third part is an explanation and summary of the proof that the vacuum expectation values (VEV's) are invariant under the complex Lorentz group. In the fourth part, the real regular points of the W-functions are characterized. The fifth and sixth parts contain the application to TCP-invariance and to spin and statistics.

444

Mathematical Part

§1. The Homogeneous Lorentz Group.

$\cdots\cdots$ *As with the Hall-Wightman theorem, Jost's mathematical development is too difficult and too long to include here.* $\cdots\cdots$

Under the $(N+1)!$ permutations of the z_k, there is an exact non-trivial relation which leads to no increase in the analyticity requirements of $\mathcal{G}_N{}'$. These state

$$\langle \psi^0_{\nu_0}(r_0)\psi^1_{\nu_1}(r_1)\cdots\psi^N_{\nu_N}(r_N)\rangle_0 = \sigma\langle \psi^N_{\nu_N}(r_N)\cdots\psi^1_{\nu_1}(r_1)\psi^0_{\nu_o}(r_0)\rangle_0, \cdots\cdots \tag{55}$$

§6. The Connection Between Spin and Statistics [31].

The value of σ in (5.5) is determined by the law of spin and statistics, which tells what σ must be for the VEV of the product of a field with its complex conjugate. From this, it follows which fields $\psi_\nu(x)$ cannot commute with $\psi_\mu(y)$ for spacelike separation.

In this section, for the first time, the positivity of the scalar product is essential. Indeed, already in Sec.5, the complex conjugate ψ^* of a field ψ was used. What was really made use of, however, was merely the existence of a *hermitian* scalar product.

There is the following

Assertion: Let ν stand for $\alpha_1\cdots\alpha_n, \beta_1\cdots\beta_m$. From the weakened commutation relations

$$\langle \psi^*_\nu(x)\psi_\mu(y)\rangle_0 = (-1)^{n+m+1}\langle \psi_\mu(y)\psi^*_\nu(x)\rangle_0 \tag{61}$$

for $(x-y)^2 < 0$, it follows that

$$\psi_\nu(x)\Omega = \psi^*_\mu(x)\Omega = 0, \tag{62}$$

where Ω is the vacuum state.

Comments: 1) (6.1) is true for a weak commutation relation because of the *Theorem* in Sec.4 that the real vectors in $\mathcal{R}'$ are identical with the spacelike vectors.

2) (6.2) is an abbreviation for the equation

$$\psi(f)\Omega = \psi^*(f)\Omega = 0, \tag{62}$$

where

$$\psi(f) = \int f^\mu(x)\psi_\mu(x)d^4x \tag{63}$$

and its hermitian conjugate for $\psi^*(f)$, with permitted test functions f.

445

Proof: From (6.1) and (5.9) follows

$$\langle \psi_\nu^*(x)\psi_\mu(y)\rangle_0 = -\langle \psi_\mu(-y)\psi_\nu^*(-x)\rangle_0 \tag{64}$$

for arbitrary x and y. One multiplies both sides by $f^{\nu*}(x)f^\mu(y)$ and integrates over x and y to get

$$|\psi(f)\Omega|^2 = -|\psi^*(g)\Omega|^2 \tag{65}$$

where $g^\mu(x) \equiv f^{\mu*}(-x)$. The assertion (6.2) follows from (6.5) [35].

Our assertion contains an essential element of the connection of spin with statistics, in that it requires a *single valued* field which satisfies

$$\langle [\psi_\nu^*(x), \psi_\mu(y)]_+ \rangle_0 = 0 \text{ for } (x - y)^2 < 0,$$

to annihilate the vacuum. The same would appear to be true for a *double valued* field, for which

$$\langle [\psi_\nu^*(x), \psi_\mu(y)]_- \rangle_0 = 0$$

for $(x - y)^2 < 0$.

However, without additional assumptions, one cannot show the desired conclusion to hold, that (6.4) requires the *vanishing* of the fields as in (6.2).

As such an assumption, one can perhaps use the postulate of *strong locality*, by which field operators at spacelike separation either commute or anti-commute. We have then the

Conclusion: From strong locality and (6.1) it follows that $\psi_\nu(x) = 0$.

Proof: First (6.2) follows and from that for the real points in $\mathcal{G}_N{}'$

$$\langle \phi_0(r_0) \cdots \psi_\mu(r_k) \cdots \phi_N(r_N)\rangle_0 = \sigma \langle \phi_0(r_0) \cdots \phi_N(r_N)\psi_\mu(r_k)\rangle_0 = 0. \tag{66}$$

σ is the signature and the ϕ are other fields of the theory. Since $(r_0 \cdots, r_N)$ is a regular point of the left side of (6.6), it follows generally that

$$\langle \phi_0(r_0) \cdots \psi_\mu(r_k) \cdots \phi_N(r_N)\rangle_0 = 0. \tag{67}$$

Therefore all VEV's which contain ψ_μ as a factor must vanish. So ψ_μ also vanishes.

Of course, one can reach these conclusions through still other assumptions. It seems to be unclear to explain (6.2) by a postulate of impossibility. On the other hand, if one is inclined to ask about the asymptotic behavior of the fields, then (6.2) represents a contradiction. One has the further benefit, that one can speak of particles and the actual conclusion has something to do with statistics.

446

We might even consider the idea, that the question whether the positive metric is consistent with strong locality should be pushed aside for a while. If one were to weaken the locality, then one has a much wider choice of spaces which, to be sure, yield no more information and permit no interesting conclusions. If one gives up the positive metric then it becomes very difficult to do anything; if one abolishes it, one runs the risk that the whole solid seeming foundation turns into a morass of options.

The Spin-Statistics Law along with strong locality and a positive metric has been used in a remarkable way as an assumption, one would like to say even as a guide. The preoccupation with it has been most fruitful - after a long time it has been the true source of even greater physics.

This article was already written in November 1958 and was intended as a gift for the sixtieth birthday of W. Pauli. The author decided not to do a revision, in light of the unexpected death of Wolfgang Pauli.

Footnotes and References.

. . . .

11) W. Pauli and V. Weisskopf, Helv. Phys. Acta **7**, 709 (1934).

12) W. Pauli, Ann. Inst. H. Poincaré **6**, 137 (1936).

13) P.A.M. Dirac, Proc. Roy. Soc. **A155**, 447 (1936).

14) M. Fierz, Helv. Phys. Acta **12**, 3 (1939); **23**, 412 (1950).

15) E. Wigner, Ann. Math., Princeton **40**, 149 (1939); V. Bargmann and E. Wigner, Proc. Nat. Acad. Sci., Wash. **34**, 211 (1940); E. Wigner, Z. Phys. **124**, 665 (1947).

16) W. Pauli, Phys. Rev. **58**, 716 (1940).

17) W. Pauli and F.J. Belinfante, Physica **7**, 177 (1940). · · ·

19) W. Pauli, Progr. Theor. Phys. **5.4**, 526 (1950).

20) W. Pauli, Rev. Mod. Phys. **15**, 145 (1943).

21) G. Källén, Encyclopedia of Physics, Vol.5/1, p.199.

22) J. Schwinger, Phys. Rev. **82**, 914 (1951). · · · · · · · · ·

24) G. Lüders, Mat.-phys. Medd. **28**, 5 (1954).

25) W. Pauli, *Niels Bohr and the Development of Physics*, p.30, 1955.

26) G. Lüders, Z. Naturf. **13a**, 254 (1958). · · · · · ·

30) R. Jost, Helv. Phys. Acta **30**, 409 (1957).

31) N. Burgoyne, Nuovo Cim. **8**, 607 (1958). · · · · · · · · ·

Chapter 19

Schwinger, Euclidean Field Theory,
Source Theory, and the
Spin-Statistics Connection

Summary: The long pursuit by Schwinger of an understanding of the spin-statistics connection, based fundamentally on Pauli's strong-reflection invariance as realized in Euclidean field theory, is reviewed.

§1. Introduction.

In a long series of papers and books [19.1] over almost thirty years, Schwinger sought to revise the logic of the Spin-Statistics Theorem *vis á vis* the **TCP**-theorem. In his approach, the fundamental role is given to neither. In a series of steps borrowing much from Pauli's original proof [19.2] - the Spin-Statistics Theorem *and* the **TCP**-theorem are deduced from " the general dynamical theory of fields, together with the specific assumption of invariance under the proper orthochronous Lorentz subgroup, and the existence of a lowest energy state".

In the first of the series (see our Ch.16), Schwinger used his recently constructed relativistic quantum field theory to demonstrate the spin-statistics connection as a result of the time-reversal invariance of Quantum Electrodynamics. Here we discuss the rest of his work devoted to the subject. Schwinger constructs a basic - but remarkably rich - theory which leads almost directly to the spin-statistics connection. The theory considers, for the moment, real Hermitian multicomponent fields χ, whose statistics and spins will be narrowed eventually to the desired result, by a succession of requirements on the dynamical behavior of the theory. The kinetic time-derivative term in the free field Lagrangian turns out to be the decisive feature which determines the allowed spin-statistics connections. It is an important feature of his proof that finite-dimensional Hermitian fields such as employed here - both tensors and spinors - can be chosen real for the Lorentz group.

The Lagrangian is written in a Dirac form involving finite dimensional numerical matrices K^μ, M and first derivatives of the field $\chi = (\chi_1 \cdots \chi_n)$,

$$\mathcal{L}(\chi) = \frac{i}{4} \left(\chi^T K^\mu \partial_\mu \chi - \partial_\mu \chi^T K^\mu \chi \right) - \frac{1}{2} \chi^T M \chi - \mathcal{H}_I(\chi), \qquad (1)$$

within an irrelevant four-divergence. Hermiticity of the Lagrangian requires that $K^\dagger = K$, $M^\dagger = M$. These separate naturally into K and M matrices which are real-symmetric or imaginary-antisymmetric. The fields χ are assumed to commute or anticommute at spacelike separation and we can deduce generic canonical equal-time (anti-)commutation relations from the Schwinger Action Principle

$$\delta\xi = -i[\xi, \delta\mathcal{A}]_-. \qquad (2)$$

The problem is more complicated when there are constraints. The necessary modifications are discussed by Johnson and Sudarshan [19.3].

The Euler-Lagrange equation

$$\partial_\mu \frac{\delta\mathcal{L}}{\delta(\partial_\mu\chi)} = \frac{\delta\mathcal{L}}{\delta\chi} \qquad (3)$$

is

$$\frac{i}{2}(K^{\mu T} - \$K^\mu)\partial_\mu\xi + \frac{1}{2}(\$M^T + M)\xi = -\delta\mathcal{H}_I/\delta\xi. \qquad (4)$$

Here the sign $\$$ is $(+1)$ if χ commutes with $\delta\chi$ and (-1) if they anticommute. The statistics are determined by the choice of K as either imaginary-antisymmetric, in which case only the sign choice $\$ = +1$ and Bose-Einstein statistics gives a nonvanishing kinematical term in the field equations; or real-symmetric, in which case the choice $\$ = -1$ and Fermi-Dirac statistics is required. The opposite choice must be made to get nonzero mass terms in the equations of motion: M real-symmetric for Bose-Einstein, imaginary-antisymmetric for Fermi-Dirac.

449

There still remains the problem of connecting the spin with the statistics. An easy way is to note that the Lagrangian must be an invariant under the rotation group and under the particular internal symmetry group of the fields. Simple examples convince us that the rotationally invariant scalar product of integral spin fields is symmetric (compare $\vec{A} \cdot \vec{B}$ and $\vec{B} \cdot \vec{A}$), and for half-integral fields is antisymmetric (compare the spin-0 combination of two spin-$\frac{1}{2}$ spinors $\alpha_1\beta_2 - \beta_1\alpha_2$). For the Lagrangian to be rotationally invariant, the term $\xi_r M_{rs} \xi_s$ must therefore be symmetric or antisymmetric for integral or half-integral spin, which requires M to be symmetric or antisymmetric, giving the connection between spin and statistics. Schwinger's argument was considerably more complicated, as we will see.

An example in zero space dimensions is the choice

$$\chi \Rightarrow \phi = \begin{pmatrix} \phi_1 \\ \phi_2 \end{pmatrix} \quad \text{and} \quad K^0 = \begin{pmatrix} 0 & +i \\ -i & 0 \end{pmatrix}, \quad M = m \begin{pmatrix} 1 & 0 \\ 0 & 1 \end{pmatrix}$$

in the Bose-Einstein case, and

$$\chi \Rightarrow \psi = \begin{pmatrix} \psi_1 \\ \psi_2 \end{pmatrix} \quad \text{and} \quad K^0 = \begin{pmatrix} 1 & 0 \\ 0 & 1 \end{pmatrix}, \quad M = m \begin{pmatrix} 0 & i \\ -i & 0 \end{pmatrix}$$

in the Fermi-Dirac case. Note that we are now dealing with "oscillators" and no determination has been made yet about the "spin" possibilities for either ϕ or ψ. The free particle equations of motion in the Bose-Einstein case are

$$\partial_t \phi_2 + m\phi_1 = 0 \text{ and } -\partial_t \phi_1 + m\phi_2 = 0, \tag{5}$$

from which we can eliminate $\phi_2 = \partial_t \phi_1$ and get the "Klein-Gordon" equation for the independent field ϕ_1

$$(\partial_t^2 + m^2)\phi_1 = 0. \tag{6}$$

The free particle equations for the Fermi-Dirac case

$$i\partial_t \phi_1 - im\psi_2 = 0 \quad \text{and} \quad i\partial_t \phi_2 + im\psi_1 = 0 \tag{7}$$

also require the "Klein-Gordon" equation. These are the Bose-Einstein and Fermi-Dirac oscillators.

Note also that $\mathcal{L}$ is symmetric under exchange of two Bose-Einstein fields $\phi_i \leftrightarrow \phi_j$, antisymmetric under the exchange of two Fermi-Dirac fields $\psi_i \leftrightarrow \psi_j$.

Schwinger states that " the development of the dynamical theory also shows that $\mathcal{H}_I$ and hence $\mathcal{L}$ are even functions of the Fermi-Dirac fields." This is a familiar result for half-integral spin fields which have to occur an even number of times to give a rotationally invariant Lagrangian, but why must this be true on the basis of Fermi-Dirac statistics alone?

Schwinger obtains the condition that $\mathcal{H}_I$ and $\mathcal{L}$ must contain even powers of the Fermi-Dirac fields, from the requirement that the left- and right-variations of $\mathcal{H}_I$ with respect to the ψ's should have opposite signs, in order to be compatible with commutation for $[\phi, \delta\psi]_-$ but anticommutation for $[\psi, \delta\psi]_+$.

Next, Schwinger introduces a trick to take account of the anticommutation within a pair of ψ's during reordering. The matrix transpose in Hermitian conjugation is replaced by a factor "i" attached to each ψ in a pair to be anticommuted. In this way, Schwinger replaces Hermitian conjugation by complex conjugation so that

$$\mathcal{L}(\phi, \psi) = \mathcal{L}^\dagger(\phi, \psi) = \mathcal{L}^{T*}(\phi, \psi) = \mathcal{L}^*(\phi, i\psi). \tag{8}$$

But it is unaesthetic to introduce antihermitian anticommuting fields. Instead here we choose $i\tau_2\psi$ which is still Hermitian. Then we write

$$\mathcal{L}(\phi, \psi) = \mathcal{L}^\dagger(\phi, \psi) = \mathcal{L}^{T*}(\phi, i\tau_2\psi). \tag{9}$$

The "$i\tau_2$" associated with the Fermi-Dirac field ψ replaces the anticommutation and multiplies every product of $2n$ ψ's by $(-1)^n$ and allows their free reordering back to their original untransposed order. Here $i\tau_2$ is a real

451

antisymmetric matrix coupling pairs of Fermi-Dirac fields. For spinor fields it could be identified with $i\sigma_2$.

Next we observe that there is a complex conjugate algebra of equal physical significance with the original. The Schwinger Action Principle which propagates a state forward in time

$$\delta\langle t_1|t_2\rangle = \langle t_1|\delta\{i\int_{t_2}^{t_1} d^4x \mathcal{L}(\phi,\psi)\}|t_2\rangle, \qquad (10)$$

for the complex conjugate algebra becomes

$$\delta\langle t_1|^*|t_2\rangle^* = \langle t_1|^*\delta\{-i\int_{t_2}^{t_1} d^4x \mathcal{L}(\phi,i\sigma_2\psi)\}|t_2\rangle^*. \qquad (11)$$

Now we proceed heuristically for the moment, for the sake of obtaining Schwinger's conclusion directly. Recall the effects of (weak-)time-reversal:

1) $T_w F = F^* T_w$ for algebraic functions; and

2) two time-reversals reverse the sign of a half-integral spin field $T_w^2 \chi_{\frac{1}{2}} = -\chi_{\frac{1}{2}}$, so the "eigenvalues" of T_w are 1 for integral spin fields, $i\sigma_2$ for half-integral spin fields. Then weak time-reversal undoes the above complex conjugation

$$
\begin{aligned}
T_w \delta\langle t_1|^*|t_2\rangle^* &= \delta\langle t_1|t_2\rangle \\
&= \langle t_1|\delta\{+i\int_{t_2}^{t_1} d^4x \mathcal{L}(\phi_T, -i\sigma_2\psi_T\}|t_2\rangle, \qquad (12)
\end{aligned}
$$

where the time-reversed fields $T_w\phi = \phi_T = \phi_{int}$ for integral spin fields and $i\sigma_2\phi_{half}$ for half-integral, and similarly for the Fermi-Dirac fields ψ_{int} and ψ_{half}. With this transformation

$$
\begin{aligned}
\delta\langle t_1|t_2\rangle &= \langle t_1|\delta\{i\int_{t_2}^{t_1} d^4x \mathcal{L}(\phi_{int}, i\tau_2\phi_{half}; -i\sigma_2\psi_{int}, \psi_{half})\}|t_2\rangle, \quad \text{and} \\
&= \langle t_1|\delta\{i\int_{t_2}^{t_1} d^4x \mathcal{L}(\phi_{int}, \phi_{half}; \psi_{int}, \psi_{half})\}|t_2\rangle, \qquad (13)
\end{aligned}
$$

from the original. Here we have Schwinger's "proof".

452

The "wrong-spin" fields have opposite sign kinetic terms in these equivalent descriptions, and would lead to different spectra, to say nothing of negative kinetic energies bilinear in the field, in violation of the invariance we started with. The conclusion is that only the "right-spin" fields are permitted. In a more detailed discussion, this is a consequence of Pauli's original strong-reflection invariance $x, t \rightarrow -x, -t$, supported with a number of supplementary assumptions as required.

Dirac-like Lagrangians at most linear in the first derivatives of the fields, which are Hermitian of some unspecified dimension, are convenient for the immediate purpose but make other problems unwieldy. The simplest case of a real scalar field ϕ requires the introduction of an auxiliary 4-vector field V_μ. The Lagrangian

$$
\begin{aligned}
\mathcal{L} &= \frac{m^2}{2}\left(\phi^2 + V_\mu V_\mu\right) - \frac{m}{2}\phi\overleftrightarrow{\partial}_\mu V_\mu, \\
&= \frac{m^2}{2}\left(\phi^2 + V_\mu V_\mu\right) - m\phi\overrightarrow{\partial}_\mu V_\mu
\end{aligned}
\tag{14}
$$

within a divergence. By inspection

$$
K = 2m \begin{pmatrix} 0 & i \\ -i & 0 \end{pmatrix} \quad \text{and} \quad M = m^2 \begin{pmatrix} 1 & 0 \\ 0 & 1 \end{pmatrix},
\tag{15}
$$

both 5×5 matrices. The equation of motion for ϕ is a constraint equation

$$
m\phi = \partial_\mu V_\mu
\tag{16}
$$

and the equation of motion for V_μ is

$$
- m\partial_\mu\phi = m^2 V_\mu,
\tag{17}
$$

leading to the Klein-Gordon equation for ϕ. The usual Klein-Gordon Lagrangian is recovered by using the equations of motion to eliminate the auxiliary field V_μ from the original.

Schwinger's discussion is not complete. The explicit construction for anticommuting fields of higher half-integral spin turns out to be impossible.

453

The relativistic Hermitian spinors have extra dimensions (as many as 24 for spin-$\frac{3}{2}$) which must be reduced to $(2(2S+1) = 8)$ independent degrees of freedom by subsidiary conditions. Johnson and Sudarshan have shown that this program is blocked by the appearance of anticommutation relations of indefinite sign where positive definite ones are required. They observe that in the anticommutation relations

$$[\psi_j, \psi_k]_+ = \sim \delta_{jk} \Rightarrow K_0^{-1}{}_{jk}, \tag{18}$$

the matrix K_0 is usually either indefinite or singular, indicating that the fields ψ_j are not independent. Projection operators must be found, a sequence of constraints imposed, and non-singular K-matrices of reduced dimension constructed. But this program is halted by the fact that the reduced K-matrix is dependent on the fields except in the case of spin-$\frac{1}{2}$, where no constraints are necessary and $K = 1$. The kinematics thus depends on the dynamics! As a result, for a charged spin-$\frac{3}{2}$ field the anticommutator also depends on the external field in such a manner that the quantization becomes inconsistent.

Johnson and Sudarshan conclude that only spin-$\frac{1}{2}$ fields can be regarded as fundamental. Higher half-integral spin fields cannot be represented by a local action principle. Their result supports the view that the spin-statistics connection need be demonstrated only for spin-0 and spin-$\frac{1}{2}$.

§2. Euclidean Field Theory.

In search of an independent justification of Pauli's strong-reflection invariance, Schwinger constructed a continuation of relativistic quantum field theory from the physical Minkowski space (x_1, x_2, x_3, t) with Lorentz metric $(+ + + -)$ to an associated Euclidean space [19.4] $(x_1, x_2, x_3, x_4 = \pm it)$ with metric $(+ + + +)$. A Lorentz boost in the x_1 direction corresponds to a rotation in the (x_1, x_4)-plane. The group of Lorentz transformations is replaced by the group of rotations in four dimensional Euclidean space. One consequence of Euclidean invariance is that strong-reflection invariance

$x_\mu \to -x_\mu, \mu = 1 \cdots 4$ is included in the invariance under ordinary rotations. We can anticipate that Pauli's proof of the Spin-Statistics Theorem, based as it was on the strong-reflection invariance of the spinor representations of the (proper-)Lorentz group, should have a presence in the Euclidean field theory.

The distinction between integral spin and half-integral spin fields appears when one insists on the existence of six imaginary antisymmetric matrices for the generators of the four dimensional infinitesimal orthogonal transformations. These are constructed from the continuation of the finite dimensional nonunitary representations of the generators of the Lorentz group using the Weyl unitary trick. Since we have been using Hermitian fields, any fermionic charge has to be introduced in terms of a 2×2 antisymmetric imaginary "charge" matrix.

Schwinger shows that the Green's functions of relativistic quantum field theory in Minkowski space-time have a unique image in a four dimensional Euclidean manifold with (x_1, x_2, x_3, t) replaced by (x_1, x_2, x_3, x_4). Continuation of the Euclidean Green's function to $x_4 \to \mp it$ gives the Feynman Green's function $G_\pm(t) = G_E(\mp i x_4)$. The Lorentz invariance properties of the original Minkowski space relativistic quantum field theory translate almost directly into four dimensional rotational invariance of the corresponding Euclidean theory. The Hermitian spin matrices $S_{\mu\nu}, \mu, \nu = 1 \cdots 4$ have $S_{kl}(k, l = 1 \cdots 3)$ antisymmetric, but S_{k4} symmetric. These must be transformed into six imaginary antisymmetric matrices describing the independent rotations in the four dimensional Euclidean space. Schwinger defines the Euclidean generators for integer spins as

$$S_{\mu\nu}^E = e^{\mp i\pi/4 R_t} S_{\mu\nu} e^{\pm i\pi/4 R_t}, \tag{19}$$

where the top sign refers to G_+, the bottom to G_-. R_t is the time-reflection matrix. It is easily verified that

$$\left(S_{\mu\nu}^E\right)^\dagger = -\left(S_{\mu\nu}^E\right)^* = -S_{\nu\mu}^E = S_{\mu\nu}^E \tag{20}$$

because $S_{kl}R_t = +R_tS_{kl}$ and $S_{k4}R_t = -R_tS_{k4}$.

There are complications in the case of half-integral spin, which Schwinger overcomes by the introduction of an auxiliary matrix l interpreted as the fermionic charge in the Hermitian field theory. The spinor representations for Euclidean space are complex, in contrast to the real spinors in Minkowski spaces. Therefore the matrix always exists. The problem arises for half-integral spin because the space reflection matrix R_s is anti-Hermitian and antisymmetric, and the time reflection matrix R_t is Hermitian and antisymmetric. In this case it is the direct product of the imaginary, antisymmetric fermionic charge matrix l with the space reflection matrix R_s which is required to construct the Euclidean rotation generators for half-integral spin fields as

$$S_{\mu\nu}^E = e^{\pm\pi/4R_s l}S_{\mu\nu}e^{\mp\pi/4R_s l}, \tag{21}$$

which are Hermitian and antisymmetric as required. The charge matrix l is also necessary in the case of half-integral spin fields to make the Euclidean time-reflection matrix $R_4 \equiv R_tR_s l$ real, symmetric and orthogonal.

With these definitions, the reflection matrices are all real, symmetric and orthogonal. They commute for integral spin fields, anticommute for half-integral spin fields. All reference to the Lorentz metric has been removed.

The demonstration of the spin-statistics connection is rather indirect and unsuitable for a proof, certainly of the sort Neuenschwander wants. Schwinger bases his demonstration on the Green's functions

$$G_+(x_1\cdots x_p) = \langle 0|T_+(\chi(x_1)\cdots\chi(x_p))|0\rangle \tag{22}$$

and

$$G_-(x_1\cdots x_p) = \langle 0|T_-(\chi(x_p)\cdots\chi(x_1))|0\rangle, \tag{23}$$

where T_+ is the time ordering operator putting later times to the left, and T_- is that putting later times to the right. The relation between the time orderings depends on the number of anticommuting Fermi-Dirac fields in the product. For our Hermitian fields, the connection between the two Green's functions is simply

$$G_-(x_1 \cdots x_p) = G_+(x_1 \cdots x_p)^\dagger. \tag{24}$$

In order to have invariance under strong reflection R_{st} for which $x_\mu \to -x_\mu$, Schwinger requires

$$
\begin{aligned}
R_{st} G_+(-x_1 \cdots - x_p) &= \langle 0 | R_{st} \chi(-x_1) \cdots | 0 \rangle \\
&= G_+(x_1 \cdots x_p)(i)^{2n'} \\
&= G_+(x_p \cdots x_1)(-)^n (i)^{2n'}.
\end{aligned}
\tag{25}
$$

A factor $(-)^n$ from reordering n Fermi-Dirac pairs (recall that Schwinger established that they occur in pairs) to get the correct time order, compensates a factor $(i)^{2n'}$ from the time-reversal of $2n'$ half-integral spin particles. Invariance of the Green's functions under strong reflection is obtained for the spin-statistics connection $n = 2n'$, identifying the Fermi-Dirac particles as those with half-integral spin.

Schwinger generalizes the Euclidean theory to include non-Hermitian fields by simply taking linear combinations

$$\phi^\pm \sim \phi_1 \pm i\phi_2. \tag{26}$$

We return to this demonstration of the spin-statistics connection in a paper to be discussed in the next section. It does seem that nothing has been gained in terms of directness or simplicity over the proofs in Minkowski space constructed by Lüders and Zumino, and by Burgoyne [19.5], which also restrict their considerations to finite component relativistic fields.

§3. Euclidean Proof of the Spin-Statistics Theorem.

Brown and Schwinger [19.6] clarify the Euclidean proof of the Spin-Statistics Theorem by specializing to the two-point Green's function of Bose-Einstein and Fermi-Dirac fields. As in Schwinger's earlier work, the discussion involves Hermitian fields $\chi_a(x), a = 1 \cdots n$ of either integral or half-integral spin, with internal (charge) degrees of freedom a usually left implicit. The proof will require covariance under the proper orthochronous Lorentz group $L_+^\uparrow$, and the existence of a unique vacuum ground state.

The vacuum expectation value of two fields - the unordered two-point Wightman function - is, by translation invariance,

$$\langle 0|\chi(x)\chi(x')|0\rangle = \langle 0|\chi(0)e^{iP\cdot(x-x')}\chi(0)|0\rangle \qquad (27)$$

where we use $\chi(x) = e^{-iP\cdot x}\chi(0)e^{+iP\cdot x'}$ and $P_\mu|0\rangle = 0$. Inserting a complete set of physical positive energy momentum eigenstates of mass$^2 = \kappa^2$, we get, assuming nonnegative energies,

$$e^{iP\cdot(x-x')} = \int d^4p \int_0^\infty d\kappa^2 e^{ip\cdot(x-x')}\eta_+(p)\delta(p^2 + \kappa^2)M(p, \kappa^2), \qquad (28)$$

with $p^2 + \kappa^2 = \vec{p}^2 + \kappa^2 - p_0^2$, and $\eta_+ = 1$ for $p_0 > 0$, $= 0$ for $p_0 < 0$.

$$M(p, \kappa^2) = \sum_\alpha |p, \kappa^2; \alpha\rangle\langle p, \kappa^2; \alpha| \qquad (29)$$

is a Hermitian projection operator. Finally, the two point function is expressed as

$$\langle 0|\chi_a(x)\chi_b(x')|0\rangle = \int d^4p \int d\kappa^2\delta(p^2 + \kappa^2)\eta_+(p)e^{ip\cdot(x-x')}m_{ab}(p, \kappa^2) \qquad (30)$$

where

$$m_{ab}(p, \kappa^2) = \langle 0|\chi_a(0)M(p, \kappa^2)\chi_b(0)|0\rangle \qquad (31)$$

is a finite dimensional positive Hermitian mass-matrix.

The Lorentz invariance of the Green's function translates to a Euclidean rotational invariance. The mass-matrix $m_{ab}(p, \kappa^2)$ is a matrix constructed from the algebra of representations of O_4, arbitrary functions of $p^2 = -\kappa^2$, and polynomials in p_μ - the four-dimensional spherical harmonics - which are homogeneous of degree $2S$ with S the highest spin value in χ.

458

Euclidean invariance includes strong reflection invariance, $p \rightarrow -p$. The continuation $p \rightarrow -p$ of the four-dimensional spherical harmonics in $m(p, \kappa^2)$ results in a sign change

$$m(-p, \kappa^2) = (-)^{2S} m(p, \kappa^2) \tag{32}$$

in the continuation of the mass-matrix from positive to negative p_0 for half-integral spin.

The commutator or anticommutator of two fields involves also

$$
\begin{aligned}
\langle 0|\chi_b(x')\chi_a(x)|0\rangle &= \int d^4p d\kappa^2 \delta(p^2 + \kappa^2) \eta_+(p) e^{ip\cdot(x'-x)} m_{ba}(p, \kappa^2) \\
&= \int d^4q d\kappa^2 \delta(q^2 + \kappa^2) \eta_-(q) e^{iq\cdot(x-x')} m_{ab}^*(-q, \kappa^2) \tag{33}
\end{aligned}
$$

where we have used the Hermiticity of the mass-matrix

$$m_{ba}(p, \kappa^2) = m_{ab}^*(p, \kappa^2), \tag{34}$$

also $\eta_+(p) = \eta_-(-p)$, and we have changed integration variables from p to $q = -p$.

Writing $\eta_\pm(p) = (1 \pm \epsilon(p))/2$ with $\epsilon(p)$ equal to the sign of p_0, gives for the (anti-)commutator

$$\langle 0|[\chi(x), \chi(x')]_\pm|0\rangle = \int d^4p d\kappa^2 \mathcal{F}_\pm(p, \kappa^2) e^{ip\cdot(x-x')} \delta(p^2 + \kappa^2), \tag{35}$$

with the auxiliary function $\mathcal{F}_\pm(p, \kappa^2)$ defined as

$$\mathcal{F}_\pm(p, \kappa^2) = \frac{1}{2} \left\{ m(p, \kappa^2) \pm m^*(-p, \kappa^2) + \epsilon(p)(m(p, \kappa^2) \mp m^*(-p, \kappa^2)) \right\}. \tag{36}$$

This expression can be factored into differential operators (with p replaced by ∇/i) acting on two standard functions which are well known from the commutators of spinless free fields. One, the causal commutator $\Delta(x - x'; \kappa^2)$ containing the factor $\epsilon(p)$, vanishes for spacelike $x - x'$ and gives no restriction. The other, $\Delta_1(x - x'; \kappa^2)$, does not vanish for spacelike

459

$x - x'$. In order to get a causal commutator or anticommutator for the fields χ, the coefficient of Δ_1 must vanish. Setting this term to zero requires the spin-statistics connection. The coefficient of the Δ_1 term which must be set to zero is

$$m(p, \kappa^2) \pm m^*(-p, \kappa^2) \Rightarrow 0. \tag{37}$$

It suffices to look just at the diagonal matrix elements of m which are real and satisfy

$$m(-p, \kappa^2) = (-)^{2S} m(p, \kappa^2). \tag{38}$$

The coefficient of Δ_1 vanishes for integral spin in the case of commutation, and for half-integral spin in the case of anticommutation. Causal anticommutation is not possible in the case of integral spin, but causal commutation is. Causal commutation is not possible in the case of half-integral spin, but causal anticommutation is.

The Spin-Statistics Theorem is proved for integral and half-integral particles in a very symmetrical way using Euclidean field theory, strong reflection invariance, and the Euclidean rotational invariance to establish the behavior of the mass-matrix under reflection.

Brown and Schwinger foreclose various possibilities. For example, the mass-matrix must be nonzero at least in its diagonal elements, or else the corresponding field χ_a must vanish. The mass matrix, viewed as a differential operator, cannot annihilate Δ_1 which is a Bessel function, because the mass-matrix is a finite polynomial in ∇ as a result of replacing the Lorentz invariance by the Euclidean rotational invariance. Brown and Schwinger also extend the proof to off-diagonal commutators, compound particles and charged particles by various auxiliary arguments.

It is worth emphasizing, as Schwinger in fact did, that the key ingredient in the proof is the strong reflection property of the mass-matrix, which is determined by the O_4 spherical harmonics; but such a continuation from Minkowski to Euclidean space is possible only if strong reflection is defined

460

in Minkowski theory. If we start from the Euclidean space and analytically continue, the strong reflection is automatically defined. The debt to the Pauli discovery of strong reflection invariance and his proof of the Spin-Statistics Theorem from the properties of the spinor representation of the Lorentz group is evident.

§4. Schwinger's Source Theory.

Schwinger's source theory [19.7] is a phenomenology exquisitely informed by his intuitive mastery of all the implications of field theory. The result is the introduction of a generating functional for S-matrix elements essentially the same as the one resulting from the Feynman path integral formulation, $S(\eta) = e^{iW(\eta)}$. Various derivatives of $S(\eta)$ evaluated at $\eta \to 0$ produce physical S-matrix elements. The argument for the spin-statistics connection involves only the phase

$$W(\eta) = \frac{1}{2} \int dx\,dx'\,\eta_\mu(x) K_{\mu\nu}(x - x')\eta_\nu(x'). \tag{39}$$

Here $\eta(x')$ is the source which emits the field propagated by $K(x - x')$ to be absorbed by the receiver $\eta(x)$. The propagator K is constructed by Schwinger using intuitive arguments for relativistic particles of arbitrary spin. Here we simply follow his arguments for spin-$\frac{1}{2}$ with

$$K(x - x') = \gamma_0 G_+(x - x') \tag{40}$$

the 4×4 matrix propagator of the Dirac equation.

Schwinger requires that the integrand in $W(\eta)$ should be symmetric in $x \leftrightarrow x'$

$$\eta_\mu(x) K_{\mu\nu}(x - x')\eta_\nu(x') = \eta_\nu(x') K_{\nu\mu}(x' - x)\eta_\mu(x). \tag{41}$$

As we will see in a moment, the Dirac propagator is antisymmetric

$$K_{\mu\nu}(x - x') = -K_{\nu\mu}(x' - x). \tag{42}$$

The immediate conclusion is that the sources of the spin-$\frac{1}{2}$ Dirac particle must be anticommuting Grassmann quantities for which

$$\eta_\mu(x)\eta_\nu(x') = -\eta_\nu(x')\eta_\mu(x). \tag{43}$$

461

To demonstrate the antisymmetry of the Dirac propagator, note that Schwinger has simplified the discussion by using Hermitian Dirac fields. This choice fixes the Dirac matrices to be in the Majorana representation, in which γ_0 is antisymmetric imaginary and squares to 1, and the $\gamma_j (j = 1, 2, 3)$ are symmetric imaginary and anticommute with γ_0. With

$$K(x - x') = \gamma_0(m + \gamma \cdot \nabla/i)\Delta_+(x - x'), \qquad (44)$$

and $\Delta_+(x - x') = \Delta_-(x' - x)$, the transpose is

$$K^T(x' - x) = -\gamma_0(m + \gamma \cdot \nabla'/i)\Delta_-(x' - x). \qquad (45)$$

This is recognized - except for the sign - as the propagator for negative energy quanta propagating backwards in time from x to x'. Time reversal invariance requires these two amplitudes to be the same. Schwinger achieves this by requiring that the sources anticommute, compensating the antisymmetry of the Dirac propagator.

Schwinger goes on to show that it is not possible to avoid the anticommutation by the device of including an antisymmetric charge matrix to reverse the sign of antiparticle amplitudes relative to particle amplitudes. He uses the same argument introduced by Feynman and analyzed by Pauli, that such a device violates unitarity by introducing an indefinite metric into the Hilbert space.

§5. Concluding Remarks.

Schwinger was forced to equivocate on the question of the spin-statistics relation in the presence of interactions. He appealed to his own precursor of "asymptotic freedom" and assumed that at sufficiently high energies the kinematic term in the Lagrangian dominated the interaction term, and that they had the same exchange symmetry for the sake of consistency. He had a similarly cavalier attitude toward higher spin particles which he took to be composite or to have wave-function components which were, and to have exchange properties which are simply the product of the constituents. This

462

attitude is supported by the conclusion of Johnson and Sudarshan already referred to above, that spin-$\frac{3}{2}$ fields cannot be fundamental. What the impact on this result is from the advances of Supersymmetry is another in a seemingly endless series of questions, but one we must leave to others.

Schwinger's arguments seem so technical and based on such arcane manipulations of field theoretic results, that one is left with little insight or understanding. They leave the spin-statistics connection and the Spin-Statistics Theorem as mysterious as ever.

Bibliography and References.

19.1) J. Schwinger, Phys. Rev. **82**, 914 (1951); see our App.16A. Also J. Schwinger, Proc. Natl. Acad. Sci. USA **44**, 223, 617 (1958).

19.2) W. Pauli, Phys. Rev. **58**, 716 (1940); see our App.14A.

19.3) K. Johnson and E.C.G. Sudarshan, Ann. Phys. **13**, 126 (1961).

19.4) J. Schwinger, Phys. Rev. **115**, 721 (1959); J. Schwinger, Proc. Natl. Acad. Sci. USA, **44**, 956 (1958).

19.5) G. Lüders and B. Zumino, Phys. Rev. **110**, 1450 (1958); N. Burgoyne, Nuovo Cim. **8**, 607 (1958); see our App.17A,B.

19.6) L.S. Brown and J. Schwinger, Progr. Theor. Phys. **26**, 917 (1961).

19.7) J. Schwinger, *Particles, Sources and Fields, Vol. I* (Addison- Wesley, Reading MA, 1989), pp.99-144.

Chapter 20

Responses to Neuenschwander's Question

Summary: The responses to Neuenschwander's challenging question - Is there a simple, straightforward, accessible, intuitive, 'physical' proof of the Spin-Statistics Theorem? - are dealt with individually, and in all cases found to lack credibility. This conclusion is a reiteration and expansion of the original evaluation of the responses by Hilborn.

§1. Introduction.

Neuenschwander's question in the marvelously stimulating, informative, and frequently entertaining 'Questions and Answers' section of the American Journal of Physics [20.1] - where he quotes Feynman [20.2] (circa 1963):

"··· Why is it that particles with half-integral spin are Fermi particles ··· whereas particles with integral spin are Bose particles ··· we cannot give you an elementary explanation. ··· Pauli ··· complicated ideas of quantum field theory and relativity ··· not ··· elementary ··· we do not have a complete understanding ···";

and then asks: "Has anyone made any progress towards an "elementary" argument for the Spin-Statistics Theorem?" - excited us, obviously, but generated a remarkably limited direct response. Two years after publication of the question, there have appeared only four replies. We deal with them directly.

In our opinion, and in agreement with the response by Hilborn [20.3], none of the direct replies to Neuenschwander, nor any of the references contained in them, contain satisfactory proofs of the Spin-Statistics Theorem, elementary or otherwise. The situation is left essentially as Feynman described it thirty years earlier, although there *had* been an evolution from Pauli's proof to those of Lüders and Zumino and of Burgoyne, which Feynman did not mention. But these proofs too are still intimately involved with

464

relativistic quantum field theory, and furthermore make essential reference to the daunting mathematical developments of the Hall-Wightman theorem. They are by no means elementary or intuitive. A proof is presented in the next chapter which we believe satisfies all of Neuenschwander's requirements. The point of view preferred there is much closer to the ideas of Schwinger but also much more simple. In this chapter, however, we deal with the various proposals brought forward in response to Neuenschwander.

§2. Bacry's Proof and Hilborn's Critique.

First we discuss the simplest proof due to Bacry [20.4]. Bacry identifies an exchange operator $\mathcal{E}$ with a suitably chosen rotation operator R. The situation described in his reply is a particularly simple one which illustrates the essential idea but also contains a basic misconception which invalidates his argument from being the sought-after "simple intuitive proof". Later we discuss a more general argument along the same line due to Broyles, which defines the exchange operator in terms of a rotation operator suitable for general situations beyond the simple configuration described by Bacry. However, Broyles' argument suffers from the same critical flaw as does Bacry's. Our conclusions are in full agreement with Hilborn's evaluation.

Bacry considers the state of an electron at $(x, y, z) = (+a, 0, 0)$ with spin component $s_z = +\frac{1}{2}$ described by a wave function

$$\psi_A = \begin{pmatrix} \delta(x - a)\delta(y)\delta(z) \\ 0 \end{pmatrix}, \tag{1}$$

and another electron at $(x, y, z) = (-a, 0, 0)$ with $s_z = -\frac{1}{2}$ described by a wave function

$$\psi_B = \begin{pmatrix} 0 \\ \delta(x + a)\delta(y)\delta(z) \end{pmatrix}. \tag{2}$$

The two electron wave function is written as

$$\Psi_{AB}(1, 2) = \psi_A(1)\psi_B(2) \pm \psi_B(1)\psi_A(2), \tag{3}$$

465

where we need to make a choice $\pm$ between a symmetric wave function or an antisymmetric wave function for the two electrons. Under the exchange operation $\mathcal{E}_{12}$ taking $1 \Leftrightarrow 2$,

$$\mathcal{E}_{12}\Psi_{AB}(1,2) \equiv \Psi_{AB}(2,1) = \pm\Psi_{AB}(1,2). \tag{4}$$

Bacry then observes that a finite rotation by π around the y-axis leaves this two particle state unchanged. This rotation is generated by the operator

$$
\begin{aligned}
R_y(\pi) &= e^{-i\pi J_y} = e^{-i\pi L_y}e^{-i\sigma_y\pi/2} \\
&= e^{-i\pi L_y}(-i\sigma_y) \\
&= e^{-i\pi L_y}\begin{pmatrix} 0 & -1 \\ 1 & 0 \end{pmatrix}.
\end{aligned}
\tag{5}
$$

Acting on the wave function ψ_A

$$R_y(\pi)\psi_A(x,y,z) = \begin{pmatrix} 0 \\ \delta(-x-a)\delta(y)\delta(-z) \end{pmatrix} = \psi_B(x,y,z), \tag{6}$$

and on ψ_B

$$R_y(\pi)\psi_B(x,y,z) = -\psi_A(x,y,z). \tag{7}$$

Acting on the two particle wave function

$$
\begin{aligned}
R_y(\pi)\Psi_{AB}(1,2) &= -(\psi_B(1)\psi_A(2) \pm \psi_A(1)\psi_B(2)) \\
&\equiv -\pm\Psi_{AB}(1,2).
\end{aligned}
\tag{8}
$$

Bacry (and in a similar context, Broyles) now makes the unjustifiable assumption which negates the proof. From the fact that the two particle state is invariant under the finite rotation $R_y(\pi)$, he concludes that the wave function is also, and requires that

$$R_y(\pi)\Psi_{AB}(1,2) \equiv \Psi_{AB}(1,2)$$

which, if true, would require the choice of the negative sign in the $\pm$ and would determine that

$$\mathcal{E}_{12}\Psi_{AB}(1,2) \equiv \Psi_{AB}(2,1) = -\Psi_{AB}(1,2),$$

the desired result.

However, the invariance of the state does *not* require the invariance of the wave function in the case of a *discrete* symmetry, which is what we have here. There is nothing to rule out the possibility of a sign change of the wave function under a rotation through π, and therefore the above argument can give no information about the choice of exchange symmetry of the wave function.

The change of sign of a wave function under a discrete symmetry transformation is a common feature of wave functions: we have only to note the invariance under a 2π rotation of a spin-$\frac{1}{2}$ particle state, but the wave function changes sign; or the invariance under reflections of a pseudoscalar state, but again the wave function changes sign.

Broyles' argument, which predates the whole Neuenschwander incident, is primarily concerned with showing that a rotation operator exists which can serve as the exchange operator for two spins not simply parallel or antiparallel (as in Bacry's case), and that therefore a rotation operator exists which exchanges the two particles for general spin states. The actual details of this operator (in Broyles' Eqn.16, a result which is somewhat contrived and artificial, in our view) need not concern us. What does concern us is his Postulate A on which he predicates his proof of the Spin-Statistics Theorem:

"Postulate A: If we write the wave function for two particles in such a way as to exhibit all of the internal quantum numbers and the spatial position of each and, furthermore, if the two sets of quantum numbers including coupling constants and spin are identical with the exception of the spin components along some axis and the spatial positions, *then this wave function must be invariant* (*Note: Italics added*) to any (Poincaré) transformation of the coordinate frame (with all physical apparatus connected to it) that produces a wave function with the same two sets of quantum numbers."

The critical part of this postulate, with which we disagree, is the phrase

467

in italics.

Broyles goes on to emphasize: "Any combination of rotations and translations of the coordinate frame that leaves the picture looking just as it did before these operations, must also leave the wave function unchanged." He does emphasize that the postulate is special to two particle wave functions for the reason cited above, that the wave function for a single particle with spin-$\frac{1}{2}$ under a rotation of 2π is an immediate exception without this restriction. We conclude that Broyles' Postulate A is *ad hoc* special pleading which has no other purpose than to construct his proof of the Spin-Statistics Theorem, plays no other role, and is not a valid postulate of quantum mechanics.

In Hilborn's words: Bacry's and similarly Broyles' "$\cdots$ argument establishes a spin-statistics connection at the expense of an additional assumption about how the wave function behaves under coordinate transformations. This assumption goes beyond the requirement that all *observables* remain unchanged and is equivalent to restricting the wave functions to the totally symmetric or totally antisymmetric representations of the permutation group or, equivalently, restricting physical states to those represented by single rays in Hilbert space."

§3. Topological Markers and Feynman's Models.

The first response to Neuenschwander's Question was from Gould [20.5] who referred to the Feynman Lectures and summarized Feynman's 1986 Dirac Lecture [20.6] on the same topic. Here, as Gould describes it, Feynman "sketched" an elementary argument for the spin-statistics connection. The Dirac Lecture reads as Feynman at his finest and most charismatic, although, sadly, approaching the final stages of a long illness. It is a monument to a remarkable person's incredible "$\cdots$ lightness of being", and an example to us all that he could maintain his youthful spirit, energy and exuberance so close to the end. Unfortunately, the fascinating scenarios that he painted in this lecture do not constitute a proof or even an explanation of the Spin-

Statistics Theorem. In another reply, von Foerster [20.7] recalled similar heuristic explanations of the result by others.

Gould summarizes Feynman's argument, in part, by recounting the paradoxical behavior of the rotation of a tethered classical object. The purpose of the classical paradox is to convince people that a 2π rotation is not just a trivial return of everything to the way it was, even classically, and that we should not be distressed by the resulting change in the sign of the wave function of a spin-$\frac{1}{2}$ particle. This demonstration - in the hands of anyone but Feynman - has surely convinced countless pained spectators at innumerable cocktail parties of only one thing: avoid physicists as overbearing bores. The demonstration is pure old fashioned snake-oil peddling. The scene is usually made somewhat palatable by visualizing a dancing girl with a wine glass (full) balanced in the palm of her hand, but an aging physicist with a coffee cup (preferably empty) will have to suffice here. The point is - grasping the handle of the cup, rotate the cup through through an angle of 2π around a vertical axis while keeping feet fixed, but at the expense of a twisted arm. A further 2π rotation through a total of 4π returns the cup (and the arm) to the original configuration. This is supposed to remind us of the sign change in the wave function of a spin-$\frac{1}{2}$ particle under a 2π rotation, and the need to rotate the spin-$\frac{1}{2}$ particle twice around through 4π to return to the original wave function. So far, so good. But no further.

Hilborn states it clearly: "$\cdots$ analogy is not an explanation. Nowhere does the spin of the object enter the discussion nor is it clear what the twist in the constraint has to do with the change in sign of the fermion's wave function. $\cdots$ why are boson wave functions unchanged $\cdots$ bosons are composed of an even number of $\cdots$ fermions $\cdots$. But photons $\cdots$ are not considered to be composites."

It was Feynman's purpose to show that "$\cdots$ the mysterious minus signs in the behavior of Fermi particles are really due to unnoticed 2π rotations!" Feynman - ever facile - produces in his Dirac Lecture two other models of

469

identical particle exchange which reproduce the spin-statistics connection. One is a nearly classical model of a spin-$\frac{1}{2}$ object which has the required change of sign under a 2π rotation, and also under an exchange of two identical particles. Feynman describes a composite object consisting of a spin-0 electric charge e and a spin-0 magnetic monopole of charge g, originally due to Saha [20.8]. The electromagnetic angular momentum

$$\vec{L} = \int \vec{r} \times (\vec{E} \times \vec{B}) d^3 r \tag{9}$$

is independent of the separation of e and g, directed along the line between them, and equal to eg, giving the Dirac relation $eg = \frac{1}{2}$ when the angular momentum assumes its minimum non-zero value.

Now suppose the electric charge e is moved in a circle around the magnetic charge g. The wave function acquires a phase

$$\phi = e \int \vec{A} \cdot d\vec{l} = e \int \vec{B} \cdot d\vec{S}. \tag{10}$$

The surface integral $\int dS$ can be deformed into an easily done integral over a hemisphere centered on the magnetic charge, giving the result

$$\phi = e \times (4\pi g)/2 = eg \times 2\pi = \pi. \tag{11}$$

As desired, the phase of the spin-$\frac{1}{2}$ object has changed by π for a rotation through 2π.

Next, Feynman considers the process of exchanging two (very compact) eg composites, call them 1 at x and 2 at y. He views this as 1 translated from $x \to y$ in the vector potential of 2, and 2 from $y \to x$ in that of 1. They are supposed to have their axes parallel and fixed in direction throughout the exchange. Then the phase acquired by the composite wave function during the exchange is

$$
\begin{aligned}
\phi &= \phi_1 + \phi_2 \\
&= e \int_x^y \vec{A_2} \cdot d\vec{l_1} + e \int_y^x \vec{A_1} \cdot d\vec{l_2}, \\
&= e \int \vec{B} \cdot d\vec{S} = \pi,
\end{aligned} \tag{12}
$$

470

just the same closed line-integral that occurs in the 2π rotation, and just what is required by the Spin-Statistics Theorem.

Finally, Feynman proposes a prescription for the exchange operator $\mathcal{E}_{12}$: rotate each particle around the other by an angle π, which is equivalent to a 2π rotation of one particle around the other. The net effect for spin-0 particles is a factor 1 and for spin-$\frac{1}{2}$ particles a factor -1 in accord with the Spin-Statistics Theorem.

As a corollary, Feynman reminds us of an argument by Finkelstein [20.9] to demonstrate that in the rigid rotation of two particles with no rotation of their internal axes, there is a relative rotation of each by π. This can be seen by attaching the two ends of a ribbon, one to each particle, and identifying the "inside edge" initially, which becomes the "outside edge" after the rigid rotation. To complete the exchange and return to the original configuration (including the ribbon) requires a further rotation of each particle around its own body axis by π, for a total rotation by 2π and a factor -1.

Feynman's verbal arguments were remarkably beautiful and inspiring but they were also as difficult to grasp, and hold onto, and take home, and possess as one's own, as are smoke rings. So it is with great humility that we quarrel with what is, after all, only our second-hand perception of Feynman's arguments, in full knowledge of the fact that if he were here to mesmerize us again, we would be fully convinced again.

§4. Critique of Topological Markers.

So what *is* wrong with these proofs of the Spin-Statistics Theorem which Feynman sketched?

The argument Feynman borrowed from Finkelstein, which endowed elementary particles with a connecting ribbon to keep track of their orientation, has to be dispensed with for the reason that there are *no* topological appendages to Cartan spinors. We need a proof that views an elementary

471

particle as a mathematical point or we need to prove that such a view is untenable, but we cannot endow elementary particles with a property which is needed for no other purpose. So a ribbon attached to spinors is not allowed, and we have no reason to identify the exchange operation as a rigid rotation (with internal axes unrotated) followed by two internal rotations through angle π to make the spinors "face" each other again, thereby producing a minus sign. Cartan spinors have no face.

The demonstration based on the charge-monopole composite - although it is an intuitive *tour de force* - suffers from the same disqualification of endowing an elementary particle with the unphysical superstructure of a magnetic field. In the ribbon case, the exchange operation is the rigid rotation of each (with internal axes held fixed) followed by two rotations by π which result in the sign change required for Fermi-Dirac statistics. In the composite case, the exchange operation is the rigid rotation of each composite (with internal axes held fixed) but with *no* apparent internal rotations. The purpose of the ribbon model is to convince us that an unintended or unnoticed or at least unmentioned rotation of 2π has actually occurred somehow.

In either case, the topological markers on the elementary particles have to be ruled out as extraneous.

Biedenharn and Louck [20.10] have discussed just these properties of rotations in Chapter 2 of their book "Angular Momentum in Quantum Physics". They describe in detail and illustrate with intricate diagrams the Dirac construction which demonstrates that "$\cdots$ *for solid bodies* a rotation by 2π is *not* equivalent to the identity, but that a rotation by 4π is $\cdots$." The solid (*i.e.*, impenetrable) body is connected to an external coordinate frame by at least three strings which become inextricably tangled after the 2π rotation but can be untangled after 4π. After a number of caveats concerning the Dirac construction, Biedenharn and Louck then make the unequivocal statement "Dirac's result must be carefully distinguished from the similar behavior of spinors under rotation. $\cdots$ spinors are *point* objects, in contrast to the objects in Dirac's construction, which must have a finite size."

The Cartan definition of a spinor associates a spinor (ξ_0, ξ_1) with an isotropic vector (x_1, x_2, x_3) defined as a three dimensional complex Euclidean vector of zero-length $x_1^2 + x_2^2 + x_3^2 = 0$. The connection is

$$x_1 = \xi_0^2 - \xi_1^2, \quad x_2 = i(\xi_0^2 + \xi_1^2), \quad x_3 = -2\xi_0\xi_1, \tag{13}$$

which are nicely expressed using the Pauli matrices as

$$\vec{x} = \xi^T \mathcal{C}\vec{\sigma}\xi \tag{14}$$

with $\mathcal{C} = i\sigma_2$. Only the relative sign of ξ_0 and ξ_1 is defined since the spinor reverses sign under a 2π rotation. This is the source of the statement that the Cartan spinor is a point spinor of zero length. It has no other identifying properties beyond its associated isotropic vector which determines two complex spinor components within a phase. There is no topological handle, as would be the case for the solid bodies of Dirac's construction and similar classical analogs.

Biedenharn and Louck give the connection between the 2 × 2 unitary unimodular matrix U which rotates the spinor $\xi' = U\xi$ and the elements of the 3 × 3 orthogonal matrix R which rotates the vector, $x' = Rx$:

$$R_{ij} = \frac{1}{2}tr(\sigma_i U \sigma_j U^\dagger), \tag{15}$$

where the σ_i are the Pauli matrices. The mapping is well behaved, with one coverage of U corresponding to two coverages of R, as one would expect from the fact that $x \sim \xi^2$ and correspondingly $R \sim UU^\dagger$.

Hilborn's reply refers also to the possibility of the braid group playing a role in understanding the spin-statistics relation by some topological generalization of the so-far restricted point elementary particles. Biedenharn and Louck introduce the braid group of order n as the crossings of n strings

473

attached at top and bottom, which run continuously downward without looping back. There are two elementary operations:

a) σ_i which crosses string i *over* string $i + 1$ (numbered from the left before (above) the operation);

b) σ_i^{-1} which crosses string i *under* string $i + 1$.

The operations form a group generated by the operators

$$\sigma_1, \sigma_2 \cdots, \sigma_{n-1}$$

for which

a) σ_0 and σ_1 generate the group through

$$\sigma_0 = \sigma_1 \sigma_2 \cdots \sigma_{n-1} \qquad \sigma_i = (\sigma_0)^{i-1} \sigma_1 (\sigma_0)^{-i+1}$$

b) $\sigma_i \sigma_j = \sigma_j \sigma_i$ for $j \geq i + 2$,

c) $\sigma_i \sigma_{i+1} \sigma_i = \sigma_{i+1} \sigma_i \sigma_{i+1}$.

Biedenharn and Louck characterize the braid group mathematically as more fundamental than the permutation group, and physically as the natural tool to analyze many path dependent problems. The braid group is different from the permutation group for 2-dimensions but they coincide for 3 or more dimensional Euclidean space.

Imbo, Imbo, and Sudarshan [20.11] have shown using braid group analysis that - far from providing an understanding of the spin-statistics connection - a topological extension of point spinors leads to a proliferation of presumably unrealized possibilities beyond Bose-Einstein, Fermi-Dirac, and even para- and θ-statistics. The so-called exotic statistics are associated with higher dimensional representations of the permutation group, in contrast to the Bose-Einstein and the Fermi-Dirac statistics which are associated with the one-dimensional totally symmetric and totally antisymmetric representations. Some of the exotic statistics play a physical role in 2-dimensional condensed matter physics, for example in the fractional quantum Hall effect, and possibly also in high-T_c superconductivity. Here we restrict our discussion to the original Bose-Einstein and Fermi-Dirac cases.

Imbo, Imbo, and Sudarshan quantize a classical system of n identical particles and elect to remove from the manifold of states M^n the subcomplex Δ where two or more particle coordinates coincide, which constitute troublesome fixed points under permutations. The permutation operator S_n acts freely on $M^n - \Delta$ and the orbit space $Q_n(M) = (M^n - \Delta)/S_n$ is a smooth manifold. Without removing the fixed points, only symmetric states and Bose-Einstein statistics would be permitted. With fixed points removed, particle world lines are not permitted to cross. The particles are now endowed with a property of solid bodies, that they are impenetrable. Their world lines become unbreakable threads trailing behind them eventually forming a tangled skein, whence the braid group and the proliferation of configurations. This example is fair warning to anyone who would too lightly change the rules of the game.

A sophisticated variant of Bacry's identification of an exchange operator with a rotation operator can be found in the work of Balachandran et al [20.12]. They obtain the spin-statistics connection without relativity or field theory, but with detailed topological assumptions which exclude coincident coordinates for many-body states. We can only refer the reader to their detailed arguments from which we cite their qualifying remarks: "one outstanding problem $\cdots$ concerns its relation to field theory $\cdots$ suggestive if as yet vague likeness to Fock space $\cdots$ whether spinorial fermions are actually admitted $\cdots$." Without a connection to field theory - which is, after all, no more than quantum mechanics - the significance of this work is an open question which we cannot pursue here.

§5. Feynman's Unitarity Argument Revisited.

In his 1986 Dirac Lecture Feynman also returned to the same argument for the spin-statistics relation that he had made some thirty-five years before in his original quantum electrodynamics paper [20.13]. Recall that Pauli [20.14] had deconstructed that analysis and showed that the sign change conjectured there for antiparticle amplitudes violated charge-conjugation invariance and amounted to a field theory with an indefinite Hilbert space metric, hence the violations of unitarity. In his lecture, Feynman ignored Pauli's comment (to which he evidently never responded) and presented the same argument in a slightly different guise.

Feynman argues that the unitarity of the S-matrix requires a cancellation of signs which arise from two sources: one sign change arises from "particles propagating backward in time" requiring two time-reversals of a Dirac spinor with a resulting sign change (analogous to that occurring in a 2π rotation), and a cancelling sign change from the anticommutation of Dirac field operators. His claim is that the time-reversal properties of the Dirac spinor require Fermi-Dirac statistics in order to avoid violations of unitarity, clearly containing seeds of Schwinger's proof [20.15] based on time-reversal invariance, and reversing the logical preeminence of the Spin-Statistics Theorem over the **TCP**-theorem which had been specifically established in the proofs of Lüders and Zumino, and of Burgoyne [20.16].

First Feynman establishes that spin-$\frac{1}{2}$ states change sign under two time-reversals. The effect of time-reversal T on spin-$\frac{1}{2}$ states must be

$$T|m_z = +\rangle = e^{i\phi}|m_z = -\rangle \quad \text{and} \quad T|m_z = -\rangle = e^{i\xi}|m_z = +\rangle. \quad (16)$$

With $TF = F^*T$ defining the effect of (weak-) time-reversal on algebraic functions F to be complex conjugation, two time-reversals give

$$T^2|m_z = +\rangle = e^{i(\xi-\phi)}|m_z = +\rangle. \quad (17)$$

But Feynman shows that we cannot choose $\phi = \xi$. For consider the time-reversal of states quantized along the x-axis. With

$$|m_x = \pm\rangle = |m_z = +\rangle \pm |m_z = -\rangle \quad (18)$$

476

then

$$T|m_x = +\rangle = e^{i\psi}|m_x = -\rangle = e^{i\psi} (|m_z = +\rangle - |m_z = -\rangle)$$
$$= e^{i\phi}|m_z = -\rangle + e^{i\xi}|m_z = +\rangle, \tag{19}$$

so we must have

$$e^{i\psi} = e^{i\phi} \quad \text{but} \quad -e^{i\psi} = e^{i\xi} \tag{20}$$

and

$$e^{i(\xi-\phi)} = -1. \tag{21}$$

The result is that two time-reversals change the sign of a spin-$\frac{1}{2}$ state, and, by superposition, any half-integral spin state. The sign reversal does not occur for integral spin states because they include the unique $M = 0$ state for which

$$T^2|M = 0\rangle = Te^{i\alpha}|M = 0\rangle = e^{-i\alpha}T|M = 0\rangle = e^{-i\alpha}e^{+i\alpha}|M = 0\rangle, \tag{22}$$

so $T^2 = 1$, a result that can be extended to all integral spin states because they are superposable with the $M = 0$ state.

Next Feynman considers the unitarity of the scattering matrix S, which is defined as the operator which evolves the quantum system from early (non-interacting) times, through the scattering interval, to late (again non-interacting) times

$$\Psi(t \to \infty) = S\Psi(t \to -\infty). \tag{23}$$

In order to maintain orthonormality of states propagated through the scattering, the S-matrix must be unitary

$$S^\dagger S = SS^\dagger = 1. \tag{24}$$

It is useful to define the transition matrix $\mathcal{T}$ by

$$S = 1 - 2i\mathcal{T}. \tag{25}$$

These matrix operators are familiar in their elementary form for individual partial waves elastically scattered by a central potential. In this case, $S =$

477

$e^{2i\delta}$, $\mathcal{T} = -e^{i\delta} \sin \delta$ and the cross-section is $|\mathcal{T}|^2 = \sin^2 \delta$ within factors of no concern here. The unitarity of the S-matrix imposes a requirement on the $\mathcal{T}$-matrix

$$S^\dagger S = (1 + 2i\mathcal{T}^\dagger)(1 - 2i\mathcal{T}) = 1 - 2i(\mathcal{T} - \mathcal{T}^\dagger) + 4\mathcal{T}^\dagger \mathcal{T}. \qquad (26)$$

For diagonal matrix elements $\mathcal{T}_{ii}$

$$Im\mathcal{T}_{ii} = -\left(\mathcal{T}^\dagger \mathcal{T}\right)_{ii} = -\sum_j |\mathcal{T}_{ji}|^2 \leq 0. \qquad (27)$$

For a given state i, the sum is over all energy and momentum conserving states j. The right hand side has a ready interpretation in terms of the total cross-section for scattering from state i to all possible states j, and it unequivocally determines the sign of the imaginary part of the diagonal $\mathcal{T}$-matrix elements, which correspond to forward elastic scattering. These results are familiar for the scattering amplitudes of individual partial waves elastically scattered by a central potential. There, $-Im\mathcal{T} = \sin^2 \delta = |\mathcal{T}|^2 = \sigma/4\pi k^2$ with σ the partial cross-section, and k the momentum.

Feynman evaluates the scattering amplitude using the Feynman rules. Consider the following somewhat artificial example, just to make his point: we imagine a toy model of spinless mesons ϕ coupled to Dirac particles ψ by an interaction $\phi^2 \bar\psi \psi$. The $\phi\phi \to \phi\phi$ scattering amplitude includes an amplitude with a $\psi, \bar\psi$ loop as intermediate state. We are instructed that the same amplitude evaluated with the $\psi, \bar\psi$ loop replaced by a loop of two spin-0 particles has the appropriate sign required to respect unitarity.

What is different about the amplitude with the $\psi, \bar\psi$ loop? Feynman says that there are two differences. One is a sign change due to the rearrangement - by three anticommutations - of Dirac field operators from the order which occurs naturally in the product of interaction Hamiltonians evaluated at the two vertices

$$\bar\psi(2)\psi(2)\bar\psi(1)\psi(1),$$

into the order

$$\psi(1)\bar\psi(2)\psi(2)\bar\psi(1).$$

The rearrangement is necessary so that we can identify the Feynman propagator

$$S_F(2 \leftarrow 1) = \langle 0|\psi(2)\bar{\psi}(1)|0\rangle \qquad (28)$$

of a particle propagating forward in time from vertex 1 to vertex 2 followed by the Feynman propagator

$$S_F(1 \leftarrow 2) = \langle 0|\psi(1)\bar{\psi}(2)|0\rangle \qquad (29)$$

of a particle propagating backward in time from 2 to 1. This triple anticommutation results in a characteristic minus sign accompanying every closed fermion loop when evaluated with the Feynman rules.

Minus sign number one.

Feynman explains away this minus sign as necessary to cancel another minus sign introduced by a double time-reversal of the "actual" antiparticle spinors at vertices 1 and 2 to the "backward in time" negative energy particle spinors. The net result is said to be consistent with the sign of the spinless loop case which is taken to be consistent with the unitarity of the S-matrix.

Lets see how this happens. The Feynman propagator

$$S_F(x_2 - x_1) = -i \int \frac{d^3p}{(2\pi)^3} \frac{m}{E} \times$$
$$\times \left[\Theta(t_2 - t_1)\Lambda_+(p)e^{-ip\cdot(x_2-x_1)} + \Theta(t_1 - t_2)\Lambda_-(p)e^{+ip\cdot(x_2-x_1)} \right] \quad (30)$$

is the amplitude for a free Dirac particle to propagate forward in time from $1 \rightarrow 2$ when the unit step function $\Theta(t_2 - t_1) = 1$ or for a free Dirac antiparticle to propagate forward in time from $2 \rightarrow 1$ when the other step function $\Theta(t_1 - t_2) = 1$. It is the great - but not entirely free - elegance of the Feynman rules to treat the antiparticle propagation as if it were negative energy particles propagating backward in time.

The projection operator $\Lambda_+(p)$ onto positive energy Dirac states, equal to the sum over free Dirac spinors $u(p)$ satisfying $(p \cdot \gamma - m)u(p) = 0$ with

$p^2 = m^2$, $p_0 > 0$ and normalized to $\bar{u}u = 1$, is

$$\Lambda_+(p) = \sum_{s=\pm} u(p;s)\bar{u}(p;s) = \frac{p \cdot \gamma + m}{2m}. \tag{31}$$

The projection operator onto negative energy states will be defined similarly in a moment.

We begin to understand Feynman's argument. It becomes clear that the Feynman Dirac propagator on the mass shell, where the virtual particles become real and contribute to the unitarity sum in the imaginary part of the amplitudes, is the product of a real, positive projection operator times a Feynman scalar propagator. The Dirac propagator must contribute with the same sign to the imaginary part as does the scalar propagator. Our only concern - still assuming the spin-0 loop is well behaved - is the overall external sign of the amplitude.

What about the "negative energy" projection operator which enters when we consider an antiparticle propagating forward in time as a negative energy particle propagating backward in time, as Feynman does in his prescription for the loop amplitude. We need to examine the particle propagator Λ_+, with momentum p continued to the reversed four-momentum $-p$. We find that it is *not* directly the (more properly called) antiparticle projection operator $\Lambda_-(p)$. An extra minus sign is needed in the continuation.

$$\begin{aligned}
\Lambda_-(p) &= \Lambda_+(-p) = \frac{-p \cdot \gamma + m}{2m} \\
&\neq \sum_s u(-p;s)\bar{u}(-p;s) \\
\text{but} &= -\sum_s u(-p;s)\bar{u}(-p;s) \\
\text{or} &= -\sum_s v(p;s)\bar{v}(p;s).
\end{aligned} \tag{32}$$

An extra minus sign must be inserted in the continuation

$$u(p)\bar{u}(p) \to -u(-p)\bar{u}(-p) = -v(p)\bar{v}(p). \tag{33}$$

Minus sign number two. Just Feynman's change of sign from the time-reversal $(p \to -p)$ of two Dirac spinors.

At last, the $\psi, \bar{\psi}$ loop is expressed in a way which makes clear that it has the same sign imaginary part as obtained for a loop with spinless particles. The difference is two minus signs - one from the original anticommutations, one from the double time reflection of the Dirac spinors; and two projection operators, one onto physical, positive energy particle states and one onto so-called "negative energy", but actually physical positive energy antiparticle states.

The perfect equivalence between particle and antiparticle propagation in the Feynman propagator is clear from the original prescription. There Feynman eliminated forward propagation of negative energy states by subtracting the negative energy contributions for all times. The subtracted term satisfies the homogeneous Dirac equation and leaves the Green's function satisfying the original defining inhomogeneous differential equation, but different boundary conditions. The Feynman Green's function becomes

$$
\begin{aligned}
S_F(x' - x) &= -i\Theta(t' - t) \sum \psi\bar{\psi} \\
&\to -i\Theta(t' - t) \sum_{+} \psi\bar{\psi} + i\Theta(t - t') \sum_{-} \psi\bar{\psi}.
\end{aligned}
\tag{34}
$$

Recall that the negative energy spinors have $\bar{\psi}_-\psi_- = \bar{v}v = -1$ and require a minus sign in the projection operator $\Lambda_-(p)$.

A nearly identical situation arises in the Feynman Green's function for the spin-0 Klein-Gordon equation. Here a sign change sneaks in because the norm of positive energy states is

$$
\sim \int \phi_+^\dagger \left(i \overleftrightarrow{\partial}_0 \right) \phi_+,
$$

but the negative of this expression for negative energy states. Bjorken and Drell [20.17] have a full description of the two cases in their "Relativistic Quantum Mechanics".

This lack of simple continuability is also obvious from the gap between the static $+m$ and $-m$ four-component Dirac spinors. We have (in an abbreviated 2×2 notation)

$$u(m) = \begin{pmatrix} 1 \\ 0 \end{pmatrix} \quad \text{and} \quad u(-m) \equiv v(m) = \begin{pmatrix} 0 \\ 1 \end{pmatrix}, \tag{35}$$

and the projection operators

$$\begin{aligned} \Lambda_+(m) &= \begin{pmatrix} 1 & 0 \\ 0 & 0 \end{pmatrix} = \sum_m u(m)\bar{u}(m) \\ &= \begin{pmatrix} 1 \\ 0 \end{pmatrix} (1 \ \ 0) \begin{pmatrix} 1 & 0 \\ 0 & -1 \end{pmatrix}, \end{aligned} \tag{36}$$

but

$$\begin{aligned} \Lambda_-(m) &= \begin{pmatrix} 0 & 0 \\ 0 & 1 \end{pmatrix} = -\sum_m u(-m)\bar{u}(-m) = -\sum_m v(m)\bar{v}(m) \\ &= -\begin{pmatrix} 0 \\ 1 \end{pmatrix} (0 \ \ 1) \begin{pmatrix} 1 & 0 \\ 0 & -1 \end{pmatrix}. \end{aligned} \tag{37}$$

The same arguments - in particular the sign of the one loop amplitudes - follow also from the Feynman Path Integral formulation of quantum mechanics. The path integral formulation achieves all the results (and more) of canonical quantum field theory without the formalism of second quantization. The device which makes this possible for Fermi-Dirac anticommuting fields is the calculus of Grassmann variables which anticommute with each other. It is a long story, and we simply accept the result that the usual Feynman rules are reproduced [20.18].

There is the same hesitation about treating scalar fields as Grassmann variables that existed for treating them as anticommuting quantum fields. It would appear that

$$\phi(x)\phi(y) = -\phi(y)\phi(x) \tag{38}$$

requires $\phi^2(x) = 0$, returning us to arguments from Pauli to deWet to Burgoyne, and all the others, which preclude this possibility. Such objects *are* useful as purely mathematical devices called "ghost" fields - scalar fields satisfying Fermi-Dirac statistics but having no matrix elements among the physical states. As ghosts they do have the wrong sign of scattering amplitude to respect unitarity. Is this really irreparable? It must be, because the propagation of positive energy particles (and antiparticles) forward in time requires the standard choice of sign of the imaginary part. Reversing the sign of the imaginary part (by replacing $m^2 - i\epsilon$ by $m^2 + i\epsilon$ in the Feynman propagators) would propagate *negative* energy particles forward in time.

Commuting Dirac fields violate unitarity as in Feynman's above diagrammatic argument. The interesting questions from this point of view are the basically unanswerable existential question of 'Why Grassmann variables?', and a familiar question in reverse - 'How do we recognize *a priori* that a Grassmann field is a Dirac field?'.

Feynman's argument - based upon the internal consistency of the Feynman Rules for perturbation theory in relativistic quantum field theory - seems to be an esoteric if not tenuous thread on which to hang a proof of the Spin-Statistics Theorem. We hazard a guess that it leaves Neuenschwander still dissatisfied.

Bibliography and References.

20.1) D.E. Neuenschwander, Am. J. Phys. **62**, 972 (1994).

20.2) R.P. Feynman, R.B. Leighton, and M. Sands, *The Feynman Lectures on Physics, Vol.3* (Addison-Wesley, Reading MA, 1965), Ch.4.

20.3) R.C. Hilborn, Am. J. Phys. **63**, 298 (1995).

20.4) H. Bacry, Am. J. Phys. **63**, 297 (1995); A.A. Broyles, Am. J. Phys. **44**, 340 (1976).

20.5) R.R. Gould, Am. J. Phys. **63**, 109 (1995).

20.6) R.P. Feynman, *Elementary Particles and the Laws of Physics. The 1986 Dirac Memorial Lecture* (Cambridge, New York, 1987), pp.2-59.

20.7) T. von Foerster, Am. J. Phys. **64**, 526 (1996).

20.8) M. Saha, Ind. J. Phys. **10**, 145 (1936); see also Phys. Rev. **75**, 1968 (1949).

20.9) D. Finkelstein, Phys. Rev. **100**, 924 (1955); also D. Finkelstein and J. Rubinstein, J. Math. Phys. **9**, 1762 (1968).

20.10) L.C. Biedenharn and J.D. Louck, *Angular Momentum in Quantum Mechanics. Theory and Application* (Addison-Wesley, Reading MA, 1981), pp.7-26.

20.11) T.D. Imbo, C.S. Imbo, and E.C.G. Sudarshan, Phys. Lett. **B234**, 103 (1990).

20.12) A.P. Balachandran, A. Daughton, Z.-C. Gu, R.D. Sorkin, G. Marmo, and A.M. Srivastava, Int. Jour. Mod. Phys. **A8**, 2993 (1993).

20.13) R.P. Feynman, Phys. Rev. **76**, 749 (1949); see our App.15A.

20.14) W. Pauli, Progr. Theor. Phys. **5**, 526 (1950); see our App.15B.

20.15) J. Schwinger, Phys. Rev. **82**, 914 (1951); see our App.16A.

20.16) G. Lüders and B. Zumino, Phys. Rev. **110**, 1450 (1958); also N. Burgoyne, Nuovo Cim. **8**, 607 (1958); see our App.17A,B.

20.17) J.D. Bjorken and S.D. Drell, *Relativistic Quantum Mechanics* (McGraw-Hill, New York, 1964), pp.95,188.

20.18) M. Kaku, *Quantum Field Theory: A Modern Introduction* (Oxford, New York, Oxford, 1993) pp.285-289.

Chapter 21

Overview and Epilog

Summary: The proofs of the Spin-Statistics Theorem are reviewed and summarized. Following this, we present a new proof of the great theorem which comes closest to satisfying Neuenschwander's requirements: the proof is simple, physical, and intuitive, but still not completely free from the complications of relativistic quantum field theory. This is as much as one can hope for the proof of a theorem whose most dramatic physical consequences are manifestly extremely non-relativistic: the Exclusion Principle effects on electrons in atoms and metals, the Bose-Einstein condensation effects in superfluids and superconductors, in lasers, and most recently in trapped atoms condensed at tenths of a micro-Kelvin. Finally, the most fundamental possible understanding of the spin-statistics connection is presented, based on the fact that classical oscillators satisfying Bose-Einstein statistics must necessarily be identified with spin-0 fields; those satisfying Fermi-Dirac statistics must necessarily be identified with spin-$\frac{1}{2}$ fields.

§1. Introduction.

The work of Pauli and his contemporaries and the more recent work of Schwinger are both based on Lagrangian field theory. They both assume that the kinematic part of the Lagrangian by itself determines the spin-statistics connection. This would be compelling if we had reason to believe perturbation theory based on the free Lagrangian. Schwinger [21.1] assumes the validity of this strategy independent of perturbation theory on the grounds that at "sufficiently high" energies, the kinematic terms dominate. This is certainly true for potential scattering where we know the first Born approximation is good at high enough energy. It is also true for the all-important non-Abelian gauge theories; but in general the validity of this assumption cannot be taken for granted.

485

The possibility of going beyond this limitation is contained in the work of Umezawa and Kamefuchi [21.2]. They show that the two-point Wightman function for a scalar field

$$F^{(2)}(x, y) = \langle 0|\phi(x)\phi(y)|0\rangle \tag{1}$$

has the spectral resolution

$$F^{(2)}(x, y) = \int_0^\infty d(m^2)\rho(m^2)F_0^{(2)}(x - y; m) \tag{2}$$

with a nonnegative weight function $\rho(m^2)$ and the free particle Wightman function $F_0^{(2)}$ [21.3]. A similar result holds for spinor fields.

Burgoyne [21.4] established the spin-statistics relation according to the Wightman axioms [21.5]. For a scalar field obeying anticommutation relations for spacelike separated field points, it follows from the structure of the Wightman function that

$$F_A^{(2)}(x, y) = 0, \quad x - y \text{ spacelike.} \tag{3}$$

But this implies that

$$\left\| \int dx\, f(x)\phi(x)|0\rangle \right\|^2 = 0 \tag{4}$$

for any suitable test function $f(x)$, so $\phi(x)$ annihilates the vacuum and must therefore be a null field. Burgoyne concludes that scalar fields cannot obey the anticommutation relation "all others zero". Only scalar fields obeying commutation relations are possible. Similar arguments can be constructed for other tensor fields.

For spin-$\frac{1}{2}$ fields assumed to commute at spacelike separations, we can show in a similar fashion that

$$\left\| \int dx\, u_r(x)\psi_r(x)|0\rangle \right\|^2 = 0 \tag{5}$$

for any test function $u_r(x)$, which shows that such $\psi_r(x)$ must vanish. Conversely, if the spinor fields anticommute the structure of the two-point

486

Wightman function imposes no constraints. These considerations can be extended to higher half-integral spin fields.

Wightman theory makes no assumptions about equations of motion, and assumes no specific Lagrangian; yet the analysis yields conclusions very similar to those obtained by Pauli and by Schwinger. The common feature of all these is the symmetry properties of the Lorentz covariants bilinear in the fields, which themselves are finite dimensional representations of the Lorentz group.

We note that all the proofs are "negative proofs": integral spin fields cannot satisfy anticommuting Fermi-Dirac statistics; half-integral spin fields cannot satisfy commuting Bose-Einstein statistics. If there are more general statistics then the spin-statistics relation is not proved, but only a set of wrong-connections is ruled out. We emphasize that relativistic invariance is only used to the extent of identifying the symmetry or the antisymmetry of terms bilinear in the fields.

Green [21.6] has shown that there are generalizations of the Bose-Einstein and of the Fermi-Dirac statistics in which the commutation relations are not bilinear but trilinear. Green also suggested an *ansatz* where each "para-Bose" field is the sum of mutually *anti-commuting* Bose-Einstein fields and each "para-Fermi" field is the sum of mutually *commuting* Fermi-Dirac fields. The Green ansatz furnishes a reducible representation of the "para-statistics" and is essentially reincarnated in modern quantum field theory as "color".

We may well ask whether Green's parastatistics is the most general. To explore this question we consider a quantum field theory over a manifold M of arbitrary but finite dimension. The wave functions should be suitable square-integrable functions over the manifold with its natural measure μ. The states of n-identical particles are described by either the wavefunctions $\psi(x_1, \cdots x_n)$, or the wavefunctions $\psi(x_1', \cdots x_n')$ where $(x_1', \cdots x_n')$ are a permutation of $(x_1, \cdots x_n)$. But since for *identical* particles these wavefunctions represent the *same* configuration, the wavefunctions are multivalued and depend on the variety of paths by which we can go from $(x_1, \cdots x_n)$ to $(x_1', \cdots x_n')$. Since those paths which can be continuously deformed into each other are to produce the same wave function, it follows that the wavefunctions are a representation of the homotopy group in the manifold over which they are defined. This is obtained as follows: for n distinguishable particles the configuration space would have been

$$\mathcal{M} \times \mathcal{M} \times \cdots \times \mathcal{M} \text{ (n times)} \equiv \mathcal{M}^n. \tag{6}$$

When the particles are indistinguishable, the configuration space must be reduced to

$$\frac{\mathcal{M}^n}{\mathcal{S}_n} \tag{7}$$

where $\mathcal{S}_n$ is the $n!$ element discrete group of permutations over n variables. But this is not a manifold unless we remove the "diagonal" Δ^{n-1} where two or more coordinates coincide. So we get the manifold [21.7]

$$\frac{\mathcal{M}^n - \Delta^{n-1}}{\mathcal{S}_n}. \tag{8}$$

The first homotopy group is the "braid group"

$$B_n(\mathcal{M}) = \Pi_1 \left(\frac{\mathcal{M}^n - \Delta^{n-1}}{\mathcal{S}_n} \right). \tag{9}$$

If the dimension of the manifold is three or greater, and M is simply connected (that is, all closed curves in M can be continuously shrunk to nothing) then

$$B_n(\mathcal{M}) = \mathcal{S}_n. \tag{10}$$

The representations of this class of braid groups is just Green's parastatistics.

When the base manifold M is the two-dimensional Euclidean space (2-space plus 1-time) there is a new possibility for the statistics: this is the "anyon" statistics in which the interchange in a definite labelling scheme of the jth-particle with the kth-particle multiplies the wave function by a phase factor $e^{i\theta}$. Note that in this case "spin" is associated with rotations in a plane for which all irreducible representations of the covering group are one-dimensional and labelled by one real number. Such wave functions are best described in terms of the complex variables

$$(z_j - z_k)^{\theta/2\pi} \tag{11}$$

multiplying any nonsingular function of the z's. In this case no connection exists between the "spin" and the "statistics".

§2. Elementary Proof of the Spin-Statistics Theorem.

We return to the situation where the spin and statistics *are* connected.

Following earlier work by Sudarshan [21.8], we show that the full complication of Lorentz invariance and of relativistic quantum field theory are not explicitly necessary in the proof of the Spin-Statistics Theorem. There still remains however a key part of the argument which must be traced to the requirement of separate relativistic kinematic Lagrangians for individual relativistic fields.

Consider the usual (3+1)-dimensional space-time. We impose four conditions on the kinematic part of the Lagrangian: it must be
1) derivable from a local Lorentz invariant field theory for fields ξ which are each a finite dimensional representation of the Lorentz group (tensor or spinor);
2) in the Hermitian field basis $\xi = \xi^\dagger$;
3) at most linear in the first derivatives of the fields; and
4) bilinear in the fields.

These embody the physical requirements that the Euler-Lagrange equations should be first order linear differential equations of the Hamiltonian form local in space and time. A wide variety of field equations is possible by the artifice of auxiliary fields which can be eliminated even at the Lagrangian stage, so the Schwinger prescription is not as limiting as it might

489

appear.

The kinematic terms in the Lagrangian have the general form

$$
\begin{aligned}
\mathcal{L}_K \;=\; & \frac{i}{2}\left(\xi_r \dot{\xi}_s - \dot{\xi}_r \xi_s\right) K^0_{rs} \\
& -\frac{i}{2}\left(\xi_r \cdot \nabla_j \xi_s - \nabla_j \xi_r \cdot \xi_s\right) K^j_{rs} + \xi_r \xi_s M_{rs},
\end{aligned} \tag{12}
$$

summed on $j = x, y, z$. The indices (r, s) label the space-time and spin components of the field ξ and are summed over also.

It is a property of the $O(3)$ group of rotations in three dimensions that representations belonging to *integral spin* have a bilinear scalar product *symmetric* in the indices of the factors: for example the scalar product of two vectors is

$$
(V_1, V_2) = \sum_{rs} V_{1r} V_{2s} \delta_{rs}, \tag{13}
$$

a familiar result. In contrast, *half-integral spin* representations have *anti-symmetric* scalar products: for spin-$\frac{1}{2}$, the scalar product is

$$
(\psi_1, \psi_2) = \sum_{rs} \psi_{1r} \psi_{2s} (i\sigma_y)_{rs}. \tag{14}
$$

This result is familiar from the rotational invariant spin-0 combination of the usual spin-$\frac{1}{2}$ spinors α and β

$$
\phi_{12}(j = 0) = (\alpha_1 \beta_2 - \beta_1 \alpha_2). \tag{15}
$$

What is less familiar is the use of Hermitian fields, especially in the case of half-integral spin.

Already we can note that the invariance of these scalar products under the exchange $1 \leftrightarrow 2$ is consistent with the spin-statistics relation.

The kinematic Lagrangian density is of the form

$$
\mathcal{L}_K = \xi_r \Lambda_{rs} \xi_s \tag{16}
$$

where the matrix Λ_{rs} contains differential operators as well as numerical matrices:

$$\Lambda = \left(\frac{i}{2}K^0 \overleftrightarrow{\partial}_t - \frac{i}{2}K_j \overleftrightarrow{\partial}_j + M\right). \tag{17}$$

The terms in the Lagrangian must be scalar invariants under the group of the indices (rs), that is they must be scalar products bilinear in the ξ_r. In order to give an invariant nonvanishing result under the exchange $r \Leftrightarrow s$, which results in

$$\xi_r \Lambda_{rs} \xi_s \to \xi_s \Lambda_{sr} \xi_r = \pm \xi_r \Lambda_{sr} \xi_s, \tag{18}$$

the simplest term M must be *symmetric* for *Bose-Einstein* (+) statistics and *antisymmetric* for *Fermi-Dirac* (-) statistics. But this is compatible only with the usual spin-statistics relation: a symmetric scalar product corresponding to Bose-Einstein statistics can only be rotationally invariant for an integral spin field; an antisymmetric scalar product corresponding to Fermi-Dirac statistics can only be rotationally invariant for a half-integral spin field. This is the essential point of Sudarshan's elementary proof.

The differential operators

$$\overleftrightarrow{\partial} \equiv \overrightarrow{\partial} - \overleftarrow{\partial} \tag{19}$$

are themselves antisymmetric, so K^0 and K_j must be *antisymmetric* for *Bose-Einstein* and *symmetric* for *Fermi-Dirac* statistics. In addition, the matrices K^0, K_j and M are restricted by the Hermiticity of the Lagrangian.

So far the result does not require Lorentz invariance, although it is consistent with Lorentz invariant theories. We will need reference to relativistic quantum field theory in order to exclude the possibility of an arbitrary antisymmetrization on two *different* fields in the kinematic Lagrangian. We do require space-dimension three in order to have multicomponent irreducible representations with symmetric or antisymmetric invariant scalar product. As a typical nonrelativistic quantum field theory we consider quantum hydrodynamics. This Galilean field theory should be quantized according to Bose-Einstein statistics. By simply appending a Pauli spin-$\frac{1}{2}$ spinor, we can change the required statistics to Fermi-Dirac.

§3. Further Comments on the Elementary Proof.

Sudarshan's proof should be compared to a related proof of Schwinger, and to the very different proof of Lüders and Zumino. Both Sudarshan and Schwinger base their proofs on Dirac-like Lagrangians, both ignore interactions, and both make use of Hermitian fields. Both recognize that the mass matrix M must be symmetric for Bose-Einstein statistics, antisymmetric for Fermi-Dirac. The great simplification in Sudarshan's proof is to recognize that the spin-statistics connection can be made directly from rotational invariance, without ever having to appeal directly, as Schwinger does, to time-reversal invariance. Schwinger's subsequent efforts to establish the spin-statistics connection (see our Eqn19.13) succeed but at the price of a much more difficult proof which has no intuitive access.

Sudarshan, with Schwinger, makes the great simplification of using Hermitian fields $\xi_r = \xi_r^\dagger$. The index r refers only to spin and space-time degrees of freedom. Any possible internal "flavor" degrees of freedom must be treated separately as we will confirm in a moment, in which case the rotationally invariant $\xi^T M \xi$ term in the Lagrangian is just the metric (Clebsch-Gordan coefficient $(sms - m | ss00)$) in the spin-space: symmetric for integral spin s, antisymmetric for half-integral spin.

A possibility is that the *a priori* fundamental Hermitian fields (ξ_1, ξ_2), each of which satisfies its own Lagrangian, occur in flavor pairs. These can be identified as the real and imaginary parts of complex fields $\psi \neq \psi^\dagger$ which satisfy a global (charge-conserving) gauge invariance under the phase transformation $\psi \to e^{i\alpha}\psi$. Then the pair of Hermitian fields is rigidly rotated by the free flavor neutral Hamiltonian leaving the norm $(\xi_1)^2 + (\xi_2)^2$ invariant. In this case, the K and M matrices are the direct product of the 2-dimensional unit matrix in flavor times the spin-metric matrix. In this way, Sudarshan's proof can be extended to include non-Hermitian fields.

One might think that a flavor singlet Lagrangian for which the spin-statistics connection had been established could then be antisymmetrized on particle flavor, for example

$$\mathcal{L} = \psi^\dagger(i\partial_t - M)\psi \Rightarrow \sum_{j,k=1}^{2} \sigma_y(j,k)\psi_j^\dagger(i\partial_t - M)\psi_k,$$

thereby reversing the original conclusion. However, the Lagrangian can be diagonalized in the flavor indices (j, k) leading to two fields with identical independent Lagrangians of opposite sign, identical but opposite spectra, and a total field energy which is unbounded below. Just this form occurs in the *difference* of the Klein-Gordon Lagrangians for a charged particle ϕ^+ and its antiparticle $\phi^- = \phi^{+\dagger}$, expressed in terms of the Hermitian fields

$$\phi_1 = (\phi^+ + \phi^-)/\sqrt{2} \quad \text{and} \quad \phi_2 = (\phi^+ - \phi^-)/\sqrt{2}i,$$

leading to an antisymmetric Lagrangian

$$\mathcal{L}_A \sim \phi^{+\dagger}\phi^+ - \phi^{-\dagger}\phi^- \sim i(\phi_1\phi_2 - \phi_2\phi_1),$$

which requires anticommutation relations for the Klein-Gordon field. The subtracted Lagrangian leads to negative energies, which can only be avoided at the price of a negative metric for the antiparticle states with all its associated problems. This quagmire is avoided in relativistic quantum field theory by postulating a Hilbert space of positive energy states with positive definite metric.

Separate flavor singlets (ξ_1, ξ_2) with flavor symmetric Lagrangians

$$\mathcal{L} \sim (\xi_1)^2 + (\xi_2)^2$$

avoid these pathologies and return us to the Sudarshan proof.

The origin of the sign in the spin-metric [21.9] is related to the invariance under rotations of the scalar product $\alpha^\dagger\alpha + \beta^\dagger\beta$, which is imposed by defining

$$\alpha_k^\dagger \equiv \alpha_{t/r;k} = -\beta_k, \quad \beta_k^\dagger \equiv \beta_{t/r;k} = +\alpha_k,$$

in terms of the time-reverse spinors. Consistency for all angular momenta requires the spin-metric to be

$$(jmj - m|jj00) = (-)^{j+m}/\sqrt{2j+1}. \tag{20}$$

Two such time-reversals result in a phase $(-)^{2j}$ consistent with Feynman's discussion (Eqn20.21), and already used in Schwinger's Euclidean proof (Eqn19.32).

The second-quantized Hermitian field ξ expressed in terms of angular momentum eigenstates and their time-reverse is qualitatively

$$\xi \sim \sum aU + b^\dagger U^{\dagger T} \to \sum aU + a^\dagger U_{t/r}. \tag{21}$$

The one-particle expectation value of the mass term in the Lagrangian is

$$\langle 1|L \sim \xi M\xi|1\rangle = \sum U_{t/r}U. \tag{22}$$

Summed over spin components it is a rotational invariant as required.

Everything but the spin-statistics connection looks complicated in terms of Schwinger's Hermitian fields satisfying a Dirac-like Lagrangian linear in field derivatives. The simplest case of the Klein-Gordon Lagrangian for a single Hermitian scalar field ϕ expressed in the Schwinger form requires an auxiliary 4-vector field $V_\mu = \partial_\mu \phi$ and 5×5 K and M matrices.

The non-relativistic Schrödinger equation for complex $\psi \neq \psi^\dagger$ can be included in the proof only by taking the limit of the Klein-Gordon or the Dirac Lagrangians. The difficulty appears because the Schrödinger Lagrangian is not directly of the Schwinger form, but contains terms like $\phi_1 \phi_2$. Without additional arguments from the relativistic theories, we cannot rule out the possibility of antisymmetrizing such terms, leaving either choice of statistics. This leads to an implicit but critical reliance on relativistic Lagrangians.

The electromagnetic field is a particularly simple case using Sudarshan's proof. Since the electromagnetic field is Hermitian, it can be understood without recourse to the Schwinger Dirac-like Lagrangian. The behavior of the electric field term in the Lagrangian (in a convenient gauge)

$$E^2 = \sum_{r,s=1,2,3} \partial_t A_r A_s g_{rs}$$

494

is sufficient to require Bose-Einstein quantization using Sudarshan's argument. The symmetry of g_{rs} and invariance under exchange of A_r and A_s is all that is required.

None of the above considerations limit the statistics of composite and nonlocal entities. Usually a composite particle is considered as a collection of point particles with integer relative angular momenta and in such simple cases, a composite consisting of an even number of fermions and any number of bosons would be a boson; with an odd number of fermions, a fermion. So He^4 would be a boson, He^3 a fermion. When excitations with topological obstructions are included the situation can change. In the Saha model, a charged particle-magnetic monopole system with half-integral electromagnetic field angular momentum obeys Fermi-Dirac statistics [21.10]. Similarly in a $(2+1)$-dimension space-time, charged particles with a minimal Chern-Simons interaction appear to change their statistics at low energies.

§4. Dirac Equation from Grassmann Theory.

In this Section we deduce the spin-statistics connection *starting* from fundamental anticommuting Grassmann variables [21.11] which are defined by the anticommutation relation

$$\xi_j \xi_k + \xi_k \xi_j = \delta_{jk}. \tag{23}$$

By a series of inferences, we show that the only possible Lagrangian for the associated field is a first order Dirac Lagrangian. From this point the Dirac equation with 4-component spinors, spin-$\frac{1}{2}$, and all the rest follows as usual. The difference is that we have *started* with an anticommuting quantum field which at the outset was required to satisfy the Pauli Exclusion Principle, and - by inference from the *only possible* classical Lagrangian for anticommuting objects - must satisfy the Dirac equation.

Schwinger's Lagrangian, linear in the first derivative of the field, suggests that we start with a Grassmann variable defined at a point, a function of time only, and construct the basic dynamics [21.12]. By embedding the result into a Lorentz invariant form, we limit the choice to a Grassmann field of spin-$\frac{1}{2}$ satisfying the Dirac equation. This program is carried out, starting with the prescription for the Lagrangian of the Grassmann oscillator

$$\mathcal{L} = \frac{i}{2} \sum_k \xi_k \dot{\xi}_k - \mathcal{H}_I(\xi). \tag{24}$$

The kinetic term $\frac{i}{2}\xi\dot{\xi}$ is Hermitian provided the ξ's anticommute and are themselves Hermitian. The simplest choice is a two component object with $k = 1, 2$. The generalized momentum

$$\Pi_k = \frac{\partial \mathcal{L}}{\partial \dot{\xi}_k} = \frac{i}{2}\xi_k \tag{25}$$

(by convention, all derivatives on anticommuting objects are from the right) leads to the canonical anticommutation relations

$$[\xi_k, \Pi_j]_+ = \frac{i}{2}\delta_{kj} \Rightarrow [\xi_k, \xi_j]_+ = \delta_{kj}. \tag{26}$$

This is positive definite as required. The prescription "all the rest zero" will be invoked where needed. The anticommutation relation is canonical except for the factor $\frac{1}{2}$ which is necessary for the Hamilton equations of motion to agree with the Euler-Lagrange equations.

The Hamiltonian is

$$\mathcal{H} = \Pi_k \dot{\xi}_k - \mathcal{L} = \mathcal{H}_I(\xi) \tag{27}$$

where the term linear in the velocity $\dot{\xi}$ disappears as usual. We choose

$$\mathcal{H}_I = \frac{1}{2}\xi^T M \xi \tag{28}$$

with $M^\dagger = M = -M^T$ in order that $\mathcal{H}^\dagger = \mathcal{H} = \mathcal{H}^T$. The specific 2×2 example we discuss will have $M = \frac{1}{2}\sigma_y$.

The antisymmetry of the 2×2 M-matrix permits embedding in a rotationally invariant spin-$\frac{1}{2}$ field.

The Euler-Lagrange equation of motion

$$\frac{d}{dt}\Pi_k = \frac{\partial \mathcal{L}}{\partial \xi_k} \Rightarrow \dot{\xi}_k = iM_{ks}\xi_s \Rightarrow \ddot{\xi} = -M^2\xi, \tag{29}$$

with the familiar solution

$$\begin{pmatrix} \xi_1(t) \\ \xi_2(t) \end{pmatrix} = \begin{pmatrix} +\cos(t/2) & +\sin(t/2) \\ -\sin(t/2) & +\cos(t/2) \end{pmatrix} \begin{pmatrix} \xi_1(0) \\ \xi_2(0) \end{pmatrix}. \tag{30}$$

Note that the solution satisfies the original time-independent anticommutation relations. The Hamilton equation of motion

$$\frac{d}{dt}\xi_k = i[\xi_k, \mathcal{H}]_- \tag{31}$$

leads to the same result.

Next we observe that the Lagrangian $\mathcal{L}$ contains only a first derivative and is reminiscent of the Dirac form. There is a minor technicality that so far the fields have been Hermitian which puts the Dirac equation into the Majorana representation [21.13] where, for example, $\gamma_0 = \gamma_0^\dagger = -\gamma_0^T$ similar to M but 4×4. We can infer

$$\mathcal{L}_D = \frac{1}{2} \int d^3x \psi^T(x,t)\gamma_0 \left(\gamma_0 E - \vec{\gamma} \cdot \vec{p} - m\right) \psi(x,t). \tag{32}$$

Here $\gamma_j = -\gamma_j^\dagger = \gamma_j^T$ for $j = x, y, z$ in the Majorana representation. Having made the embedding, we can go to a general representation with complex Dirac spinors by a unitary transformation and return if we wish to the familiar standard representation. We have generalized the summation on k to include an integration over the spatial positions so that the anticommutation relations are generalized to $\sim \delta_{jk}\delta^3(x - x')$ at spacelike points when we invoke the "all others zero" prescription. Also the two 2-dimensional static Grassmann variables ξ_k, ξ_k' defined independently at each point and satisfying a pointlike Schrödinger equation must be embedded in an irreducible (that is, not separable) way in a covariant 4-dimensional structure as a direct product

$$\psi = \begin{pmatrix} \xi \\ \xi' \end{pmatrix}. \tag{33}$$

The 2×2 matrices K_2 and M_2 are embedded in the 4×4 matrices $K_4 \equiv \gamma_0^2 = 1_4$ and $M_4 \equiv \gamma_0$ as

$$K_4 = \begin{pmatrix} K_2 & 0 \\ 0 & K_2 \end{pmatrix} \quad \text{but} \quad M_4 = \begin{pmatrix} 0 & M_2 \\ M_2 & 0 \end{pmatrix}, \tag{34}$$

off-diagonal for irreducibility.

The *second-quantized* Dirac equation and all its consequences, most particularly spin-$\frac{1}{2}$, follow immediately.

§5. No Bose-Einstein Dirac Lagrangian.

The question is whether there can be a commuting field which has a Dirac Lagrangian. Clearly there can if we permit auxiliary fields, as we mentioned for the case of the Klein-Gordon scalar field which required a scalar field and an auxiliary four vector field in the Schwinger formulation. What we seek to eliminate here is an elementary field without any auxiliary fields with only first time-derivatives. In this case, from Eqn.48 in §1, the Schwinger Lagrangian must have $K = K^\dagger = -K^T$ imaginary-antisymmetric and $M = M^\dagger = M^T$ real-symmetric in order to give non-vanishing terms in the Euler-Lagrange equations of motion. The generalized momentum

$$\Pi_k = \frac{\partial \mathcal{L}}{\partial \dot{\xi}_k} = \frac{i}{2} \xi_j K_{jk} \tag{35}$$

and the canonical commutation relations become

$$[\xi_r, \Pi_s]_- = \frac{i}{2} \delta_{rs} \Rightarrow [\xi_r, \xi_s]_- = K_{rs}^{-1}. \tag{36}$$

Now we look for various finite dimensional representations. We easily see that there can be no one-dimensional representation because K is antisymmetric with no diagonal element. For a two dimensional representation, $K = \sigma_y$. When we try to embed the theory in three dimensional space and require it to be relativistically invariant we eliminate the 2-dimensional representation on the now familiar grounds of Dirac's algebra. A candidate K matrix for a three-dimensional representation is L_y which satisfies

$L_y = L_y^\dagger = -L_y^T$ but has no inverse and fails to give sensible commutation relations for the fields [21.14], as well as failing the Dirac algebra.

It seems impossible for a Bose-Einstein field to have a Dirac Lagrangian but we go on to explore the spectrum of the Hamiltonian for representations which can be embedded relativistically. The Hamiltonian is

$$\mathcal{H} = \Pi^T \dot{\xi} - \mathcal{L} = \frac{1}{2}\xi^T M \xi \qquad (37)$$

with M real-symmetric. The trivial choice $M = 1$, the unit matrix, corresponds to a product of all covariant or all contravariant representations of the group $O(4)$ which cannot be embedded into representations of the Lorentz group. In order to get products of covariant and contravariant representations, of dimension at least 4×4, we must choose a nontrivial $M = M^\dagger = M^T$ such as

$$M = \begin{pmatrix} 0 & \sigma_z \\ \sigma_z & 0 \end{pmatrix}. \qquad (38)$$

But this M has equal numbers of $+1$ and -1 eigenvalues and corresponds to a Hamiltonian with negative energies. These are just the signs that are reversed by the anticommuting Grassmann variables in the allowed Grassmann-Dirac theory, and serve to eliminate the possibility of the relativistic embedding of a Bose-Einstein Dirac Lagrangian .

§6. No Fermi-Dirac Klein-Gordon Lagrangian.

A different Lagrangian which we might consider for the Grassmann variables is the Klein-Gordon form

$$\mathcal{L} = \frac{1}{2}\dot{\xi}^T K \dot{\xi} - \frac{1}{2}\xi^T M \xi. \qquad (39)$$

Now $K^\dagger = K = -K^T$ and $M^\dagger = M = -M^T$ are required for $\mathcal{L}^\dagger = \mathcal{L} = \mathcal{L}^T$. The generalized momentum

$$\Pi_s = \frac{\partial \mathcal{L}}{\partial \dot{\xi}_s} = -K_{sr}\dot{\xi}_r = \dot{\xi}_r K_{rs} \qquad (40)$$

defines the canonical anticommutator

$$[\xi_s, \Pi_r]_+ = i\delta_{sr} \Rightarrow [\xi_s, \dot{\xi}_r]_+ = iK_{sr}^{-1}. \tag{41}$$

Here we are reminded of deWet's theorem [21.15] from his 1939 Princeton PhD thesis. The usual prescription "all others zero" includes

$$[\xi_s, \xi_s]_+ = 0 \tag{42}$$

and would require $\xi_s \equiv 0$. The conclusion is that Grassmann fields cannot have a Klein-Gordon Lagrangian. The conclusion is somewhat trivial for one-dimensional Grassmann variables which do not exist anyway.

If we persist in solving the model for a 2-dimensional Grassmann variable, we find a problem. The Euler-Lagrange equations of motion are

$$\ddot{\xi} = -\left(K^{-1}M\right)\xi. \tag{43}$$

Choosing

$$K = \frac{M}{\omega^2} = \begin{pmatrix} 0 & i \\ -i & 0 \end{pmatrix} \tag{44}$$

in our two dimensional model, we get $\ddot{\xi} = -\omega^2\xi$. The solutions

$$\begin{pmatrix} \xi_1(t) \\ \xi_2(t) \end{pmatrix} = \begin{pmatrix} +\cos\omega t & +\sin\omega t \\ -\sin\omega t & +\cos\omega t \end{pmatrix} \begin{pmatrix} \xi_1(0) \\ \xi_2(0) \end{pmatrix} \tag{45}$$

satisfy the equations of motion and the anticommutation relations if

$$[\xi_r, \xi_s]_+ = \delta_{rs}, \tag{46}$$

and seemingly evade deWet's theorem for a 2-component Grassmann variable. But this should not be a surprise, because deWet's theorem requires Lorentz invariance, and Lorentz invariance requires both 2-dimensional representations $\mathcal{D}(\frac{1}{2}, 0)$ and $\mathcal{D}(0, \frac{1}{2})$.

A 4×4 choice for M and K is

$$M = \begin{pmatrix} 0 & i\sigma_x \\ -i\sigma_x & 0 \end{pmatrix}, \quad K = \begin{pmatrix} 0 & i\sigma_z \\ -i\sigma_z & 0 \end{pmatrix}, \tag{47}$$

giving

$$K^{-1}M = \begin{pmatrix} i\sigma_y & 0 \\ 0 & i\sigma_y \end{pmatrix}. \tag{48}$$

Solutions of the form

$$\xi(t) \sim e^{\pm i\omega t} \tag{49}$$

exist only if the eigenvalues ω^2 of $K^{-1}M$ are real and positive. This is clearly not the case in our example where they are $\pm i$. Other choices for M and K either lead to the same result or are the trivial choice $K^{-1}M = 1$ corresponding to the direct product of two decoupled 2×2 representations which cannot be relativistically embedded.

We conclude that no relativistic Grassmann field can be a Klein-Gordon field. Of course the Grassmann field does satisfy the Klein-Gordon equation as a result of the Dirac equation, but that is distinct from having a Klein-Gordon action (not allowed) rather than a Dirac action (allowed).

§7. Concluding Remarks.

Sudarshan's arguments based on rotational invariance lead to the simplest, most transparent, and most elementary proof of the Spin-Statistics Theorem, refining and greatly simplifying a previous proof due to Schwinger based on time-reversal invariance. Sudarshan's proof eliminates the explicit dependence of the proof on relativistic quantum field theory, but a critical implicit dependence still remains to limit the kinematic Lagrangian to separate symmetric flavor singlets with antisymmetrization on flavor indices forbidden only by arguments which have basic relativistic roots.

A fundamental understanding of the spin-statistics connection is obtained in the derivation of the Dirac equation as the only possible relativistic embedding of the Lagrangian theory of the simplest point Grassmann oscillator. The basic field is defined at the outset as an anticommuting quantum field and by deWet's arguments is found to satisfy the Dirac equation for spin-$\frac{1}{2}$. The arguments of Johnson and Sudarshan rule out the possibility of fundamental fields having half-integral spin greater than $\frac{1}{2}$, so the fundamental connection between Grassmann variables and Dirac spinors is established and is complete. Schwinger's arguments for composite fields are sufficient in other cases. The Klein-Gordon Lagrangian with canonical anti-commutation relations is ruled out for anticommuting Grassmann fields by analogs of deWet's theorem: Intrinsically positive anticommutators turn out to be negative. Commuting Bose-Einstein fields cannot have a Dirac action, which would lead to negative energies.

Understanding the puzzle of the spin-statistics connection requires that we admit the existence of the most elementary (two component) Grassmann oscillators, which anticommute and must relativistically embed in the spin-$\frac{1}{2}$ Dirac equation. Commuting fields cannot satisfy the Dirac action and relativity and have a positive definite Hamiltonian, an old result. Conversely, a Klein-Gordon action for an anticommuting field leads to null fields, another old result. Commuting fields satisfy the Klein-Gordon action without contradiction, again an old and familiar result.

Clearly, a unifying point of view for understanding the spin-statistics connection presents itself. Start with two fundamental oscillator fields: a commuting one, which must must have a Klein-Gordon action and spin-0; and an anti-commuting one which must have a Dirac action, and consequently spin-$\frac{1}{2}$.

The role of the pre-existing proofs of the Spin-Statistics Theorem is to demonstrate the consistency of the relativistic field theory constructed from the fundamental quantum oscillators whose spin-statistics relation is already determined by Sudarshan's nonrelativistic argument.

§8. Final Summary.

Finally, to emphasize our point one last time, we summarize the history and the current status of the Spin-Statistics Theorem:

For sixty years, following the direction set forth by Pauli, people have struggled to understand the spin-statistics connection as the consequence of consistency requirements dictated by relativistic quantum field theory. This struggle has been particularly frustrating and exclusionary for the vast majority who have no other need for relativistic quantum field theory, and are deterred from learning it for this sole purpose, or from mastering the further arcana of the Hall-Wightman theorem. Even the vast majority of practising physicists have been excluded from an intimate understanding of, and required to take on faith, what is surely one of the most important facts of our world.

We simplify the problem in two steps. The first step is Sudarshan's demonstration that rotational invariance of the Lagrangian requires the Spin-Statistics Theorem in an apparently non-relativistic way, which does however still depend implicitly on positivity requirements of relativistic quantum field theory. In the second step, finally, we make the *spin-statistics* connection *understandable* by reversing the question to that of the *statistics-spin* connection. We show that ordinary classical commuting Bose-Einstein number-valued oscillators embed naturally into relativistic quantum field theoretic Klein-Gordon fields of spin-0; conversely, not-so-ordinary classical anticommuting Fermi-Dirac Grassmann-valued oscillators embed naturally into relativistic quantum field theoretic Dirac fields with spin-$\frac{1}{2}$.

What remains to be understood in more fundamental terms is the existence of the two types of oscillator, number-valued and Grassmann-valued.

Bibliography and References.

21.1) J. Schwinger, Proc. Nat. Ac. Sci. USA **44**, 223, 617 (1958).

21.2) H. Umezawa and S. Kamefuchi, Progr. Theor. Phys. **4**, 543 (1951); see also G. Källen, Helv. Phys. Acta **25**, 417 (1952); and H. Lehman, Nuovo Cim. **11**, 342 (1954).

21.3) J.D. Bjorken and S.D. Drell, *Relativistic Quantum Fields* (McGraw-Hill, New York, 1965), pp.139,215,390.

21.4) N. Burgoyne, Nuovo Cim. **8**, 607 (1958); see our App.17B.

21.5) A.S. Wightman, Phys. Rev. **101**, 860 (1956); see our App.18A.

21.6) H.S. Green, Phys. Rev. **90**, 270 (1953).

21.7) T.D. Imbo, C.S. Imbo, and E.C.G. Sudarshan, Phys. Lett. **B234**, 103 (1990).

21.8) E.C.G. Sudarshan, in: *Proc. Nobel Symposium 8* (Almquist and Wiksell, Stockholm, 1968), ed. N. Svartholm, p.379; also, E.C.G. Sudarshan, Statistical Physics Supplement: Journ. Ind. Inst. Sci., June, 123 (1975).

21.9) A.R. Edmonds, *Angular Momentum in Quantum Mechanics* (Princeton, Princeton, 1957), pp.30,46; A. March, *Quantum Mechanics of Particles and Wave Fields* (Wiley, New York, 1951), pp.140,150.

21.10) M.N. Saha, Ind. J. Phys. **10**, 145 (1936), also Phys. Rev. **75**, 1968 (1949); see also R.P. Feynman, *Elementary Particles and the Laws of Physics. The 1986 Dirac Memorial Lectures* (Cambridge, New York, 1987), p.56.

21.11) F.A. Berezin, *Introduction to Superanalysis* (Reidel, Boston, 1988), ed. A.A. Kirilov, p.374.

21.12) J.L. Martin, Proc. Roy. Soc. **A251**, 536, 543 (1959).

21.13) P. Roman, *Theory of Elementary Particles* (North Holland, 1960), p.125.

21.14) K. Johnson and E.C.G. Sudarshan, Ann. Phys. **13**, 126 (1961).

21.15) J.S. deWet, Phys. Rev. **57**, 646 (1940); see our App.13B.

Index

BY PERSONAL NAME (first refer-
    ence in Chapter):

507

508

Wien's Law 75
Wightman A.S. 1, 331, 406, 425
    axioms, 486
    function, 425, 458, 486
Wigner E.P. 8, 168, 235, 330

Zumino B. 3, 235, 283, 310, 331, 350,
    370, 393, 405, 425, 457, 464

BY SUBJECT (first reference in Chapter):

Analytic continuation 427
    function, 426
Annihilation 368
    operator, 8, 151, 170, 234, 256
Anomalous Zeeman effect 23, 52
Anomalous moment 23, 52 , 204
Anticommutation relation 5, 168, 256,
    282, 305, 331, 347, 390, 405,
    449, 486
Anticommutator 10, 168, 298, 333, 366,
    406, 454
Anticommutation 371
Anticommuting operator 172
Anticommuting field 451
Antisymmetric state 110, 170

Braid group 473

Canonical variables 151
    quantization, 232
Charge-conjugation invariance 301, 334,
    405
    transformation, 390
Color 487
Commutation relation 5, 149, 168, 233,
    256, 279, 305, 347, 374, 405,
    449, 486
Commutator 16, 168, 281, 333, 348, 394,
    406, 459
Conserved current 280
Continuity equation 278
Creation operator 8, 151, 170, 234, 256,
    374

Discrete symmetry 467

Dotted indices 279, 347

Electrodynamics 234, 368 ,390, 448, 475
Electron spin 22, 50
Energy-momentum tensor 278, 347
Exchange operator 465

Fermionic charge 455
First quantization 233
Future tube 430

Indefinite metric 374
Irreducible spinor 282

Lightcone 426

Magnetic monopole 470

Nucleon 310

Occupation number 151
    operator, 170
    representation, 149
    state, 150

Periodic Table 21

Quantum field theory 151, 235, 256, 346,
    390, 411, 431, 448, 465, 487

Reciprocity theorem 26
Relativistic doublets 23, 52
Rotation group 450
    operator, 465
Rotational invariance 414, 450, 489

Second quantization 150, 233, 370
Solid bodies 472
Space reflection 301, 391
Spin-Statistics Theorem 1, 235, 256, 277,
    301, 330, 345, 368, 390, 393,
    405, 425, 448, 464
Spin-3/2 fields 454
Spinor calculus 278
    classes, 347
    double-valued, 349
    indices, 282, 349
    irreducible, 347
    single-valued, 349

Tabelle 1. Ursprüngliches Bohrsches Schema der Edelgaskonfiguration.

Element	Atom-Nr.	Anzahl der n_k-Elektronen														
		1_1	2_1	2_2	3_1	3_2	3_3	4_1	4_2	4_3	4_4	5_1	5_2	5_3	6_1	6_2
Helium	2	2	—	—	—	—	—	—	—	—	—	—	—	—	—	—
Neon	10	2	4	4	—	—	—	—	—	—	—	—	—	—	—	—
Argon	18	2	4	4	4	4	—	—	—	—	—	—	—	—	—	—
Krypton	36	2	4	4	6	6	6	4	4	—	—	—	—	—	—	—
Xenon	54	2	4	4	6	6	6	6	6	6	—	4	4	—	—	—
Emanation	86	2	4	4	6	6	6	8	8	8	8	6	6	6	4	4

Tabelle 2. Schema der Edelgaskonfigurationen nach Stoner.

Element	Atom-Nr.	Anzahl der n_{k_1,k_2}-Elektronen														
		1_1	2_1	$2_2(1+2)$	3_1	$3_2(1+2)$	$3_3(2+3)$	4_1	$4_2(1+2)$	$4_3(2+3)$	$4_4(3+4)$	5_1	$5_2(1+2)$	$5_3(2+3)$	6_1	$6_2(1+2)$
Helium	2	2	—	—	—	—	—	—	—	—	—	—	—	—	—	—
Neon	10	2	2	2+4	—	—	—	—	—	—	—	—	—	—	—	—
Argon	18	2	2	2+4	2	2+4	—	—	—	—	—	—	—	—	—	—
Krypton	36	2	2	2+4	2	2+4	4+6	2	2+4	—	—	—	—	—	—	—
Xenon	54	2	2	2+4	2	2+4	4+6	2	2+4	4+6	—	2	2+4	—	—	—
Emanation	86	2	2	2+4	2	2+4	4+6	2	2+4	4+6	6+8	2	2+4	4+6	2	2+4

PAULI   1940